Geometry and Computing

Christian Perwass

Geometric Algebra with Applications in Engineering

With 62 Figures

Springer

Dr. habil. Christian Bernd Ulrich Perwaß
Department of Computer Science
Christian-Albrechts-Universität zu Kiel
Germany
christian@perwass.de

ISBN 978-3-642-10032-1 e-ISBN 978-3-540-89068-3

Springer Series in Geometry and Computing

ACM Computing Classification (1998): F.2.2, I.3.5, I.4, G.1.3, I.1.2, G.4, I.2.9

Mathematics Subjects Classification (2000): 14-01, 51-01, 65D, 68U, 53A

Cover design: deblik, Berlin
Printed on acid-free paper

9 8 7 6 5 4 3 2 1

springer.com

To Susanne

Preface

Geometry is a pervasive mathematical concept that appears in many places and in many disguises. The representation of geometric entities, of their unions and intersections, and of their transformations is something that we, as human beings, can relate to particularly well, since these concepts are omnipresent in our everyday life. Being able to describe something in terms of geometric concepts lifts a purely formal description to something we can relate to. The application of "common geometrical sense" can therefore be extremely helpful in understanding many mathematical concepts. This is true not only for elements directly related to geometry such as the fundamental matrix, but also for polynomial curves such as Pythagorean hodograph curves, and analysis-related aspects such as the Cauchy–Riemann equations [127], to mention a few examples.

An algebra for geometry is therefore a desirable mathematical tool, and indeed many such algebras have been developed. The first ones were probably Grassmann's algebra of extensive entities, Hamilton's quaternions, and complex numbers, which were all combined by W. K. Clifford into a single algebraic system, *Clifford algebra.* The aim of this text is not only to show that Clifford algebra is *the* geometric algebra, but also to demonstrate its advantages for important engineering applications.

This text is split into two main parts: a theoretical and an application part. In the theoretical part, the mathematical foundations are laid and a methodology is presented for the representation and numerical implementation of geometric constraints, with possibly uncertain geometric entities. The application part demonstrates with a number of examples how this methodology can be applied, but also demonstrates that the representative power of geometric algebra can lead to interesting new results.

This text originates from my work in the Cognitive Systems Group at the Christian-Albrechts-University in Kiel, Germany, led by Professor Dr. Gerald Sommer. First and foremost, I am indebted to him for his constant support of my research in geometric algebra. I would also like to thank Professor Dr. Wolfgang Förstner from the University of Bonn, Germany, and Professor Dr.

Rida Farouki from the University of California at Davis (UCD), USA, for inviting me to work with them. The work on random multivector variables and the Gauss–Helmert model originated from a collaboration with Professor Förstner in 2004. The work on Pythagorean hodographs is the result of a two-month stay at UCD in 2005, where I worked with Professor Farouki.

Of course, in addition to Professor Sommer, my coworkers and students at the Cognitive Systems Group in Kiel were an invaluable help in forming, discussing, and implementing new ideas. In particular, I would like to thank Dr. Vladimir Banarer, with whom I had many insightful discussions, not just about hypersphere neurons; Christian Gebken who I worked with on an implementation of geometric algebra on an FPGA, on the estimation of uncertain geometric entities, and on pose estimation with projective and catadioptric cameras; Anti Tolvanen, for his collaboration on central catadioptric cameras; Dr. Sven Buchholz, for his pure-mathematical assessments; and Professor Dr. Bodo Rosenhahn, for collaboration on pose estimation and coupled twists. I am also grateful for the technical assistance of Gerd Diesner and Henrik Schmidt, and the administrative support provided by Francoise Maillard.

Last but not least, I thank my wife Susanne and my parents Heide and Ulrich for their help and encouragement, which made this work possible.

Stuttgart, *Christian B. U. Perwaß*
September 2008

Contents

1 **Introduction** ... 1
- 1.1 History ... 3
- 1.2 Geometry ... 5
- 1.3 Outlook ... 6
- 1.4 Overview of This Text ... 8
  - 1.4.1 CLUCalc ... 9
  - 1.4.2 Algebra ... 9
  - 1.4.3 Geometries ... 10
  - 1.4.4 Numerics ... 10
  - 1.4.5 Uncertain Geometric Entities and Operators ... 13
  - 1.4.6 The Inversion Camera Model ... 13
  - 1.4.7 Monocular Pose Estimation ... 14
  - 1.4.8 Versor Functions ... 14
  - 1.4.9 Random Variable Space ... 15
- 1.5 Overview of Geometric Algebra ... 15
  - 1.5.1 Basics of the Algebra ... 16
  - 1.5.2 General Vectors ... 17
  - 1.5.3 Geometry ... 19
  - 1.5.4 Transformations ... 21
  - 1.5.5 Outermorphism ... 23

2 **Learning Geometric Algebra with CLUCalc** ... 25
- 2.1 Background ... 26
- 2.2 The Software ... 27
  - 2.2.1 Editor Window ... 28
  - 2.2.2 Visualization Window ... 29
  - 2.2.3 Output Window ... 29
  - 2.2.4 Command Line Parameters ... 30
- 2.3 The Scripting Language ... 30
  - 2.3.1 Basics ... 30
  - 2.3.2 Visualizing Geometry ... 32

2.3.3 User Interaction ... 39
2.3.4 Animation ... 40
2.3.5 Annotating Graphics ... 42
2.3.6 Multivector Calculations ... 45
2.4 Summary ... 48

Part I Theory

3 Algebra ... 51
3.1 Basics ... 52
3.1.1 Axioms ... 52
3.1.2 Basic Properties ... 54
3.1.3 Algebraic Basis ... 56
3.1.4 Involutions ... 59
3.1.5 Duality ... 60
3.1.6 Inner and Outer Product ... 61
3.2 Blades ... 64
3.2.1 Geometric Product ... 64
3.2.2 Outer Product ... 66
3.2.3 Scalar Product ... 68
3.2.4 Reverse ... 69
3.2.5 Conjugate ... 69
3.2.6 Norm ... 70
3.2.7 Inner Product ... 72
3.2.8 Duality ... 79
3.2.9 Inverse ... 81
3.2.10 Projection ... 82
3.2.11 Rejection ... 83
3.2.12 Meet and Join ... 84
3.2.13 Regressive Product ... 89
3.3 Versors ... 89
3.3.1 Definitions ... 90
3.3.2 Properties ... 91
3.4 Linear Functions ... 93
3.4.1 Determinant ... 94
3.4.2 Determinant Product ... 96
3.4.3 Inverse ... 96
3.4.4 Summary ... 96
3.5 Reciprocal Bases ... 97
3.5.1 Definition ... 98
3.5.2 Example ... 98
3.6 Differentiation ... 99
3.6.1 Vector Derivative ... 99
3.6.2 Multivector Differentiation ... 101
3.6.3 Tensor Representation ... 103

3.7 Algorithms ... 104
3.7.1 Basis Orthogonalization ... 105
3.7.2 Factorization of Blades ... 105
3.7.3 Evaluation of the Join ... 106
3.7.4 Versor Factorization ... 107
3.8 Related Algebras ... 109
3.8.1 Gibbs's Vector Algebra ... 109
3.8.2 Complex Numbers ... 110
3.8.3 Quaternions ... 111
3.8.4 Grassmann Algebra ... 114
3.8.5 Grassmann–Cayley Algebra ... 115

4 Geometries ... 119
4.1 Euclidean Space ... 121
4.1.1 Outer-Product Representations ... 122
4.1.2 Geometric Interpretation of the Inner Product ... 123
4.1.3 Inner Product Representation ... 124
4.1.4 Reflections ... 127
4.1.5 Rotations ... 130
4.1.6 Mean Rotor ... 133
4.2 Projective Space ... 134
4.2.1 Definition ... 134
4.2.2 Outer-Product Representations ... 138
4.2.3 Inner-Product Representations ... 140
4.2.4 Reflections in Projective Space ... 142
4.2.5 Rotations in Projective Space ... 143
4.3 Conformal Space ... 145
4.3.1 Stereographic Embedding of Euclidean Space ... 146
4.3.2 Homogenization of Stereographic Embedding ... 147
4.3.3 Geometric Algebra on $\mathbb{R}^{n+1,1}$ ... 150
4.3.4 Inner-Product Representations in $\mathbb{G}_{4,1}$ ... 152
4.3.5 Outer-Product Representations in $\mathbb{G}_{4,1}$ ... 158
4.3.6 Summary of Representations ... 161
4.3.7 Stratification of Spaces ... 161
4.3.8 Reflections in $\mathbb{G}_{n+1,1}$ ... 163
4.3.9 Inversions in $\mathbb{G}_{n+1,1}$ ... 164
4.3.10 Translations in $\mathbb{G}_{n+1,1}$ ... 168
4.3.11 Rotations in $\mathbb{G}_{n+1,1}$ ... 170
4.3.12 Dilations in $\mathbb{G}_{n+1,1}$ ... 171
4.3.13 Summary of Operator Representations ... 172
4.3.14 Incidence Relations ... 173
4.3.15 Analysis of Blades ... 176
4.4 Conic Space ... 179
4.4.1 Polynomial Embedding ... 180
4.4.2 Symmetric-Matrix Vector Space ... 181

4.4.3 The Geometric Algebra $\mathbb{G}_6$ ... 183
4.4.4 Rotation Operator ... 184
4.4.5 Analysis of Conics ... 186
4.4.6 Intersecting Lines and Conics ... 189
4.4.7 Intersection of Conics ... 190
4.5 Conformal Conic Space ... 193
4.5.1 The Vector Space ... 193
4.5.2 The Geometric Algebra $\mathbb{G}_{5,3}$ ... 194

5 Numerics ... 197
5.1 Tensor Representation ... 198
5.1.1 Component Vectors ... 199
5.1.2 Example: Geometric Product in $\mathbb{G}_2$ ... 200
5.1.3 Subspace Projection ... 201
5.1.4 Example: Reduced Geometric Product ... 202
5.1.5 Change of Basis ... 203
5.2 Solving Linear Geometric Algebra Equations ... 203
5.2.1 Inverse of a Multivector ... 204
5.2.2 Versor Equation ... 205
5.2.3 Example: Inverse of a Multivector in $\mathbb{G}_2$ ... 207
5.3 Random Multivectors ... 208
5.3.1 Definition ... 208
5.3.2 First-Order Error Propagation ... 210
5.3.3 Bilinear Functions ... 212
5.3.4 Summary ... 214
5.4 Validity of Error Propagation ... 215
5.4.1 Non-Gaussivity ... 215
5.4.2 Error Propagation Bias ... 217
5.4.3 Conclusions ... 220
5.5 Uncertainty in Projective Space ... 221
5.5.1 Mapping ... 221
5.5.2 Random Homogeneous Vectors ... 223
5.5.3 Conditioning ... 224
5.6 Uncertainty in Conformal Space ... 228
5.6.1 Blades and Operators ... 230
5.7 Uncertainty in Conic Space ... 233
5.8 The Gauss–Markov Model ... 234
5.8.1 Linearization ... 235
5.8.2 Constraints on Parameters Alone ... 236
5.8.3 Least-Squares Estimation ... 236
5.8.4 Numerical Calculation ... 238
5.8.5 Generalization ... 238
5.9 The Gauss–Helmert Model ... 239
5.9.1 The Constraints ... 239
5.9.2 Least-Squares Minimization ... 242

5.9.3 Derivation of the Covariance Matrix $\Sigma_{\Delta p,\Delta p}$ 244
5.9.4 Numerical Evaluation 246
5.9.5 Generalization 248
5.10 Applying the Gauss–Markov and Gauss–Helmert Models 248
5.10.1 Iterative Application of Gauss–Helmert Method 249

Part II Applications

6 Uncertain Geometric Entities and Operators 255
6.1 Construction 255
6.1.1 Geometric Entities in Conformal Space 256
6.1.2 Geometric Entities in Conic Space 258
6.1.3 Operators in Conformal Space 259
6.2 Estimation 263
6.2.1 Estimation of Geometric Entities 263
6.2.2 Versor Equation 265
6.2.3 Projective Versor Equation 267
6.2.4 Constraint Metrics 268
6.2.5 Estimation of a 3D Circle 272
6.2.6 Estimation of a General Rotor 274
6.3 Hypothesis Testing 275

7 The Inversion Camera Model 277
7.1 The Pinhole Camera Model 279
7.2 Definition of the Inversion Camera Model 281
7.3 From Pinhole to Lens 282
7.3.1 Mathematical Formulation 283
7.3.2 Relationship Between Focal Length and Lens Distortion 288
7.4 Fisheye Lenses 293
7.5 Catadioptric Camera 294
7.6 Extensions 296

8 Monocular Pose Estimation 299
8.1 Initial Pose 301
8.2 Formulation of the Problem in CGA 305
8.3 Solution Method 307
8.3.1 Tensor Form 308
8.3.2 Jacobi Matrices 310
8.3.3 Constraints on Parameters 311
8.3.4 Iterative Estimation 313
8.4 Experiments 316
8.4.1 Setup 317
8.4.2 Execution 320
8.4.3 Results 321
8.5 Conclusions 322

9 Versor Functions ... 325
9.1 Coupled Motors ... 326
9.1.1 Cycloidal Curves ... 326
9.1.2 Fourier Series ... 327
9.1.3 Space Curves ... 330
9.2 Pythagorean-Hodograph Curves ... 330
9.2.1 Relation to Versor Equation ... 332
9.2.2 Pythagorean-Hodograph Curves ... 333
9.2.3 Relation Between the Rotation and Reflection Forms ... 336
9.2.4 Pythagorean-Hodograph Quintic Hermite Interpolation 339
9.2.5 Degrees of Freedom ... 340
9.2.6 Curves of Constant Length ... 342
9.2.7 Pythagorean-Hodograph Curves in $\mathbb{R}^n$ ... 347
9.2.8 Summary ... 347
9.2.9 Proof of Lemma 9.5 ... 348

10 Random-Variable Space ... 351
10.1 A Random-Variable Vector Space ... 352
10.1.1 Probability Space ... 352
10.1.2 Continuous Random Variables ... 352
10.1.3 Multiple Random Variables ... 354
10.2 A Hilbert Space of Random Variables ... 356
10.2.1 The Norm ... 356
10.2.2 The Scalar Product ... 357
10.2.3 The Dirac Delta Distribution ... 358
10.3 Geometric Algebra over Random Variables ... 360
10.3.1 The Norm ... 361
10.3.2 General Properties ... 362
10.3.3 Correlation ... 363
10.3.4 Normal Random Variables ... 365

Notation ... 369

References ... 371

Index ... 381

Chapter 1
Introduction

Geometric algebra is currently not a widespread mathematical tool in the fields of computer vision, robot vision, and robotics within the engineering sciences, where standard vector analysis, matrix algebra, and, at times, quaternions are mainly used. The prevalent reason for this state of affairs is probably the fact that geometric algebra is typically not taught at universities, let alone at high-school level, even though it appears to be *the* mathematical language for geometry. This unfortunate situation seems to be due to two main aspects. Firstly, geometric algebra combines many mathematical tools that were developed separately over the past 200-odd years, such as the standard vector analysis, Grassmann's algebra, Hamilton's quaternions, complex numbers, and Pauli matrices. To a certain extent, teaching geometric algebra therefore means teaching all of these concepts at once. Secondly, most applications in two- and three-dimensional space, which are the most common spaces in engineering applications, can be dealt with using standard vector analysis and matrix algebra, without the need for additional tools. The goal of this text is thus to demonstrate that geometric algebra, which combines geometric transformations with the construction and intersection of geometric entities in a single framework, can be used advantageously in the analysis and solution of engineering applications.

Matrix algebra, or linear algebra in general, probably represents the most versatile mathematical tool in the engineering sciences. In fact, any geometric algebra can be represented in matrix form, or, more to the point, geometric algebra is a particular subalgebra of general tensor algebra (see e.g. [148]). However, this constraint can be an advantage, as for example in the case of quaternions, which form a subalgebra of geometric algebra. While rotations about an axis through the origin in 3D space can be represented by 3×3 matrices, it is a popular method to use quaternions instead, because their components are easier to interpret (direction of rotation axis and rotation angle) and, with only four parameters, they are a nearly minimal parameterization. Given the four components of a quaternion, the corresponding (scaling) rotation is uniquely determined, whereas the three Euler angles from

Geometry and Computing.

which a rotation matrix can be constructed do not suffice by themselves. It is also important to define in which order the three rotations about the three basis axes are executed, in order to obtain the correct rotation matrix. It is also not particularly intuitive what type of rotation three Euler angles represent. This is the reason why, in computer graphics software libraries such as OpenGL [23, 22], rotations are always given in terms of a rotation axis and a rotation angle. Internally, this is then transformed into the corresponding rotation matrices.

Apart from the obvious interpretative advantages of quaternions, there are clear numerical advantages, at the cost that only rotations can be described, whereas matrices can represent any linear function. For example, two rotations are combined by multiplying two quaternions or two rotation matrices. The product of two quaternions can be represented by the product of a 4×4 matrix with a 4×1 vector, while in the case of a matrix representation two 3×3 matrices have to be multiplied. That is, the former operation consists of 16 multiplications and 12 additions, while the latter needs 27 multiplications and 18 additions. Furthermore, when one is solving for a rotation matrix, two additional constraints need to be imposed on the nine matrix components: matrix orthogonality and scale. In the case of quaternions, the orthogonality constraint is implicit in the algebraic structure and thus does not have to be imposed explicitly. The only remaining constraint is therefore the quaternion scale.

The quaternion example brings one of the main advantages of geometric algebra to the fore: by reducing the representable transformations from all (multi)linear functions to a particular subset, for example rotation, a more optimal parameterization can be achieved and certain constraints on the transformations are embedded in the algebraic structure. In other words, the group structure of a certain set of matrices is made explicit in the algebra. On the downside, this implies that only a subset of linear transformations is directly available. Clearly, any type of function, including linear transformations, can still be defined on algebraic entities, but not all functions profit in the same way from the algebraic structure.

The above discussion gives an indication of the type of problems where geometric algebra tends to be particularly beneficial: situations where only a particular subset of transformations and/or geometric entities are present. The embedding of appropriate constraints in the algebraic structure can then lead to descriptive representations and optimized numerical constraints.

One area where the embedding of constraints in the algebraic structure can be very valuable is the field of artificial neural networks, or classification algorithms in general. Any classification algorithm has to make some assumptions about the structure of the feature space or, rather, the form of the separation boundaries between areas in the feature space that belong to different classes. Choosing the best basis functions (kernels) for such a separation can improve the classification results considerably. geometric algebra offers, through its algebraic structure, methods to advantageously implement such basis functions, in particular for geometric constraints. This has been shown, for example, by Buchholz and Sommer [27, 29, 28] and by Bayro-Corrochano and Buchholz [17]. Banarer, Perwass, and Sommer have shown, furthermore, that hyperspheres as represented in the geometric algebra of conformal space are effective basis functions for classifiers [134, 14, 15]. In this text, however, these aspects will not be detailed further.

1.1 History

Before discussing further aspects of geometric algebra, it is helpful to look at its roots. Geometric algebra is basically just another name for Clifford algebra, a name that was introduced by David Hestenes to emphasize the geometric interpretation of the algebra. He first published his ideas in a refined version of his doctoral thesis in 1966 [87]. A number of books extending this initial "proof of concept" (as he called it) followed in the mid 1980s and early 1990s [91, 88, 92]. Additional information on his current projects can be found on the website [90].

Clifford algebra itself is about 100 years older. It was developed by William K. Clifford (1845–1879) in 1878 [37, 35]. A collection of all his papers can be found in [36]. Clifford's main idea was that the *Ausdehnungslehre* of Hermann G. Grassmann (1809–1877) and the quaternion algebra of William R. Hamilton (1805–1865) could be combined in a single geometric algebra, which was later to be known as *Clifford algebra*. Unfortunately, Clifford died very young and had no opportunity to develop his algebra further. In around 1881 the physicist Josiah W. Gibbs (1839–1903) developed a method that made it easier to deal with vector analysis, which advanced work on Maxwell's electrodynamics considerably. This was probably one of the reasons why Gibbs's vector analysis became the standard mathematical tool for physicists and engineers, instead of the more general but also more elaborate Clifford algebra.

Within the mathematics community, the Clifford algebra of a quadratic module was a well-established theory by the end of the 1970s. See, for example, the work of O'Meara [129] on the theory over fields or the work of Baeza [13] on the theory over rings. The group structure of Clifford algebra was detailed by, for example, Porteous [148] and Gilbert and Murray [79] in the 1990s. The latter also discussed the relation between Clifford algebra and Dirac operators, which is one of the main application areas of Clifford algebra in physics.

The Dirac equation of quantum mechanics has a natural representation in Clifford algebra, since the Pauli matrices that appear in it form an algebra that is isomorphic to a particular Clifford algebra. Some other applications of Clifford algebra in physics are to Maxwell's equations of electrodynamics, which have a very concise representation, and to a flat-space theory of gravity

[46, 106]. An introduction to geometric algebra for physicists was published by Doran and Lasenby in 2003 [45].

The applications of Clifford algebra or geometric algebra in engineering were initially based on the representation of geometric entities in projective space, as introduced by Hestenes and Ziegler in [92]. These were, for example, applications to projective invariants (see e.g. [107, 108]) and multiple-view geometry (see e.g. [18, 145, 144]). Initial work in the field of robotics used the representation of rotation operators in geometric algebra (see e.g. [19, 161]). The first collections of papers discussing applications of geometric algebra in engineering appeared in the mid 1990s [16, 48, 164, 165]. The first design of a geometric algebra coprocessor and its implementation in a field-programmable gate array (FPGA) was developed by Perwass, Gebken, and Sommer in 2003 [137].

In 2001 Hongbo Li, David Hestenes, and Alyn Rockwood published three articles in [164] introducing the *conformal model* [116], spherical conformal geometry [117], and a universal model for conformal geometries [118]. These articles laid the foundation for a whole new class of applications that could be treated with geometric algebra. Incidentally, Pierre Anglés had already developed this representation of the conformal model independently in the 1980s [6, 7, 8] but, apparently, it was not registered by the engineering community. His work on conformal space can also be found in [9].

The conformal model is based on work by Friedrich L. Wachter (1792–1817), a student of J. Carl F. Gauss (1777–1855), who showed that a certain surface in hyperbolic geometry was metrically equivalent to Euclidean space. Forming a geometric algebra over a homogeneous embedding of this space extends the representation of points, lines, planes, and rotations about axes through the origin to point pairs, circles, spheres, and conformal transformations, which include all Euclidean transformations. Note that the representation of Euclidean transformations in the conformal model is closely related to biquaternions, which had been investigated by Clifford himself in 1873 [34], a couple of years before he developed his algebra. The extended set of basic geometric entities in the conformal model and the additionally available transformations made geometric algebra applicable to a larger set of application areas, for example pose estimation [155, 152], a new type of artificial neural network [15], and the description of space groups [89].

Although the foundations of the conformal model were laid by Li, Hestenes, and Rockwood in 2001, its various facets, properties, and extensions are still a matter of ongoing research (see e.g. [49, 105, 140, 160]). One of the aims of this text is to present the conformal model in the context of other geometric algebra models and to give a detailed derivation of the model itself, as well as in-depth discussions of its geometric entities and transformation operators. Even though the conformal model is particularly powerful, it is presented as one special case of a general method for representing geometry and transformations with geometric algebra. One result of this more general outlook is the geometric algebra of conic space, which is published here in all of its details for the first time. In this geometric algebra, the algebraic entities represent all types of projective conic sections.

1.2 Geometry

One of the novel features of the discussion of geometric algebra in this text is the explicit separation of the algebra from the representation of geometry through algebraic entities. In the author's opinion this is an advantageous view, since it clarifies the relation between the various geometric models and indicates how geometric algebra may be developed for new geometric models. The basic idea that blades represent linear subspaces through their null space has already been noted by Hestenes. However, the consequent explicit application of this notion to the discussion of different geometric models and algebraic operations was first developed in [140] and is extended in this text. Note that the notion of representing geometry through null spaces is directly related to *affine varieties* in algebraic geometry [38].

One conclusion that can be drawn from this view of geometric algebra is that there exist only three fundamental operations in the algebra, all based on the same algebraic product: "addition", "subtraction", and reflection of linear subspaces. All geometric entities, such as points, lines, planes, spheres, circles, and conics, are represented through linear subspaces, and all transformations, such as rotation, inversion, translation, and dilation, are combinations of reflections of linear subspaces. Nonlinear geometric entities such as spheres and nonlinear transformations such as inversions stem from a particular embedding of Euclidean space in a higher-dimensional embedding space, such that linear subspaces in the embedding space represent nonlinear subspaces in the Euclidean space, and combinations of reflections in the embedding space represent nonlinear transformations in the Euclidean space. This aspect is detailed further in Chap. 4, where a number of geometries are discussed in detail.

The fact that all geometric entities and all transformation operations are constructed through fundamental algebraic operations leads to two key properties of geometric algebra:

1. Geometric entities and transformation operators are constructed in exactly the same way, independent of the dimension of the space they are constructed in.
2. The intersection operation and the transformation operators are the same for all geometric entities in all dimensions.

For example, it is shown in Sect. 4.2 that in the geometric algebra of the projective space of Euclidean 3D space $\mathbb{R}^3$, a vector $\boldsymbol{A}$ represents a point. The *outer product* (see Sect. 3.2.2) of two vectors $\boldsymbol{A}$ and $\boldsymbol{B}$ in this space, denoted by $\boldsymbol{A} \wedge \boldsymbol{B}$, then represents the line through $\boldsymbol{A}$ and $\boldsymbol{B}$. If $\boldsymbol{A}$ and $\boldsymbol{B}$

are vectors in the projective space of $\mathbb{R}^{10}$, say, they still represent points and $\boldsymbol{A} \wedge \boldsymbol{B}$ still represents the line through these points.

The intersection operation in geometric algebra is called the *meet* and is denoted by $\vee$. If $\boldsymbol{A}$ and $\boldsymbol{B}$ represent any two geometric entities in any dimension, then their intersection is always determined with the meet operation as $\boldsymbol{A} \vee \boldsymbol{B}$. Note that even if the geometric entities represented by $\boldsymbol{A}$ and $\boldsymbol{B}$ have no intersection, the meet operation results in an algebraic entity. Typically this entity then represents an imaginary geometric object or one that lies at infinity. This property of the intersection operation has the major advantage that no additional checks have to be performed before the intersection of two entities is computed.

The application of transformations is similarly dimension-independent. For example, if $\boldsymbol{R}$ is an algebraic entity that represents a rotation and $\boldsymbol{A}$ represents any geometric entity, then $\boldsymbol{R}\,\boldsymbol{A}\,\boldsymbol{R}^{-1}$ represents the rotated geometric entity, independent of its type and dimension. Note that juxtaposition of two entities denotes the algebra product, the *geometric product*.

To a certain extent, it can therefore be said that geometric algebra allows a coordinate-free representation of geometry. Hestenes emphasized this property by developing the algebra with as little reference to a basis as possible in [87, 91]. However, since in those texts the geometric algebra over real-valued vector spaces is discussed, a basis can always be found, but it is not essential for deriving the properties of the algebra. The concept of a basis-independent geometric algebra was extended even further by Frank Sommen in 1997. He constructed a Clifford algebra completely without the notion of a vector space [162, 163]. In this approach abstract basic entities, called *vector-variables*, are considered, which are not entities of a particular vector space. Instead, these vector-variables are characterized purely through their algebraic properties. An example of an application of this *radial algebra* is the construction of Clifford algebra on super-space [21].

The applications considered in this text, however, are all related to particular vector spaces, which allows a much simpler construction of geometric algebra. The manipulation of analytic expressions in geometric algebra is still mostly independent of the dimensionality of the underlying vector space or any particular coordinates.

1.3 Outlook

From a mathematical point of view, geometric algebra is an elegant and analytically powerful formalism to describe geometry and geometric transformations. However, the particular aim of the engineering sciences is the development of solutions to problems in practical applications. The tools that are used to achieve that aim are only of interest insofar as the execution speed and the accuracy of solutions are concerned. Nevertheless, this does not preclude research into promising mathematical tools, which could lead to a substantial gain in application speed and accuracy or even new application areas. At the end of the day, though, a mathematical tool will be judged by its applicability to practical problems. In this context, the question has to be asked:

What are the advantages of geometric algebra?

Or, more to the point, *when should geometric algebra be used?* There appears to be no simple, generally applicable answer to this question. Earlier in this introduction, some pointers were given in this context. Basically, geometric algebra has three main properties:

1. Linear subspaces can be represented in arbitrary dimensions.
2. Subspaces can be added, subtracted, and intersected.
3. Reflections of subspaces in each other can be performed.

Probably the most important effects that these fundamental properties have are the following:

- With the help of a non-linear embedding of Euclidean space, it is possible to represent non-linear subspaces and transformations. This allows a (multi)linear representation of circles and spheres, and non-linear transformations such as inversions.
- The combination of the basic reflection operations results in more complex transformations, such as rotation, translation, and dilation. The corresponding transformation operators are nearly minimal parameterizations of the transformation. Additional constraints, such as the orthogonality of a rotation matrix, are encoded in the algebraic structure.
- Owing to the dimension-independent representation of geometric entities, only a single intersection operation is needed to determine the intersections between arbitrary combinations of entities. This can be a powerful analytical tool for geometrical constructions.
- The dimension-independent representation of geometric entities and reflections also has the effect that a transformation operator can be applied to any element in the algebra in any dimension, be it a geometric entity or another transformation.
- Since all transformations are represented as multilinear operations, the uncertainty of Gaussian distributed transformation operators can be effectively represented by covariance matrices. That is, *the uncertainty of geometric entities and transformations can be represented in a unified fashion.*

These properties are used in Chap. 6 to construct and estimate uncertain geometric entities and transformations. In Chap. 7, the availability of a linearized inversion operator leads to a unifying camera model. Chapter 8 exploits the encoding of transformation constraints in the algebraic structure. In Chap. 9, the dimension independence of the reflection operation leads to an immediate extension of Pythagorean-hodograph curves to arbitrary dimensions. And Chap. 10 demonstrates the usefulness of representing linear subspaces in the Hilbert space of random variables.

It is hoped that the subjects discussed in this text will help researchers to identify advantageous applications of geometric algebra in their field.

1.4 Overview of This Text

This section gives an overview of the main contributions of this text in the context of research in geometric algebra and computer vision. The main aim of this text is to give a detailed presentation of and, especially, to develop new tools for the three aspects that are necessary to apply geometric algebra to engineering problems:

1. **Algebra**, the mathematical formalism.
2. **Geometry**, the representation of geometry and transformations.
3. **Numerics**, the implementation of numerical solution methods.

This text gives a thorough description of geometric algebra, presents the geometry of the geometric algebra of Euclidean, projective, conformal, and a novel conic space in great detail, and introduces a novel numerical calculation method for geometric algebra that incorporates the notion of random algebra variables. The methodology presented combines the representative power of the algebra and its effective algebraic manipulations with the demands of real-life applications, where uncertain data is unavoidable.

A number of applications where this methodology is used are presented, which include the description of uncertain geometric entities and transformations, a novel camera model, and monocular pose estimation with uncertain data. In addition, applications of geometric algebra to some special polynomial curves (Pythagorean-hodograph curves), and the geometric algebra over the Hilbert space of random variables are presented.

In addition to the mathematical contributions, the software tool CLUCalc, developed by the author, is introduced. CLUCalc is a stand-alone software program that implements geometric algebra calculations and, more importantly, can visualize the geometric content of algebraic entities automatically. It is therefore an ideal tool to help in the learning and teaching of geometric algebra.

In the remainder of this section, the main aspects of all chapters are detailed.

1.4.1 CLUCalc

The software tool CLUCALC (Chap. 2) [133] was developed by the author with the aim of furthering the understanding, supporting the teaching, and implementing applications of geometric algebra. For this purpose, a whole new programming language, called CLUSCRIPT, was developed, which was honed for the programming of geometric algebra expressions. For example, most operator symbols typically used in geometric algebra are available in CLUSCRIPT. However, probably the most useful feature in the context of geometric algebra is the automatic visualization of the geometric content of multivectors. That is, the user need not know what a multivector represents in order to draw it. Instead, the meaning of multivectors and the effect of algebraic operations can be *discovered* interactively. Since geometric algebra is all about geometry, CLUSCRIPT offers simple access to powerful visualization features, such as transparent objects, lighting effects, texture mapping, animation, and user interaction. It also supports the annotation of drawings using LaTeX text, which can also be mapped onto arbitrary surfaces. Note that virtually all of the figures in this text were created with CLUCALC, and the various applications presented were implemented in CLUSCRIPT.

1.4.2 Algebra

One aspect that has been mentioned before is the clear separation of algebraic entities and their geometric interpretation. The foundation for this view is laid in Chap. 3 by introducing the inner- and outer-product null spaces. This concept is extended to the *geometric* inner- and outer-product null spaces in Chap. 4 on geometries, to give a general methodology of how to represent geometry by geometric algebra. Note that this concept is very similar to *affine varieties* in algebraic geometry [38].

Another important aspect that is treated explicitly in Chap. 3 is that of null vectors and null blades, which is something that is often neglected. Through the definition of an algebra conjugation, a *Euclidean scalar product* is introduced, which, together with a corresponding definition of the magnitude of an algebraic entity, or *multivector*, allows the definition of a Hilbert space over a geometric algebra. While this aspect is not used directly, algebra conjugation is essential in the general definition of subspace addition and subtraction, which eventually leads to factorization algorithms for blades and versors that are also valid for null blades and null versors. Furthermore, a *pseudoinverse* of null blades is introduced. The relevance of these operations is very high when working with the conformal model, since geometric entities are represented by blades of null vectors. In order to determine the meet, i.e. the general intersection operation, between arbitrary blades of null vectors, a factorization algorithm for such blades has to be available.

Independent of the null-blade aspect, a novel set of algorithms that are essential when implementing geometric algebra on a computer are presented. This includes, in particular, the evaluation of the *join* of blades, which is necessary for the calculation of the meet. In addition, the versor factorization algorithm is noteworthy. This factorizes a general transformation into a set of reflections or, in the case of the conformal model, inversions.

1.4.3 Geometries

While the representation of geometric objects and transformations through algebraic entities is straightforward, extracting the geometric information from the algebraic entities is not trivial. For example, in the conformal model, two vectors $\boldsymbol{S}_1$ and $\boldsymbol{S}_2$ can represent two spheres and their outer product $\boldsymbol{C} = \boldsymbol{S}_1 \wedge \boldsymbol{S}_2$ the intersection circle of the two spheres (see Sect. 4.3.4). Extracting the circle parameters center, normal, and radius from the algebraic entity $\boldsymbol{C}$ is not straightforward. Nevertheless, a knowledge of how this can be done is essential if the algebra is to be used in applications.

Therefore, the analysis of the geometric interpretation of algebraic entities that represent elements of 3D Euclidean space is discussed in some detail. A somewhat more abstract discussion of the geometric content of algebraic entities in the conformal model of arbitrary dimension can be found in [116]. Confining the discussion in this text to the conformal model of 3D Euclidean space simplifies the formulas considerably.

Another important aspect of Chap. 4 is a discussion of how incidence relations between geometric entities are represented through algebraic operations. This knowledge is pivotal when one is expressing geometric constraints in geometric algebra.

A particularly interesting contribution is the introduction of the conic space, which refers to the geometric algebra over the vector space of reduced symmetric matrices. In the geometric algebra of conic space, the outer product of five vectors represents the conic section that passes through the corresponding five points. The outer product of four points represents a point quadruplet, which is also the result of the meet of two five-blades, i.e. the intersection of two conic sections. The discussion of conic space starts out with an even more general outlook, whereby the conic and conformal spaces are particular subspaces of the general polynomial space.

1.4.4 Numerics

The essential difference in the treatment of numerical calculation with geometric algebra between this text and the standard approach introduced by Hestenes is that algebraic operations in geometric algebra are regarded here as bilinear functions and expressed through tensor contraction. This approach was first introduced by the author and Sommer in [146]. While the standard approach has its merits in the analytical description of derivatives as algebraic entities, the tensorial approach allows the direct application of (multi)linear optimization algorithms. At times, the tensorial approach also leads to solution methods that cannot be easily expressed in algebraic terms.

An example of the latter case is the versor equation (see Sect. 5.2.2). Without delving into all the details, the problem comes down to solving for the multivector $\boldsymbol{V}$, given multivectors $\boldsymbol{A}$ and $\boldsymbol{B}$, the equation

$$\boldsymbol{V}\boldsymbol{A} - \boldsymbol{B}\boldsymbol{V} = 0 .$$

The problem is that it is impossible to solve for $\boldsymbol{V}$ through algebraic manipulations, since the various multivectors typically do not commute. However, in the tensorial representation of this equation, it is straightforward to solve for the components of $\boldsymbol{V}$ by evaluating the null space of a matrix.

In the standard approach a function $\boldsymbol{C}(\boldsymbol{V})$, say, would be defined as

$$\boldsymbol{C} : \boldsymbol{V} \mapsto \boldsymbol{V}\boldsymbol{A} - \boldsymbol{B}\boldsymbol{V} .$$

The solution to $\boldsymbol{V}$ is then the vector $\widehat{\boldsymbol{V}}$, that minimizes $\Delta(\widehat{\boldsymbol{V}}) := \boldsymbol{C}(\widehat{\boldsymbol{V}}) \cdot \widetilde{\boldsymbol{C}}(\widehat{\boldsymbol{V}})$. To evaluate $\widehat{\boldsymbol{V}}$, the derivative of $\Delta(\widehat{\boldsymbol{V}})$ with respect to $\widehat{\boldsymbol{V}}$ has to be calculated, which can be done with the multivector derivative introduced by Hestenes [87, 91] (see Sect. 3.6). Then a standard gradient descent method or something more effective can be used to find $\widehat{\boldsymbol{V}}$. For an example of this approach, see [112]. It is shown in Sect. 5.2.2, however, that the tensorial approach also minimizes $\Delta(\widehat{\boldsymbol{V}})$; this method is much easier to apply.

Another advantage of the tensorial approach is that Gaussian distributed random multivector variables can be treated directly. This use of the tensorial approach was developed by the author in collaboration with W. Förstner and presented at a Dagstuhl workshop in 2004 [136]. The present text extends this and additional publications [76, 138, 139] to give a complete and thorough discussion of the subject area. The two main aspects with respect to random multivector variables, whose foundations are laid in Chap. 5, are the construction and estimation of uncertain geometric entities and uncertain *transformations* from uncertain data. For example, an uncertain circle can be constructed from the outer product of three uncertain points, and an uncertain rotation operator may be constructed through the geometric product of two uncertain reflection planes.

The representation of uncertain transformation operators is certainly a major advantage of geometric algebra over a matrix representation. This is demonstrated in Sect. 5.6.1, where it is shown that the variation of a random transformation multivector in the conformal model, such as a rotation operator, lies (almost) in a linear subspace. A covariance matrix is therefore well suited to representing the uncertainty of a Gaussian distributed random transformation multivector variable. It appears that the representation of uncertain transformations with matrices is more problematic. Heuel investigated an approach whereby transformation matrices are written as column vectors and their uncertainty as an associated covariance matrix [93]. He notes that this method is numerically not very stable.

A particularly convincing example is the representation of rotations about the origin in 3D Euclidean space. Corresponding transformation operators in the conformal model lie in a linear subspace, which also forms a subalgebra, and a subgroup of the Clifford group. Rotation matrices, which are elements of the orthogonal group, do not form a subalgebra at the same time. That is, the sum of two rotation matrices does not, in general, result in a rotation matrix. A covariance matrix on a rotation matrix can therefore only represent a tangential uncertainty.

Instead of constructing geometric entities and transformation operators from uncertain data, they can also be *estimated* from a set of uncertain data. For this purpose, a linear least-squares estimation method, the Gauss–Helmert model, is presented (see Sect. 5.9), which accounts for uncertainties in measured data. Although this is a well-known method, a detailed derivation is given here, to hone the application of this method to typical geometric-algebra problems.

In computer vision, the measured data consists typically of the locations of image points. Owing to the digitalization in CCD chips and/or the point spread function (PSF) of the imaging system, there exists an unavoidable uncertainty in the position measurement. Although this uncertainty is usually small, such small variations can lead to large deviations in a complex geometric construction. Knowing the final uncertainty of the elements that constitute the data used in an estimation may be essential to determining the reliability of the outcome.

One such example is provided by a catadioptric camera, i.e. a camera with a 360 degree view that uses a standard projective camera which looks at a parabolic mirror. While it can be assumed that the uncertainties in the position of all pixels in the camera are equal, the uncertainty in the corresponding projection rays reflected at the parabolic mirror varies considerably depending on where they hit the mirror. Owing to the linearization of the inversion operation in the conformal model, which can be used to model reflection in a parabolic mirror (see Sect. 7.2), simple error propagation can be used here to evaluate the final uncertainty of the projection rays. These uncertain rays may then form the data that is used to solve a pose estimation problem (see Chap. 8).

1.4.5 Uncertain Geometric Entities and Operators

In Chap. 6, the tools developed in Chap. 5 are applied to give examples of the construction and estimation of uncertain geometric entities and transformation operators. Here the construction of uncertain lines, circles, and conic sections from uncertain points is visualized, to show that the standard-deviation envelopes of such geometric entities are not simple surfaces. The use of a covariance matrix should always be favored over simple approximations such as a tube representing the uncertainty of a line.

Furthermore, the effect of an uncertain reflection and rotation on an ideal geometric entity is shown. This has direct practical relevance, for example, in the evaluation of the uncertainty induced in a light ray which is reflected off an uncertain plane.

With respect to the estimation of geometric entities and transformation operators, standard problems that occur in the application of geometric algebra to computer vision problems are presented, and it is shown how the methods introduced in Chap. 5 can be used to solve them. In addition, the metrics used implicitly in these problems are investigated. This includes the first derivation of the point-to-circle metric and the versor equation metric.

The quality of the estimation methods presented is demonstrated in two experiments: the estimation of a circle and of a rotation operator. Both experiments demonstrate that the Gauss–Helmert method gives better results than a simple null-space estimation. The estimation of the rotation operator is also compared with a standard method, where it turns out that the Gauss–Helmert method gives better and more stable results.

Another interesting aspect is hypothesis testing, where questions such as "does a point lie on a line?" are answered in a statistical setting. Here, the first visualization of what this means for the question "does a point lie on a circle?" in the conformal model is given (see Fig. 6.10).

1.4.6 The Inversion Camera Model

The inversion camera model is a novel camera model that was first published by Perwass and Sommer in [147]. In Chap. 7 an extended discussion of this camera model is given, which is applied in Chap. 8 to monocular pose estimation. The inversion camera model combines the pinhole camera model, a lens distortion model, and the catadioptric-camera model for the case of a parabolic mirror. All these configurations can be obtained by varying the position of a focal point and an inversion sphere. Since inversion can be represented through a linear operator in the conformal model, the geometric algebra of conformal space offers an ideal framework for this camera model. The unification of three camera models that are usually treated separately can lead to generalized constraint equations, as is the case for monocular pose estimation. Note that the camera model is represented by a transformation operator in the geometric algebra of conformal space, which implies that it can be easily associated with a covariance matrix that represents its uncertainty.

1.4.7 Monocular Pose Estimation

In Chap. 8 an application is presented that uses all aspects of the previously presented methodology: a geometric setup is translated into geometric-algebra expressions, algebraic operations are used to express a geometric constraint, and a solution is found using the tensorial approach, which incorporates uncertain data. While the pose estimation problem itself is well known, a number of novel aspects are introduced in this chapter:

- a simple and robust method to find an initial pose,
- the incorporation of the inversion camera model into the pose constraint,
- a pose constraint equation that is quadratic in the components of the pose operator without making any approximations, and
- the covariance matrix for the pose operator and the inversion camera model operator.

The pose estimation method presented is thoroughly tested against ground truth pose data generated with a robotic arm for a number of different imaging systems.

1.4.8 Versor Functions

In Chap. 9 some instances of versor functions are discussed, that is, functions of the type $\mathcal{F} \;:\; t \mapsto \boldsymbol{A}(t)\,\boldsymbol{N}\,\widetilde{\boldsymbol{A}}(t)$. One geometric interpretation of this type of function is that a *preimage* $\boldsymbol{A}(t)$ scales and rotates a vector $\boldsymbol{N}$. It is shown that cycloidal curves generated by coupled motors, Fourier series of complex-valued functions, and Pythagorean-hodograph (PH) curves are all related to this form.

In the context of PH curves, a new representation based on the reflection of vectors is introduced, and it is shown that this is equivalent to the standard quaternion representation in the case of cubic and quintic PH curves. This novel representation has the advantage that it can be immediately extended to arbitrary dimensions, and it gives a geometrically more intuitive representation of the degrees of freedom. This also leads to the identification of parameter subsets that generate PH curves of constant length but of different shape. The work on PH curves resulted from a collaboration of the author with R. Farouki [135].

1.4.9 Random Variable Space

Chap. 10 gives an example of a geometric algebra over a general Hilbert space, instead of a real-valued vector space. The Hilbert space chosen here is that of random variables. This demonstrates how the geometric concepts of geometric algebra can be applied to function spaces. Some fundamental properties of random variables are presented in this context. This leads to a straightforward derivation of the Cauchy–Schwarz inequality and an extension of the correlation coefficient to an arbitrary number of random variables. The geometric insight into geometric algebra operations gained for the Euclidean space can be applied directly to random variables, which gives properties such as the expectation value, the variance, and the correlation a direct geometric meaning.

1.5 Overview of Geometric Algebra

In this section, a mathematical overview of some of the most important aspects of geometric algebra is presented. The purpose is to give those readers who are not already familiar with geometric algebra a gentle introduction to the main concepts before delving into the detailed mathematical analysis. To keep it simple, only the geometric algebra of Euclidean 3D space is used in this introduction and formal mathematical proofs are avoided. A detailed introduction to geometric algebra is given in Chap. 3.

One problem in introducing geometric algebra is that depending on the reader's background, different introductions are best suited. For pure mathematicians the texts on Clifford algebra by Porteous [148], Gilbert and Murray [79], Lounesto [119], Riesz [149], and Ablamowicz et al. [3], to name just a few, are probably most instructive. In this context, the text by Lounesto presenting counterexamples of theorems in Clifford algebra should be of interest [120].

The field of Clifford analysis is only touched upon when basic multivector differentiation is introduced in Sect. 3.6. This is sufficient to deal with the simple functions of multivectors that are encountered in this text. Thorough treatments of Clifford analysis can be found in [24, 41, 42].

This text is geared towards the use of geometric algebra in engineering applications and thus stresses more those aspects that are related to the representation of geometry and Euclidean transformations. It may thus not treat all aspects that the reader is interested in. As always, the best way is to read a number of different introductions to geometric algebra. After the books by Hestenes [87, 88, 91], there are a number of papers and books geared towards various areas, in particular physics and engineering. For physics-related introductions see, for example [45, 83, 110], and for engineering-related introductions [47, 48, 111, 140, 50, 172].

1.5.1 Basics of the Algebra

The real-valued 3D Euclidean vector space is denoted by $\mathbb{R}^3$, with an orthonormal basis $e_1, e_2, e_3 \in \mathbb{R}^3$. That is,

$$e_i * e_j = \delta_{ij} , \qquad \delta_{ij} := \begin{cases} 1 : i = j , \\ 0 : i \neq j , \end{cases}$$

where $*$ denotes the standard scalar product. The geometric algebra of $\mathbb{R}^3$ is denoted by $\mathbb{G}(\mathbb{R}^3)$, or simply $\mathbb{G}_3$. Its algebra product is called the *geometric product*, and is denoted by the juxtaposition of two elements. This is just as in matrix algebra, where the matrix product of two matrices is represented by juxtaposition of two matrix symbols. The effect of the geometric product on the basis vectors of $\mathbb{R}^3$ is

$$e_i\, e_j = \begin{cases} e_i * e_j : i = j , \\ \quad e_{ij} : i \neq j . \end{cases} \tag{1.1}$$

One of the most important points to note for readers who are new to geometric algebra is the case when $i \neq j$, where $e_{ij} \equiv e_i\, e_j$ represents a *new algebraic element*, sometimes also denoted by e_{ij} for brevity. Similarly, if $i, j, k \in \{1, 2, 3\}$ are three different indices, $e_{ijk} := e_i\, e_j\, e_k$ is yet another new algebraic entity. That this process of creating new entities cannot be continued indefinitely is ensured by defining the geometric product to be associative and to satisfy $e_i\, e_i = 1$. The latter is also called the *defining equation* of the algebra. For example,

$$(e_i\, e_j\, e_k)\, e_k = (e_i\, e_j)\, (e_k\, e_k) = (e_i\, e_j)\, 1 = e_i\, e_j\ .$$

In Chap. 3 it is shown that (1.1) together with the standard axioms of an associative algebra suffices to show that

$$e_i\, e_j = -e_j\, e_i \qquad \text{if} \quad i \neq j\ . \tag{1.2}$$

This implies, for example, that

$$(e_i\, e_j\, e_k)\, e_j = (e_i\, e_j)\, (e_k\, e_j) = -(e_i\, e_j)\, (e_j\, e_k) = -e_i\, (e_j\, e_j)\, e_k = -e_i\, e_k\ .$$

From these basic rules, the basis of $\mathbb{G}_3$ can be found to be

$$\overline{\mathbb{G}}_3 := \left\{ 1,\, e_1,\, e_2,\, e_3,\, e_{12},\, e_{13},\, e_{23},\, e_{123} \right\} . \tag{1.3}$$

In general, the dimension of the geometric algebra of an n-dimensional vector space is 2^n.

1.5.2 General Vectors

The operations introduced in the previous subsection are valid only for a set of orthonormal basis vectors. For general vectors of $\mathbb{R}^3$, the algebraic operations have somewhat more complex properties, which can be related to the properties of the basis vectors by noting that any vector $\boldsymbol{a} \in \mathbb{R}^3$ can be written as

$$\boldsymbol{a} := a^1\, \boldsymbol{e}_1 + a^2\, \boldsymbol{e}_2 + a^3\, \boldsymbol{e}_3 \,.$$

Note that the scalar components of the vector $\boldsymbol{a}$, a^1, a^2, a^3, are indexed by a superscript index. This notation will be particularly helpful when algebra operations are expressed as tensor contractions. In particular, the *Einstein summation convention* can be used, which states that a subscript index repeated as a superscript index within a product implies a summation over the range of the index. That is,

$$\boldsymbol{a} := a^i\, \boldsymbol{e}_i \equiv \sum_{i=1}^{3} a^i\, \boldsymbol{e}_i \,.$$

It is instructive to consider the geometric product of two vectors $\boldsymbol{a}, \boldsymbol{b} \in \mathbb{R}^3$ with $\boldsymbol{a} := a^i\, \boldsymbol{e}_i$ and $\boldsymbol{b} := b^i\, \boldsymbol{e}_i$. Using the multiplication rules of the geometric product between orthonormal basis vectors as introduced in the previous subsection, it is straightforward to find

$$\begin{aligned} \boldsymbol{a}\,\boldsymbol{b} = \;& (a^1\, b^1 + a^2\, b^2 + a^3\, b^3) \\ &+ (a^2\, b^3 - a^3\, b^2)\, \boldsymbol{e}_{23} \\ &+ (a^3\, b^1 - a^1\, b^3)\, \boldsymbol{e}_{31} \\ &+ (a^1\, b^2 - a^2\, b^1)\, \boldsymbol{e}_{12} \,. \end{aligned} \tag{1.4}$$

Recall that $\boldsymbol{e}_{23}$, $\boldsymbol{e}_{31}$ and $\boldsymbol{e}_{12}$ are new basis elements of the geometric algebra $\mathbb{G}_3$. Because these basis elements contain two basis vectors from the vector space basis of $\mathbb{R}^3$, they are said to be of *grade* 2. Hence, the expression

$$(a^2\, b^3 - a^3\, b^2)\, \boldsymbol{e}_{23} + (a^3\, b^1 - a^1\, b^3)\, \boldsymbol{e}_{31} + (a^1\, b^2 - a^2\, b^1)\, \boldsymbol{e}_{12}$$

is said to be a vector of grade 2, because it is a linear combination of the basis elements of grade 2.

The sum of scalar products in (1.4) is clearly the standard scalar product of the vectors $\boldsymbol{a}$ and $\boldsymbol{b}$, i.e. $\boldsymbol{a} * \boldsymbol{b}$. The grade 2 part can also be evaluated separately using the *outer product*. The outer product is denoted by $\wedge$ and defined as

$$\boldsymbol{e}_i \wedge \boldsymbol{e}_j = \begin{cases} 0 : i = j \\ \boldsymbol{e}_i\, \boldsymbol{e}_j : i \neq j \end{cases} \tag{1.5}$$

Since $\boldsymbol{e}_i \wedge \boldsymbol{e}_j = \boldsymbol{e}_i \, \boldsymbol{e}_j$ if $i \neq j$, it is clear that $\boldsymbol{e}_i \wedge \boldsymbol{e}_j = -\boldsymbol{e}_j \wedge \boldsymbol{e}_i$. Also, note that the expression $\boldsymbol{e}_i \wedge \boldsymbol{e}_i = 0$ is the defining equation for the Grassmann, or exterior, algebra (see Sect. 3.8.4). Therefore, by using the outer product, the Grassmann algebra is recovered in geometric algebra.

Using this definition of the outer product, it is not too difficult to show that

$$\boldsymbol{a} \wedge \boldsymbol{b} = (a^2 \, b^3 - a^3 \, b^2) \, \mathrm{e}_{23} + (a^3 \, b^1 - a^1 \, b^3) \, \mathrm{e}_{31} + (a^1 \, b^2 - a^2 \, b^1) \, \mathrm{e}_{12} \; ,$$

which is directly related to the standard vector cross product of $\mathbb{R}^3$, since

$$\boldsymbol{a} \times \boldsymbol{b} = (a^2 \, b^3 - a^3 \, b^2) \, \mathrm{e}_1 + (a^3 \, b^1 - a^1 \, b^3) \, \mathrm{e}_2 + (a^1 \, b^2 - a^2 \, b^1) \, \mathrm{e}_3 \; .$$

To show how one expression can be transformed into the other, the concept of the *dual* has to be introduced. To give an example, the dual of e_1 is $\mathrm{e}_2 \, \mathrm{e}_3$, that is, the geometric product of the remaining two basis vectors. This can be interpreted in geometric terms by saying that the dual of the subspace parallel to e_1 is the subspace perpendicular to e_1, which is spanned by e_2 and e_3.

A particularly powerful feature of geometric algebra is that this dual operation can be expressed through the geometric product. For this purpose, the *pseudoscalar* $\boldsymbol{I}$ of $\mathbb{G}_3$ is defined as the basis element of highest grade, i.e. $\boldsymbol{I} := \mathrm{e}_1 \, \mathrm{e}_2 \, \mathrm{e}_3$. Using again the rules of the geometric product, it follows that the inverse pseudoscalar is given by $\boldsymbol{I}^{-1} = \mathrm{e}_3 \, \mathrm{e}_2 \, \mathrm{e}_1$, such that $\boldsymbol{I} \, \boldsymbol{I}^{-1} = \boldsymbol{I}^{-1} \, \boldsymbol{I} = 1$. The dual of some multivector $\boldsymbol{A} \in \mathbb{G}_3$ is denoted by $\boldsymbol{A}^*$ and can be evaluated via

$$\boldsymbol{A}^* = \boldsymbol{A} \, \boldsymbol{I}^{-1} \; .$$

It now follows that

$$(\boldsymbol{a} \wedge \boldsymbol{b})^* = (\boldsymbol{a} \wedge \boldsymbol{b}) \, \boldsymbol{I}^{-1} = \boldsymbol{a} \times \boldsymbol{b} \; .$$

What does this show? First of all, it shows that the important vector cross product can be recovered in geometric algebra. Even more importantly, it implies that the vector cross product is only a special case, for a three-dimensional vector space, of a much more general operation: the outer product. Whereas the expression $\boldsymbol{a} \wedge \boldsymbol{b}$ is a valid operation in any dimension greater than or equal to two, the vector cross product is defined only in three dimensions.

From the geometric interpretation of the dual, the geometric relation between the vector cross product and the outer product can be deduced. The vector cross product of $\boldsymbol{a}$ and $\boldsymbol{b}$ represents a vector perpendicular to those two vectors. Hence, the outer product $\boldsymbol{a} \wedge \boldsymbol{b}$ represents the plane spanned by $\boldsymbol{a}$ and $\boldsymbol{b}$. It may actually be shown that with an appropriate definition of the magnitude of multivectors, the magnitude $\|\boldsymbol{a} \wedge \boldsymbol{b}\|$ is the area of the parallelogram spanned by $\boldsymbol{a}$ and $\boldsymbol{b}$, as illustrated in Fig. 1.1 (see Sect. 4.1.1).

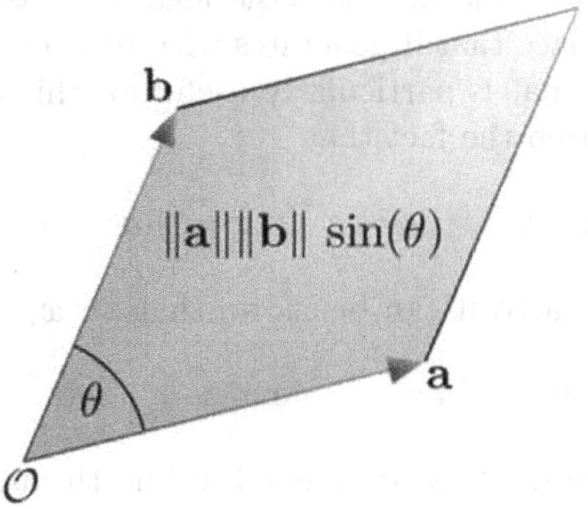

Fig. 1.1 The magnitude of a blade $\boldsymbol{a} \wedge \boldsymbol{b}$ is the area of the parallelogram spanned by $\boldsymbol{a}$ and $\boldsymbol{b}$

One important equation that follows from the above analysis is the relation between the geometric, the scalar and the outer product:

$$\boldsymbol{a}\,\boldsymbol{b} = \boldsymbol{a} * \boldsymbol{b} + \boldsymbol{a} \wedge \boldsymbol{b} \,. \tag{1.6}$$

That is, the geometric product of two vectors results in the sum of a scalar and a grade 2 vector (also called a *bivector*). Note that this is true only for vectors of grade 1. While it may appear strange at first that this is a sum of different types of elements (scalars and bivectors), it is simply a linear combination of elements of the algebraic basis. Adding elements of different types is quite natural, for example, when working with complex numbers, where real and imaginary numbers are added.

It is interesting to note that, in fact, certain subalgebras of geometric algebra are isomorphic to complex numbers. This can be seen quite easily, by first evaluating the square of the bivector $\boldsymbol{e}_1\,\boldsymbol{e}_2 \in \mathbb{G}_3$:

$$(\boldsymbol{e}_{12})^2 = \boldsymbol{e}_{12}\,\boldsymbol{e}_{12} = -(\boldsymbol{e}_1\,\boldsymbol{e}_2)\,(\boldsymbol{e}_2\,\boldsymbol{e}_1) = -\boldsymbol{e}_1\,(\boldsymbol{e}_2\,\boldsymbol{e}_2)\,\boldsymbol{e}_1 = -(\boldsymbol{e}_1\,\boldsymbol{e}_1) = -1 \,.$$

This shows that bivectors in geometric algebra square to minus 1. In fact, there are many such entities. Using the simple rules of the geometric product for orthonormal basis vectors, it can be easily shown that algebraic entities of the type $\boldsymbol{w} = a + b\,\boldsymbol{e}_{12}$ behave in the same way as complex numbers $w = a + \mathsf{i}\,b$, where $\mathsf{i} := \sqrt{-1}$ denotes the imaginary unit (see Sect. 3.8.2).

1.5.3 Geometry

So far, mainly the algebraic properties of geometric algebra have been discussed. In this subsection, the notion of how geometry is represented through algebraic entities is introduced. The basic idea is to associate an algebraic entity with the null space that it generates with respect to a particular operation. One operation that is particularly useful for this purpose is the outer product; this stems from the fact that

$$\boldsymbol{x} \wedge \boldsymbol{a} = 0 \qquad \Longleftrightarrow \qquad \boldsymbol{x} = u\,\boldsymbol{a}\,,\ \forall\, u \in \mathbb{R}\,,$$

where $\boldsymbol{x}, \boldsymbol{a} \in \mathbb{R}^3$. Similarly, it can be shown that for $\boldsymbol{x}, \boldsymbol{a}, \boldsymbol{b} \in \mathbb{R}^3$,

$$\boldsymbol{x} \wedge \boldsymbol{a} \wedge \boldsymbol{b} = 0 \qquad \Longleftrightarrow \qquad \boldsymbol{x} = u\,\boldsymbol{a} + v\,\boldsymbol{b}\,,\ \forall\, u, v \in \mathbb{R}\,.$$

In this way, $\boldsymbol{a}$ can be used to represent the line through the origin in the direction of $\boldsymbol{a}$ and $\boldsymbol{a} \wedge \boldsymbol{b}$ to represent the plane through the origin spanned by $\boldsymbol{a}$ and $\boldsymbol{b}$. Later on in the text, this will be called the *outer-product null space* (see Sect. 3.2.2). This concept is extended in Chap. 4 to allow the representation of arbitrary lines, planes, circles, and spheres.

For example, in the geometric algebra of the projective space of $\mathbb{R}^3$, vectors represent points in Euclidean space and the outer product of two vectors represents the line passing through the points represented by those vectors. Similarly, the outer product of three vectors represents the plane through the three corresponding points (see Sect. 4.2).

In the geometric algebra of conformal space, there exist two special points: the point at infinity, denoted by $\boldsymbol{e}_\infty$, and the origin $\boldsymbol{e}_o$. Vectors in this space again represent points in Euclidean space. However, the outer product of two vectors $\boldsymbol{A}$ and $\boldsymbol{B}$, say, represents the corresponding point pair. The line through the points $\boldsymbol{A}$ and $\boldsymbol{B}$ is represented by $\boldsymbol{A} \wedge \boldsymbol{B} \wedge \boldsymbol{e}_\infty$, that is, the entity that passes through the points $\boldsymbol{A}$ and $\boldsymbol{B}$ and infinity. Furthermore, the outer product of three vectors represents the circle through the three corresponding points, and similarly for spheres (see Sect. 4.3).

Another important product that has not been mentioned yet is the inner product, denoted by $\cdot$. The inner product of two vectors $\boldsymbol{a}, \boldsymbol{b} \in \mathbb{R}^3$ is equivalent to the scalar product, i.e.

$$\boldsymbol{a} \cdot \boldsymbol{b} = \boldsymbol{a} * \boldsymbol{b}\,.$$

However, the inner product of a grade 1 and a grade 2 vector results in a grade 1 vector and not a scalar. That is, given three vectors $\boldsymbol{a}, \boldsymbol{b}, \boldsymbol{c} \in \mathbb{R}^3$, then

$$\boldsymbol{x} := \boldsymbol{a} \cdot (\boldsymbol{b} \wedge \boldsymbol{c})$$

is a grade 1 vector. It is shown in Sect. 3.2.7 that this equation can be expanded into

$$\boldsymbol{x} = (\boldsymbol{a} \cdot \boldsymbol{b})\,\boldsymbol{c} - (\boldsymbol{a} \cdot \boldsymbol{c})\,\boldsymbol{b} = (\boldsymbol{a} * \boldsymbol{b})\,\boldsymbol{c} - (\boldsymbol{a} * \boldsymbol{c})\,\boldsymbol{b}\,.$$

Using this expansion, it is easy to verify that

$$a * x = 0 \iff a \perp x \,,$$

i.e. $\boldsymbol{a}$ is perpendicular to $\boldsymbol{x}$. The geometric meaning of this property is that the inner product removes a linear subspace. That is, while the outer product combines linear subspaces, the inner product subtracts them. Using these operations, a general intersection operation, the *meet*, can be defined (see Sect. 3.2.12). Using the meet, the intersection between any pair of geometric entities can be evaluated.

1.5.4 Transformations

It was mentioned earlier that the fundamental transformation available in geometric algebra is reflection. All other transformations, such as rotation, inversion, and translation, are represented through combinations of this basic transformation. In Sect. 3.3 it is shown that given vectors $\boldsymbol{a}, \boldsymbol{n} \in \mathbb{R}^3$, the vector

$$\boldsymbol{b} := \boldsymbol{n}\,\boldsymbol{a}\,\boldsymbol{n}^{-1}$$

is the reflection of $\boldsymbol{a}$ in $\boldsymbol{n}$, as indicated in Fig. 1.2.

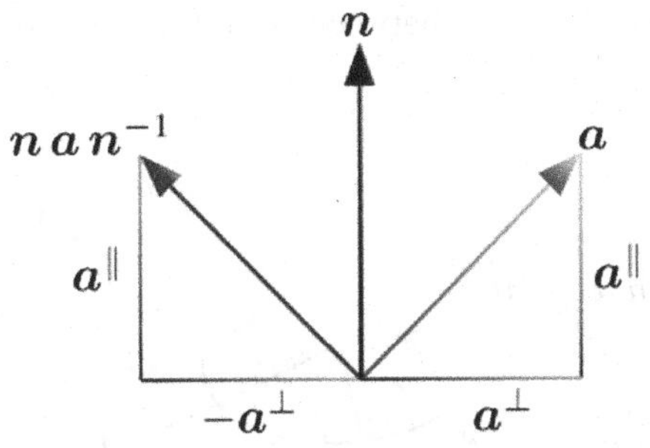

Fig. 1.2 Reflection of $\boldsymbol{a}$ in $\boldsymbol{n}$

The conformal space of the Euclidean space $\mathbb{R}^1$ is a good example of how the reflection operation can represent a non-linear transformation. The first step in constructing the conformal space is to embed the Euclidean space $\mathbb{R}^1$ as the unit circle $\mathbb{S}^1 \subset \mathbb{R}^2$ in $\mathbb{R}^2$. Owing to the particular embedding of $\mathbb{R}^1$, a reflection in $\mathbb{R}^2$ has the effect of an inversion in $\mathbb{R}^1$, as indicated in Fig. 1.3 (see Sect. 4.3.9). Here $\boldsymbol{e}_1$ and $\boldsymbol{e}_+$ denote orthonormal basis vectors of $\mathbb{R}^2$, and $\mathcal{S}$ denotes the embedding operator from $\mathbb{R}^1$ into $\mathbb{R}^2$. That is,

$$x^{-1} = \mathcal{S}^{-1}\Big(\, \boldsymbol{e}_1\, \mathcal{S}(x)\, \boldsymbol{e}_1 \,\Big)\,.$$

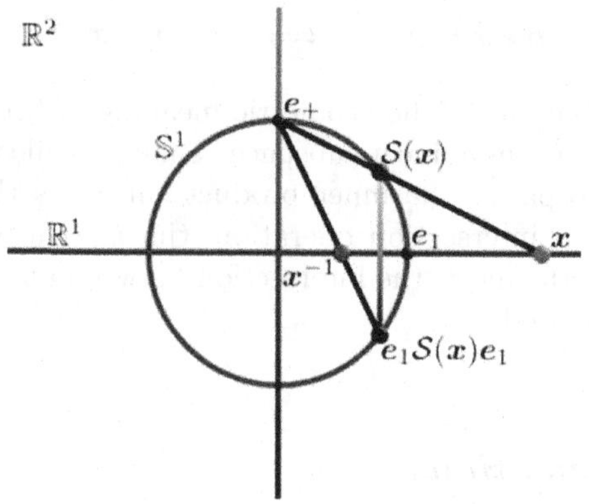

Fig. 1.3 Reflection of $\mathcal{S}(\boldsymbol{x})$ in $\boldsymbol{e}_1$ represents inversion of $\boldsymbol{x}$

Returning again to Euclidean space, given unit vectors $\hat{\boldsymbol{n}}, \hat{\boldsymbol{m}} \in \mathbb{S}^2 \subset \mathbb{R}^3$, where $\mathbb{S}^2$ denotes the unit sphere in $\mathbb{R}^3$, the consecutive reflection of $\boldsymbol{a}$ in $\hat{\boldsymbol{n}}$ and $\hat{\boldsymbol{m}}$ can be evaluated via

$$\boldsymbol{b} = \hat{\boldsymbol{m}}\,\hat{\boldsymbol{n}}\,\boldsymbol{a}\,\hat{\boldsymbol{n}}\,\hat{\boldsymbol{m}}\,.$$

The effect of this double reflection is a rotation of $\boldsymbol{a}$ in the plane spanned by $\hat{\boldsymbol{n}}$ and $\hat{\boldsymbol{m}}$ by twice the angle between $\hat{\boldsymbol{n}}$ and $\hat{\boldsymbol{m}}$, as shown in Fig. 1.4 (see Sect. 4.1.5).

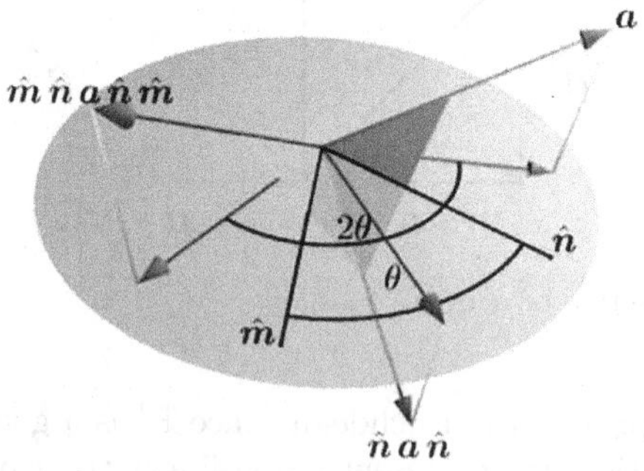

Fig. 1.4 Rotation of $\boldsymbol{a}$ by consecutive reflection in $\hat{\boldsymbol{n}}$ and $\hat{\boldsymbol{m}}$

The product $\hat{\boldsymbol{m}}\,\hat{\boldsymbol{n}}$ may therefore be interpreted as a rotation operator, which is typically called a *rotor*. Defining $\boldsymbol{R} := \hat{\boldsymbol{m}}\,\hat{\boldsymbol{n}}$, the rotation of $\boldsymbol{a}$ can be written as $\boldsymbol{b} = \boldsymbol{R}\,\boldsymbol{a}\,\widetilde{\boldsymbol{R}}$ where $\widetilde{\boldsymbol{R}}$ denotes the *reverse* of $\boldsymbol{R}$, which is defined as $\widetilde{\boldsymbol{R}} := \hat{\boldsymbol{n}}\,\hat{\boldsymbol{m}}$. From the relation between the geometric product and the scalar and outer products as given in (1.6), it follows that

$$\boldsymbol{R} = \hat{\boldsymbol{m}}\,\hat{\boldsymbol{n}} = \hat{\boldsymbol{m}} * \hat{\boldsymbol{n}} + \hat{\boldsymbol{m}} \wedge \hat{\boldsymbol{n}} = \cos\theta + \sin\theta\,\frac{\hat{\boldsymbol{m}} \wedge \hat{\boldsymbol{n}}}{\|\hat{\boldsymbol{m}} \wedge \hat{\boldsymbol{n}}\|}\,, \tag{1.7}$$

because $\|\hat{\boldsymbol{m}} \wedge \hat{\boldsymbol{n}}\| = \sin\theta$, if $\theta = \angle(\hat{\boldsymbol{m}}, \hat{\boldsymbol{n}})$ (see Sect. 4.1.1). It was shown earlier that a bivector $\boldsymbol{e}_1\,\boldsymbol{e}_2$, for example, squares to -1. The same is true for any unit bivector such as $\hat{\boldsymbol{U}} := (\hat{\boldsymbol{m}} \wedge \hat{\boldsymbol{n}})/\|\hat{\boldsymbol{m}} \wedge \hat{\boldsymbol{n}}\|$, i.e. $\hat{\boldsymbol{U}}^2 = -1$. Identifying $\hat{\boldsymbol{U}} \cong \mathsf{i}$, where i denotes the imaginary unit $\mathsf{i} = \sqrt{-1}$, it is clear that (1.7) can be written as

$$\cos\theta + \sin\theta\,\hat{\boldsymbol{U}} \cong \cos\theta + \sin\theta\,\mathsf{i} = \exp(\theta\,\mathsf{i})\,.$$

Extending the definition of the exponential function to geometric algebra, it may thus be shown that the rotor given in (1.7) can be written as

$$\boldsymbol{R} = \exp(\theta\,\hat{\boldsymbol{U}})\,.$$

Just as a rotation operator in Euclidean space is a combination of reflections, the available transformations in conformal space are combinations of inversions, which generates the group of conformal transformations. This includes, for example, dilation and translation. Hence, rotors in the conformal embedding space can represent dilations and translations in the corresponding Euclidean space.

1.5.5 Outermorphism

The *outermorphism* property of transformation operators, not to be confused with an *automorphism*, plays a particularly important role, and is one of the reasons why geometric algebra is such a powerful language for the description of geometry. The exact mathematical definition of the outermorphism of transformation operators is given in Sect. 3.3. The basic idea is as follows: if $\boldsymbol{R}$ denotes a rotor and $\boldsymbol{a}, \boldsymbol{b} \in \mathbb{R}^3$ are two vectors, then

$$\boldsymbol{R}\,(\boldsymbol{a} \wedge \boldsymbol{b})\,\widetilde{\boldsymbol{R}} = (\boldsymbol{R}\,\boldsymbol{a}\,\widetilde{\boldsymbol{R}}) \,\wedge\, (\boldsymbol{R}\,\boldsymbol{b}\,\widetilde{\boldsymbol{R}})\,.$$

Since geometric entities are represented by the outer product of a number of vectors, called a *blade*, the above equation implies that the rotation of a blade is equivalent to the outer product of the rotation of the constituent vectors. That is, a rotor rotates any geometric entity, unlike rotation matrices, which differ for different geometric entities. Of course, all transformation operators that are constructed from the geometric product of vectors satisfy the outermorphism property.

Chapter 2
Learning Geometric Algebra with CLUCalc

Before the theory of geometric algebra is discussed in detail in the following chapters, the software tool CLUCALC [133] is presented, which can help one to understand the algebra's algebraic and geometric properties. CLUCALC allows the user to enter geometric algebra expressions in much the same way as they are written in this text, and to investigate and visualize their outcome. In fact, a whole programming language, called CLUSCRIPT, has been developed by the author that combines structured programming with an intuitive representation of mathematical operations and simple methods to generate high-quality visualizations.

In particular, readers who are interested in the geometric representation of multivectors, as discussed in Chap. 4, will find that CLUCALC facilitates their understanding considerably; this will lead to a quicker and more intuitive insight into the structure of geometric algebra. However, CLUCALC can also be used to implement complex algorithms, to generate illustrations for publications, and to give presentations including animated and user-interactive embedded 3D visualizations. All drawings shown in this text, for example, were generated using CLUCALC. Some additional interesting features of CLUCALC are:

- automatic analysis and visualization of multivectors with respect to their geometric content;
- user-interactive and animated visualizations;
- rendering of arbitrary LaTeX text[1] for the annotation of graphics and for texture mapping onto arbitrary surfaces;
- support for solving (multi)linear multivector equations;
- full support for the tensor representation of multivector equations, as discussed in Chap. 5;
- support for error propagation in algebraic operations;
- reading, writing, and texture mapping of images;
- control of external hardware through a serial port.

[1] This requires LaTeX and Ghostscript to be installed on the system.

C. Perwass, *Geometric Algebra with Applications in Engineering.* Geometry and Computing.

These features indicate that CLUCalc is more than a visualization tool for geometric algebra. In fact, it has been used in various applications. For example, the monocular pose estimation algorithm presented in Chap. 8 was implemented and tested with CLUCalc. CLUCalc was also used for the development and implementation of an inverse kinematic algorithm, which was used to control a robotic arm [95, 96]. CLUScripts that implement these algorithms can be found on [94]. Another application is the visualization of point groups, and wallpaper groups with CLUCalc [97, 142, 99, 98, 143, 100]. The visualizations scripts are freely available from [141]. The Space Group Visualizer, which is the first tool to visualize all 230 space groups in 3D space, is also based on CLUCalc. There are many CLUScripts available that visualize physical and geometrical constellations. For example, there are scripts that simulate the gravitational three-body problem, visualize parts of the solar system, and demonstrate the constraints of the fundamental matrix. All of these can be found in the distribution of CLUCalc and on [133]. The current, freely available versions of CLUCalc for various operating systems, as well as a complete documentation of all its features, can be found at that site.

The aim of this chapter is to show the potential that CLUCalc offers for learning and teaching geometric algebra, but also anything else that has a geometric representation. In the next section, the motivation for the development of CLUCalc and some other software tools are briefly discussed. Then the basics of CLUCalc and the programming language CLUScript are introduced, which should help the reader to use CLUCalc alongside this text as a learning tool.

2.1 Background

Probably the first automatic visualization of elements of geometric algebra was implemented in a MATLAB toolbox called GABLE, which allowed the user to visualize such elements in 3D Euclidean space [51]. Unfortunately, elements in projective or conformal space could not be visualized. Therefore, the development of a C++ software library was started in 2001 by the author, which could automatically interpret elements of a geometric algebra in terms of their geometric representation and visualize them. The visualization was done using the OpenGL 3D graphics library and worked for 3D Euclidean space and the corresponding projective and conformal spaces. The software library, called CLUDraw, was made available in August 2001 under the GNU Public License agreement as open source software. CLUDraw itself was based on a C++ implementation of geometric algebra by the author, which had been under development since 1996.

The drawback of this library was that a user had to know how to program in C++ in order to use it. Furthermore, every time the program was changed, it had to be recompiled in order to visualize the new result. Many people who might have been interested in geometric algebra could therefore not use this tool, and even for those who did know how to program in C++, it was quite tedious to constantly recompile a program when a small change was made to a variable. This is the context in which the idea was born of developing a stand-alone program, with an integrated parser, such that formulas could be typed in and visualized right away. This was the birth of CLUCALC, whose first version was made available for download in February 2002.

The basic philosophy behind the development of CLUCALC was that the user had a geometric-algebra formula and would like to know what that formula implied geometrically. To simplify this process as much as possible, the programming language CLUSCRIPT was developed from scratch. In this way, the symbols representing the various products in geometric algebra could be chosen such that they were as close as possible to the ones used on paper. For example, the formula

$Y = (A \wedge B) \cdot X$ becomes `Y = (A ^ B) . X`

in CLUSCRIPT.

In addition to CLUCALC a number of programming libraries and stand-alone programs in a variety of languages are available. Probably the first implementation of geometric algebra on a computer was produced by a group from the Helsinki University of Technology, lead by Pertti Lounesto, around 1982. This software tool, called CLICAL, can quickly perform calculations in non-degenerate algebras $\mathbb{G}_{p,q}$ for any signature with $p + q < 10$. The latest version of CLICAL is available from [121].

There are packages for the symbolic computer algebra system Maple [1, 2], the C++ software library GluCat [114], the C++ software library generator Gaigen [72], the visualization tool GAViewer [73], and the Java library Clados [44], to name just a few. A more in-depth overview of current software implementations of geometric algebra can be found in [4].

2.2 The Software

In this section, the basic usage of the CLUCALC software program is described. A detailed description can be found in the online manual at [133]. Figure 2.1 shows a screenshot of CLUCALC after it has been started. There are three main windows: the editor, the output and the visualization window. Note that the menus differ somewhat from window to window. If one of the windows cannot be seen, it can be made active by pressing the following key combinations:

- `SHIFT + CTRL + e` activates the editor window.
- `SHIFT + CTRL + v` activates the visualization window.

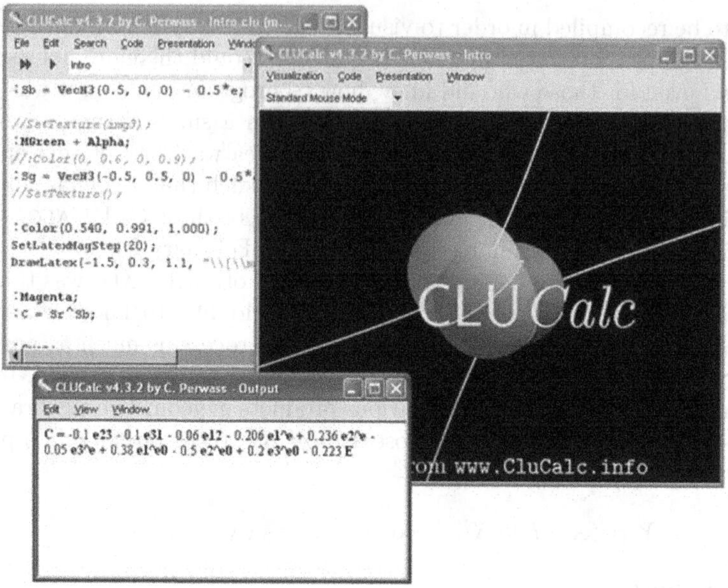

Fig. 2.1 Screenshot of CLUCALC, showing the editor, output and visualization windows

- SHIFT + CTRL + o activates the output window.

CLUSCRIPTs are written in the editor window. When executed the script draws graphical output in the visualization window and outputs text in the output window. Should an error occur during parsing or execution of a script, a corresponding error text is shown in the output window and the editor window displays the appropriate code line.

2.2.1 Editor Window

The editor implements syntax highlighting to facilitate the programming of CLUSCRIPTs. It offers the standard functionality of an editor, with copy, cut, paste, and search and replace routines. A script can be executed (parsed) by pressing `CTRL + p`.

Within the editor window, a number of scripts can be opened concurrently. To switch between the scripts, either the combo box at the top of the editor window or the key combinations `CTRL` + ← and `CTRL` + → can be used. When one switches between scripts, they are not automatically executed again. This is helpful when one is working on an "include" file of a main file while visualizing only the main file. Independently of which script is currently displayed in the editor, it is always possible to reparse and execute the script that is currently visualized by selecting from the `Code` menu the command `Parse Main`. Alternatively, the key combination `CTRL + m` is available.

When the right mouse button is clicked in the editor window, a context menu pops up that allows the insertion of standard CLUScript constants into the script text.

2.2.2 Visualization Window

The visualization window displays anything that is drawn by the script. Independently of what has been drawn, the user can rotate and move the visualization using the mouse. This is done by placing the mouse anywhere inside the visualization window, holding down the left or right mouse button, and moving the mouse. When the left mouse button is pressed, the visualization is rotated, whereas holding the right mouse button translates it. Rotations and translations along different axes can be achieved by holding the `SHIFT` button at the same time. Note that selecting `Local Rotation` in the `Visualization` menu allows a different type of rotation, which may be better suited for certain visualizations. Another useful feature is available when `Animation` and `Mouse animation` are selected in the `Visualization` menu, where the visualization can be given a rotation impulse with the mouse, and the rotation is then continued.

In addition to rotating and translating the whole visualization, the user can also interact with the visualization or, rather, the visualization script using the mouse. This is controlled via `Mouse Modes`, which can be selected in a combo box in the visualization window, or by using the keys `CTRL + 0` to `CTRL + 9`. Mouse mode zero is the standard mouse mode, which, when selected, allows the user to rotate and translate the whole visualization. If one of the other mouse modes is selected, moving the mouse in the visualization window with a mouse button pressed varies internal variables and reexecutes the script at each mouse position change. Within the script, these internal variables can then be used to influence the visualization. In this way, highly interactive scripts can be created.

Note that the visualization window can be toggled between full-screen view and a normal window with the key combination `CTRL + f`.

2.2.3 Output Window

The output window simply displays the values of variables. Using the `View` menu, the variable types can be chosen to be displayed, and it is possible to toggle between a simple and a more complex display. The `Edit` menu allows the user to copy the content of the output window as HTML code into the clipboard.

2.2.4 Command Line Parameters

CLUCalc has a number of command line parameters that can be quite helpful for various purposes. In general, the filename of a CLUScript, when used as a command line parameter, will automatically load and execute the script when CLUCalc is started. Furthermore, the following parameters are available:

- `--main-path=[path]`. Here, a path can be specified which has as subdirectories the `ExampleScripts` and `Documentation` directories. This option allows CLUCalc to find the documentation and example scripts, even if it is started from a different directory.
- `--no-intro`. This starts CLUCalc with an empty script and tiles the windows such that all of them can be seen. This is a useful setting when one is working with CLUCalc.
- `--viz-only`. This starts CLUCalc with only the visualization window. That is, the user can neither edit nor see the script or the text output. This can be useful when one is giving presentations with some other program and using CLUCalc for visualizations in between.

2.3 The Scripting Language

In this section, an overview of the CLUScript programming language is given. A detailed programming and reference manual is included in the CLUCalc distribution and is also available from [133].

2.3.1 Basics

CLUScript is somewhat similar to the C programming language. Just as in C, every program line has to be ended by a semicolon. The advantage of this convention is that a program line can be extended over a number of text lines. Comments can be included in the script in the same way as in C and C++. For example,

```
// This is a single-line comment
/*
        and here
   is a block comment */
```

```
A = 1 + 2;
// The above line gives the same result as
A =
        1
    + 2;
```

Such a script can be executed with the key combination `CTRL + p`. The end of a script does not have to be signaled with a particular keyword. By the way, a question mark at the beginning of a line prints the result in the output window after evaluation of the corresponding line.

It was mentioned in the introduction to this chapter that one aim in the development of the CLUScript language was to map the operator symbols of geometric algebra one-to-one, as far as possible, to the programming language. The resultant syntax is shown in Table 2.1.

Lists are a very useful feature of CLUCalc. A list is generated using square brackets `[]`, as in `lA = [1,2,3];`, or by employing the function `List`. For example, `lA = List(3);` generates a list with three entries and sets them to zero. An empty list can be generated by writing `lA = [];`. Lists can also be nested and they can contain elements of arbitrary type. For example, `lA = [1, "Hello", [2,3]];` is also a valid list, which contains an integer, a string, and a sublist. Elements in a list can be accessed using *round* brackets `()`. For example, `lA(2)` returns the second element in a list `lA`. The second element in a list can be replaced with a new value via `lA(2) = 1;`.

A particularly useful feature of CLUScript is the automatic expansion of operations on lists to all elements in the list. For example,

```
lA = [1,2,3] + 1;
```

adds 1 to all elements in the list and stores the result in `lA`. This is also possible between two lists. For example, the expression `lA = [1,2] + [1,2];` generates the nested list `[[2,3],[3,4]]`. By the way, this can be transformed into a matrix variable using the function `Matrix`. There also exist "point operators" that expand componentwise between lists. For example, `lA = [1,2] .+ [1,2]` results in the list `[2,4]`. That is, the first and the second components are added separately. For more details, see the CLUCalc reference manual. The following script is a short example, which uses this expansion of operations on lists to rotate a list of vectors in a single command line:

```
R = RotorE3(0,1,0, Pi/4);
lV = R * [ VecE3(1,0,0), VecE3(0,1,0), VecE3(0,0,1) ] * ~R;
```

Table 2.1 Operator symbols in CLUSCRIPT

Operation	Formula	CLUSCRIPT
Geometric product	$\boldsymbol{A}\boldsymbol{B}$	`A * B`
Inner product	$\boldsymbol{A} \cdot \boldsymbol{B}$	`A . B`
Outer product	$\boldsymbol{A} \wedge \boldsymbol{B}$	`A ^ B`
Addition	$\boldsymbol{A} + \boldsymbol{B}$	`A + B`
Subtraction	$\boldsymbol{A} - \boldsymbol{B}$	`A - B`
Join	$\boldsymbol{A} \dot{\wedge} \boldsymbol{B}$	`A | B`
Meet	$\boldsymbol{A} \vee \boldsymbol{B}$	`A & B`
Inverse	$\boldsymbol{A}^{-1}$	`!A`
Reverse	$\widetilde{\boldsymbol{A}}$	`~A`
Dual	$\boldsymbol{A}^*$	`*A`
Grade projection	$\langle \boldsymbol{A} \rangle_k$	`A°k`
Integer power	$\boldsymbol{A}^k$	`A^^k`

2.3.2 Visualizing Geometry

In order to describe geometry, the geometric algebra of 3D Euclidean space or the corresponding projective and conformal spaces are typically considered. In each space, vectors and blades usually represent different geometric entities and therefore have to be analyzed differently. CLUCALC allows the user to work in all of these three spaces concurrently and to transfer vectors from any space to any other.

The basis vectors of each of the spaces can be defined by calling the functions `DefVarsE3()`, `DefVarsP3()`, and `DefVarsN3()`, for the Euclidean space $\mathbb{R}^3$ and for the corresponding projective and conformal spaces, respectively. The variables and their algebraic contents defined by these functions are listed in table 2.2. Note that in addition to these three spaces, the conic space discussed in Sect. 4.4 is implemented. See the reference manual for more details.

Vectors in the various spaces can be generated either through linear combination of the basis vectors or with the functions `VecE3()`, `VecP3()`, and `VecN3()`. These functions also transform vectors from one space into another. Here is an example script:

```
?Ae = VecE3(1,2,0); // Create vector in E3 at (1,2,0)
?Ap = VecP3(Ae);    // Embed vector Ae in projective space
?An = VecN3(Ap);    // Embed proj. vector in conformal space
```

The question mark at the beginning of each line is an operator that prints the contents of the element to its right in the output window. The output of the above script is

Table 2.2 Variables defined by the functions `DefVarsE3()`, `DefVarsP3()`, and `DefVarsN3()` and their algebraic meaning

DefVarsE3()		DefVarsP3()		DefVarsN3()	
Entity	Var.	Entity	Var.	Entity	Var.
1	id	1	id	1	id
$\boldsymbol{e}_1$	e1	$\boldsymbol{e}_1$	e1	$\boldsymbol{e}_1$	e1
$\boldsymbol{e}_2$	e2	$\boldsymbol{e}_2$	e2	$\boldsymbol{e}_2$	e2
$\boldsymbol{e}_3$	e3	$\boldsymbol{e}_3$	e3	$\boldsymbol{e}_3$	e3
		$\boldsymbol{e}_4$	e4	$\boldsymbol{e}_+$	ep
				$\boldsymbol{e}_-$	em
				$\boldsymbol{e}_\infty \equiv \boldsymbol{e}_+ + \boldsymbol{e}_-$	einf
				$\boldsymbol{e}_o \equiv \frac{1}{2}(\boldsymbol{e}_- - \boldsymbol{e}_+)$	e0
				$\boldsymbol{n} \equiv \boldsymbol{e}_\infty$	n
				$\bar{\boldsymbol{n}} \equiv \boldsymbol{e}_o$	nb
$\boldsymbol{e}_1 \wedge \boldsymbol{e}_2 \wedge \boldsymbol{e}_3$	I	$\boldsymbol{e}_1 \wedge \boldsymbol{e}_2 \wedge \boldsymbol{e}_3 \wedge \boldsymbol{e}_4$	I	$\boldsymbol{e}_1 \wedge \boldsymbol{e}_2 \wedge \boldsymbol{e}_3 \wedge \boldsymbol{e}_\infty \wedge \boldsymbol{e}_o$	I

Ae = 1 **e1** + 2 **e2**
Ap = 1 **e1** + 2 **e2** + 1 **e4**
An = 1 **e1** + 2 **e2** + 2.5 **e** + 1 **e0**

In projective space, the homogeneous dimension is denoted by `e4`. In conformal space, `e` and `einf` denote $\boldsymbol{e}_\infty$ and `e0` denotes $\boldsymbol{e}_o$. Moving between the different spaces is particularly useful for showing the different geometric interpretations of blades in the different spaces. However, note that blades cannot be transferred directly from one space to another. This has to be done via the constituent vectors. Here is an example:

```
Ae = VecE3(1,0,0);      // Create unit vector along x-dir.
Be = VecE3(0,1,0);      // Create unit vector along y-dir.

:Red;       // Switch current color to red.
:Ae^Be;     // Visualize the outer product of Ae and Be

:Blue;                   // Switch current color to blue.
:VecP3(Ae)^VecP3(Be);    // Visualize outer prod. of Ae and Be
                         // when embedded in projective space

:Green;                  // Switch current color to green.
:VecN3(Ae)^VecN3(Be);    // Visualize outer prod. of Ae and Be
                         // when embedded in conformal space
```

The colon is an operator that tries to visualize the element to its right. Figure 2.2(a) shows the resultant visualization of this script. The disk is the result of `Ae^Be`, the line is the result of `VecP3(Ae)^VecP3(Be)`, and the two points are the result of `VecN3(Ae)^VecN3(Be)`. The disk represents

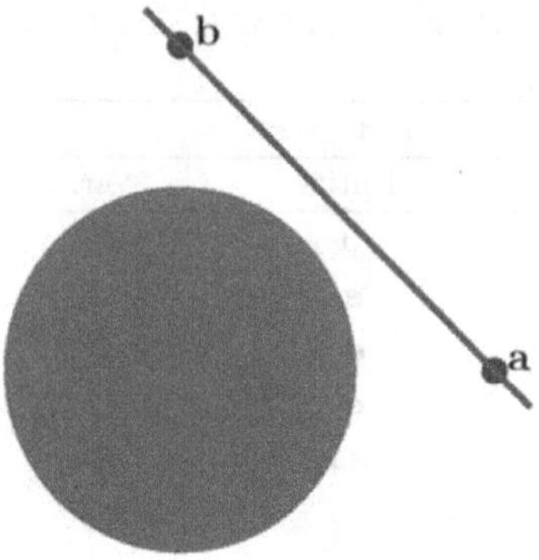

(a) Geometric interpretation of the outer product of two Euclidean vectors (the disk) and the geometric interpretation of the outer product when the vectors are embedded in the corresponding projective space (the line) and conformal space (the point pair)

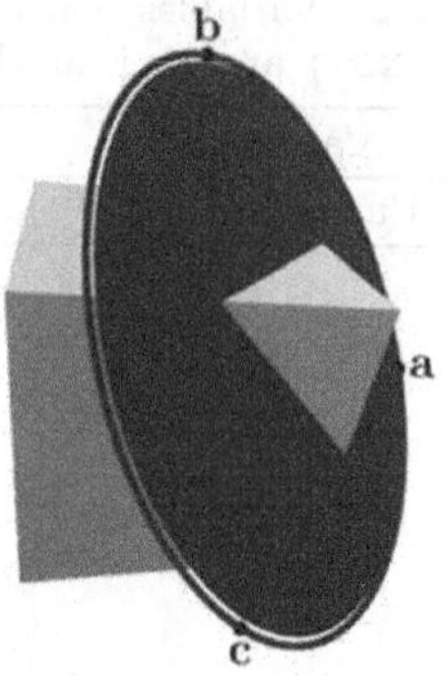

(b) Geometric interpretation of the outer product of three vectors in the various spaces. The Euclidean, projective, and conformal interpretations are now the whole space, the disk and the circle, respectively

Fig. 2.2 Geometric interpretation of blades in different spaces

the subspace spanned by `Ae` and `Be`, where the area of the disk is the magnitude of the blade `Ae^Be`. In projective space the outer product of the embedded points `VecP3(Ae)` and `VecP3(Be)` represents the line passing through the points `Ae` and `Be` in Euclidean space. The length of the line gives the magnitude of the corresponding blade. The outer product of the two vectors when embedded in conformal space (`VecN3(Ae)^VecN3(Be)`) represents the point pair (`Ae`, `Be`) in Euclidean space and is visualized accordingly. Note that such a point pair is in fact a one-dimensional sphere.

Figure 2.2(b) shows the same for the outer product of three vectors $\boldsymbol{a}$, $\boldsymbol{b}$, and $\boldsymbol{c}$, when regarded as Euclidean vectors and embedded in the corresponding projective and conformal spaces. The outer product of three Euclidean vectors in 3D Euclidean space represents the whole space. This is represented by the cube. The volume of the cube is equal to the magnitude of the blade. When these three vectors are embedded in projective space, their outer product represents a plane in Euclidean space, which is represented as a disk. Again, the area of the disk is the magnitude of the blade. The outer product of the three vectors, when embedded in conformal space, represents a circle through these three points in Euclidean space. Hence, the circle is drawn through the three points. The radius of the circle can be extracted from the blade, but is not directly related to the magnitude of the blade in conformal space.

Note that the script shown above does not automatically generate the annotations seen in Fig. 2.2. This has to be done with the help of additional commands, as will be explained later on. Nonetheless, the images in Fig. 2.2 were generated directly with CLUCALC, and it is possible to use arbitrary LaTeX code for the annotations.

Intersections of geometric entities can also be evaluated and visualized quite easily. In geometric algebra, the *meet* of two blades gives the blade representing the largest subspace that the two blades have in common (see Sect. 3.2.12). Geometrically speaking, this is an intersection. In CLUCALC, the operator `&` is used to evaluate the meet of two multivectors. Note that when `&` is applied to two integer values, it evaluates their bitwise logical AND. In the following example script, two spheres represented in conformal space are intersected and the resulting circle is intersected again with a plane:

```
SetPointSize(6); // Increase size of points
SetLineWidth(4); // Increase width of lines

DefVarsN3(); // Define standard variables for conformal space
:N3_SOLID; // Display spheres as solid objects
:DRAW_POINT_AS_SPHERE; // Draw points as small spheres

//      red   green  blue   alpha = transparency
:Color(1.000, 0.378, 0.378, 0.8); // Use a transparent color
// Make center of first sphere user-interactive
:S1 = SphereN3(VecE3(1), 1);

:MBlue; // Medium blue
:S2 = SphereN3(0.5,0,0, 1); // Another sphere of radius 1

:Green;
:C = S1 & S2; // The intersection of the two spheres

:Orange;
// Create plane, where 'e' is a predefined variable denoting
// the point at infinity. 'e' is defined through the function
// DefVarsN3() which was called in the fourth line.
// The scalar factor affects the size
// of the plane visualization.
:P = 10*VecN3(0,0,0) ^ VecN3(1,0,0) ^ VecN3(0,0,1) ^ e;

:Magenta;
:X = P & C; // Evaluate intersection of circle and plane
```

Figure 2.3 shows the result of the above script. Note that all intersections were evaluated with the same operation: the meet. The meet operator, as implemented in CLUCALC, considers only the algebraic properties of the multivectors and knows nothing about their geometric interpretation. It is, in fact, a direct implementation of the mathematical definition of the meet. The center of the sphere `S1` in the above script can be changed interactively by the user with the mouse when mouse mode 1 is selected. This will be explained in more detail in Sect. 2.3.3. If `S1` is moved by the user such that

(a) The meet of two spheres resulting in a circle. The meet of the circle with the plane results in a point pair

(b) The meet of two spheres that do not intersect, resulting in a circle with an imaginary radius. The meet of the imaginary circle with the plane results in an imaginary point pair

Fig. 2.3 Intersection of spheres

the spheres do not intersect anymore, the meet operation returns an algebraic object that can be interpreted as a circle with an imaginary radius. This is visualized by a dotted circle, as shown in Fig. 2.3(b). Imaginary spheres and point pairs exist as well and are visualized as transparent spheres and two points connected by a dotted line, respectively.

Typically the visualization of functions is also of high interest. With CLUCALC, scalar-valued, vector-valued, and certain multivector-valued functions can be drawn. One particularly interesting feature is the visualization of circle-valued functions. Since a circle can be represented in conformal space as a trivector, i.e. the outer product of three vectors, a circle-valued function is synonymous with a trivector-valued function. The following script gives an example of this. Figure 2.4 shows the visualization result of this script. In Fig. 2.4(a) the circles are superimposed on the surface while in figure 2.4(b) an image has been loaded and mapped onto the surface. Note that any image could have been used for this purpose.

```
// Preliminary settings
_BGColor = White;
EnableSmoothLine(1);
SetLineWidth(2);

// Define the standard variables for
// conformal space.
DefVarsN3();

// Generate a standard plane passing through
// (0,0,1), (0,1,0), (0,0,-1) and infinity.
P = VecN3(0,0,1) ^ VecN3(0,1,0) ^ VecN3(0,0,-1) ^ e;

// Now define a function which generates
```

```
// circles.
MyFunc =
{
    a = _P(1); // first parameter of function
    Plane = _P(2); // second parameter of function

    // Generate radius of circle from 'a'
    r = pow(cos(1.5*a), 2) + 0.1;
    // Generate center of circle from 'a'
    d = 0.5*pow(cos(1*a), 2) + 0.1;

    // Generate a sphere
    S = SphereN3(0,0,d, r);
    // Generate circle as intersection of
    // sphere 'S' and plane 'P'.
    C = S & Plane;

    // Generate a rotor, rotating about angle 'a'.
    R = RotorN3(0,1,0, a);

    // Generate a color depending on the circle's radius
    Col = Color(r - 0.1, 0, 1.1 - r);

    // Evaluate the return value of this function
    // as the rotated circle and the color.
    // No semicolon here since this is the return value.
    [(R * C * ~R), Col]
}

// Preparations for a loop that generates a list
// of circles by evaluating the function 'MyFunc'
// for consecutive values of its parameter.
phi = 0;
Circle_list = []; //MyFunc(phi);
Color_list = []; //Color(1,0,0);
// Here the loop starts
loop
{
    // Check the loop-end criterion
    if (phi > 2*Pi)
        break;

    // Store return values in variables
    [Circ, Col] = MyFunc(phi, P);

    // Add a circle to the circle list
    Circle_list << Circ;
    // Add a corresponding color to the color list.
    Color_list << Col;

    // Increase the counting variable.
    phi = phi + Pi / 40;
}
```

(a) Surface with superimposed circles

(b) Same surface with texture mapping

Fig. 2.4 Surface swept out by a circle-valued function

```
// Create a check-box tool, which allows the
// user to decide whether to draw the actual
// circles or not.
if (CheckBox("Show circles", 0))
    :Circle_list;

// Create a check-box tool, which allows the user
// to decide whether to use a texture or not.
if (CheckBox("Use Texture", 0))
{
    // Load image 'checkers.png' and set it as texture
    SetTexture( ReadImg( "checkers.png" ) );
    // Repeat texture with 1/0.2 times over whole surface
    SetTextureRepeat(0.2);
}

// Now draw the surface spanned by the circles
// using the corresponding colors.
DrawCircleSurface(Circle_list, Color_list);

// Reset texture mapping
SetTexture();
```

To make such surfaces look even better, it is possible to add additional lights to the scene. Basically all of the lighting features offered by OpenGL can be accessed directly by CLUCalc. For details on how to use lights, please refer to the documentation included with a CLUCalc distribution or the online manual on [133].

2.3.3 User Interaction

Although static visualizations of geometric algebra formulas are already quite useful, some properties can be understood much better with interactive visualizations. Making a visualization user-interactive is done quite easily with CLUCalc. For example, a vector whose position can be changed by the user is generated with `VecE3(1)`. The number passed to the function `VecE3` denotes the mouse mode mentioned earlier. This mouse mode can be set via a menu in CLUCalc. If, in this example, the user sets the mouse mode to 1, and then places the mouse pointer in the visualization window, holds down the right mouse button and moves the mouse, the return value of `VecE3(1)` is changed and the whole script is reexecuted. Therefore, anything that depends on the return value of `VecE3(1)` will also be changed and redisplayed. The scripts presented so far may thus be made user-interactive by simply changing the initial `VecE3` functions from fixed values (e.g. `VecE3(1,0,0)`) to mouse modes (e.g. `VecE3(1)`).

Apart from generating user-interactive vectors, it is also possible to extract user-interactive scalar values which are related to certain mouse movements in given mouse modes. The function which does this is called `Mouse`. For example, the return value of the function call `Mouse(1,2,1)` depends on the movement of the mouse along its x-axis when mouse mode 1 is selected and the right mouse button is pressed. The first parameter gives the mouse mode, the second one the mouse button (1, left; 2, right), and the third parameter gives the axis (1, x-axis; 2, y-axis; 3, z-axis). By default, moving the mouse with the right mouse button pressed changes the values for the x- and z-axes. When the `SHIFT` key is pressed at the same time, the values for the x- and y-axes are changed. The `Mouse` function for the left mouse button, i.e. when the second parameter is set to 1, works very similarly. The only difference is that the values returned lie in the range $[0, 2\pi[$. This is very useful when making rotors interactive. Here is an example script:

```
// Use mouse mode 1 and left mouse button
?a = Mouse(1,1,1);  // Value changed by movement in x-dir.
?b = Mouse(1,1,2);  // Value changed by pressing "shift"
                    // and moving in y-dir.
?c = Mouse(1,1,3);  // Value changed by NOT pressing "shift"
                    // and movement in y-dir.

:Red;                      // Set color to red
:Ry = RotorE3(0,1,0, a);   // Rotor about y-axis with angle 'a'

:Color(0.1, 0.2, 0.8);     // Set color to given (r,g,b) values
:Rz = RotorE3(0,0,1, b);   // Rotor about z-axis with angle 'b'

:Color(0.2, 0.8, 0.1);     // Set color to given (r,g,b) values
:Rx = RotorE3(1,0,0, c);   // Rotor about x-axis with angle 'c'
```

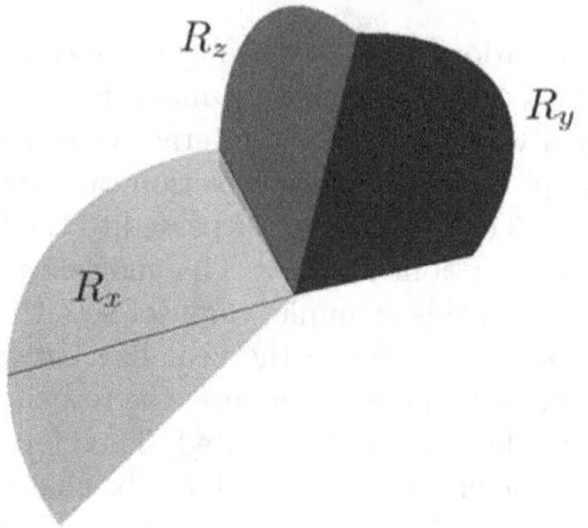

Fig. 2.5 Three rotors about the x-, y- and z-axes

The text output window of CLUCALC will now display the values of x, y, and z, and the visualization window displays the rotors. Figure 2.5 shows an example visualization. A rotor is always visualized as a rotation axis with a partial disk perpendicular to it, representing the rotation plane and the rotation angle. By switching to mouse mode 1, holding down the left mouse button, and moving the mouse, the user can continually change the visualization according to the movement of the mouse.

2.3.4 Animation

Besides user-interactive visualizations, animation can be very helpful. The two features can also be mixed, such that the user can interact with some animated features. Again, it is very simple to animate a script with CLUCALC. First, CLUCALC has to be told that a script is to be animated. This is done with the script line `EnableAnimate( true );`. Once the script has been parsed, it is executed continually, with a maximum of 25 executions per second. The number of executions that can be achieved per second depends on the script to be executed and the computer used.

In order to generate animated visualizations, two variables are now available: `Time` and `dTime`. The former gives the time in seconds that has elapsed since the start of the animation, and the latter gives the time that has elapsed since the last execution of the script. Although the time variables give the time in seconds, their precision lies in the area of one millisecond. Here is an example script that uses animation:

```
// Enable animation of this script.
```

```
EnableAnimate( true );

// Output time since start of animation in text output window.
?Time;

DegPerSec = 45; // Rotate with 45 degrees per second

// Evaluate current rotation angle. The variable 'RadPerDeg'
// is predefined and gives radians per degree. The variable
// 'Pi' is also predefined with the value of pi.
// The operator '%' evaluates the modulus.
?angle = (Time * DegPerSec * RadPerDeg) % (2*Pi);

:Black;
// This will become the user-interactive rotation axis with
// initial value (0,1,0), i.e. the y-axis.
:w = VecE3(1) + VecE3(0,1,0);

:Blue;
:R = RotorP3(w, angle); // The actual rotor about axis 'w'

:Red;
// The plane through points (1,0,0), (1,1,0), (1,0,1).
A = VecP3(1,0,0)^VecP3(1,1,0)^VecP3(1,0,1);

// The plane A rotated by R.
// Note that ~R denotes the reverse of R.
:B = R * A * ~R;
```

The animation feature may also be used to simulate physical effects such as springs or gravitation. In this case it is usually necessary to initialize a set of variables during the first run of a script. Subsequent runs should then only adapt the variables' current values. This can be done in CLUCALC via the predefined variable `ExecMode`. The value of this variable may differ in each execution of a script and depends on the reason why the script was executed. To check for a particular execution mode, a bitwise AND operation of `ExecMode` with one of a set of predefined variables has to be perfomed. For example, if the script is executed because the user has changed it, then `ExecMode & EM_CHANGE` is non-zero, where `&` is the operator for a bitwise AND. This is used in the following example script, where a mass feels a gravitational pull towards the origin:

```
EnableAnimate( true ); // Make this script animated

// Check whether script has just been loaded
// or has been changed.
if (ExecMode & EM_CHANGE)
{
    // if true: initialize variables.
    TimeFactor = 0.01; // Factor for simulation timestep
    G = 6.67e-2; // Gravitational constant (in some units)
```

```
    pos = VecP3(0,1,0);     // Position vector in proj. space
    vel = DirVecP3(6,0,0); // Direction vector in proj. space

    Mass = 1; // Mass of object at origin
    O = VecP3(0,0,0); // the origin
}

?deltaT = dTime * TimeFactor; // Current time step

dist = abs(pos - O); // Distance of mass to origin
dir = (O - pos) / dist; // Normalized direction to origin
acc = G * Mass / (dist * dist); // Current acceleration
vel = vel + acc * dir; // Adapt velocity

:Red;
:pos = pos + vel * deltaT; // Adapt position and draw it
:Blue;
:O;
```

2.3.5 Annotating Graphics

It was mentioned before that it is possible to annotate graphics generated with CLUCALC using arbitrary LaTeX code. This is a very useful feature, because visualizations can become much easier to understand when the various elements are labeled. The advantage of using LaTeX code is, clearly, that virtually all mathematical symbols are available. It is also convenient for producing illustrations for printed texts, since the same symbols can be used in the text and the illustration. In order for CLUCALC to be able to render LaTeX code, the LaTeX software environment and some other helper programs have to be installed. For the exact details, see [133].

LaTeX text can be rendered in CLUCALC using the command `DrawLatex`. The text always has to be drawn relative to a point. For example, to draw the text "Hello World" at position $(1,1,1)$, one can write

```
DrawLatex(1,1,1, "Hello World").
```

Instead of passing a string containing the LaTeX text, it is also possible to give the filename of a LaTeX file, which is rendered instead.

The text is rendered as a bitmap and then drawn with the bottom left corner at the position given. Since the text is drawn as a bitmap, it is not changed perspectively when moved around in the visualization space. However, it is possible to use texture mapping to draw LaTeX text on arbitrary surfaces. This is demonstrated by the following short script. The visualization generated by this script is shown in Fig. 2.6.

```
// Set Latex rendering size
SetLatexMagStep( 20 );
// Generate image of Latex text in white.
```

Fig. 2.6 Texture mapping of LaTeX text on an arbitrary surface

```
:White;
imgText = GetLatexImg( "$\\int$\\LaTeX\\, dt", "latex" );

// Set Color to red.
:Red;
// Set image of Latex text as texture
SetTexture( imgText );
// Repeat texture 5 times in each direction
ScaleFrame( "texture", 5, 5, 1 );
// Plot surface with texture map
:Plot( VecE3(x, sin(0.5*Pi*x)*cos(0.5*Pi*y), y),
[x, -1, 1, 20], [y, -1, 1, 20]);
// Reset texture mapping
SetTexture();
```

It is also possible to make an arbitrary adjustment of the text bitmap relative to the point given (see the help on `SetLatexAlign` within CLUCalc).

The process of rendering a bitmap from LaTeX code is not particularly fast. Therefore, rendered text bitmaps are cached, so that they only have to be rendered once when the script is loaded. However, this would mean that anybody who wanted to use a script that contained LaTeX annotations would need to have LaTeX installed. Furthermore, it can be quite annoying to always have to wait for a couple of seconds before all the LaTeX text for a script is rendered. For all of these reasons, the rendered LaTeX bitmaps can be stored in files and are then readily available when a script is loaded. For example,

```
DrawLatex(1,1,1, "Hello World", "text1")
```

renders the text "Hello World" and stores the rendered bitmap with a filename that is constructed from the name of the script and "text1". For more details on how to render LaTeX text, see [133] again.

Besides annotating geometric entities in a visualization, it is also of practical value to have text that does not move with the visualization, for example a title for the visualization. This is also possible with CLUCalc, by drawing text in an overlay. Note that anything, not just text, can be drawn in an overlay, which allows a static foreground or background. The following script gives an example of this feature. The corresponding visualization is shown in Fig. 2.7.

```
_BGColor = White; // Set White as background color

DefVarsE3();              // Define variables for E3
// Set Latex magnification. This is chosen very high here,
// since this script was used to generate the
// corresponding figure for this text.
SetLatexMagStep(20);
SetPointSize(8);          // Size in which points are drawn.
:DRAW_POINT_AS_SPHERE;    // Draw points as small spheres.
:E3_DRAW_VEC_AS_ARROW;    // Draw vectors in E3 as arrows.

:e1 :Red;         // Draw basis vector e1 in red.
:e2 :Green;       // Draw basis vector e2 in green.
:e3 :Blue;        // Draw basis vector e3 in blue.

:Black;           // Draw the following text in black
SetImgAlign(0, 0.5); // Align text left/centered
SetImgPos( e1 );
:GetLatexImg("$\\vec{e}_1$", "e1"); // Draw text

SetImgAlign(0.5, 0); // Align text centered/bottom
SetImgPos( e2 );
:GetLatexImg("$\\vec{e}_2$", "e2"); // Draw text

SetImgAlign(1, 0.5); // Align text right/centered
SetImgPos( e3 );
:GetLatexImg("$\\vec{e}_3$", "e3"); // Draw text

:E3_DRAW_VEC_AS_POINT;  // Draw vectors as points
:Orange;          // Set color to orange
:A = VecE3(1);   // Draw user-interactive vector
:Black;                       // Set color to black
SetImgAlign(-0.05,0);       // Align text
SetImgPos( A );
:GetLatexImg(  // Draw formula at position of 'A'
  "\\[\\oint_{\\mathcal{S}}\\,f(\\vec{x})\\,d\\vec{x}\\]",
  "formula");

// Now start an overlay to draw text
// that does not move when the user
// rotates or translates the above visualization.
scTitle = Scene( "Title" );
EnableSceneResetFrame( scTitle, true );
SetSceneOverlay( scTitle, 0, 1, 0, 1, -1, 1, false );
DrawToScene( scTitle );
SetLatexMagStep(26); // Size of Latex textS
SetImgAlign(0,1);  // Align text left/top.
SetImgPos( 0, 1, 0 );
// Draw Title of slide
:GetLatexImg("\\sl The Title", "title");
```

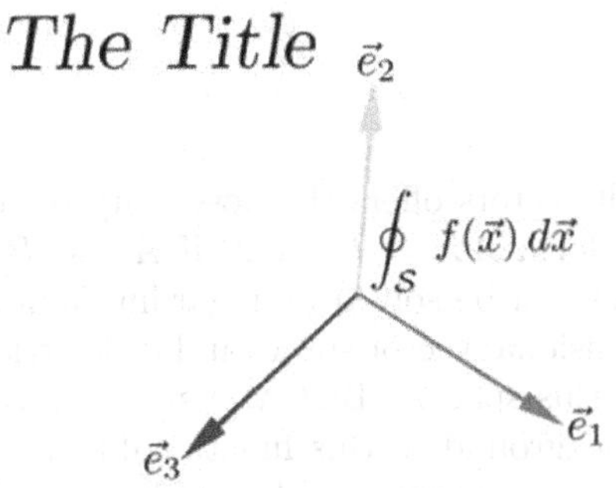

Fig. 2.7 LaTeX text in a visualization

```
DrawToScene();          // End overlay here
// Draw Title
:scTitle;
```

With the overlay feature, basically all prerequisites to give presentations with CLUCalc are available. Indeed, the other elements that are needed, such as a full-screen mode, stepping through a list of slides, and using a white background, are also provided by CLUCalc. In the CLUCalc distribution, an example presentation is included, which could serve as a template for your own presentations. Of course, preparing presentations in CLUCalc is not as easy as using a tool such as OpenOffice or PowerPoint, but CLUCalc allows you to combine text with interactive and/or animated 3D graphics.

2.3.6 Multivector Calculations

Besides the visualization features, CLUCalc also supports advanced calculations with multivectors. A very useful operation is the inversion of a multivector, if the multivector has an inverse. The inversion is obtained with the operator `!`. If a multivector has no inverse, then zero is returned. For example, the script

```
DefVarsE3();                // Define variables for E3

?M = 1 + e1 + e2^e3;        // Define some multivector
?iM = !M;                   // Evaluate inverse of M
?"M * iM = " + M*iM;        // Check that iM is inverse of M

?W = 1 + e1;                // A non-invertible multivector
?iW = !W;                   // The inversion
```

results in the output

M = 1 + 1 **e1** + 1 **e23**
iM = 0.2 + 0.2 **e1** + -0.6 **e23** + 0.4 **I**

Constant = M * iM = 1
W = 1 + 1 **e1**
iW = 0

The inversion of multivectors offers the possibility to solve simple multivector equations of the form $AX = B$ for X if A and B are known. More complex linear equations can be solved by regarding multivectors as simple vectors in a higher-dimensional vector space, and geometric-algebra products as bilinear functions in this space. CLUCALC supports this way of looking at multivectors. The background to this functionality is discussed in detail in Chap. 5. The main ideas are presented briefly in the following.

The algebraic basis of a geometric algebra over an n-dimensional vector space has dimension 2^n. Denoting the algebraic basis of $\mathbb{G}_n$ by $\{\boldsymbol{E}_i\}$, then an arbitrary multivector $\boldsymbol{A} \in \mathbb{G}_n$ can be written as

$$\boldsymbol{A} = \sum_{i=1}^{2^n} a^i\, \boldsymbol{E}_i,$$

where the $\{a^i\} \subset \mathbb{R}$ are scalars. The inner, outer, and geometric products of geometric algebra between two multivectors may then be written as

$$\boldsymbol{A} \circ \boldsymbol{B} = \sum_{i=1}^{2^n} \sum_{j=1}^{2^n} \sum_{k=1}^{2^n} a^i\, b^j\, \Gamma^k{}_{ij}\, \boldsymbol{E}_k, \tag{2.1}$$

where $\boldsymbol{B} = \sum_{i=1}^{2^n} b^i\, \boldsymbol{E}_i$, $\circ$ is a placeholder for one of the products, and $\Gamma^k{}_{ij}$ is a tensor whose entries are either 1, -1, or zero. This tensor encodes the particular properties of the product used. Therefore, this tensor is different for each product. If a third multivector $\boldsymbol{C} \in \mathbb{G}(\mathbb{R}^n)$ is written as $\boldsymbol{C} = \sum_{i=1}^{2^n} c^i\, \boldsymbol{E}_i$ and $\boldsymbol{C} = \boldsymbol{A} \circ \boldsymbol{B}$, then the relation between the $\{a^i\}$, $\{b^j\}$, and $\{c^k\}$ follows from (2.1) as

$$c^k = \sum_{i=1}^{2^n} \sum_{j=1}^{2^n} \alpha^i\, \beta^j\, \Gamma^k{}_{ij}\,. \tag{2.2}$$

If $\boldsymbol{A}$ and $\boldsymbol{C}$ are known and $\boldsymbol{B}$ is to be evaluated, the sum over i on the right-hand is evaluated, which then gives $c^k = \sum_{j=1}^{2^n} b^j\, H^k{}_j$, where $H^k{}_j := \sum_{i=1}^{2^n} a^i\, \Gamma^k{}_{ij}$. By writing the $\{c^k\}$ and $\{b^j\}$ as column vectors c and b, respectively, and $H^k{}_j$ as a matrix H, the above equation becomes $\mathsf{c} = \mathsf{H}\,\mathsf{b}$, which may be solved for b by inverting H. This is exactly what is done by CLUCALC internally when the inverse of a multivector is evaluated.

If, instead of $\boldsymbol{A}$ and $\boldsymbol{C}$, in the above example, $\boldsymbol{B}$ and $\boldsymbol{C}$ are known and $\boldsymbol{A}$ is to be evaluated, then first the sum over j has to be evaluated. Taking care of the order of the indices i and j of $\Gamma^k{}_{ij}$ is important, since the products are in general not commutative, i.e. $\boldsymbol{A} \circ \boldsymbol{B} \neq \boldsymbol{B} \circ \boldsymbol{A}$. In CLUCALC the function `GetMVProductMatrix` is used to evaluate the sum of the components of a multivector with the tensor $\Gamma^k{}_{ij}$ for a given product. The functions `MV2Matrix` and `Matrix2MV` transform multivectors to a vector representation and vice versa. Here is a simple example script that evaluates the geometric product of two multivectors using this vector representation:

```
DefVarsE3();    // Define variables for E3
A = e1;         // Define multivector A
B = e2;         // Define multivector B

// Evaluate the matrix for the product A * B of the
// components of A summed over index i of the
// tensor g^k_ij representing the geometric product.
Agp = GetMVProductMatrix(A, MVOP_GP, 1 /* from left  */);
// Do the same for product B * A with components of a
gpA = GetMVProductMatrix(A, MVOP_GP, 0 /* from right */);

// Transform multivector B in matrix representation.
// Bm is a column vector.
Bm = MV2Matrix(B);

// The operator '*' between two matrices evaluates
// the standard matrix product.
C1m = Agp * Bm; // Evaluate A * B
C2m = gpA * Bm; // Evaluate B * A

?C1 = Matrix2MV(C1m); // Transform C1m back to a multivector
?C2 = Matrix2MV(C2m); // Transform C2m back to a multivector
```

This script produces the text output

C1 = [1 **e12**]
C2 = [-1 **e12**]

Many more things can be done with these functions. It is, for example, possible to map not a whole multivector to a vector, but only certain components. In this way, it is possible to solve for a rotor numerically, while ensuring that the result can have only scalar and bivector components. This is discussed in more detail in [133].

Another advanced feature of CLUCalc is the evaluation of products in geometric algebra with error propagation. This means that for each multivector a covariance matrix can be defined representing the uncertainty with which the multivector is known (see Chap. 5). It is then possible to evaluate the geometric, inner, and outer products of multivectors with associated covariance matrices and propagate their uncertainties. Since intersections of geometric objects can in principle be evaluated using the inner, outer, and/or geometric product, it is possible to perform uncertain geometric reasoning with CLUCalc. Furthermore, since circles and spheres are represented in conformal space in a linear fashion, error propagation extends readily to these objects. One can, for example, evaluate a circle with an associated covariance matrix from three uncertain points simply by evaluating their outer product with error propagation. The mathematical aspects of this are discussed in more detail in Chap. 6. How error propagation can be used in CLUCALC is, again, presented in [133].

There are many more aspects of CLUCALC that would be worth mentioning. However, what has been presented so far should suffice to allow the reader to use CLUCALC as a helpful software tool to facilitate the understanding of the remainder of this text.

2.4 Summary

The goal of this chapter was to give the reader a first impression of the features of CLUCALC and CLUSCRIPT, and to show how this software may be used to support the learning and teaching of geometry and other subjects that profit from geometric visualizations. To summarize, probably the most important features of CLUCALC are the following:

1. CLUCALC is a tool that analyzes entities of geometric algebra with respect to their geometric meaning and visualizes them. This can be very useful when learning about the geometric algebras of Euclidean, projective, and conformal space, since it is possible to *discover* what the inner, outer, and geometric products mean geometrically.
2. CLUCALC can be used to write scripts that evaluate complex algorithms, and offers a platform to easily visualize results as interactive and/or animated 3D graphics. This is certainly a useful aspect when one is doing research with geometric algebra.
3. CLUCALC allows the user to produce images for printed publications that use the same LaTeX fonts as the printed text. By allowing transparent surfaces and user-defined lighting, it allows the user to greatly increase the visual impact of 3D graphics in a printed medium.
4. All visualizations can be used immediately in presentations. In particular, the combination of fixed text and interactive and animated 3D graphics in a slide can further the understanding of complex geometric relations and is also bound to catch the attention of an audience.

Part I
Theory

Chapter 3
Algebra

In this chapter, a detailed introduction to the algebraic properties of geometric algebra is given. The aim is to not only give an axiomatic derivation but also to discuss the calculation rules that are needed to deal effectively with geometric-algebra equations. The chapter starts with an axiomatic discussion of geometric algebra in terms of the elements of a canonical vector space basis. Since, in this text, only geometric algebras over vector spaces are used and, for any vector space, a basis can be found, this approach does not constitute a loss of generality.

After the fundamental properties of *basis blades* are introduced in Sect. 3.1, these are extended to general *blades* in Sect. 3.2. The properties derived here are those that are most often used in later chapters. While blades represent linear subspaces and thus geometric entities, Sect. 3.3 discusses *versors*, which represent transformation operations, such as reflection and rotation. In this context, the *Clifford* group and its subgroups the *pin* and the *spin* group are discussed. A more general type of transformation of multivectors is presented in Sect. 3.4, where general linear functions are considered. These are directly related to matrix algebra, and certain properties of determinants are easily derived in this context. Section 3.5 introduces the concept of reciprocal bases, which play an important role in a number of applications, for example the representation of pinhole cameras in geometric algebra (see Sect. 7.1). They are also directly related to linear functions, since they can be used to generate basis transformation matrices.

After the discussion of various transformations of multivectors, a brief introduction to multivector differentiation in Sect. 3.6 completes the picture. In this section, differentiation and integration are also discussed with respect to the tensor representation of geometric-algebra operations. This representation, which is described in detail in Sect. 5.1, basically expresses geometric-algebra operations as bilinear functions, for which differentiation and integration are well defined.

C. Perwass, *Geometric Algebra with Applications in Engineering.*
Geometry and Computing.

Section 3.7 introduces fundamental algorithms operating on multivectors. In particular, the algorithm to evaluate the *join* of two blades plays an important role, since it is essential to evaluate the meet between arbitrary blades.

This chapter concludes with an overview of related "geometric" algebras in Sect. 3.8. Here the relation to Gibbs's vector algebra, complex numbers, quaternions, Grassmann algebra, and Grassmann–Cayley algebra is discussed. Grassmann–Cayley algebra is also used for applications in computer vision and robotics. Therefore, the principal differences and similarities between geometric algebra and Grassmann–Cayley algebra are presented here.

3.1 Basics

3.1.1 Axioms

Since only finite-dimensional geometric algebras over the reals are used in this text, the following axioms are specialized to this case.

Let $\mathbb{R}^{p,q}$ denote a $(p+q)$-dimensional vector space over the reals $\mathbb{R}$. Furthermore, let a commutative scalar product be defined as $* : \mathbb{R}^{p,q} \times \mathbb{R}^{p,q} \to \mathbb{R}$. That is, for $\boldsymbol{a}, \boldsymbol{b} \in \mathbb{R}^{p,q}$,

$$\boldsymbol{a} * \boldsymbol{b} = \boldsymbol{b} * \boldsymbol{a} \quad \in \mathbb{R}.$$

Definition 3.1 (Canonical Vector Basis). The canonical basis of $\mathbb{R}^{p,q}$, denoted by $\overline{\mathbb{R}}^{p,q}$, is defined as the totally ordered set

$$\overline{\mathbb{R}}^{p,q} := \{\boldsymbol{e}_1, \ldots, \boldsymbol{e}_p, \boldsymbol{e}_{p+1}, \ldots, \boldsymbol{e}_{p+q}\} \subset \mathbb{R}^{p,q},$$

where the $\{\boldsymbol{e}_i\}$ have the property

$$\boldsymbol{e}_i * \boldsymbol{e}_j = \begin{cases} +1, \; 1 \leq i = j \leq p, \\ -1, \; p < i = j \leq p+q, \\ \;\;0, \; i \neq j. \end{cases} \tag{3.1}$$

The combination of a vector space with a scalar product is called a *quadratic space*. Quadratic spaces form the basis for the construction of geometric algebras. While the quadratic space $(\mathbb{R}^{p,q}, *)$ plays a central role in applications, geometric algebras can be constructed over any type of quadratic space, one example of which is Hilbert spaces. It is therefore possible to construct a geometric algebra over a finite Fourier basis or a finite random-variable space (see Sect. 10).

Axiom 3.1 (Geometric Algebra) *Let $\mathbb{A}(\mathbb{R}^{p,q})$ denote the associative algebra over the quadratic space $(\mathbb{R}^{p,q}, *)$ and let $\circ$ denote the algebraic product.*

Note that the field $\mathbb{R}$ and the vector space $\mathbb{R}^{p,q}$ can both be regarded as subspaces of $\mathbb{A}(\mathbb{R}^{p,q})$. The algebra $\mathbb{A}(\mathbb{R}^{p,q})$ is said to be a geometric algebra if for $\boldsymbol{a} \in \mathbb{R}^{p,q} \subset \mathbb{A}(\mathbb{R}^{p,q})$, $\boldsymbol{a} \circ \boldsymbol{a} = \boldsymbol{a} * \boldsymbol{a}$.

The geometric algebra over $\mathbb{R}^{p,q}$ is denoted by $\mathbb{G}(\mathbb{R}^{p,q})$ or simply $\mathbb{G}_{p,q}$ and the algebra product is called the *Clifford* or *geometric* product. Although the geometric product is denoted here by $\circ$, it is represented later by juxtaposition for brevity.

In the following, all axioms of a geometric algebra are given explicitly. First of all, the elements of $\mathbb{G}_{p,q}$, which are called *multivectors*, satisfy the axioms of a vector space over the field $\mathbb{R}$.

Axiom 3.2 *The following two operations exist in $\mathbb{G}_{p,q}$:*

1. **Multivector addition.** *For any two elements $\boldsymbol{A}, \boldsymbol{B} \in \mathbb{G}_{p,q}$ there exists an element $\boldsymbol{C} = \boldsymbol{A} + \boldsymbol{B} \in \mathbb{G}_{p,q}$, their sum.*
2. **Scalar multiplication.** *For any element $\boldsymbol{A} \in \mathbb{G}_{p,q}$ and any scalar $\alpha \in \mathbb{R}$, there exists an element $\alpha\boldsymbol{A} \in \mathbb{G}_{p,q}$, the α-multiple of $\boldsymbol{A}$.*

Axiom 3.3 (Vector Space) *Let $\boldsymbol{A}, \boldsymbol{B}, \boldsymbol{C} \in \mathbb{G}_{p,q}$ and $\alpha, \beta \in \mathbb{R}$.*

1. **Associativity** *of multivector addition:*
$$(\boldsymbol{A} + \boldsymbol{B}) + \boldsymbol{C} = \boldsymbol{A} + (\boldsymbol{B} + \boldsymbol{C}).$$
2. **Commutativity**:
$$\boldsymbol{A} + \boldsymbol{B} = \boldsymbol{B} + \boldsymbol{A}.$$
3. **Identity element of addition.** *There exists an element $0 \in \mathbb{G}_{p,q}$, the zero element, such that*
$$\boldsymbol{A} + 0 = \boldsymbol{A}$$
4. **Associativity** *of scalar multiplication:*
$$\alpha(\beta\boldsymbol{A}) = (\alpha\beta)\boldsymbol{A}.$$
5. **Commutativity** *of scalar multiplication:*
$$\alpha\boldsymbol{A} = \boldsymbol{A}\alpha.$$
6. **Identity element of scalar multiplication.** *The identity element $1 \in \mathbb{R}$ satisfies*
$$1\,\boldsymbol{A} = \boldsymbol{A}.$$
7. **Distributivity** *of multivector sums:*
$$\alpha\,(\boldsymbol{A} + \boldsymbol{B}) = \alpha\boldsymbol{A} + \alpha\boldsymbol{B}.$$

8. **Distributivity** *of scalar sums:*

$$(\alpha + \beta)\, \boldsymbol{A} = \alpha\, \boldsymbol{A} + \beta\, \boldsymbol{A}.$$

It follows from these axioms that for each $\boldsymbol{A} \in \mathbb{G}_{p,q}$ there exists an element $-\boldsymbol{A} := (-1)\boldsymbol{A}$ such that

$$\boldsymbol{A} - \boldsymbol{A} := \boldsymbol{A} + (-\boldsymbol{A}) = \boldsymbol{A} + (-1)\boldsymbol{A} = \big(1 + (-1)\big)\, \boldsymbol{A} = 0\, \boldsymbol{A} = 0.$$

Axiom 3.4 *The axioms related to the algebraic product, i.e. the geometric product, are as follows. Let $\boldsymbol{A}, \boldsymbol{B}, \boldsymbol{C} \in \mathbb{G}_{p,q}$ and $\alpha, \beta \in \mathbb{R}$.*

1. **The algebra is closed** *under the geometric product:*

$$\boldsymbol{A} \circ \boldsymbol{B} \in \mathbb{G}_{p,q}.$$

2. **Associativity***:*

$$(\boldsymbol{A} \circ \boldsymbol{B}) \circ \boldsymbol{C} = \boldsymbol{A} \circ (\boldsymbol{B} \circ \boldsymbol{C}).$$

3. **Distributivity***:*

$$\boldsymbol{A} \circ (\boldsymbol{B} + \boldsymbol{C}) = \boldsymbol{A} \circ \boldsymbol{B} + \boldsymbol{A} \circ \boldsymbol{C} \quad \textit{and} \quad (\boldsymbol{B} + \boldsymbol{C}) \circ \boldsymbol{A} = \boldsymbol{B} \circ \boldsymbol{A} + \boldsymbol{C} \circ \boldsymbol{A}.$$

4. **Scalar multiplication***:*

$$\alpha \circ \boldsymbol{A} = \boldsymbol{A} \circ \alpha = \alpha \boldsymbol{A}.$$

All axioms given so far define an associative algebra. What actually separates Clifford algebra from other algebras is the *defining equation.*

Axiom 3.5 *Let $\boldsymbol{a} \in \mathbb{R}^{p,q} \subset \mathbb{G}_{p,q}$; then*

$$\boldsymbol{a} \circ \boldsymbol{a} = \boldsymbol{a} * \boldsymbol{a} \; \in \mathbb{R}. \tag{3.2}$$

That is, the geometric product of a vector (*not* a multivector in general) with itself maps to an element of the field $\mathbb{R}$.

3.1.2 Basic Properties

All properties of the geometric algebra $\mathbb{G}_{p,q}$ can be derived from the axioms given in the previous subsection. In this subsection, basic observations are presented, which are extended in later subsections.

The defining equation of geometric algebra states that the geometric product of a vector $\boldsymbol{a} \in \mathbb{R}^{p,q} \subset \mathbb{G}_{p,q}$ with itself results in a scalar in $\mathbb{R}$. From this property, the representation of the scalar product of different vectors in terms of the geometric product can be derived. Let $\boldsymbol{a}, \boldsymbol{b} \in \mathbb{R}^{p,q} \subset \mathbb{G}_{p,q}$; then

$$\begin{aligned}
&(\boldsymbol{a}+\boldsymbol{b}) \circ (\boldsymbol{a}+\boldsymbol{b}) = (\boldsymbol{a}+\boldsymbol{b}) * (\boldsymbol{a}+\boldsymbol{b})\\
\iff\ &\boldsymbol{a}\circ\boldsymbol{a}+\boldsymbol{a}\circ\boldsymbol{b}+\boldsymbol{b}\circ\boldsymbol{a}+\boldsymbol{b}\circ\boldsymbol{b} = \boldsymbol{a}*\boldsymbol{a}+2\,\boldsymbol{a}*\boldsymbol{b}+\boldsymbol{b}*\boldsymbol{b}\\
\iff\ &\frac{1}{2}(\boldsymbol{a}\circ\boldsymbol{b}+\boldsymbol{b}\circ\boldsymbol{a}) = \boldsymbol{a}*\boldsymbol{b}.
\end{aligned} \tag{3.3}$$

The expression $\frac{1}{2}(\boldsymbol{a}\circ\boldsymbol{b}+\boldsymbol{b}\circ\boldsymbol{a})$ is called the *anticommutator product.* In this text, the operator symbol $\overline{\times}$ is used to represent the anticommutator product, i.e.

$$\boldsymbol{a}\overline{\times}\boldsymbol{b} := \frac{1}{2}(\boldsymbol{a}\circ\boldsymbol{b}+\boldsymbol{b}\circ\boldsymbol{a}). \tag{3.4}$$

Similarly, the *commutator product* is denoted by $\underline{\times}$, such that

$$\boldsymbol{a}\underline{\times}\boldsymbol{b} := \frac{1}{2}(\boldsymbol{a}\circ\boldsymbol{b}-\boldsymbol{b}\circ\boldsymbol{a}). \tag{3.5}$$

In the literature, the commutator product of two multivectors $\boldsymbol{A},\boldsymbol{B}\in\mathbb{G}_{p,q}$ is usually written as $[\boldsymbol{A},\boldsymbol{B}]$ and the anticommutator product as $\{\boldsymbol{A},\boldsymbol{B}\}$. In this text, however, the symbols introduced above are used to emphasize the operator quality of these products. By applying the properties of the geometric product, it can be seen immediately that the geometric product of two multivectors can be written as the sum of the commutator and anticommutator products:

$$\boldsymbol{A}\circ\boldsymbol{B} = \boldsymbol{A}\overline{\times}\boldsymbol{B} + \boldsymbol{A}\underline{\times}\boldsymbol{B}. \tag{3.6}$$

Two vectors $\boldsymbol{a},\boldsymbol{b}\in\mathbb{R}^{p,q}$ are called *orthogonal* if $\boldsymbol{a}*\boldsymbol{b}=0$, or, in terms of the geometric product, $\boldsymbol{a}\overline{\times}\boldsymbol{b}=0$. The elements of $\overline{\mathbb{R}}^{p,q}$ therefore have the properties

$$\boldsymbol{e}_i\overline{\times}\boldsymbol{e}_i \neq 0 \quad\text{and}\quad \boldsymbol{e}_i\overline{\times}\boldsymbol{e}_j = 0\,,\ i\neq j. \tag{3.7}$$

Hence, for $i\neq j$,

$$\begin{aligned}
\boldsymbol{e}_i\circ\boldsymbol{e}_j &= \boldsymbol{e}_i\overline{\times}\boldsymbol{e}_j + \boldsymbol{e}_i\underline{\times}\boldsymbol{e}_j = \boldsymbol{e}_i\underline{\times}\boldsymbol{e}_j\,,\\
\boldsymbol{e}_j\circ\boldsymbol{e}_i &= \boldsymbol{e}_j\overline{\times}\boldsymbol{e}_i + \boldsymbol{e}_j\underline{\times}\boldsymbol{e}_i = \boldsymbol{e}_j\underline{\times}\boldsymbol{e}_i\,,
\end{aligned} \tag{3.8}$$

and, since $\boldsymbol{A}\underline{\times}\boldsymbol{B} = -\boldsymbol{B}\underline{\times}\boldsymbol{A}$, by definition,

$$\boldsymbol{e}_i\circ\boldsymbol{e}_j = -\boldsymbol{e}_j\circ\boldsymbol{e}_i. \tag{3.9}$$

Note that the entity $\boldsymbol{e}_i\circ\boldsymbol{e}_j\in\mathbb{G}_{p,q}$ cannot be reduced to an element of $\mathbb{R}^{p,q}$. Instead, it represents an additional basis element of $\mathbb{G}_{p,q}$.

Consider the geometric product of two vectors $\boldsymbol{a},\boldsymbol{b}\in\mathbb{G}(\mathbb{R}^2)$ defined as $\boldsymbol{a} = a^1\,\mathrm{e}_1 + a^2\,\mathrm{e}_2$ and $\boldsymbol{b} = b^1\,\mathrm{e}_1 + b^2\,\mathrm{e}_2$, where $\{\,\mathrm{e}_1,\,\mathrm{e}_2\,\}\subset\mathbb{R}^2$ is an orthonormal basis of $\mathbb{R}^2$:

$$
\begin{aligned}
\boldsymbol{a} \circ \boldsymbol{b} &= (a^1\mathbf{e}_1 + a^2\mathbf{e}_2)\,(b^1\mathbf{e}_1 + b^2\mathbf{e}_2) \\
&= \left(a^1 b^1\, \mathbf{e}_1 \circ \mathbf{e}_1 + a^2 b^2\, \mathbf{e}_2 \circ \mathbf{e}_2\right) + \left(a^1 b^2\, \mathbf{e}_1 \circ \mathbf{e}_2 + a^2 b^1\, \mathbf{e}_2 \circ \mathbf{e}_1\right) \\
&= \left(a^1 b^1 + a^2 b^2\right) + \left(a^1 b^2 - a^2 b^1\right) \mathbf{e}_1 \circ \mathbf{e}_2\,, \\
\boldsymbol{b} \circ \boldsymbol{a} &= (b^1\mathbf{e}_1 + b^2\mathbf{e}_2)\,(a^1\mathbf{e}_1 + a^2\mathbf{e}_2) \\
&= \left(b^1 a^1\, \mathbf{e}_1 \circ \mathbf{e}_1 + b^2 a^2\, \mathbf{e}_2 \circ \mathbf{e}_2\right) + \left(b^1 a^2\, \mathbf{e}_1 \circ \mathbf{e}_2 + b^2 a^1\, \mathbf{e}_2 \circ \mathbf{e}_1\right) \\
&= \left(a^1 b^1 + a^2 b^2\right) - \left(a^1 b^2 - a^2 b^1\right) \mathbf{e}_1 \circ \mathbf{e}_2.
\end{aligned}
\tag{3.10}
$$

It therefore follows that

$$
\begin{aligned}
\boldsymbol{a} \overline{\times} \boldsymbol{b} &= \frac{1}{2}(\boldsymbol{a} \circ \boldsymbol{b} + \boldsymbol{b} \circ \boldsymbol{a}) = a^1 b^1 + a^2 b^2, \\
\boldsymbol{a} \underline{\times} \boldsymbol{b} &= \frac{1}{2}(\boldsymbol{a} \circ \boldsymbol{b} - \boldsymbol{b} \circ \boldsymbol{a}) = \left(a^1 b^2 - a^2 b^1\right) \mathbf{e}_1 \circ \mathbf{e}_2.
\end{aligned}
\tag{3.11}
$$

As previously shown, $\boldsymbol{a} \overline{\times} \boldsymbol{b} \in \mathbb{R}$, and $\boldsymbol{a} \underline{\times} \boldsymbol{b}$ generates an element of $\mathbb{G}_2$ which is an element neither of $\mathbb{R}^2$ nor of $\mathbb{R}$. Thus, the geometric product of two vectors results in a sum of a scalar and an additional element of the algebra.

If $\boldsymbol{a}$ and $\boldsymbol{b}$ are unit vectors, it follows directly from (3.11), that

$$
\boldsymbol{a} \overline{\times} \boldsymbol{b} = \cos\theta \qquad \text{and} \qquad (\boldsymbol{a} \underline{\times} \boldsymbol{b}) \circ (\mathbf{e}_2 \circ \mathbf{e}_1) = \sin\theta, \tag{3.12}
$$

where θ is the angle spanned by $\boldsymbol{a}$ and $\boldsymbol{b}$. That is, $\boldsymbol{a} \overline{\times} \boldsymbol{b}$ gives the length of the projection of one vector onto the other, and $\boldsymbol{a} \underline{\times} \boldsymbol{b}$ gives the length of the *rejection* of the two vectors. Therefore, if $\boldsymbol{a}$ and $\boldsymbol{b}$ are general vectors,

$$
(\boldsymbol{a} \underline{\times} \boldsymbol{b}) \circ (\mathbf{e}_2 \circ \mathbf{e}_1) = \|\boldsymbol{a}\|\,\|\boldsymbol{b}\|\,\sin\theta, \tag{3.13}
$$

which is the area of the parallelogram spanned by $\boldsymbol{a}$ and $\boldsymbol{b}$. This will be discussed again later on.

3.1.3 Algebraic Basis

In this subsection it is shown how an algebraic basis of $\mathbb{G}_{p,q}$ can be constructed. From now on, the geometric product will be denoted by juxtaposition of symbols. For example, the geometric product of $\boldsymbol{A}, \boldsymbol{B} \in \mathbb{G}_{p,q}$ is now written as $\boldsymbol{A}\,\boldsymbol{B}$.

In the following, let $\mathbb{A}[i]$ denote the ith element of an ordered set $\mathbb{A}$. That is, if $\mathbb{A} := \{2, 3, 1\}$, then $\mathbb{A}[2] = 3$. Also, if the product operator $\prod$ is applied to elements of a geometric algebra, it refers to the geometric product of the operands. That is, $\prod_{i=1}^{3} \mathbf{e}_i = \mathbf{e}_1\,\mathbf{e}_2\,\mathbf{e}_3$, with $\{\mathbf{e}_i\} \subset \overline{\mathbb{R}}^{p,q}$.

Definition 3.2 (Basis blade). A *basis blade* in $\mathbb{G}_{p,q}$ is the geometric product of a number of different elements of the canonical basis $\overline{\mathbb{R}}^{p,q}$. Let $\mathbb{A} \subset \{1, \ldots, p+q\}$; then $e_{\mathbb{A}}$ denotes the basis blade

$$e_{\mathbb{A}} := \prod_{i=1}^{|\mathbb{A}|} \overline{\mathbb{R}}^{p,q}\left[\mathbb{A}[i]\right].$$

For example, if $\mathbb{A} := \{2, 3, 1\}$, then $e_{\mathbb{A}} = e_2\, e_3\, e_1$.

Definition 3.3 (Grade). The *grade* of a basis blade $e_{\mathbb{A}} \in \mathbb{G}_{p,q}$, with $\mathbb{A} \subset \{1, \ldots, p+q\}$, is denoted by $\mathrm{gr}(e_{\mathbb{A}})$ and is defined as $\mathrm{gr}(e_{\mathbb{A}}) := |\mathbb{A}|$. Note that $e_{\emptyset} = 1$, and thus $\mathrm{gr}(1) = 0$. Furthermore, the following definitions are made:

$$\begin{aligned}
\mathrm{gr}_+(e_{\mathbb{A}}) &:= \left|\{\, a \in \mathbb{A} \;:\; 1 \leq a \leq p \,\}\right|, \\
\mathrm{gr}_-(e_{\mathbb{A}}) &:= \left|\{\, a \in \mathbb{A} \;:\; p < a \leq p+q \,\}\right|.
\end{aligned}$$

That is, gr_+ and gr_- give the numbers of basis vectors in a basis blade that square to $+1$ and -1, respectively.

Given a vector space $\mathbb{R}^{p,q}$ with a canonical basis

$$\overline{\mathbb{R}}^{p,q} = \{e_1, \ldots, e_p, e_{p+1}, \ldots, e_{p+q}\},$$

there are 2^n $(n = p+q)$ ways to combine the $\{\, e_i \,\}$ with the geometric product such that no two of these products are linearly dependent, i.e. there exist 2^n linearly independent basis blades. The collection of 2^n linearly independent basis blades forms an algebraic basis of $\mathbb{G}_{p,q}$.

Recall that the geometric product is associative. Hence, $(e_i\, e_j)\, e_k$ may be written as $e_i\, e_j\, e_k$. Recall also that $e_i\, e_j = -e_j\, e_i$ for $i \neq j$. Therefore, using a different order for the constituent $\{\, e_i \,\}$ in the basis blades can at most change the signs of the basis blades. While this implies that there exist a great number of algebra bases for $\mathbb{G}_{p,q}$, the choice of basis is arbitrary, because the algebras formed by all of these different bases are isomorphic. Nevertheless, it is useful to define a *canonical algebraic basis* of $\mathbb{G}_{p,q}$, which is closely related to an ordered power set.

Definition 3.4 (Ordered Power Set). Let $\mathbb{I} := \{\, 1, \ldots, n \,\} \subset \mathbb{N}$ with cardinality $n = |\mathbb{I}|$ and denote by $\mathcal{P}(\mathbb{I})$ the power set of $\mathbb{I}$, such that $|\mathcal{P}(\mathbb{I})| = 2^n$. The *ordered power set* of $\mathbb{I}$, denoted by $\mathcal{P}_O(\mathbb{I})$, is a totally ordered set of the following type:

1. The elements of $\mathcal{P}_O(\mathbb{I})$, which are subsets of $\mathbb{I}$, are ordered by cardinality in ascending order.
2. The members of each element of $\mathcal{P}_O(\mathbb{I})$ are ordered in ascending order.
3. The elements of $\mathcal{P}_O(\mathbb{I})$ of equal cardinality are ordered in lexicographical order.

Furthermore, we define $\mathcal{P}_O^k(\mathbb{I}) := \{\, \mathbb{A} \in \mathcal{P}_O(\mathbb{I}) \,:\, |\mathbb{A}| = k \,\}$, whose elements are also ordered lexicographically.

For example, if $\mathbb{I} := \{\,1, 2, 3\,\}$, then

$$\mathcal{P}_O(\mathbb{I}) = \Big\{ \{\emptyset\},\ \{1\},\ \{2\},\ \{3\},\ \{1,2\},\ \{1,3\},\ \{2,3\},\ \{1,2,3\} \Big\}.$$

This index set is then used to construct a canonical algebraic basis.

Definition 3.5 (Canonical Algebraic Basis). The canonical algebraic basis of $\mathbb{G}_{p,q}$, denoted by $\overline{\mathbb{G}}_{p,q}$, is constructed as follows. Let the canonical vector basis of $\mathbb{R}^{p,q}$ be given by

$$\overline{\mathbb{R}}^{p,q} = \{e_1,\, \ldots,\, e_p,\, e_{p+1},\, \ldots,\, e_{p+q}\},$$

and let $\mathbb{I} := \{\,1, \ldots, p+q\,\}$. The *canonical algebraic basis* of $\mathbb{G}_{p,q}$ is then given by the ordered set

$$\overline{\mathbb{G}}_{p,q} := \Big\{\ e_{\mathbb{A}} \ :\ \mathbb{A} \in \mathcal{P}_O(\mathbb{I})\ \Big\}.$$

The order of the elements of $\overline{\mathbb{G}}_{p,q}$ is the same as that of $\mathcal{P}_O(\mathbb{I})$, i.e. if for $\mathbb{A}, \mathbb{B} \in \mathcal{P}_O(\mathbb{I})$, $\mathbb{A} < \mathbb{B}$, then $e_{\mathbb{A}} < e_{\mathbb{B}}$. Furthermore, we define

$$\overline{\mathbb{G}}^k_{p,q} := \Big\{\ e_{\mathbb{A}} \ :\ \mathbb{A} \in \mathcal{P}_O^k(\mathbb{I})\ \Big\}.$$

For example, consider the vector space $\mathbb{R}^3$ with a canonical basis $\overline{\mathbb{R}}^3 = \{\, e_1, e_2, e_3 \,\}$. The canonical algebraic basis of $\mathbb{G}_3$ is then given by

$$\overline{\mathbb{G}}_3 := \{1,\ e_1,\ e_2,\ e_3,\ e_1\, e_2,\ e_1\, e_3,\ e_2\, e_3,\ e_1\, e_2\, e_3\}. \tag{3.14}$$

Let $\boldsymbol{E}_i := \overline{\mathbb{G}}_{p,q}[i]$; then a general multivector of $\mathbb{G}_{p,q}$ may be written as

$$\boldsymbol{A} = a^i\, \boldsymbol{E}_i\,, \quad i \in \{\,1, 2, \ldots, 2^n\,\}, \tag{3.15}$$

where $\{\, a^i \,\} \subset \mathbb{R}$ and a summation over the range of i is implied. That is,

$$a^i\, E_i \;\equiv\; \sum_{i=1}^{2^n} a^i\, E_i\,; \quad i \in \{\,1, 2, \ldots, 2^n\,\}. \tag{3.16}$$

Multivectors that are linear combinations of basis blades of the same grade play an important role later on. It is therefore useful to have the following definition.

Definition 3.6 (k-Vector Space). The k-vector space of $\mathbb{G}_{p,q}$, denoted by $\mathbb{G}^k_{p,q}$, is the vector space spanned by $\overline{\mathbb{G}}^k_{p,q}$. The dimension of $\mathbb{G}^k_{p,q}$ is $\binom{p+q}{k}$.

Note that in $\mathbb{G}_{p,q}$ the highest-dimensional k-vector space has $k = p + q$, and is of dimension 1. Since the 0-vector space, i.e. the scalars, is also of dimension 1, an element of $\mathbb{G}^{(p+q)}_{p,q}$ is called a *pseudoscalar*.

Definition 3.7 (Pseudoscalar). The element of highest grade of the canonical basis of $\mathbb{G}_{p,q}$, i.e. $\overline{\mathbb{G}}_{p,q}[2^{p+q}]$, is said to be the *pseudoscalar* of $\mathbb{G}_{p,q}$.

3.1.4 Involutions

An involution is, by definition, an operation that maps an operand to itself when applied twice. The involution that will be used most often in the applications of geometric algebra in this book is the *reverse*.

Definition 3.8 (Reverse). Let $\mathbb{A} := \{i_1, i_2 \ldots, i_k\} \subset \{1, \ldots, p+q\}$, such that $\boldsymbol{e}_{\mathbb{A}} \in \mathbb{G}_{p,q}$. Then the reverse of $\boldsymbol{e}_{\mathbb{A}}$, denoted by $\widetilde{\boldsymbol{e}}_{\mathbb{A}}$, is defined as

$$\widetilde{\boldsymbol{e}}_{\mathbb{A}} := \prod_{j=1}^{k} \overline{\mathbb{R}}^{p,q}\big[\mathbb{A}[k-j+1]\big].$$

For example, if $\boldsymbol{E}_i := \overline{\mathbb{G}}_3[i]$, then $\boldsymbol{E}_8 = \boldsymbol{e}_1\, \boldsymbol{e}_2\, \boldsymbol{e}_3$, and $\widetilde{\boldsymbol{E}}_8 = \boldsymbol{e}_3\, \boldsymbol{e}_2\, \boldsymbol{e}_1$. It is clear that the reverse of a reverse is again the initial element. It also follows from the definition of the reverse that

$$(\boldsymbol{E}_i\, \boldsymbol{E}_j)^{\sim} = \widetilde{\boldsymbol{E}}_j\, \widetilde{\boldsymbol{E}}_i. \tag{3.17}$$

Since a basis blade differs from its reverse only in the order of its constituent basis vectors, the two can differ only by a sign. It may be shown that for $\boldsymbol{E}_i := \overline{\mathbb{G}}_{p,q}[i]$ with grade $k = \text{gr}(\boldsymbol{E}_i)$,

$$\widetilde{\boldsymbol{E}}_i = (-1)^{k(k-1)/2}\, \boldsymbol{E}_i. \tag{3.18}$$

The reverse operation is distributive, and thus the reverse of a general multivector is obtained by applying the reverse operation to the constituent basis blades of the multivector. That is, for $\boldsymbol{A} \in \mathbb{G}_{p,q}$ with $\boldsymbol{A} := a^i\, \boldsymbol{E}_i$, $\widetilde{\boldsymbol{A}} = a^i\, \widetilde{\boldsymbol{E}}_i$.

The second important involution in geometric algebra is closely related to the reverse. In this text, it is called the *conjugate* of geometric algebra.

Definition 3.9 (Conjugate). Let $\mathbb{A} \subset \{1, \ldots, p+q\}$, such that $\boldsymbol{e}_{\mathbb{A}} \in \mathbb{G}_{p,q}$. The *conjugate* of $\boldsymbol{e}_{\mathbb{A}}$, denoted by $\boldsymbol{e}^{\dagger}_{\mathbb{A}}$, is defined as

$$\boldsymbol{e}^{\dagger}_{\mathbb{A}} := (-1)^r\, \widetilde{\boldsymbol{e}}_{\mathbb{A}}, \qquad r := \text{gr}_{-}(\boldsymbol{e}_{\mathbb{A}}).$$

The relation between a basis blade $\boldsymbol{E}_i$ and its conjugate $\boldsymbol{E}^{\dagger}_i$ is therefore

$$\boldsymbol{E}_i^{\dagger} = (-1)^r \, (-1)^{k(k-1)/2} \, \boldsymbol{E}_i, \tag{3.19}$$

where $k = \text{gr}(\boldsymbol{E}_i)$ and $r = \text{gr}_{-}(\boldsymbol{E}_i)$. Clearly, the conjugate is equivalent to the reverse in $\mathbb{G}_p$, since the corresponding $\overline{\mathbb{R}}^p$ contains no basis vectors that square to -1. The conjugate has the useful property that for any basis blade $\boldsymbol{E}_i := \overline{\mathbb{G}}_{p,q}[i]$,

$$\boldsymbol{E}_i \, \boldsymbol{E}_i^{\dagger} = +1, \tag{3.20}$$

which is *not* necessarily the case for $\boldsymbol{E}_i \, \widetilde{\boldsymbol{E}}_i$. Hence, the conjugate of a basis blade is its inverse, i.e. $\boldsymbol{E}_i^{-1} = \boldsymbol{E}_i^{\dagger}$.

3.1.5 Duality

Duality is a simple but also very important concept in geometric algebra. Let $\boldsymbol{E}_i := \overline{\mathbb{G}}_{p,q}[i]$; then the dual of a basis blade $\boldsymbol{E}_i$ is denoted by $\boldsymbol{E}_i^*$ and defined as

$$\boldsymbol{E}_i^* := \boldsymbol{E}_i \, \boldsymbol{I}^{-1}, \tag{3.21}$$

where $\boldsymbol{I} := \overline{\mathbb{G}}_{p,q}^{(p+q)}$ denotes the unit pseudoscalar of $\mathbb{G}_{p,q}$ and $\boldsymbol{I}^{-1}$ denotes its inverse. In $\mathbb{G}_3$,

$$\boldsymbol{I} := \mathbf{e}_1\mathbf{e}_2\mathbf{e}_3 \quad \text{and} \quad \boldsymbol{I}^{-1} = \mathbf{e}_3\mathbf{e}_2\mathbf{e}_1. \tag{3.22}$$

By employing the associativity of the geometric product, it can easily be seen that $\boldsymbol{I}\boldsymbol{I}^{-1} = 1$ when the above definitions are used:

$$\begin{aligned} \boldsymbol{I}\boldsymbol{I}^{-1} &= (\mathbf{e}_1\mathbf{e}_2\mathbf{e}_3)(\mathbf{e}_3\mathbf{e}_2\mathbf{e}_1) \\ &= (\mathbf{e}_1\mathbf{e}_2)(\mathbf{e}_3\mathbf{e}_3)(\mathbf{e}_2\mathbf{e}_1) \; ; \quad \mathbf{e}_3\mathbf{e}_3 = 1 \\ &= \mathbf{e}_1(\mathbf{e}_2\mathbf{e}_2)\mathbf{e}_1 \; ; \quad \mathbf{e}_2\mathbf{e}_2 = 1 \\ &= \mathbf{e}_1\mathbf{e}_1 \; ; \quad \mathbf{e}_1\mathbf{e}_1 = 1 \\ &= 1. \end{aligned} \tag{3.23}$$

It is also straightforward to see that $\boldsymbol{I}^{-1} = -\boldsymbol{I}$ in $\mathbb{G}_3$. From the associativity of the geometric product and the property $\mathbf{e}_i\mathbf{e}_j = -\mathbf{e}_j\mathbf{e}_i$ for $i \neq j$, it follows that

$$\begin{aligned} \boldsymbol{I}^{-1} &= \mathbf{e}_3(\mathbf{e}_2\mathbf{e}_1) = -\mathbf{e}_3(\mathbf{e}_1\mathbf{e}_2) = -(\mathbf{e}_3\mathbf{e}_1)\mathbf{e}_2 \\ &= (\mathbf{e}_1\mathbf{e}_3)\mathbf{e}_2 = \mathbf{e}_1(\mathbf{e}_3\mathbf{e}_2) = -\mathbf{e}_1(\mathbf{e}_2\mathbf{e}_3) \\ &= -\boldsymbol{I}. \end{aligned} \tag{3.24}$$

Therefore, $\boldsymbol{I}\boldsymbol{I} = -\boldsymbol{I}\boldsymbol{I}^{-1} = -1$. The meaning of duality from a geometric point of view can be elucidated by considering the following equation:

$$\boldsymbol{E}_i\boldsymbol{E}_i^* = \boldsymbol{E}_i\boldsymbol{E}_i\boldsymbol{I}^{-1} \simeq \boldsymbol{I}, \tag{3.25}$$

since $\boldsymbol{E}_i \, \boldsymbol{E}_i \in \mathbb{R}$. Here "$\simeq$" denotes equality up to a scalar constant. This shows that a basis blade multiplied by its dual gives the pseudoscalar up to a scalar constant. For example, when we use the order of basis blades given in (3.14) for $\mathbb{G}_3$, it follows that

$$\boldsymbol{E}_2^* = \boldsymbol{E}_2 \boldsymbol{I}^{-1} = -\boldsymbol{E}_2 \boldsymbol{I} = -\boldsymbol{e}_1(\boldsymbol{e}_1\boldsymbol{e}_2\boldsymbol{e}_3) = -\boldsymbol{e}_2\boldsymbol{e}_3 = -\boldsymbol{E}_5. \tag{3.26}$$

Therefore, $\boldsymbol{E}_2 \boldsymbol{E}_2^* = -\boldsymbol{E}_2 \boldsymbol{E}_5 = -\boldsymbol{I}$. Note that applying the dual operation twice does not necessarily result in the original basis blade. For example,

$$(\boldsymbol{E}_2^*)^* = -\boldsymbol{E}_5^* = -\boldsymbol{E}_2. \tag{3.27}$$

On the other hand, applying the dual operation four times in a row will always return a basis blade to itself. Whether the dual operation has to be applied twice or four times to return a basis blade to itself depends on whether the pseudoscalar squares to $+1$ or -1.

The dual operation can also be applied to arbitrary multivectors. In this case the dual is applied to every basis blade contained in the multivector. In other words, the dual operation is distributive. For example, let $\boldsymbol{A} \in \mathbb{G}_{p,q}$ be defined as $\boldsymbol{A} := a^i \, \boldsymbol{E}_i$. Then $\boldsymbol{A}^* = a^i \, \boldsymbol{E}_i^*$.

3.1.6 Inner and Outer Product

The inner and the outer product are two special operations in Clifford algebra which are of particular importance. It is important to note at this point that the inner product referred to here is an algebraic operation that need not result in a scalar, and is also not positive definite. This differs from typical definitions of the inner product.

Initially, the inner and the outer product will be defined on basis blades, and will then be extended to multivectors. These basic definitions form the basis for the derivation of a number of important identities in the next section.

Before defining the inner and the outer product, the *grade projection* bracket has to be introduced.

Definition 3.10 (Grade Projection). Let $\boldsymbol{E}_i := \overline{\mathbb{G}}_{p,q}[i]$; then the *grade projection* of $\boldsymbol{E}_i$ onto grade k is written as $\langle \boldsymbol{E}_i \rangle_k$ and is defined as

$$\langle \boldsymbol{E}_i \rangle_k = \begin{cases} \boldsymbol{E}_i, \ \mathrm{gr}(\boldsymbol{E}_i) = k \,, \\ 0, \ \ \mathrm{gr}(\boldsymbol{E}_i) \neq k \,. \end{cases} \tag{3.28}$$

The grade projection operator is distributive and, for some $a \in \mathbb{R}$ and $\boldsymbol{A} \in \mathbb{G}_{p,q}$, $\langle a\boldsymbol{A} \rangle_k = a \, \langle \boldsymbol{A} \rangle_k$. Hence, for $\boldsymbol{A} \in \mathbb{G}_{p,q}$ with $\boldsymbol{A} = a^i \, \boldsymbol{E}_i$, it follows that

$$\langle \boldsymbol{A} \rangle_k = a^i \, \langle \boldsymbol{E}_i \rangle_k \,. \tag{3.29}$$

Definition 3.11 (Inner Product). Let $\boldsymbol{E}_i := \overline{\mathbb{G}}_{p,q}[i]$; then the *inner product* of two basis blades $\boldsymbol{E}_i$ and $\boldsymbol{E}_j$, with $k = \mathrm{gr}(\boldsymbol{E}_i)$, $l = \mathrm{gr}(\boldsymbol{E}_j)$, is defined as

$$\boldsymbol{E}_i \cdot \boldsymbol{E}_j := \begin{cases} \langle \boldsymbol{E}_i \boldsymbol{E}_j \rangle_{|k-l|} & , i,j > 0\,, \\ 0\,, & i = 0 \text{ and/or } j = 0\,. \end{cases} \tag{3.30}$$

That is, if the grade of the geometric product of $\boldsymbol{E}_i$ and $\boldsymbol{E}_j$ is $|k-l|$, then $\boldsymbol{E}_i \cdot \boldsymbol{E}_j = \boldsymbol{E}_i \boldsymbol{E}_j$. Otherwise, the inner product of the two basis blades is zero. Note that this definition implies that the inner product of a scalar with any element of the algebra is identically zero. The outer product is defined as follows.

Definition 3.12 (Outer Product). Let $\boldsymbol{E}_i := \overline{\mathbb{G}}_{p,q}[i]$; then the *outer product* of two basis blades $\boldsymbol{E}_i$ and $\boldsymbol{E}_j$, with $k = \mathrm{gr}(\boldsymbol{E}_i)$, $l = \mathrm{gr}(\boldsymbol{E}_j)$, is defined as

$$\boldsymbol{E}_i \wedge \boldsymbol{E}_j := \langle \boldsymbol{E}_i \boldsymbol{E}_j \rangle_{k+l}\,. \tag{3.31}$$

That is, if the grade of the geometric product of $\boldsymbol{E}_i$ and $\boldsymbol{E}_j$ is $k+l$, then $\boldsymbol{E}_i \wedge \boldsymbol{E}_j = \boldsymbol{E}_i \boldsymbol{E}_j$. Otherwise, the outer product is zero. Note that the outer product of a scalar with any element of the algebra is equivalent to the geometric product of the two elements. Here are two examples for the application of the inner and the outer product:

$$(\boldsymbol{e}_1\boldsymbol{e}_2) \cdot \boldsymbol{e}_2 = \langle \boldsymbol{e}_1\boldsymbol{e}_2\boldsymbol{e}_2 \rangle_{|2-1|} = \langle \boldsymbol{e}_1 \rangle_1 = \boldsymbol{e}_1, \tag{3.32}$$

$$(\boldsymbol{e}_1\boldsymbol{e}_2) \wedge \boldsymbol{e}_3 = \langle \boldsymbol{e}_1\boldsymbol{e}_2\boldsymbol{e}_3 \rangle_{2+1} = \langle \boldsymbol{e}_1\boldsymbol{e}_2\boldsymbol{e}_3 \rangle_3 = \boldsymbol{e}_1\boldsymbol{e}_2\boldsymbol{e}_3. \tag{3.33}$$

The definitions of the outer and the inner product make it necessary to be very careful when writing down geometric-algebra equations. Quite often, people like to write the symbol $\cdot$ to denote the product of a scalar with anything else. In geometric algebra, this can lead to considerable confusion and incorrect results, since $\cdot$ denotes the inner product, and the inner product of a scalar with any other element, and also with another scalar, is *zero* by definition.

Lemma 3.1. *Let* $\mathbb{A} \subseteq \{1, \ldots, p+q\}$ *and* $i \in \{1, \ldots, p+q\}$, *such that* $\boldsymbol{e}_{\mathbb{A}}, \boldsymbol{e}_i \in \mathbb{G}_{p,q}$. *Then*

$$\boldsymbol{e}_i\, \boldsymbol{e}_{\mathbb{A}} = \begin{cases} (-1)^{|\mathbb{A}|}\; \boldsymbol{e}_{\mathbb{A}}\, \boldsymbol{e}_i,\; i \notin \mathbb{A}\,, \\ (-1)^{|\mathbb{A}|+1}\; \boldsymbol{e}_{\mathbb{A}}\, \boldsymbol{e}_i,\; i \in \mathbb{A}\,. \end{cases}$$

Proof. This follows directly from the associativity of the geometric product, and the fact that $\boldsymbol{e}_i\, \boldsymbol{e}_j = -\boldsymbol{e}_j\, \boldsymbol{e}_i$ if $i \neq j$.
□

Lemma 3.2. *Let* $\mathbb{B} \subseteq \mathbb{A} \subseteq \{1, \ldots, p+q\}$, *such that* $\boldsymbol{e}_{\mathbb{A}}, \boldsymbol{e}_{\mathbb{B}} \in \mathbb{G}_{p,q}$. *Then*

$$\boldsymbol{e}_{\mathbb{A}}\, \boldsymbol{e}_{\mathbb{B}} = (-1)^{|\mathbb{B}|\,(|\mathbb{A}|-1)}\; \boldsymbol{e}_{\mathbb{B}}\, \boldsymbol{e}_{\mathbb{A}}.$$

Proof. This follows directly from the associativity of the geometric product and Lemma 3.1.
□

Lemma 3.3. *Let* $\mathbb{A} \subseteq \{1, \ldots, p+q\}$ *and* $i \in \{1, \ldots, p+q\}$*, such that* $\boldsymbol{e}_{\mathbb{A}}, \boldsymbol{e}_i \in \mathbb{G}_{p,q}$*. Then*

$$\boldsymbol{e}_i \wedge \boldsymbol{e}_{\mathbb{A}} = \frac{1}{2}\big(\boldsymbol{e}_i\,\boldsymbol{e}_{\mathbb{A}} + (-1)^{|\mathbb{A}|}\,\boldsymbol{e}_{\mathbb{A}}\,\boldsymbol{e}_i\big).$$

Proof. If $i \in \mathbb{A}$, then $\boldsymbol{e}_i \wedge \boldsymbol{e}_{\mathbb{A}} = \langle \boldsymbol{e}_i\,\boldsymbol{e}_{\mathbb{A}} \rangle_{1+|\mathbb{A}|} = 0$ and from Lemma 3.1 $\boldsymbol{e}_i\,\boldsymbol{e}_{\mathbb{A}} = (-1)^{|\mathbb{A}|+1}\,\boldsymbol{e}_{\mathbb{A}}\,\boldsymbol{e}_i$, whence $\boldsymbol{e}_i\,\boldsymbol{e}_{\mathbb{A}} + (-1)^{|\mathbb{A}|}\,\boldsymbol{e}_{\mathbb{A}}\,\boldsymbol{e}_i = 0$. If $i \notin \mathbb{A}$, then $\boldsymbol{e}_i \wedge \boldsymbol{e}_{\mathbb{A}} = \langle \boldsymbol{e}_i\,\boldsymbol{e}_{\mathbb{A}} \rangle_{1+|\mathbb{A}|} = \boldsymbol{e}_i\,\boldsymbol{e}_{\mathbb{A}}$ and, from Lemma 3.1, $\boldsymbol{e}_i\,\boldsymbol{e}_{\mathbb{A}} = (-1)^{|\mathbb{A}|}\,\boldsymbol{e}_{\mathbb{A}}\,\boldsymbol{e}_i$, whence $\frac{1}{2}\big(\boldsymbol{e}_i\,\boldsymbol{e}_{\mathbb{A}} + (-1)^{|\mathbb{A}|}\,\boldsymbol{e}_{\mathbb{A}}\,\boldsymbol{e}_i\big) = \boldsymbol{e}_i\,\boldsymbol{e}_{\mathbb{A}}$.
□

Lemma 3.4. *Let* $\mathbb{A} \subseteq \{1, \ldots, p+q\}$ *and* $i \in \{1, \ldots, p+q\}$*, such that* $\boldsymbol{e}_{\mathbb{A}}, \boldsymbol{e}_i \in \mathbb{G}_{p,q}$*. Then*

$$\boldsymbol{e}_i \cdot \boldsymbol{e}_{\mathbb{A}} = \frac{1}{2}\big(\boldsymbol{e}_i\,\boldsymbol{e}_{\mathbb{A}} - (-1)^{|\mathbb{A}|}\,\boldsymbol{e}_{\mathbb{A}}\,\boldsymbol{e}_i\big).$$

Proof. The proof follows along the same lines as for Lemma 3.3. Note that this equality also holds if $\mathbb{A} = \emptyset$.
□

In order to apply the inner and the outer product to general multivectors, the above definitions have to be extended.

Axiom 3.6 *Let* $a, b \in \mathbb{R}$ *and* $\boldsymbol{A}, \boldsymbol{B}, \boldsymbol{C} \in \mathbb{G}_{p,q}$*; then*

$$\begin{aligned} \boldsymbol{A} \cdot (\boldsymbol{B} + \boldsymbol{C}) &= \boldsymbol{A} \cdot \boldsymbol{B} + \boldsymbol{A} \cdot \boldsymbol{C}, \\ (a\boldsymbol{A}) \cdot (b\boldsymbol{B}) &= ab\,(\boldsymbol{A} \cdot \boldsymbol{B}). \end{aligned}$$

Axiom 3.7 *Let* $a, b \in \mathbb{R}$ *and* $\boldsymbol{A}, \boldsymbol{B}, \boldsymbol{C} \in \mathbb{G}_{p,q}$*; then*

$$\begin{aligned} \boldsymbol{A} \wedge (\boldsymbol{B} + \boldsymbol{C}) &= \boldsymbol{A} \wedge \boldsymbol{B} + \boldsymbol{A} \wedge \boldsymbol{C}, \\ (a\boldsymbol{A}) \wedge (b\boldsymbol{B}) &= ab\,(\boldsymbol{A} \wedge \boldsymbol{B}). \end{aligned}$$

Therefore, the inner and the outer product of two multivectors $\boldsymbol{A}, \boldsymbol{B} \in \mathbb{G}_{p,q}$, with $\boldsymbol{A} := a^i \boldsymbol{E}_i$ and $\boldsymbol{B} := b^i \boldsymbol{E}_i$, can be evaluated as follows:

$$\boldsymbol{A} \cdot \boldsymbol{B} = \sum_{i,j} a^i b^j\,(\boldsymbol{E}_i \cdot \boldsymbol{E}_j), \tag{3.34}$$

$$\boldsymbol{A} \wedge \boldsymbol{B} = \sum_{i,j} a^i b^j\,(\boldsymbol{E}_i \wedge \boldsymbol{E}_j). \tag{3.35}$$

It may also be shown that the outer product of multivectors is associative.

Lemma 3.5. *Let* $\boldsymbol{A}, \boldsymbol{B}, \boldsymbol{C} \in \mathbb{G}_{p,q}$*; then*

$$\boldsymbol{A} \wedge (\boldsymbol{B} \wedge \boldsymbol{C}) = (\boldsymbol{A} \wedge \boldsymbol{B}) \wedge \boldsymbol{C}.$$

Proof. Let $\boldsymbol{E}_i := \overline{\mathbb{G}}_{p,q}[i]$, $\boldsymbol{A} = a^i\, \boldsymbol{E}_i$, $\boldsymbol{B} = b^i\, \boldsymbol{E}_i$ and $\boldsymbol{C} = c^i\, \boldsymbol{E}_i$. Consider initially three basis blades with grades $r = \operatorname{gr}(\boldsymbol{E}_i)$, $s = \operatorname{gr}(\boldsymbol{E}_j)$, and $t = \operatorname{gr}(\boldsymbol{E}_k)$; then

$$\boldsymbol{E}_i \wedge (\boldsymbol{E}_j \wedge \boldsymbol{E}_k) = \langle \boldsymbol{E}_i\, \boldsymbol{E}_j\, \boldsymbol{E}_k \rangle_{r+s+t} = (\boldsymbol{E}_i \wedge \boldsymbol{E}_j) \wedge \boldsymbol{E}_k.$$

Since the outer product is distributive, associativity also holds for $\boldsymbol{A} \wedge (\boldsymbol{B} \wedge \boldsymbol{C})$, which is a linear combination of basis blades.
□

Although these definitions are sufficient to evaluate the inner and the outer product of two arbitrary multivectors, the definitions alone are not particularly helpful when working with geometric-algebra expressions. What is needed is a set of identities to manipulate expressions without writing everything in terms of basis blades first.

3.2 Blades

An important concept in Clifford algebra is that of a *blade*. A blade is defined to be the outer product of a number of 1-vectors.

Definition 3.13 (Blade). Let $\{\, \boldsymbol{a}_i \,\} \subset \mathbb{R}^{p,q}$ be a set of $n \geq k$ linearly independent 1-vectors in $\mathbb{G}_{p,q}$. Then the outer product of these vectors is called a *k-blade* or a blade of grade k. A blade of grade k is denoted by $\boldsymbol{A}_{\langle k \rangle}$. For example,

$$\boldsymbol{A}_{\langle k \rangle} = \boldsymbol{a}_1 \wedge \boldsymbol{a}_2 \wedge \ldots \wedge \boldsymbol{a}_k =: \bigwedge_{i=1}^{k} \boldsymbol{a}_i.$$

Clearly, a k-blade is a linear combination of basis blades of grade k. However, not every linear combination of basis blades of grade k is a k-blade. For example, $\boldsymbol{e}_1\boldsymbol{e}_2 + \boldsymbol{e}_1\boldsymbol{e}_3$ is a 2-vector and a 2-blade, since $\boldsymbol{e}_1\boldsymbol{e}_2 + \boldsymbol{e}_1\boldsymbol{e}_3 = \boldsymbol{e}_1 \wedge (\boldsymbol{e}_2 + \boldsymbol{e}_3)$. On the other hand, $\boldsymbol{e}_1\boldsymbol{e}_2 + \boldsymbol{e}_3\boldsymbol{e}_4$ is also a 2-vector but it cannot be expressed as a 2-blade.

3.2.1 Geometric Product

Before the inner and outer products of blades are discussed, consider again the geometric product. Clearly, the geometric product of blades $\boldsymbol{A}_{\langle k \rangle}, \boldsymbol{B}_{\langle l \rangle} \in \mathbb{G}_{p,q}$ can be written as

$$A_{\langle k\rangle}B_{\langle l\rangle} = \sum_{r=0}^{n} \left\langle A_{\langle k\rangle}B_{\langle l\rangle}\right\rangle_r \,. \tag{3.36}$$

$A_{\langle k\rangle}$ and $B_{\langle l\rangle}$ are linear combinations of basis blades of grade k and l, respectively. If two basis blades $E_{\langle k\rangle}, E_{\langle l\rangle} \in \overline{\mathbb{G}}_{p,q}$ have m basis vectors in common, then

$$E_{\langle k\rangle}E_{\langle l\rangle} = \left\langle E_{\langle k\rangle}E_{\langle l\rangle}\right\rangle_{k+l-2m} \,. \tag{3.37}$$

For example,

$$(e_1e_2)(e_2e_3) = e_1(e_2e_2)e_3 = e_1e_3. \tag{3.38}$$

It therefore follows that the sum in (3.36) can be written as follows:

$$\begin{aligned} A_{\langle k\rangle}B_{\langle l\rangle} = \left\langle A_{\langle k\rangle}B_{\langle l\rangle}\right\rangle_{|k-l|} &+ \left\langle A_{\langle k\rangle}B_{\langle l\rangle}\right\rangle_{|k-l|+2} + \dots \\ &+ \left\langle A_{\langle k\rangle}B_{\langle l\rangle}\right\rangle_{k+l-2} + \left\langle A_{\langle k\rangle}B_{\langle l\rangle}\right\rangle_{k+l} \,. \end{aligned} \tag{3.39}$$

The element of lowest grade ($|k-l|$) is the inner product, and the element of highest grade ($k+l$) is the outer product. Note that not all elements of this sum are necessarily present for any particular choice of blades. Nevertheless, it can be seen that the inner and the outer product are simply two particularly interesting products, but not the only ones that could have been defined.

Given two vectors $a, b \in \mathbb{G}^1_{p,q}$, it follows from (3.39) that

$$\begin{aligned} a\,b &= \langle a\,b\rangle_{|1-1|} + \langle a\,b\rangle_{1+1} \\ &= a\cdot b + a\wedge b. \end{aligned} \tag{3.40}$$

Blades of grade 2, or *bivectors*, are also an interesting special case. For two bivectors $A_{\langle 2\rangle}, B_{\langle 2\rangle} \in \mathbb{G}_{p,q}$, it follows from (3.39) that

$$A_{\langle 2\rangle}B_{\langle 2\rangle} = \left\langle A_{\langle 2\rangle}B_{\langle 2\rangle}\right\rangle_0 + \left\langle A_{\langle 2\rangle}B_{\langle 2\rangle}\right\rangle_2 + \left\langle A_{\langle 2\rangle}B_{\langle 2\rangle}\right\rangle_4 \,. \tag{3.41}$$

That is, in addition to the inner and the outer product, there is another product which returns a bivector. It turns out that this is the commutator product, which has already been introduced. For example,

$$\begin{aligned} (e_1e_2)\underline{\times}(e_2e_3) &= \frac{1}{2}\left[(e_1e_2e_2e_3) - (e_2e_3e_1e_2)\right] \\ &= \frac{1}{2}\left[(e_1e_3) - (e_3e_1)\right] \\ &= \frac{1}{2}\left[(e_1e_3) + (e_1e_3)\right] \\ &= e_1e_3. \end{aligned} \tag{3.42}$$

Clearly, $(e_1e_2)\underline{\times}(e_1e_2) = 0$ and also $(e_1e_2)\underline{\times}(e_3e_4) = 0$.

Since the geometric product of two multivectors $A, B \in \mathbb{G}_{p,q}$ can be written as $A\,B = A\overline{\times}B + A\underline{\times}B$, for bivectors it follows that

$$A_{\langle 2\rangle} \overline{\times} B_{\langle 2\rangle} = A_{\langle 2\rangle} \cdot B_{\langle 2\rangle} + A_{\langle 2\rangle} \wedge B_{\langle 2\rangle}. \tag{3.43}$$

3.2.2 Outer Product

Owing to the distributivity of the outer product, it follows that for $A_{\langle k\rangle}, B_{\langle l\rangle} \in \mathbb{G}_{p,q}$,

$$A_{\langle k\rangle} \wedge B_{\langle l\rangle} = \left\langle A_{\langle k\rangle} \, B_{\langle l\rangle} \right\rangle_{k+l} . \tag{3.44}$$

Hence, the outer product of these blades either is zero or results in a blade of grade $k+l$. Hence, if $k+l > p+q$, then $A_{\langle k\rangle} \wedge B_{\langle l\rangle} \equiv 0$.

Consider now the outer product of a 1-vector $a \in \mathbb{G}^1_{p,q}$ and a k-blade $B_{\langle k\rangle} \in \mathbb{G}^k_{p,q}$.

Lemma 3.6. *Let $a \in \mathbb{G}^1_{p,q}$ and $B_{\langle k\rangle} \in \mathbb{G}^k_{p,q}$; then*

$$a \wedge B_{\langle k\rangle} = \frac{1}{2}\Big(a \, B_{\langle k\rangle} + (-1)^k \, B_{\langle k\rangle} \, a\Big).$$

Proof. Let $e_i := \overline{\mathbb{G}}^1_{p,q}[i]$ and $E_i := \overline{\mathbb{G}}^k_{p,q}[i]$, and define $a := \mathsf{a}^i \, e_i$, $B_{\langle k\rangle} := \mathsf{b}^i \, E_i$, where $\{\mathsf{a}^i\}, \{\mathsf{b}^i\} \subset \mathbb{R}$. Then

$$a \wedge B_{\langle k\rangle} = \mathsf{a}^i \, \mathsf{b}^j \, (e_i \wedge E_j).$$

It follows from Lemma 3.3 that $e_i \wedge E_j = \frac{1}{2}(e_i \, E_j + (-1)^k \, E_j \, e_i)$. Hence, the proposition follows by linearity.

□

Lemma 3.6 implies that given two 1-vectors $a, b \in \mathbb{G}^1_{p,q}$,

$$a \wedge b = -b \wedge a \qquad \Rightarrow \qquad a \wedge a = 0. \tag{3.45}$$

The property $a \wedge a = 0$ is of pivotal importance. From a geometric point of view, it states that the outer product of two vectors is zero if they are parallel. This is in contrast to the scalar product of two vectors, which is zero if the vectors are perpendicular. The following lemma generalizes this notion.

Lemma 3.7. *Let $\{a_1, \ldots, a_k\} \subset \mathbb{G}^1_{p,q}$ define a basis of a k-dimensional vector space ($1 \leq k \leq p+q$) and let $x := \mathsf{a}^i \, a_i$ be a vector in this vector space with $\{\mathsf{a}^i\} \subset \mathbb{R}$. Then the blade $A_{\langle k\rangle} := a_1 \wedge \ldots \wedge a_k$ satisfies the equation $A_{\langle k\rangle} \wedge x = 0$.*

Proof. The proof is done by construction. It employs the facts that the outer product is associative and that $a_i \wedge a_j = -a_j \wedge a_i$:

$$\begin{aligned} \boldsymbol{A}_{\langle k \rangle} \wedge \boldsymbol{x} &= \sum_i \mathsf{a}^i \, (\boldsymbol{a}_1 \wedge \boldsymbol{a}_2 \wedge \ldots \wedge \boldsymbol{a}_k) \wedge \boldsymbol{a}_i \\ &= \sum_i (-1)^{k-i} \, \mathsf{a}^i \, \left(\bigwedge_{j=1}^{i-1} \boldsymbol{a}_j\right) \wedge \left(\bigwedge_{j=i+1}^{k} \boldsymbol{a}_j\right) \wedge \boldsymbol{a}_i \wedge \boldsymbol{a}_i \\ &= 0, \end{aligned} \tag{3.46}$$

since $\boldsymbol{a}_i \wedge \boldsymbol{a}_i = 0 \; \forall i \in \{1, \ldots, k\}$.
□

That is, the outer product of a k-blade with a 1-vector which is linearly dependent on the constituent 1-vectors of the k-blade is zero. In this respect, it may be said that the k-blade $\boldsymbol{A}_{\langle k \rangle}$ represents the k-dimensional subspace spanned by $\{ \boldsymbol{a}_i \}$. In fact, this subspace may be defined using $\boldsymbol{A}_{\langle k \rangle}$.

Definition 3.14 (Outer-Product Null Space). The *outer-product null space* (OPNS) of a blade $\boldsymbol{A}_{\langle k \rangle} \in \mathbb{G}^k_{p,q}$, denoted by $\mathbb{NO}(\boldsymbol{A}_{\langle k \rangle})$, is defined as

$$\mathbb{NO}(\boldsymbol{A}_{\langle k \rangle}) := \left\{ \boldsymbol{x} \in \mathbb{G}^1_{p,q} \; : \; \boldsymbol{x} \wedge \boldsymbol{A}_{\langle k \rangle} = 0 \right\}.$$

Using this notation, $\mathbb{NO}(\boldsymbol{A}_{\langle k \rangle}) = \text{span}\{ \boldsymbol{a}_i \}$. Furthermore, it follows that the outer product of two blades $\boldsymbol{A}_{\langle k \rangle}, \boldsymbol{B}_{\langle l \rangle} \in \mathbb{G}_{p,q}$ is non-zero only if the respective subspaces spanned by $\boldsymbol{A}_{\langle k \rangle}$ and $\boldsymbol{B}_{\langle l \rangle}$ are disjoint. That is,

$$\boldsymbol{A}_{\langle k \rangle} \wedge \boldsymbol{B}_{\langle l \rangle} = 0 \quad \Longleftrightarrow \quad \mathbb{NO}(\boldsymbol{A}_{\langle k \rangle}) \cap \mathbb{NO}(\boldsymbol{B}_{\langle l \rangle}) \neq \emptyset. \tag{3.47}$$

Definition 3.15 (Direct Sum). Let $\mathbb{A}, \mathbb{B} \subseteq \mathbb{R}^{p,q}$; then their *direct sum* is defined as

$$\mathbb{A} \oplus \mathbb{B} := \left\{ \boldsymbol{a} + \boldsymbol{b} \; : \; \boldsymbol{a} \in \mathbb{A}, \; \boldsymbol{b} \in \mathbb{B} \right\}.$$

With this definition in mind, it follows that for $\boldsymbol{A}_{\langle k \rangle}, \boldsymbol{B}_{\langle l \rangle} \in \mathbb{G}_{p,q}$ with $\boldsymbol{A}_{\langle k \rangle} \wedge \boldsymbol{B}_{\langle l \rangle} \neq 0$,

$$\mathbb{NO}(\boldsymbol{A}_{\langle k \rangle} \wedge \boldsymbol{B}_{\langle l \rangle}) = \mathbb{NO}(\boldsymbol{A}_{\langle k \rangle}) \oplus \mathbb{NO}(\boldsymbol{B}_{\langle l \rangle}). \tag{3.48}$$

The following important lemma will be used in many proofs later on.

Lemma 3.8. *Let $\boldsymbol{A}_{\langle k \rangle} \in \mathbb{G}^k_{p,q}$; then there exists an* orthogonal *set of k vectors $\{ \boldsymbol{a}_i \} \subset \mathbb{G}^1_{p,q}$, i.e. $\boldsymbol{a}_i \wedge \boldsymbol{a}_j \neq 0$ and $\boldsymbol{a}_i * \boldsymbol{a}_j = 0$ for $i \neq j$, such that $\boldsymbol{A}_{\langle k \rangle} = \bigwedge_{i=1}^{k} \boldsymbol{a}_i$.*

Proof. From the definition of the OPNS of a blade, it follows that all blades with the same OPNS can differ only by a scalar factor. Furthermore, since $\mathbb{NO}(\boldsymbol{A}_{\langle k \rangle})$ is a k-dimensional vector space, there exists an orthonormal basis $\{ \boldsymbol{n}_i \} \subset \mathbb{G}^1_{p,q}$ of $\mathbb{NO}(\boldsymbol{A}_{\langle k \rangle})$. If we define $\boldsymbol{N}_{\langle k \rangle} := \bigwedge_{i=1}^{k} \boldsymbol{n}_i$, then there must exist a scalar $\alpha \in \mathbb{R}$ such that $\boldsymbol{A}_{\langle k \rangle} = \alpha \, \boldsymbol{N}_{\langle k \rangle}$.
□

Note that if $\{ \boldsymbol{a}_i \} \in \mathbb{G}^1_{p,q}$ is an orthogonal set of vectors, then $\boldsymbol{a}_i \, \boldsymbol{a}_j = \boldsymbol{a}_i \wedge \boldsymbol{a}_j$, for $i \neq j$. Hence, the blade $\boldsymbol{A}_{\langle k \rangle} := \bigwedge_{i=1}^{k} \boldsymbol{a}_i$ can also be constructed by taking the geometric product of the $\{ \boldsymbol{a}_i \}$, i.e.

$$A_{\langle k \rangle} = \bigwedge_{i=1}^{k} a_i = \prod_{i=1}^{k} a_i. \tag{3.49}$$

Lemma 3.6 also implies, for $a \in \mathbb{G}^1_{p,q}$ and $B_{\langle k \rangle} \in \mathbb{G}^k_{p,q}$, that

$$a \wedge B_{\langle k \rangle} = (-1)^k \, B_{\langle k \rangle} \wedge a. \tag{3.50}$$

It follows, for $A_{\langle k \rangle}, B_{\langle l \rangle} \in \mathbb{G}_{p,q}$, that

$$A_{\langle k \rangle} \wedge B_{\langle l \rangle} = (-1)^{k\,l} \, B_{\langle l \rangle} \wedge A_{\langle k \rangle}. \tag{3.51}$$

The above analysis suggests that the outer product can be regarded as an operator that combines disjoint (linear) subspaces. As will be seen next, the inner product "subtracts" (linear) subspaces.

3.2.3 Scalar Product

So far, the scalar product has been defined only for 1-vectors. This definition can be extended to blades in the following way. Let $A_{\langle k \rangle}, B_{\langle l \rangle} \in \mathbb{G}_{p,q}$; then

$$A_{\langle k \rangle} * B_{\langle l \rangle} := \big\langle A_{\langle k \rangle} \, B_{\langle l \rangle} \big\rangle_0 \; . \tag{3.52}$$

If $k = l \neq 0$, the scalar product is therefore equal to the inner product, and if $k \neq l$, then $A_{\langle k \rangle} * B_{\langle l \rangle} = 0$. From 3.81) it follows for $k = l$ that

$$A_{\langle k \rangle} * B_{\langle k \rangle} = (-1)^{k\,(k+1)} \, B_{\langle k \rangle} * A_{\langle k \rangle} = B_{\langle k \rangle} * A_{\langle k \rangle}, \tag{3.53}$$

since $k\,(k+1)$ is an even number for all $k \in \mathbb{N}^+$. An important identity related to the scalar product is described in the following lemma.

Lemma 3.9. *Let* $A_{\langle r \rangle}, B_{\langle s \rangle}, C_{\langle t \rangle} \in \mathbb{G}_{p,q}$*; then* $\big\langle A_{\langle r \rangle} \, B_{\langle s \rangle} \, C_{\langle t \rangle} \big\rangle_0 = \langle C_{\langle t \rangle} \, A_{\langle r \rangle}$ $B_{\langle s \rangle} \rangle 0$.

Proof. Since only the scalar component of the geometric product of the three blades is of interest,

$$\big\langle A_{\langle r \rangle} \, B_{\langle s \rangle} \, C_{\langle t \rangle} \big\rangle_0 = \big\langle \, \big\langle A_{\langle r \rangle} \, B_{\langle s \rangle} \big\rangle_t \;\; C_{\langle t \rangle} \, \big\rangle_0 = \big\langle A_{\langle r \rangle} \, B_{\langle s \rangle} \big\rangle_t * C_{\langle t \rangle}.$$

However, because of (3.53),

$$\big\langle A_{\langle r \rangle} \, B_{\langle s \rangle} \big\rangle_t * C_{\langle t \rangle} = C_{\langle t \rangle} * \big\langle A_{\langle r \rangle} \, B_{\langle s \rangle} \big\rangle_t = \big\langle C_{\langle t \rangle} \, A_{\langle r \rangle} \, B_{\langle s \rangle} \big\rangle_0 \; .$$

□

Lemma 3.10. *Let* $A_{\langle k \rangle} \in \mathbb{G}^k_{p,q}$*; then* $A_{\langle k \rangle} \, A_{\langle k \rangle} = A_{\langle k \rangle} * A_{\langle k \rangle}$.

Proof. Let $\boldsymbol{A}_{\langle k\rangle} := \bigwedge_{i=1}^{k} \boldsymbol{a}_i$, with $\{\boldsymbol{a}_i\} \subset \mathbb{G}^1_{p,q}$. From Lemma 3.8, it follows that there exists a set $\{\boldsymbol{n}_i\} \subset \mathbb{G}^1_{p,q}$ of orthogonal vectors such that $\boldsymbol{A}_{\langle k\rangle} = \prod_{i=1}^{k} \boldsymbol{n}_i$. Now,

$$\begin{aligned}\boldsymbol{A}_{\langle k\rangle}\,\boldsymbol{A}_{\langle k\rangle} &= (-1)^{k(k-1)/2}\,\boldsymbol{A}_{\langle k\rangle}\,\widetilde{\boldsymbol{A}}_{\langle k\rangle}\\ &= (-1)^{k(k-1)/2}\,(\boldsymbol{n}_1\cdots\boldsymbol{n}_{k-1}\,(\boldsymbol{n}_k\,\boldsymbol{n}_k)\,\boldsymbol{n}_{k-1}\cdots\boldsymbol{n}_1)\,.\end{aligned}$$

Since $(\boldsymbol{n}_i)^2 \in \mathbb{R}$ for all i, $\boldsymbol{A}_{\langle k\rangle}\,\boldsymbol{A}_{\langle k\rangle} \in \mathbb{R}$, and thus

$$\boldsymbol{A}_{\langle k\rangle}\,\boldsymbol{A}_{\langle k\rangle} = \langle \boldsymbol{A}_{\langle k\rangle}\,\boldsymbol{A}_{\langle k\rangle}\rangle_0 = \boldsymbol{A}_{\langle k\rangle} * \boldsymbol{A}_{\langle k\rangle}\,,$$

which proves the proposition.

□

3.2.4 Reverse

The reverse was introduced in Sect. 3.1.4 for basis blades. The properties described there carry over to blades. The reverse of a blade $\boldsymbol{A}_{\langle k\rangle} \in \mathbb{G}^k_{p,q}$, where $\boldsymbol{A}_{\langle k\rangle} = \bigwedge_{i=1}^{k} \boldsymbol{a}_i$, $\{\boldsymbol{a}_i\} \subset \mathbb{G}^1_{p,q}$, is given by

$$\widetilde{\boldsymbol{A}}_{\langle k\rangle} = (\boldsymbol{a}_1 \wedge \boldsymbol{a}_2 \wedge \ldots \wedge \boldsymbol{a}_k)^{\sim} = \boldsymbol{a}_k \wedge \boldsymbol{a}_{k-1} \wedge \ldots \wedge \boldsymbol{a}_1 = \bigwedge_{i=k}^{1} \boldsymbol{a}_i\,. \tag{3.54}$$

Since the reverse changes only the order of the constituent vectors, a blade and its reverse can differ only by a sign. That is,

$$\widetilde{\boldsymbol{A}}_{\langle k\rangle} = (-1)^{k(k-1)/2}\,\boldsymbol{A}_{\langle k\rangle}\,. \tag{3.55}$$

Furthermore, for two blades $\boldsymbol{A}_{\langle k\rangle}, \boldsymbol{B}_{\langle l\rangle} \in \mathbb{G}_{p,q}$,

$$(\boldsymbol{A}_{\langle k\rangle} \wedge \boldsymbol{B}_{\langle l\rangle})^{\sim} = \widetilde{\boldsymbol{B}}_{\langle l\rangle} \wedge \widetilde{\boldsymbol{A}}_{\langle k\rangle} \quad \text{and} \quad (\boldsymbol{A}_{\langle k\rangle}\,\boldsymbol{B}_{\langle l\rangle})^{\sim} = \widetilde{\boldsymbol{B}}_{\langle l\rangle}\,\widetilde{\boldsymbol{A}}_{\langle k\rangle}\,. \tag{3.56}$$

3.2.5 Conjugate

The properties of the conjugate, as discussed in Sect. 3.1.4, can also be extended to blades. That is, the conjugate of a blade $\boldsymbol{A}_{\langle k\rangle} \in \mathbb{G}^k_{p,q}$, where $\boldsymbol{A}_{\langle k\rangle} = \bigwedge_{i=1}^{k} \boldsymbol{a}_i$, $\{\boldsymbol{a}_i\} \subset \mathbb{G}^1_{p,q}$, is given by

$$\boldsymbol{A}^{\dagger}_{\langle k\rangle} = (\boldsymbol{a}_1 \wedge \boldsymbol{a}_2 \wedge \ldots \wedge \boldsymbol{a}_k)^{\dagger} = \boldsymbol{a}^{\dagger}_k \wedge \boldsymbol{a}^{\dagger}_{k-1} \wedge \ldots \wedge \boldsymbol{a}^{\dagger}_1 = \bigwedge_{i=k}^{1} \boldsymbol{a}^{\dagger}_i\,. \tag{3.57}$$

Note that while $\widetilde{\boldsymbol{a}}_i = \boldsymbol{a}_i$, $\boldsymbol{a}_i^\dagger$ is not necessarily equal to $\boldsymbol{a}_i$. This is because for $\boldsymbol{e}_i := \overline{\mathbb{G}}^1_{p,q}$, $\boldsymbol{e}_i^\dagger = -\boldsymbol{e}_i$ if $p < i \leq p+q$, by definition. Therefore, it is not possible to give a general relation between $\boldsymbol{A}_{\langle k \rangle}$ and $\boldsymbol{A}^\dagger_{\langle k \rangle}$ as in (3.55) for the reverse. However, for two blades $\boldsymbol{A}_{\langle k \rangle}, \boldsymbol{B}_{\langle l \rangle} \in \mathbb{G}_{p,q}$, it also holds that

$$(\boldsymbol{A}_{\langle k \rangle} \wedge \boldsymbol{B}_{\langle l \rangle})^\dagger = \boldsymbol{B}^\dagger_{\langle l \rangle} \wedge \boldsymbol{A}^\dagger_{\langle k \rangle} \quad \text{and} \quad (\boldsymbol{A}_{\langle k \rangle}\, \boldsymbol{B}_{\langle l \rangle})^\dagger = \boldsymbol{B}^\dagger_{\langle l \rangle}\, \boldsymbol{A}^\dagger_{\langle k \rangle}. \tag{3.58}$$

3.2.6 Norm

In Euclidean space $\mathbb{R}^p$, the magnitude of a vector $\boldsymbol{a} \in \mathbb{R}^p$ is typically given by the L_2-norm, which is defined in terms of the scalar product as

$$\|\boldsymbol{a}\| := \sqrt{\boldsymbol{a} * \boldsymbol{a}}. \tag{3.59}$$

In order to extend this to multivectors of $\mathbb{G}_{p,q}$, a product between multivectors is needed such that the product of a multivector with itself results in a *positive* scalar. Note that the scalar product as defined in Sect. 3.2.3 does not satisfy this requirement, since it can result also in negative values and zero. For example, if $\boldsymbol{e}_i := \overline{\mathbb{G}}^1_{p,q}[i]$, then $\boldsymbol{e}_{p+1} * \boldsymbol{e}_{p+1} = -1$ and $(\boldsymbol{e}_1 + \boldsymbol{e}_{p+1}) * (\boldsymbol{e}_1 + \boldsymbol{e}_{p+1}) = 0$. For these reasons, it is useful to introduce yet another scalar product of blades, which is positive definite.

3.2.6.1 Euclidean Scalar Product

Definition 3.16 (Euclidean Scalar Product). Let $\boldsymbol{A}_{\langle k \rangle}, \boldsymbol{B}_{\langle l \rangle} \in \mathbb{G}_{p,q}$; we then define the *Euclidean scalar product* as

$$\boldsymbol{A}_{\langle k \rangle} \star \boldsymbol{B}_{\langle l \rangle} := \boldsymbol{A}_{\langle k \rangle} * \boldsymbol{B}^\dagger_{\langle l \rangle} = \left\langle \boldsymbol{A}_{\langle k \rangle}\, \boldsymbol{B}^\dagger_{\langle l \rangle} \right\rangle_0 .$$

Let $\boldsymbol{E}_i := \overline{\mathbb{G}}^k_{p,q}[i]$, $\boldsymbol{A}_{\langle k \rangle} := \mathsf{a}^i\, \boldsymbol{E}_i$, and $\boldsymbol{B}_{\langle k \rangle} := \mathsf{b}^i\, \boldsymbol{E}_i$, with $\{\mathsf{a}^i\}, \{\mathsf{b}^i\} \subset \mathbb{R}$. If $k \neq l$, then $\boldsymbol{A}_{\langle k \rangle} \star \boldsymbol{B}_{\langle l \rangle} = 0$, but if $k = l$, then

$$\boldsymbol{A}_{\langle k \rangle} \star \boldsymbol{B}_{\langle l \rangle} = \mathsf{a}^i\, \mathsf{b}^j \left\langle \boldsymbol{E}_i\, \boldsymbol{E}^\dagger_j \right\rangle_0 = \mathsf{a}^i\, \mathsf{b}^j\, \delta_{ij} = \sum_i (\mathsf{a}^i\, \mathsf{b}^i), \tag{3.60}$$

where $\left\langle \boldsymbol{E}_i\, \boldsymbol{E}^\dagger_j \right\rangle_0 = \delta_{ij}$ has been used, which follows from the definition of the basis blades and (3.20). The symbol δ_{ij} denotes the Kronecker delta. Hence, if $\boldsymbol{A}_{\langle k \rangle} \neq 0$, then

$$\boldsymbol{A}_{\langle k \rangle} \star \boldsymbol{A}_{\langle k \rangle} = \sum_i (\mathsf{a}^i)^2 > 0. \tag{3.61}$$

In particular, this is also true for null-blades, which are defined later in Definition 3.18. In short, null-blades are blades that square to zero. The Euclidean scalar product, however, treats all basis blades as if they had been constructed from a Euclidean basis.

Furthermore, for general multivectors $\boldsymbol{A}, \boldsymbol{B} \in \mathbb{G}_{p,q}$ defined by $\boldsymbol{A} := \mathsf{a}^i \boldsymbol{E}_i$ and $\boldsymbol{B} := \mathsf{b}^i \boldsymbol{E}_i$, with $\boldsymbol{E}_i := \overline{\mathbb{G}}_{p,q}[i]$,

$$\boldsymbol{A} \star \boldsymbol{B} = \mathsf{a}^i \mathsf{b}^j \left\langle \boldsymbol{E}_i \boldsymbol{E}_j^{\dagger} \right\rangle_0 = \mathsf{a}^i \mathsf{b}^j \delta_{ij} = \sum_i (\mathsf{a}^i \mathsf{b}^i), \tag{3.62}$$

just as for blades. Hence, if $\boldsymbol{A} \neq 0$, then

$$\boldsymbol{A} \star \boldsymbol{A} = \sum_i (\mathsf{a}^i)^2 > 0. \tag{3.63}$$

With the help of the Euclidean scalar product, it is possible to distinguish two different types of orthogonality: *geometric orthogonality* and *self-orthogonality*. Two vectors $\boldsymbol{a}, \boldsymbol{b} \in \mathbb{G}^1_{p,q}$ are geometrically orthogonal if $\boldsymbol{a} \star \boldsymbol{b} = 0$. In this case $\boldsymbol{a}$ and $\boldsymbol{b}$ are said to be *perpendicular*. A vector $\boldsymbol{a} \in \mathbb{G}^1_{p,q}$ is self-orthogonal if $\boldsymbol{a} * \boldsymbol{a} = 0$.

3.2.6.2 Magnitude

The magnitude of a multivector is therefore defined as follows.

Definition 3.17. The magnitude of a multivector $\boldsymbol{A} \in \mathbb{G}_{p,q}$ is denoted by $\|\boldsymbol{A}\|$ and defined as

$$\|\boldsymbol{A}\| := \sqrt{\boldsymbol{A} \star \boldsymbol{A}}.$$

It follows that $\|\cdot\|$ defines a norm on $\mathbb{G}_{p,q}$. Note that for a non-null-blade $\boldsymbol{A}_{\langle k \rangle} \in \mathbb{G}^{\varnothing\, k}_{p,q}$, $\boldsymbol{A}^{\dagger}_{\langle k \rangle} = \widetilde{\boldsymbol{A}}_{\langle k \rangle}$ and thus

$$\boldsymbol{A}_{\langle k \rangle} \star \boldsymbol{A}_{\langle k \rangle} = \boldsymbol{A}_{\langle k \rangle} * \widetilde{\boldsymbol{A}}_{\langle k \rangle} = \boldsymbol{A}_{\langle k \rangle} \widetilde{\boldsymbol{A}}_{\langle k \rangle}, \tag{3.64}$$

where the last equality is due to Lemma 3.10.

From Lemma 3.8, it is clear that any blade can be written as the geometric product of a set of orthogonal vectors. For example, let $\{\, \boldsymbol{a}_i \,\} \subset \mathbb{G}^{\varnothing\, 1}_{p,q}$ denote a set of orthogonal vectors and define $\boldsymbol{A}_{\langle k \rangle} := \prod_{i=0}^{k} \boldsymbol{a}_i \in \mathbb{G}^{\varnothing\, k}_{p,q}$. Then

$$\begin{aligned} \|\boldsymbol{A}_{\langle k \rangle}\| &= \sqrt{\boldsymbol{A}_{\langle k \rangle} * \widetilde{\boldsymbol{A}}_{\langle k \rangle}} \\ &= \sqrt{\boldsymbol{a}_1 \ldots \boldsymbol{a}_k\, \boldsymbol{a}_k \ldots \boldsymbol{a}_1} \\ &= \prod_{i=1}^{k} \|\boldsymbol{a}_i\|. \end{aligned} \tag{3.65}$$

Consider a blade $\boldsymbol{A}_{\langle n \rangle} \in \mathbb{G}^n_{p,q}$ of the highest grade $n = p + q$. Such a blade is always proportional to the pseudoscalar and may thus be written in terms of the basis 1-vectors $e_i := \overline{\mathbb{G}}^1_{p,q}[i]$ as

$$A_{\langle n\rangle} = \alpha \prod_{i=1}^{n} e_i = \alpha\, I,$$

where $I := \overline{\mathbb{G}}_{p,q}[2^n]$ is the pseudoscalar and $\alpha \in \mathbb{R}$. Therefore, $A_{\langle n\rangle}$ cannot be a null-blade. Furthermore, if $A_{\langle n\rangle} \in \mathbb{G}^n_n$ is a blade in a Euclidean geometric algebra, then $I\,\widetilde{I} > 0$ and thus $A_{\langle n\rangle}\,\widetilde{A}_{\langle n\rangle} > 0$, which also follows from (3.64). If, as before, $\{\, a_i \,\} \subset \mathbb{G}^1_n$ denotes a set of orthogonal vectors such that $A_{\langle n\rangle} = \prod_{i=1}^{n} a_i$, then $\|A_{\langle n\rangle}\| = \prod_{i=1}^{n} \|a_i\|$.

If the $\{\, a_i \,\}$ are written as rows in a square matrix A, the determinant of this orthogonal matrix is also given by the product of the magnitudes of the row vectors, i.e.

$$A_{\langle n\rangle} = \det \mathsf{A}\; I. \tag{3.66}$$

3.2.7 Inner Product

3.2.7.1 Basics

Just as for the outer product, it follows from the distributivity of the inner product that for $A_{\langle k\rangle}, B_{\langle l\rangle} \in \mathbb{G}_{p,q}$,

$$A_{\langle k\rangle} \cdot B_{\langle l\rangle} = \big\langle A_{\langle k\rangle}\, B_{\langle l\rangle} \big\rangle_{|k-l|}\,. \tag{3.67}$$

The inner product therefore always reduces the grades of the constituent blades. An important difference between the inner and the outer product is that the inner product is not associative. Consider now the inner product of a 1-vector $a \in \mathbb{G}^1_{p,q}$ and a k-blade $B_{\langle k\rangle} \in \mathbb{G}^k_{p,q}$.

Lemma 3.11. *Let $a \in \mathbb{G}^1_{p,q}$ and $B_{\langle k\rangle} \in \mathbb{G}^k_{p,q}$; then*

$$a \cdot B_{\langle k\rangle} = \frac{1}{2}\Big(a\, B_{\langle k\rangle} - (-1)^k\, B_{\langle k\rangle}\, a\Big).$$

Proof. The proof is similar to that of Lemma 3.6. Let $e_i := \overline{\mathbb{G}}^1_{p,q}[i]$ and $E_i := \overline{\mathbb{G}}^k_{p,q}[i]$, and define $a := \mathsf{a}^i\, e_i$, $B_{\langle k\rangle} := \mathsf{b}^i\, E_i$, where $\{\mathsf{a}^i\}, \{\mathsf{b}^i\} \subset \mathbb{R}$. Then

$$a \cdot B_{\langle k\rangle} = \mathsf{a}^i\, \mathsf{b}^j\, (e_i \cdot E_j).$$

It follows from Lemma 3.4 that $e_i \cdot E_j = \frac{1}{2}\,(e_i\, E_j - (-1)^k\, E_j\, e_i)$. Hence, the proposition follows by linearity.

□

It follows directly that

$$a \cdot B_{\langle l \rangle} = (-1)^{(l-1)}\, B_{\langle l \rangle} \cdot a. \tag{3.68}$$

In particular, for two 1-vectors $a, b \in \mathbb{G}^1_{p,q}$, $a \cdot b = b \cdot a$. Furthermore, if we let $b_i := \overline{\mathbb{G}}^1_{p,q}$, $a := \mathsf{a}^i\, e_i$, and $b := \mathsf{b}^i\, e_i$, then

$$a \cdot b = \mathsf{a}^i\, \mathsf{b}^j\, \langle e_i\, e_j \rangle_0 \; = \mathsf{a}^i\, \mathsf{b}^i\, e_i^2, \tag{3.69}$$

where e_i^2 is either $+1$ or -1. That is, $a \cdot b = \sum_i \mathsf{a}^i\, \mathsf{b}^i\, e_i^2 = a * b$. Note that in spaces with a mixed signature, i.e. $\mathbb{G}_{p,q}$ where $p \neq 0$ and $q \neq 0$, there exist null-vectors $a \in \mathbb{G}^1_{p,q}$ with $a \neq 0$ and $a * a = 0$. More generally, there exist *null-blades*, which are defined in the following.

Definition 3.18 (Null Blade). A blade $A_{\langle k \rangle} \in \mathbb{G}^k_{p,q}$ is said to be a *null-blade* if $A_{\langle k \rangle} \cdot A_{\langle k \rangle} = 0$. The subset of null-blades in $\mathbb{G}^k_{p,q}$ is denoted by $\mathbb{G}^{\circ\, k}_{p,q} \subset \mathbb{G}^k_{p,q}$ and the subset of non-null-blades by $\mathbb{G}^{\varnothing\, k}_{p,q} \subset \mathbb{G}^k_{p,q}$. That is,

$$\mathbb{G}^{\varnothing\, k}_{p,q} := \left\{\, A_{\langle k \rangle} \in \mathbb{G}^k_{p,q} \;:\; A_{\langle k \rangle} \cdot A_{\langle k \rangle} \neq 0 \,\right\},$$
$$\mathbb{G}^{\circ\, k}_{p,q} := \left\{\, A_{\langle k \rangle} \in \mathbb{G}^k_{p,q} \;:\; A_{\langle k \rangle} \cdot A_{\langle k \rangle} = 0 \,\right\}.$$

Note that the sets $\mathbb{G}^{\circ\, k}_{p,q}$ and $\mathbb{G}^{\varnothing\, k}_{p,q}$ are not vector spaces.

Null-blades exist only in mixed-signature algebras. In Euclidean geometric algebras ($\mathbb{G}_p$) and anti-Euclidean geometric algebras ($\mathbb{G}_{0,q}$), no null-blades exist. Many of the properties of blades derived in the following are not valid for null-blades.

Analogously to the outer-product null space, an *inner-product null space* of blades can also be defined.

Definition 3.19 (Inner-Product Null Space). The *inner-product null space* (IPNS) of a blade $A_{\langle k \rangle} \in \mathbb{G}^k_{p,q}$, denoted by $\mathbb{NI}(A_{\langle k \rangle})$, is defined as

$$\mathbb{NI}(A_{\langle k \rangle}) := \left\{\, x \in \mathbb{G}^1_{p,q} \;:\; x \cdot A_{\langle k \rangle} = 0 \,\right\},$$

irrespective of whether the blade is a null-blade or not.

3.2.7.2 Properties

Using Lemmas 3.6 and 3.11, and some general identities involving the commutator and anticommutator products, the following equation can be derived, as shown in [130]. Let $a, b \in \mathbb{G}_{p,q}$ and $A_{\langle k \rangle} \in \mathbb{G}^k_{p,q}$; then

$$(A_{\langle k \rangle} \wedge a) \cdot b = A_{\langle k \rangle}\, (a \cdot b) - (A_{\langle k \rangle} \cdot b) \wedge a. \tag{3.70}$$

By applying this equation recursively, the following expression may be obtained. Let $\{\, b_i \,\} \subset \mathbb{G}^1_{p,q}$ be a set of l linearly independent vectors and let

$\boldsymbol{B}_{\langle l \rangle} = \bigwedge_{i=1}^{l} \boldsymbol{b}_i$; then

$$\boldsymbol{a} \cdot \boldsymbol{B}_{\langle l \rangle} = \sum_{i=1}^{l} (-1)^{i+1} \, (\boldsymbol{a} \cdot \boldsymbol{b}_i) \, [\boldsymbol{B}_{\langle l \rangle} \backslash \boldsymbol{b}_i], \tag{3.71}$$

where

$$[\boldsymbol{B}_{\langle l \rangle} \backslash \boldsymbol{b}_i] := \Big(\bigwedge_{r=1}^{i-1} \boldsymbol{b}_r \Big) \wedge \Big(\bigwedge_{r=i+1}^{l} \boldsymbol{b}_r \Big). \tag{3.72}$$

Note that the result of $\boldsymbol{a} \cdot \boldsymbol{B}_{\langle l \rangle}$ is again a blade, albeit of grade $l - 1$. That is, the inner product reduces the grade of the constituent elements.

An example will be helpful for understanding the above equations better. Let $\boldsymbol{a}, \boldsymbol{b}_1, \boldsymbol{b}_2 \in \mathbb{G}^1_{p,q}$; then

$$\boldsymbol{a} \cdot (\boldsymbol{b}_1 \wedge \boldsymbol{b}_2) = (\boldsymbol{a} \cdot \boldsymbol{b}_1) \, \boldsymbol{b}_2 - (\boldsymbol{a} \cdot \boldsymbol{b}_2) \, \boldsymbol{b}_1. \tag{3.73}$$

Since the inner product of two vectors is a scalar, and scalars commute with all elements of $\mathbb{G}_{p,q}$, $(\boldsymbol{a} \cdot \boldsymbol{b}_1)\boldsymbol{b}_2 = \boldsymbol{b}_2(\boldsymbol{a} \cdot \boldsymbol{b}_1)$. By convention, the inner product takes precedence over the geometric product. Therefore, (3.73) may also be written as

$$\boldsymbol{a} \cdot (\boldsymbol{b}_1 \wedge \boldsymbol{b}_2) = \boldsymbol{a} \cdot \boldsymbol{b}_1 \, \boldsymbol{b}_2 - \boldsymbol{a} \cdot \boldsymbol{b}_2 \, \boldsymbol{b}_1.$$

Here is another of equation (3.71). Let $\boldsymbol{a}, \boldsymbol{b}_1, \boldsymbol{b}_2, \boldsymbol{b}_3 \in \mathbb{G}^1_{p,q}$; then

$$\begin{aligned} \boldsymbol{a} \cdot (\boldsymbol{b}_1 \wedge \boldsymbol{b}_2 \wedge \boldsymbol{b}_3) = \quad & (\boldsymbol{a} \cdot \boldsymbol{b}_1)(\boldsymbol{b}_2 \wedge \boldsymbol{b}_3) \\ & - (\boldsymbol{a} \cdot \boldsymbol{b}_2)(\boldsymbol{b}_1 \wedge \boldsymbol{b}_3) \\ & + (\boldsymbol{a} \cdot \boldsymbol{b}_3)(\boldsymbol{b}_1 \wedge \boldsymbol{b}_2). \end{aligned} \tag{3.74}$$

The general formula for the inner product of two blades is somewhat more complicated and will not be derived here. Let $\boldsymbol{A}_{\langle k \rangle}, \boldsymbol{B}_{\langle l \rangle} \in \mathbb{G}_{p,q}$, $0 < k \leq l \leq p + q$, and define $\mathbb{I} := \{ 1, \ldots, l \}$, $\mathbb{A} := \mathcal{P}^k_O(\mathbb{I})$, and $\mathbb{A}^c[i] := \mathbb{I} \backslash \mathbb{A}[i]$. That is, $\mathbb{A}$ is an ordered set of all subsets of k elements of $\mathbb{I}$, and $\mathbb{A}^c$ is its complement with respect to $\mathbb{I}$. Furthermore, define $\mathbb{U}[i] := \mathbb{A}[i] \cup_O \mathbb{A}^c[i]$, where $\cup_O$ denotes an order-preserving union. If $\mathbb{B}$ is a permutation of $\mathbb{I}$, then

$$\epsilon_{\mathbb{B}} := \begin{cases} +1, \ \mathbb{B} \text{ is an even permutation of } \mathbb{I}\,, \\ -1, \ \mathbb{B} \text{ is an odd permutation of } \mathbb{I}\,. \end{cases} \tag{3.75}$$

Note that $\mathbb{B}$ is an even permutation of $\mathbb{I}$ if the elements of $\mathbb{B}$ can be brought into the same order as in $\mathbb{I}$ by an even number of swaps of neighboring elements of $\mathbb{B}$. Finally, if $\mathbb{B} \subseteq \mathbb{I}$, then $\boldsymbol{b}_{\mathbb{B}} := \bigwedge_{i=1}^{|\mathbb{B}|} \boldsymbol{b}_{\mathbb{B}[i]}$.

For example, if $\mathbb{I} := \{ 1, 2, 3 \}$ and $k = 2$, then $\mathbb{A} = \{ \{ 1, 2 \}, \{ 1, 3 \}, \{ 2, 3 \} \}$, $\mathbb{A}^c = \{ \{ 3 \}, \{ 2 \}, \{ 1 \} \}$, and $\mathbb{U} = \{ \{ 1, 2, 3 \}, \{ 1, 3, 2 \}, \{ 2, 3, 1 \} \}$. Then $\epsilon_{\mathbb{U}[1]} = +1$, $\epsilon_{\mathbb{U}[2]} = -1$, $\epsilon_{\mathbb{U}[3]} = +1$, and $\boldsymbol{b}_{\mathbb{U}[1]} = \boldsymbol{b}_1 \wedge \boldsymbol{b}_2 \wedge \boldsymbol{b}_3$. The inner product of two blades may now be written as

$$A_{\langle k\rangle} \cdot B_{\langle l\rangle} = \sum_{i=1}^{|\mathbb{U}|} \epsilon_{\mathbb{U}[i]} \left(A_{\langle k\rangle} \cdot b_{\mathbb{A}[i]}\right) b_{\mathbb{A}^c[i]}. \tag{3.76}$$

An example will make this equation more accessible. Let $a_1, a_2, b_1, b_2, b_3 \in \mathbb{G}^1_{p,q}$; then

$$\begin{aligned}(a_1 \wedge a_2) \cdot (b_1 \wedge b_2 \wedge b_3) = \quad & \left((a_1 \wedge a_2) \cdot (b_1 \wedge b_2)\right) b_3 \\ - & \left((a_1 \wedge a_2) \cdot (b_1 \wedge b_3)\right) b_2 \\ + & \left((a_1 \wedge a_2) \cdot (b_2 \wedge b_3)\right) b_1.\end{aligned} \tag{3.77}$$

The inner product of two blades of equal grade is always a scalar (see (3.39)). Therefore, the inner product of a 2-blade and a 3-blade results in a 1-blade, as expected.

There exist two additional important identities, which are proved in the following. The first important identity is essential in many derivations. It is a generalization of (3.70).

Lemma 3.12. *Let* $A_{\langle r\rangle}, B_{\langle s\rangle}, C_{\langle t\rangle} \in \mathbb{G}_{p,q}$ *with* $1 \le r, s, t \le n$ *and* $t \ge r + s$, *then*

$$\left(A_{\langle r\rangle} \wedge B_{\langle s\rangle}\right) \cdot C_{\langle t\rangle} = A_{\langle r\rangle} \cdot \left(B_{\langle s\rangle} \cdot C_{\langle t\rangle}\right).$$

Proof. The proof can be done by construction, using the definitions of the inner and the outer product. First of all, the outer and the inner product can be written in terms of grade projections:

$$\left(A_{\langle r\rangle} \wedge B_{\langle s\rangle}\right) \cdot C_{\langle t\rangle} = \left\langle \, \left\langle A_{\langle r\rangle}\, B_{\langle s\rangle}\right\rangle_{r+s} \; C_{\langle t\rangle} \, \right\rangle_{|(r+s)-t|}$$

From (3.39), it is clear that

$$A_{\langle r\rangle}\, B_{\langle s\rangle} = \sum_{i=0}^{m} \left\langle A_{\langle r\rangle}\, B_{\langle s\rangle}\right\rangle_{|r-s|+2i}\,, \qquad m = \frac{1}{2}\,(r + s - |r - s|),$$

and thus

$$\left\langle A_{\langle r\rangle}\, B_{\langle s\rangle}\right\rangle_{r+s} = A_{\langle r\rangle}\, B_{\langle s\rangle} - \sum_{i=0}^{m-1} \left\langle A_{\langle r\rangle}\, B_{\langle s\rangle}\right\rangle_{|r-s|+2i}\,.$$

Therefore,

$$\begin{aligned}&\left(A_{\langle r\rangle} \wedge B_{\langle s\rangle}\right) \cdot C_{\langle t\rangle} \\ &\quad = \left\langle A_{\langle r\rangle}\, B_{\langle s\rangle}\, C_{\langle t\rangle}\right\rangle_{|(r+s)-t|} \\ &\qquad - \left\langle \sum_{i=0}^{m-1} \left\langle A_{\langle r\rangle}\, B_{\langle s\rangle}\right\rangle_{|r-s|+2i} \; C_{\langle t\rangle} \right\rangle_{|(r+s)-t|}\,.\end{aligned}$$

If $t \ge r + s$, then the grade projection bracket that is subtracted has to be zero, since the geometric product $\left\langle A_{\langle r\rangle}\, B_{\langle s\rangle}\right\rangle_{|r-s|+2i} \; C_{\langle t\rangle}$ has only a term of grade $t - r - s$ if $i = m$. Furthermore,

$$\big\langle \boldsymbol{A}_{\langle r\rangle}\,\boldsymbol{B}_{\langle s\rangle}\,\boldsymbol{C}_{\langle t\rangle}\big\rangle_{t-s-r} = \Big\langle\, \boldsymbol{A}_{\langle r\rangle} \sum_{i=0}^{n} \big\langle \boldsymbol{B}_{\langle s\rangle}\,\boldsymbol{C}_{\langle t\rangle}\big\rangle_{|t-s|+2i} \,\Big\rangle_{t-s-r},$$

where $n = \frac{1}{2}\,(t+s-|t+s|)$. The geometric product on the right-hand side has only terms of grade $t-s-r$ for $i=0$. Hence,

$$\Big\langle\, \boldsymbol{A}_{\langle r\rangle} \sum_{i=0}^{n} \big\langle \boldsymbol{B}_{\langle s\rangle}\,\boldsymbol{C}_{\langle t\rangle}\big\rangle_{|s-t|+2i} \,\Big\rangle_{t-s-r} = \Big\langle\, \boldsymbol{A}_{\langle r\rangle}\, \big\langle \boldsymbol{B}_{\langle s\rangle}\,\boldsymbol{C}_{\langle t\rangle}\big\rangle_{|s-t|} \,\Big\rangle_{|r-|t-s||},$$

which proves the proposition.

□

The second identity is a restricted law of associativity for the inner product.

Lemma 3.13. *Let $\boldsymbol{A}_{\langle r\rangle}, \boldsymbol{B}_{\langle s\rangle}, \boldsymbol{C}_{\langle t\rangle} \in \mathbb{G}_{p,q}$ with $1 \le r,s,t \le n$ and $s \ge r+t$, then*

$$(\boldsymbol{A}_{\langle r\rangle} \cdot \boldsymbol{B}_{\langle s\rangle}) \cdot \boldsymbol{C}_{\langle t\rangle} = \boldsymbol{A}_{\langle r\rangle} \cdot (\boldsymbol{B}_{\langle s\rangle} \cdot \boldsymbol{C}_{\langle t\rangle}).$$

Proof. The proof is done in much the same way as for Lemma 3.12, which is why some of the details are left out here.

$$\begin{aligned}(\boldsymbol{A}_{\langle r\rangle} \cdot \boldsymbol{B}_{\langle s\rangle}) \cdot \boldsymbol{C}_{\langle t\rangle} &= \Big\langle\, \big\langle \boldsymbol{A}_{\langle r\rangle}\,\boldsymbol{B}_{\langle s\rangle}\big\rangle_{|r-s|}\; \boldsymbol{C}_{\langle t\rangle} \,\Big\rangle_{||r-s|-t|} \\ &= \big\langle \boldsymbol{A}_{\langle r\rangle}\,\boldsymbol{B}_{\langle s\rangle}\,\boldsymbol{C}_{\langle t\rangle}\big\rangle_{||r-s|-t|}\;.\end{aligned}$$

This follows since $\big\langle \boldsymbol{A}_{\langle r\rangle}\,\boldsymbol{B}_{\langle s\rangle}\big\rangle_{|r-s|}$ is the only term of $\boldsymbol{A}_{\langle r\rangle}\,\boldsymbol{B}_{\langle s\rangle}$ that contributes to the outer grade projection bracket. If $s \ge r+t$, then $||r-s|-t| = |r-|s-t||$ and hence

$$\begin{aligned}\big\langle \boldsymbol{A}_{\langle r\rangle}\,\boldsymbol{B}_{\langle s\rangle}\,\boldsymbol{C}_{\langle t\rangle}\big\rangle_{||r-s|-t|} &= \big\langle \boldsymbol{A}_{\langle r\rangle}\,\boldsymbol{B}_{\langle s\rangle}\,\boldsymbol{C}_{\langle t\rangle}\big\rangle_{|r-|s-t||} \\ &= \Big\langle\, \boldsymbol{A}_{\langle r\rangle}\, \big\langle \boldsymbol{B}_{\langle s\rangle}\,\boldsymbol{C}_{\langle t\rangle}\big\rangle_{|s-t|} \,\Big\rangle_{|r-|s-t||}\;.\end{aligned}$$

The second equality follows since $\big\langle \boldsymbol{B}_{\langle s\rangle}\,\boldsymbol{C}_{\langle t\rangle}\big\rangle_{|s-t|}$ is the only component of $\boldsymbol{B}_{\langle s\rangle}\,\boldsymbol{C}_{\langle t\rangle}$ that contributes to the outer grade projection bracket. Because

$$\Big\langle\, \boldsymbol{A}_{\langle r\rangle}\, \big\langle \boldsymbol{B}_{\langle s\rangle}\,\boldsymbol{C}_{\langle t\rangle}\big\rangle_{|s-t|} \,\Big\rangle_{|r-|s-t||} = \boldsymbol{A}_{\langle r\rangle} \cdot (\boldsymbol{B}_{\langle s\rangle} \cdot \boldsymbol{C}_{\langle t\rangle}),$$

the proposition is proven.

□

Lemma 3.14. *Let $\boldsymbol{A}_{\langle k\rangle} \in \mathbb{G}^{k}_{p,q}$ and let $\boldsymbol{I} := \overline{\mathbb{G}}_{p,q}[2^{p+q}]$ be the pseudoscalar of $\mathbb{G}_{p,q}$. Then $\boldsymbol{A}_{\langle k\rangle}\,\boldsymbol{I} = \boldsymbol{A}_{\langle k\rangle} \cdot \boldsymbol{I}$.*

Proof. We write $\boldsymbol{A}_{\langle k\rangle} = \mathsf{a}^{i}\,\boldsymbol{E}_i$, with $\boldsymbol{E}_i := \overline{\mathbb{G}}^{k}_{p,q}[i]$ and $\{\mathsf{a}^{i}\} \subset \mathbb{R}$. Then $\boldsymbol{A}_{\langle k\rangle}\boldsymbol{I} = \mathsf{a}^{i}\,\boldsymbol{E}_i\,\boldsymbol{I}$. Since $\boldsymbol{I}$ is the geometric product of all basis vectors in $\overline{\mathbb{G}}^{1}_{p,q}$, $\boldsymbol{E}_i$ and $\boldsymbol{I}$ have k basis vectors in common. Hence, $\boldsymbol{E}_i\,\boldsymbol{I} = \langle \boldsymbol{E}_i\,\boldsymbol{I}\rangle_{|k-p+q|} = \boldsymbol{E}_i \cdot \boldsymbol{I}$ for all i. The proposition thus follows immediately.

□

Some examples will clarify the use of these identities. Recall that the dual of a vector $\boldsymbol{a} \in \mathbb{G}^1_{p,q}$ is given by $\boldsymbol{a}^* = \boldsymbol{a}\boldsymbol{I}^{-1}$, where $\boldsymbol{I} := \overline{\mathbb{G}}_{p,q}[2^{p+q}]$ is the pseudoscalar of $\mathbb{G}_{p,q}$, with $p+q \geq 2$. Then

$$\boldsymbol{a}^* \cdot \boldsymbol{b} = (\boldsymbol{a} \cdot \boldsymbol{I}^{-1}) \cdot \boldsymbol{b} = \boldsymbol{a} \cdot (\boldsymbol{I}^{-1} \cdot \boldsymbol{b}) = (-1)^{n+1}\, \boldsymbol{a} \cdot (\boldsymbol{b} \cdot \boldsymbol{I}^{-1}) = (-1)^{n+1}\, \boldsymbol{a} \cdot \boldsymbol{b}^* . \quad (3.78)$$

Here Lemma 3.13 has been employed, since the grade of the pseudoscalar is greater than or equal to 2. The sign change when the order of $\boldsymbol{I}$ and $\boldsymbol{b}$ is changed follows from (3.68). Using the same symbols as in the previous example, here is an example where Lemma 3.12 is used:

$$(\boldsymbol{a} \wedge \boldsymbol{b})^* = (\boldsymbol{a} \wedge \boldsymbol{b}) \cdot \boldsymbol{I}^{-1} = \boldsymbol{a} \cdot (\boldsymbol{b} \cdot \boldsymbol{I}^{-1}) = \boldsymbol{a} \cdot \boldsymbol{b}^* . \quad (3.79)$$

This equation is quite interesting, since it shows that the outer product is in some sense dual to the inner product. It is often also useful to use Lemma 3.12 in the other direction. Let $\boldsymbol{a} \in \mathbb{G}^1_{p,q}$ and $\boldsymbol{A}_{\langle k \rangle} \in \mathbb{G}^k_{p,q}$, $2 \leq k \leq p+q$; then

$$\boldsymbol{a} \cdot (\boldsymbol{a} \cdot \boldsymbol{A}_{\langle k \rangle}) = (\boldsymbol{a} \wedge \boldsymbol{a}) \cdot \boldsymbol{A}_{\langle k \rangle} = 0. \quad (3.80)$$

Instead of evaluating the inner product of $\boldsymbol{a}$ and $\boldsymbol{A}_{\langle k \rangle}$, it can be seen immediately that the whole expression is zero.

Using Lemmas 3.12 and 3.13, it can be shown that

$$\boldsymbol{A}_{\langle k \rangle} \cdot \boldsymbol{B}_{\langle l \rangle} = (-1)^{k(l+1)}\, \boldsymbol{B}_{\langle l \rangle} \cdot \boldsymbol{A}_{\langle k \rangle} . \quad (3.81)$$

With the help of Lemma 3.12, it may be shown that for $\boldsymbol{A}_{\langle k \rangle} := \boldsymbol{a}_1 \wedge \ldots \wedge \boldsymbol{a}_k \in \mathbb{G}^k_{p,q}$ and $\boldsymbol{B}_{\langle l \rangle} := \boldsymbol{b}_i \wedge \ldots \wedge \boldsymbol{b}_l \in \mathbb{G}^l_{p,q}$,

$$\boldsymbol{A}_{\langle k \rangle} \cdot \boldsymbol{B}_{\langle l \rangle} = \boldsymbol{a}_1 \cdot \Big(\boldsymbol{a}_2 \cdot \big(\cdots (\boldsymbol{a}_k \cdot \boldsymbol{B}_{\langle l \rangle})\big)\Big). \quad (3.82)$$

Using (3.71), it follows that if there exists just one $\boldsymbol{a}_i$ such that $\boldsymbol{a}_i \cdot \boldsymbol{b}_j = 0$ for all $j \in \{1, \ldots, l\}$, then $\boldsymbol{A}_{\langle k \rangle} \cdot \boldsymbol{B}_{\langle l \rangle} = 0$.

3.2.7.3 Subspaces

It was shown in Sect. 3.2.2 that the outer product of two blades represents the direct sum of the OPNSs of the blades, if it is not zero. In a similar way, the inner product represents their direct difference.

Definition 3.20 (Direct Difference). Let $\mathbb{A}, \mathbb{B} \subseteq \mathbb{R}^{p,q}$; then their *direct difference* is defined as

$$\mathbb{A} \ominus \mathbb{B} := \{\, \boldsymbol{a} \in \mathbb{A} \,:\, \boldsymbol{a} * \boldsymbol{b}^\dagger = 0 \;\forall\, \boldsymbol{b} \in \mathbb{B} \,\}.$$

Recall that $\boldsymbol{a} * \boldsymbol{b}^\dagger \equiv \boldsymbol{a} \star \boldsymbol{b}$ is zero only if $\boldsymbol{a}$ and $\boldsymbol{b}$ are geometrically orthogonal (perpendicular) (cf. Sect. 3.2.6).

Consider first of all the case of an inner product between a vector $\boldsymbol{a} \in \mathbb{G}^{\varnothing 1}_{p,q}$ and a blade $\boldsymbol{A}_{\langle k \rangle} \in \mathbb{G}^{\varnothing k}_{p,q}$, $k \geq 2$, such that $\boldsymbol{B}_{\langle l \rangle} := \boldsymbol{a} \cdot \boldsymbol{A}_{\langle k \rangle}$, where $l = k - 1$. It follows from (3.80) that $\boldsymbol{a} \cdot \boldsymbol{B}_{\langle l \rangle} = \boldsymbol{a} \cdot (\boldsymbol{a} \cdot \boldsymbol{A}_{\langle k \rangle}) = 0$, which implies that $\boldsymbol{a}$ is perpendicular to the OPNS of $\boldsymbol{B}_{\langle l \rangle}$. Furthermore, if $k = 1$, then $\boldsymbol{a} \cdot \boldsymbol{A}_{\langle 1 \rangle} \in \mathbb{R}$ and thus $\boldsymbol{a} \cdot (\boldsymbol{a} \cdot \boldsymbol{A}_{\langle 1 \rangle}) = 0$ by definition. Therefore, if $k \geq 1$ and $\boldsymbol{a} \cdot \boldsymbol{A}_{\langle k \rangle} \neq 0$,

$$\mathbb{NO}(\boldsymbol{a} \cdot \boldsymbol{A}_{\langle k \rangle}) = \mathbb{NO}(\boldsymbol{A}_{\langle k \rangle}) \ominus \mathbb{NO}(\boldsymbol{a}). \tag{3.83}$$

Note that the OPNS represented by a scalar is the empty space $\emptyset$, since the outer product of a 1-vector with a non-zero scalar is never zero.

Now consider two blades $\boldsymbol{A}_{\langle k \rangle}, \boldsymbol{B}_{\langle l \rangle} \in \mathbb{G}^{\varnothing}_{p,q}$ with $k \leq l$ and define $\boldsymbol{A}_{\langle k \rangle} := \bigwedge_{i=1}^{k} \boldsymbol{a}_i$, where $\{ \boldsymbol{a}_i \} \subset \mathbb{G}^1_{p,q}$. It then follows from (3.82) that if $\boldsymbol{A}_{\langle k \rangle} \cdot \boldsymbol{B}_{\langle l \rangle} \neq 0$,

$$\begin{aligned} &\mathbb{NO}(\boldsymbol{A}_{\langle k \rangle} \cdot \boldsymbol{B}_{\langle l \rangle}) \\ &\quad = \left(\Big(\big(\mathbb{NO}(\boldsymbol{B}_{\langle l \rangle}) \ominus \mathbb{NO}(\boldsymbol{a}_k)\big) \ominus \mathbb{NO}(\boldsymbol{a}_{k-1}) \Big) \ominus \ldots \right) \ominus \mathbb{NO}(\boldsymbol{a}_1), \end{aligned} \tag{3.84}$$

which is equivalent to

$$\mathbb{NO}(\boldsymbol{A}_{\langle k \rangle} \cdot \boldsymbol{B}_{\langle l \rangle}) = \mathbb{NO}(\boldsymbol{B}_{\langle l \rangle}) \ominus \mathbb{NO}(\boldsymbol{A}_{\langle k \rangle}). \tag{3.85}$$

This shows that the inner and outer products of blades can be used to add and subtract subspaces.

3.2.7.4 Null Blades

The existence of null-blades in geometric algebras of mixed signature complicates many properties of blades that relate to the inner product. Consider, for example, a null-vector $\boldsymbol{a} \in \mathbb{G}^{\circ 1}_{p,q}$, i.e. $\boldsymbol{a}\,\boldsymbol{a} = 0$. The geometric product of $\boldsymbol{a}$ with its conjugate $\boldsymbol{a}^\dagger$ is not a scalar, as can be seen in the following way. Let $\boldsymbol{a} := \boldsymbol{p} + \boldsymbol{q}$, with $\boldsymbol{p}^2 > 0$, $\boldsymbol{q}^2 < 0$, and $\boldsymbol{p} \cdot \boldsymbol{q} = 0$, such that $\boldsymbol{a}^2 = \boldsymbol{p}^2 + \boldsymbol{q}^2 = 0$. Then $\boldsymbol{a}^\dagger = \boldsymbol{p} - \boldsymbol{q}$, and thus

$$\boldsymbol{a}\,\boldsymbol{a}^\dagger = \boldsymbol{p}^2 - \boldsymbol{q}^2 + \boldsymbol{q}\,\boldsymbol{p} - \boldsymbol{p}\,\boldsymbol{q} = \boldsymbol{p}^2 - \boldsymbol{q}^2 + 2\,\boldsymbol{q} \wedge \boldsymbol{p}. \tag{3.86}$$

Similarly, the geometric product of a null-blade $\boldsymbol{U}_{\langle k \rangle} \in \mathbb{G}^{\circ k}_{p,q}$ with its conjugate does not result in a scalar, i.e. $\boldsymbol{U}_{\langle k \rangle}\,\boldsymbol{U}^\dagger_{\langle k \rangle} \notin \mathbb{R}$. This follows since $\boldsymbol{U}_{\langle k \rangle}$ can be written as a geometric product of perpendicular vectors, where at least one vector is a null-vector. That is, $\boldsymbol{U}_{\langle k \rangle} := \prod_{i=1}^{k} \boldsymbol{u}_i$, where $\{ \boldsymbol{u}_i \} \in \mathbb{G}^1_{p,q}$ is a set of perpendicular vectors, which may be self-orthogonal. Hence, $\boldsymbol{u}_i \star \boldsymbol{u}_j = 0$ if $i \neq j$.

Suppose that $\boldsymbol{u}_i^2 = 0$; then $\boldsymbol{u}_i \cdot \boldsymbol{U}_{\langle k \rangle} = 0$, which means that here the inner product does not return the blade of grade $k-1$ which is geometrically orthogonal to $\boldsymbol{u}_i$, as is the case for non-null-blades.

Lemma 3.15. *Let* $\boldsymbol{a} \in \mathbb{G}^1_{p,q}$ *and* $\boldsymbol{U}_{\langle k \rangle} \in \mathbb{G}^k_{p,q}$, *where* $\boldsymbol{a}$ *and* $\boldsymbol{U}_{\langle k \rangle}$ *may be null blades; then*

$$\mathbb{NO}(\boldsymbol{a}^\dagger \cdot \boldsymbol{U}_{\langle k \rangle}) = \mathbb{NO}(\boldsymbol{U}_{\langle k \rangle}) \ominus \mathbb{NO}(\boldsymbol{a}).$$

Proof. From (3.71), it follows that

$$\boldsymbol{a}^\dagger \cdot \boldsymbol{U}_{\langle k \rangle} = \sum_{i=1}^{k} (-1)^{i+1} \, (\boldsymbol{a}^\dagger \cdot \boldsymbol{u}_i) \, [\boldsymbol{U}_{\langle k \rangle} \backslash \boldsymbol{u}_i].$$

Since $\boldsymbol{a}^\dagger \cdot \boldsymbol{u}_i = \boldsymbol{u}_i \star \boldsymbol{a}$, these terms cannot be zero, because of self-orthogonality. Hence, the same properties apply as for (3.84).
□

As before, it also follows that for $\boldsymbol{A}_{\langle k \rangle} \in \mathbb{G}^k_{p,q}$,

$$\mathbb{NO}(\boldsymbol{A}^\dagger_{\langle l \rangle} \cdot \boldsymbol{U}_{\langle k \rangle}) = \mathbb{NO}(\boldsymbol{U}_{\langle k \rangle}) \ominus \mathbb{NO}(\boldsymbol{A}_{\langle l \rangle}). \tag{3.87}$$

It also follows from the definition of the direct sum that

$$\mathbb{NO}(\boldsymbol{U}_{\langle k \rangle}) = \mathbb{NO}\big(\boldsymbol{A}_{\langle l \rangle} \wedge (\boldsymbol{A}^\dagger_{\langle l \rangle} \cdot \boldsymbol{U}_{\langle k \rangle})\big). \tag{3.88}$$

Consider, for example, a vector $\boldsymbol{a} \in \mathbb{G}^1_{p,q}$ with $\mathbb{NO}(\boldsymbol{a}) \subset \mathbb{NO}(\boldsymbol{U}_{\langle k \rangle})$; then $\boldsymbol{a} \wedge \boldsymbol{U}_{\langle k \rangle} = 0$ and, with the help of (3.70),

$$(\boldsymbol{U}_{\langle k \rangle} \cdot \boldsymbol{a}^\dagger) \wedge \boldsymbol{a} = \boldsymbol{U}_{\langle k \rangle} \underbrace{(\boldsymbol{a}^\dagger \cdot \boldsymbol{a})}_{=\|\boldsymbol{a}\|^2} - \underbrace{(\boldsymbol{U}_{\langle k \rangle} \wedge \boldsymbol{a})}_{=0} \cdot \boldsymbol{a}^\dagger = \|\boldsymbol{a}\|^2 \, \boldsymbol{U}_{\langle k \rangle}. \tag{3.89}$$

3.2.8 Duality

The effect of the dual operation on blades can best be seen in relation to the OPNS and IPNS of the blade. Let $\boldsymbol{x} \in \mathbb{G}^{\varnothing 1}_{p,q}$ and $\boldsymbol{A}_{\langle k \rangle} \in \mathbb{G}^{\varnothing k}_{p,q}$ with $k \geq 1$, and let $\boldsymbol{I} := \overline{\mathbb{G}}_{p,q}[2^{p+q}]$ denote the pseudoscalar of $\mathbb{G}_{p,q}$; then, just as in (3.78),

$$(\boldsymbol{x} \wedge \boldsymbol{A}_{\langle k \rangle})^* = (\boldsymbol{x} \wedge \boldsymbol{A}_{\langle k \rangle}) \cdot \boldsymbol{I}^{-1} = \boldsymbol{x} \cdot \boldsymbol{A}^*_{\langle k \rangle},$$

Hence,

$$\boldsymbol{x} \wedge \boldsymbol{A}_{\langle k \rangle} = 0 \qquad \Longleftrightarrow \qquad \boldsymbol{x} \cdot \boldsymbol{A}^*_{\langle k \rangle} = 0,$$

and thus

$$\mathbb{NO}(\boldsymbol{A}_{\langle k \rangle}) = \mathbb{NI}(\boldsymbol{A}^*_{\langle k \rangle}). \tag{3.90}$$

This shows that the OPNS and IPNS of a blade are directly related by the dual operation. Furthermore,

$$\boldsymbol{A}_{\langle k \rangle} \wedge \boldsymbol{A}^*_{\langle k \rangle} = \big\langle \boldsymbol{A}_{\langle k \rangle} \, \boldsymbol{A}_{\langle k \rangle} \, \boldsymbol{I}^{-1} \big\rangle_{k+|n-k|} = (\boldsymbol{A}_{\langle k \rangle} \cdot \boldsymbol{A}_{\langle k \rangle}) \, \boldsymbol{I}^{-1}, \tag{3.91}$$

where $n = p+q$ and thus $k+|n-k| = k-k+n = n$. Hence, the outer product of a blade with its dual results in a scalar multiple of the pseudoscalar. Note that (3.91) does not hold for $k = 0$, hence, the initial assumption $k \geq 1$. In terms of the OPNS of the blades,

$$\mathbb{NO}(\boldsymbol{A}_{\langle k\rangle} \wedge \boldsymbol{A}^*_{\langle k\rangle}) = \mathbb{NO}(\boldsymbol{A}_{\langle k\rangle}) \oplus \mathbb{NO}(\boldsymbol{A}^*_{\langle k\rangle}) = \mathbb{NO}(\boldsymbol{I}^{-1}). \tag{3.92}$$

Since $\boldsymbol{I}^{-1}$ is an algebraic basis element of highest grade, $\mathbb{NO}(\boldsymbol{I}^{-1}) = \mathbb{G}^1_{p,q}$. Therefore, the relation between the OPNS of a blade and its dual is

$$\mathbb{NO}(\boldsymbol{A}^*_{\langle k\rangle}) = \mathbb{G}^1_{p,q} \ominus \mathbb{NO}(\boldsymbol{A}_{\langle k\rangle}); \tag{3.93}$$

that is, their OPNSs are complementary.

3.2.8.1 Null-Blades

For null-blades, some aspects of the dual are changed. For $\boldsymbol{x} \in \mathbb{G}^1_{p,q}$ and $\boldsymbol{A}_{\langle k\rangle} \in \mathbb{G}^k_{p,q}$, it is still true that $(\boldsymbol{x} \cdot \boldsymbol{A}_{\langle k\rangle})^* = \boldsymbol{x} \cdot \boldsymbol{A}^*_{\langle k\rangle}$ and $\boldsymbol{x} \wedge \boldsymbol{A}_{\langle k\rangle} = 0$ implies $\boldsymbol{x} \cdot \boldsymbol{A}^*_{\langle k\rangle} = 0$. However, $\boldsymbol{x} \cdot \boldsymbol{A}^*_{\langle k\rangle} = 0$ does not imply $\boldsymbol{x} \wedge \boldsymbol{A}_{\langle k\rangle} = 0$, since $\boldsymbol{x}$ may be a null vector which is also part of $\boldsymbol{A}^*_{\langle k\rangle}$. Furthermore,

$$\boldsymbol{A}_{\langle k\rangle} \wedge \boldsymbol{A}^*_{\langle k\rangle} = \big\langle \boldsymbol{A}_{\langle k\rangle}\, \boldsymbol{A}_{\langle k\rangle}\, \boldsymbol{I}^{-1} \big\rangle_{k-|n-k|} \tag{3.94}$$

is zero if $\boldsymbol{A}_{\langle k\rangle}$ is a null-blade. Hence, the complement of $\boldsymbol{A}_{\langle k\rangle}$ is its conjugate. That is,

$$\boldsymbol{A}^{\dagger}_{\langle k\rangle} \wedge \boldsymbol{A}^*_{\langle k\rangle} = \Big\langle \boldsymbol{A}^{\dagger}_{\langle k\rangle}\, \boldsymbol{A}_{\langle k\rangle}\, \boldsymbol{I}^{-1} \Big\rangle_{k-|n-k|} = (\boldsymbol{A}^{\dagger}_{\langle k\rangle} \cdot \boldsymbol{A}_{\langle k\rangle})\, \boldsymbol{I}^{-1} = \boldsymbol{I}^{-1}. \tag{3.95}$$

3.2.8.2 Vector Cross Product

The properties of the dual also lead to a relation between the standard vector cross product in $\mathbb{R}^3$ and the outer product in $\mathbb{G}_3$. The vector cross product of two vectors in $\mathbb{R}^3$ results in a vector that is perpendicular to the two given vectors, whereas the outer product of two vectors results in a blade whose OPNS is the subspace spanned by the two vectors. Hence, these two operations result in blades whose OPNSs are complementary to each other. That is, for two vectors $\boldsymbol{a}, \boldsymbol{b} \in \mathbb{G}^1_3$,

$$\mathbb{NO}(\boldsymbol{a} \wedge \boldsymbol{b}) = \mathbb{NI}\big((\boldsymbol{a} \wedge \boldsymbol{b})^*\big) = \mathbb{NI}(\boldsymbol{a} \times \boldsymbol{b}). \tag{3.96}$$

It may also be shown by construction that

$$\boldsymbol{a} \times \boldsymbol{b} = (\boldsymbol{a} \wedge \boldsymbol{b})^*. \tag{3.97}$$

Note that the vector cross product is defined only in $\mathbb{R}^3$, whereas the outer product of two vectors represents the subspace spanned by the two vectors in any embedding dimension.

3.2.9 Inverse

In contrast to general multivectors $\mathbb{G}_{p,q}$, nonnull blades of $\mathbb{G}_{p,q}$ always have an inverse.

Lemma 3.16. *Let* $\boldsymbol{A}_{\langle k\rangle} \in \mathbb{G}^{\varnothing\, k}_{p,q}$; *then the* inverse *of* $\boldsymbol{A}_{\langle k\rangle}$ *is given by*

$$\boldsymbol{A}^{-1}_{\langle k\rangle} = \frac{\widetilde{\boldsymbol{A}}_{\langle k\rangle}}{\boldsymbol{A}_{\langle k\rangle}\,\widetilde{\boldsymbol{A}}_{\langle k\rangle}}.$$

Proof. It follows directly from Lemma 3.10 that $\boldsymbol{A}_{\langle k\rangle}\,\widetilde{\boldsymbol{A}}_{\langle k\rangle} = \boldsymbol{A}_{\langle k\rangle} \cdot \widetilde{\boldsymbol{A}}_{\langle k\rangle} \in \mathbb{R}$. Furthermore, since $\boldsymbol{A}_{\langle k\rangle} \cdot \widetilde{\boldsymbol{A}}_{\langle k\rangle} = (-1)^{k(k+1)}\,\widetilde{\boldsymbol{A}}_{\langle k\rangle} \cdot \boldsymbol{A}_{\langle k\rangle}$ and $(-1)^{k(k+1)} = +1$, for all $k \geq 0$,

$$\boldsymbol{A}_{\langle k\rangle}\,\boldsymbol{A}^{-1}_{\langle k\rangle} = \boldsymbol{A}^{-1}_{\langle k\rangle}\,\boldsymbol{A}_{\langle k\rangle} = 1.$$

□

3.2.9.1 Null Blades

It is easy to see that null-blades do not have an inverse.

Lemma 3.17. *Let* $\boldsymbol{A}_{\langle k\rangle} \in \mathbb{G}^{\circ\, k}_{p,q}$ *be a null-blade; then there exists no blade* $\boldsymbol{B}_{\langle k\rangle} \in \mathbb{G}^{k}_{p,q}$ *such that* $\boldsymbol{A}_{\langle k\rangle}\,\boldsymbol{B}_{\langle k\rangle} = 1$.

Proof. If $\boldsymbol{A}_{\langle k\rangle}$ is a null-blade, then $\boldsymbol{A}_{\langle k\rangle}\,\boldsymbol{A}_{\langle k\rangle} = 0$. Suppose $\boldsymbol{A}_{\langle k\rangle}$ had an inverse $\boldsymbol{B}_{\langle k\rangle}$ such that $\boldsymbol{A}_{\langle k\rangle}\,\boldsymbol{B}_{\langle k\rangle} = 1$. In this case $\boldsymbol{A}_{\langle k\rangle}\,\boldsymbol{A}_{\langle k\rangle}\,\boldsymbol{B}_{\langle k\rangle} = 0$ and thus $\boldsymbol{A}_{\langle k\rangle} = 0$, which contradicts the assumption that $\boldsymbol{A}_{\langle k\rangle} \in \mathbb{G}^{k}_{p,q}$.
□

Even though a null-blade has no inverse with respect to the geometric product, it is still possible to define an inverse with respect to the inner product.

Definition 3.21 (Pseudoinverse of Blade). The *pseudoinverse* of a blade $\boldsymbol{A}_{\langle k\rangle} \in \mathbb{G}^{k}_{p,q}$ with $1 \leq k \leq p+q$, denoted by $\boldsymbol{A}^{+}_{\langle k\rangle}$, is defined as

$$\boldsymbol{A}^{+}_{\langle k\rangle} := \frac{\boldsymbol{A}^{\dagger}_{\langle k\rangle}}{\boldsymbol{A}_{\langle k\rangle} \cdot \boldsymbol{A}^{\dagger}_{\langle k\rangle}} = \frac{\boldsymbol{A}^{\dagger}_{\langle k\rangle}}{\|\boldsymbol{A}_{\langle k\rangle}\|^2}.$$

The pseudoinverse of $\boldsymbol{A}_{\langle k\rangle} \in \mathbb{G}^{k}_{p,q}$ therefore satisfies

$$\boldsymbol{A}_{\langle k\rangle} \cdot \boldsymbol{A}^{+}_{\langle k\rangle} = \boldsymbol{A}^{+}_{\langle k\rangle} \cdot \boldsymbol{A}_{\langle k\rangle} = +1.$$

However, the geometric product $\boldsymbol{A}_{\langle k\rangle}\, \boldsymbol{A}^{+}_{\langle k\rangle}$ results in $+1$ only if $\boldsymbol{A}_{\langle k\rangle}$ is not a null-blade.

3.2.10 Projection

The projection of a vector $\boldsymbol{a} \in \mathbb{G}^{\varnothing 1}_{p,q}$ onto a vector $\boldsymbol{n} \in \mathbb{G}^{\varnothing 1}_{p,q}$ is the component of $\boldsymbol{a}$ in the direction of $\boldsymbol{n}$. This projection can be evaluated via

$$(\boldsymbol{a} \cdot \boldsymbol{n})\, \boldsymbol{n}^{-1} = (\boldsymbol{a} \cdot \hat{\boldsymbol{n}})\, \hat{\boldsymbol{n}} = \|\boldsymbol{a}\| \cos\theta\, \hat{\boldsymbol{n}}, \tag{3.98}$$

where $\hat{\boldsymbol{n}} = \boldsymbol{n}/\|\boldsymbol{n}\|$ and $\hat{\boldsymbol{n}}^{-1} = \boldsymbol{n}/\|\boldsymbol{n}\|^2$. This is generalized to blades as follows.

3.2.10.1 Definitions

Definition 3.22. The *projection* of a blade $\boldsymbol{A}_{\langle k\rangle} \in \mathbb{G}^{\varnothing k}_{p,q}$ onto a blade $\boldsymbol{N}_{\langle l\rangle} \in \mathbb{G}^{\varnothing l}_{p,q}$, with $1 \leq k \leq l \leq p+q$, is defined as

$$\mathcal{P}_{\boldsymbol{N}_{\langle l\rangle}}\big(\boldsymbol{A}_{\langle k\rangle}\big) := \big(\boldsymbol{A}_{\langle k\rangle} \cdot \boldsymbol{N}^{-1}_{\langle l\rangle}\big)\, \boldsymbol{N}_{\langle l\rangle}.$$

Consider the projection of a vector $\boldsymbol{a} \in \mathbb{G}^{\varnothing 1}_{p,q}$ onto a blade $\boldsymbol{N}_{\langle l\rangle} \in \mathbb{G}^{\varnothing l}_{p,q}$. In general, $\boldsymbol{a}$ will have a component $\boldsymbol{a}^{\parallel}$ that lies in the subspace represented by $\boldsymbol{N}_{\langle l\rangle}$. That is, if $\{\boldsymbol{n}_1, \ldots, \boldsymbol{n}_l\} \subset \mathbb{G}^{\varnothing 1}_{p,q}$ and $\boldsymbol{N}_{\langle l\rangle} := \bigwedge_{i=1}^{l} \boldsymbol{n}_i$, then $\boldsymbol{a}^{\parallel} \in \operatorname{span}\{\boldsymbol{n}_1, \ldots, \boldsymbol{n}_l\}$ and thus $\boldsymbol{a}^{\parallel} \wedge \boldsymbol{N}_{\langle l\rangle} = 0$. The remainder of $\boldsymbol{a}$ is denoted by $\boldsymbol{a}^{\perp}$, i.e. $\boldsymbol{a} = \boldsymbol{a}^{\parallel} + \boldsymbol{a}^{\perp}$. Since $\boldsymbol{a}^{\perp} \notin \operatorname{span}\{\boldsymbol{n}_1, \ldots, \boldsymbol{n}_l\}$ by definition, $\boldsymbol{a}^{\perp} \cdot \boldsymbol{N}_{\langle l\rangle} = 0$. Since $\boldsymbol{N}^{-1}_{\langle l\rangle} \propto \boldsymbol{N}_{\langle l\rangle}$, it follows that

$$\mathcal{P}_{\boldsymbol{N}_{\langle l\rangle}}(\boldsymbol{a}) = \big((\boldsymbol{a}^{\parallel} + \boldsymbol{a}^{\perp}) \cdot \boldsymbol{N}^{-1}_{\langle l\rangle}\big)\, \boldsymbol{N}_{\langle l\rangle} = \boldsymbol{a}^{\parallel}\, \boldsymbol{N}^{-1}_{\langle l\rangle}\, \boldsymbol{N}_{\langle l\rangle} = \boldsymbol{a}^{\parallel}. \tag{3.99}$$

Here the fact is used that since $\boldsymbol{a}^{\parallel} \wedge \boldsymbol{N}_{\langle l\rangle} = 0$, $\boldsymbol{a}^{\parallel} \cdot \boldsymbol{N}_{\langle l\rangle} = \boldsymbol{a}^{\parallel}\, \boldsymbol{N}_{\langle l\rangle}$. Hence, $\mathcal{P}_{\boldsymbol{N}_{\langle l\rangle}}(\boldsymbol{a})$ gives the component of $\boldsymbol{a}$ that lies in the subspace represented by $\boldsymbol{N}_{\langle l\rangle}$. Furthermore, it may be shown that, given $\{\boldsymbol{a}_1, \ldots, \boldsymbol{a}_k\} \subset \mathbb{G}^{\varnothing 1}_{p,q}$ and $\boldsymbol{N}_{\langle l\rangle} \in \mathbb{G}^{\varnothing l}_{p,q}$,

$$\mathcal{P}_{\boldsymbol{N}_{\langle l\rangle}}\left(\bigwedge_{i=1}^{k} \boldsymbol{a}_i\right) = \bigwedge_{i=1}^{k} \mathcal{P}_{\boldsymbol{N}_{\langle l\rangle}}(\boldsymbol{a}_i)\,. \tag{3.100}$$

This property of the projection operator is called *outermorphism*, not to be confused with *automorphism*. Note that $\mathcal{P}_{\boldsymbol{N}_{\langle l \rangle}}(\boldsymbol{A}_{\langle k \rangle})$ results again in a blade of grade k, or is zero if the subspaces represented by $\boldsymbol{A}_{\langle k \rangle}$ and $\boldsymbol{N}_{\langle l \rangle}$ have no k-dimensional subspace in common.

3.2.10.2 Null Blades

In order to consider the projection of null-blades, the representation of the direct difference of blades given in Lemma 3.15 has to be used. That is, instead of removing the linear dependence of a vector $\boldsymbol{a} \in \mathbb{G}^1_{p,q}$ from a blade $\boldsymbol{N}_{\langle l \rangle} \in \mathbb{G}^l_{p,q}$ by evaluating the inner product $\boldsymbol{a} \cdot \boldsymbol{N}_{\langle l \rangle}$, it is necessary to write $\boldsymbol{a}^\dagger \cdot \boldsymbol{N}_{\langle l \rangle}$. Note that

$$\left(\boldsymbol{a}^\dagger \cdot \boldsymbol{N}_{\langle l \rangle}\right)^\dagger = \boldsymbol{N}^\dagger_{\langle l \rangle} \cdot \boldsymbol{a} = (-1)^{l+1}\, \boldsymbol{a} \cdot \boldsymbol{N}^\dagger_{\langle l \rangle}.$$

Furthermore, the direct difference of the OPNSs of the blades $\boldsymbol{A}_{\langle k \rangle} \in \mathbb{G}^k_{p,q}$ and $\boldsymbol{N}_{\langle l \rangle}$ can be evaluated from the inner product $\boldsymbol{A}^\dagger_{\langle k \rangle} \cdot \boldsymbol{N}_{\langle l \rangle}$. The appropriate representation of the projection operator for general blades is therefore as follows.

Definition 3.23. Let $\boldsymbol{A}_{\langle k \rangle} \in \mathbb{G}^k_{p,q}$ and $\boldsymbol{N}_{\langle l \rangle} \in \mathbb{G}^l_{p,q}$ be two general blades with $1 \le k \le l \le p+q$; then the projection of $\boldsymbol{A}_{\langle k \rangle}$ onto $\boldsymbol{N}_{\langle l \rangle}$, denoted by $\mathcal{P}_{\boldsymbol{N}_{\langle l \rangle}}(\boldsymbol{A}_{\langle k \rangle})$, is defined as

$$\mathcal{P}_{\boldsymbol{N}_{\langle l \rangle}}(\boldsymbol{A}_{\langle k \rangle}) := (\boldsymbol{A}_{\langle k \rangle} \cdot \boldsymbol{N}^+_{\langle l \rangle}) \cdot \boldsymbol{N}_{\langle l \rangle}.$$

Note that this definition of the projection reduces to the previous definition if the blades are not null-blades. This is related to the direct difference of the OPNSs by

$$\begin{aligned}(\boldsymbol{A}_{\langle k \rangle} \cdot \boldsymbol{N}^+_{\langle l \rangle}) \cdot \boldsymbol{N}_{\langle l \rangle} &= \frac{1}{\|\boldsymbol{N}_{\langle l \rangle}\|^2}\, (\boldsymbol{A}_{\langle k \rangle} \cdot \boldsymbol{N}^\dagger_{\langle l \rangle}) \cdot \boldsymbol{N}_{\langle l \rangle} \\ &= \frac{(-1)^{k(l+1)}}{\|\boldsymbol{N}_{\langle l \rangle}\|^2}\, (\boldsymbol{A}^\dagger_{\langle k \rangle} \cdot \boldsymbol{N}_{\langle l \rangle})^\dagger \cdot \boldsymbol{N}_{\langle l \rangle}.\end{aligned}$$

3.2.11 Rejection

The *rejection* of a vector $\boldsymbol{a} \in \mathbb{G}^{\varnothing 1}_{p,q}$ from a vector $\boldsymbol{n} \in \mathbb{G}^{\varnothing 1}_{p,q}$ is the component of $\boldsymbol{a}$ that has no component parallel to $\boldsymbol{n}$. Therefore, the rejection can be defined in terms of projection. Here is the general definition for blades.

Definition 3.24. The *rejection* of a blade $\boldsymbol{A}_{\langle k\rangle} \in \mathbb{G}^k_{p,q}$ from a blade $\boldsymbol{N}_{\langle l\rangle} \in \mathbb{G}^l_{p,q}$, with $1 \leq k \leq l \leq p+q$, is defined as

$$\mathcal{P}^{\perp}_{\boldsymbol{N}_{\langle l\rangle}}(\boldsymbol{A}_{\langle k\rangle}) := \boldsymbol{A}_{\langle k\rangle} - \mathcal{P}_{\boldsymbol{N}_{\langle l\rangle}}(\boldsymbol{A}_{\langle k\rangle}) \,.$$

Note that the rejection of a blade $\boldsymbol{A}_{\langle k\rangle}$ also results in a blade of grade k. In contrast to the projection operator, however, there exists no outermorphism for the rejection operator.

3.2.12 Meet and Join

It was shown in Sects. 3.2.2 and 3.2.7, when we discussed the outer and inner products, that a blade can be regarded as representing a linear subspace, through either its outer-product or its inner-product null space. In this respect, the outer and inner products of blades have the effect of adding and subtracting subspaces, respectively. However, there are some restrictions on this. The outer product of two blades that represent partially overlapping subspaces results in zero. Similarly, the inner product of two blades whose OPNSs have a non-empty intersection also results in zero.

To remove these restrictions, the operations *meet* and *join*, which have an effect similar to the intersection and union of sets, are defined in geometric algebra. Note that both of these operations are defined only on blades.

3.2.12.1 Definitions

Definition 3.25 (Join). Let $\boldsymbol{A}_{\langle k\rangle}, \boldsymbol{B}_{\langle l\rangle} \in \mathbb{G}_{p,q}$; then their join is written as $\boldsymbol{A}_{\langle k\rangle} \dot{\wedge} \boldsymbol{B}_{\langle l\rangle}$ and is defined as the blade $\boldsymbol{J}_{\langle m\rangle}$ such that

$$\mathbb{NO}(\boldsymbol{J}_{\langle m\rangle}) = \mathbb{NO}(\boldsymbol{A}_{\langle k\rangle}) \oplus \mathbb{NO}(\boldsymbol{B}_{\langle l\rangle}), \quad \|\boldsymbol{J}_{\langle m\rangle}\| = 1.$$

Hence, if $\mathbb{NO}(\boldsymbol{A}_{\langle k\rangle}) \cap \mathbb{NO}(\boldsymbol{B}_{\langle l\rangle}) = \emptyset$, then $\boldsymbol{A}_{\langle k\rangle} \dot{\wedge} \boldsymbol{B}_{\langle l\rangle} = \boldsymbol{A}_{\langle k\rangle} \wedge \boldsymbol{B}_{\langle l\rangle}$. However, in general there is no simple algebraic operation that returns the join of two blades. An algorithm that evaluates the join of two blades is given in Sect. 3.7.3.

It follows from the definition of the join that if $\boldsymbol{J}_{\langle m\rangle}$ is a join, then so is $-\boldsymbol{J}_{\langle m\rangle}$. This non-uniqueness is typically no problem, since the join of two blades is only evaluated when a representative blade of the direct sum of the OPNS of the blades is needed. The actual scalar factor of the join blade is of no importance in this case.

Definition 3.26. The *meet* of two blades $\boldsymbol{A}_{\langle k\rangle}, \boldsymbol{B}_{\langle l\rangle} \in \mathbb{G}_{p,q}$ is denoted by $\boldsymbol{A}_{\langle k\rangle} \vee \boldsymbol{B}_{\langle l\rangle}$ and is defined as

$$\mathbb{NO}(\boldsymbol{A}_{\langle k\rangle} \vee \boldsymbol{B}_{\langle l\rangle}) = \mathbb{NO}(\boldsymbol{A}_{\langle k\rangle}) \cap \mathbb{NO}(\boldsymbol{B}_{\langle l\rangle}).$$

In the following, a closed-form expression is derived that evaluates the meet of two blades, given their join. This is done by applying de Morgan's law for the intersection of subspaces.

Definition 3.27 (De Morgan's Laws). Let $\mathbb{A}, \mathbb{B} \subseteq \mathbb{J} \subseteq \mathbb{R}^{p,q}$, and denote by $\mathbb{A}^c$ the complement of $\mathbb{A}$ with respect to $\mathbb{J}$. That is, $\mathbb{A}^c \cap \mathbb{A} = \emptyset$ and $\mathbb{A}^c \oplus \mathbb{A} = \mathbb{J}$. Then *de Morgan's laws* can be written as follows:

$$1.\ \mathbb{A} \oplus \mathbb{B} = \left(\mathbb{A}^c \cap \mathbb{B}^c\right)^c.$$
$$2.\ \mathbb{A} \cap \mathbb{B} = \left(\mathbb{A}^c \oplus \mathbb{B}^c\right)^c.$$

It was shown in Sect. 3.2.8, on duality, that the OPNSs of a blade and its dual are complementary with respect to the whole space. When evaluating the meet of two blades, it is sufficient to consider the join of the OPNSs of the blades as the whole space, since the meet has to lie in the OPNS of the join.

Let $\boldsymbol{A}_{\langle k\rangle}, \boldsymbol{B}_{\langle l\rangle} \in \mathbb{G}^{\varnothing}_{p,q}$ and $\boldsymbol{J}_{\langle m\rangle} := \boldsymbol{A}_{\langle k\rangle} \dot{\wedge} \boldsymbol{B}_{\langle l\rangle}$; then $\mathbb{NO}(\boldsymbol{A}_{\langle k\rangle})$ and $\mathbb{NO}(\boldsymbol{A}_{\langle k\rangle}\, \boldsymbol{J}^{-1}_{\langle m\rangle})$ have to be complementary spaces with respect to $\mathbb{NO}(\boldsymbol{J}_{\langle m\rangle})$, since

$$\mathbb{NO}(\boldsymbol{A}_{\langle k\rangle}\, \boldsymbol{J}^{-1}_{\langle m\rangle}) = \mathbb{NO}(\boldsymbol{A}_{\langle k\rangle} \cdot \boldsymbol{J}^{-1}_{\langle m\rangle}) = \mathbb{NO}(\boldsymbol{J}^{-1}_{\langle m\rangle}) \ominus \mathbb{NO}(\boldsymbol{A}_{\langle k\rangle}).$$

That is, if we define $\mathbb{A} := \mathbb{NO}(\boldsymbol{A}_{\langle k\rangle})$ and $\mathbb{B} := \mathbb{NO}(\boldsymbol{B}_{\langle l\rangle})$, their complements $\mathbb{A}^c$ and $\mathbb{B}^c$ may be defined as

$$\mathbb{A}^c := \mathbb{NO}(\boldsymbol{A}_{\langle k\rangle}\, \boldsymbol{J}^{-1}_{\langle m\rangle}) \quad \text{and} \quad \mathbb{B}^c := \mathbb{NO}(\boldsymbol{B}_{\langle l\rangle}\, \boldsymbol{J}^{-1}_{\langle m\rangle}). \tag{3.101}$$

Hence, $\mathbb{A}^c \cap \mathbb{B}^c = \emptyset$, and thus it follows from (3.47) that

$$\mathbb{A}^c \oplus \mathbb{B}^c = \mathbb{NO}\Big((\boldsymbol{A}_{\langle k\rangle}\, \boldsymbol{J}^{-1}_{\langle m\rangle}) \wedge (\boldsymbol{B}_{\langle l\rangle}\, \boldsymbol{J}^{-1}_{\langle m\rangle})\Big).$$

Therefore,

$$\boldsymbol{A}_{\langle k\rangle} \vee \boldsymbol{B}_{\langle l\rangle} = \big((\boldsymbol{A}_{\langle k\rangle}\, \boldsymbol{J}^{-1}_{\langle m\rangle}) \wedge (\boldsymbol{B}_{\langle l\rangle}\, \boldsymbol{J}^{-1}_{\langle m\rangle})\big)\, \boldsymbol{J}_{\langle m\rangle}. \tag{3.102}$$

A right multiplication by $\boldsymbol{J}_{\langle m\rangle}$ instead of by $\boldsymbol{J}^{-1}_{\langle m\rangle}$ is used in order to avoid an additional sign change. Recall that since, by definition, $\mathbb{NO}(\boldsymbol{A}_{\langle k\rangle}) \subseteq \mathbb{NO}(\boldsymbol{J}_{\langle m\rangle})$ and $\mathbb{NO}(\boldsymbol{B}_{\langle l\rangle}) \subseteq \mathbb{NO}(\boldsymbol{J}_{\langle m\rangle})$, $\boldsymbol{A}_{\langle k\rangle} \boldsymbol{J}_{\langle m\rangle} = \boldsymbol{A}_{\langle k\rangle} \cdot \boldsymbol{J}_{\langle m\rangle}$ and $\boldsymbol{B}_{\langle l\rangle} \boldsymbol{J}_{\langle m\rangle} = \boldsymbol{B}_{\langle l\rangle} \cdot \boldsymbol{J}_{\langle m\rangle}$, as long as $k, l \geq 1$, which is assumed in the following. Furthermore, since the sum of the grades of $(\boldsymbol{A}_{\langle k\rangle}\, \boldsymbol{J}^{-1}_{\langle m\rangle})$ and $(\boldsymbol{B}_{\langle l\rangle}\, \boldsymbol{J}^{-1}_{\langle m\rangle})$ is less than or equal to the grade of $\boldsymbol{J}_{\langle m\rangle}$, Lemma 3.12 can be used to write (3.102) as

$$\begin{aligned} A_{\langle k\rangle} \vee B_{\langle l\rangle} &= (A_{\langle k\rangle} \cdot J_{\langle m\rangle}^{-1}) \cdot \big((B_{\langle l\rangle} \cdot J_{\langle m\rangle}^{-1}) \cdot J_{\langle m\rangle}\big) \\ &= (A_{\langle k\rangle} \cdot J_{\langle m\rangle}^{-1}) \cdot \big(B_{\langle l\rangle}(J_{\langle m\rangle}^{-1} J_{\langle m\rangle})\big) \\ &= (A_{\langle k\rangle} \cdot J_{\langle m\rangle}^{-1}) \cdot B_{\langle l\rangle}\,. \end{aligned} \tag{3.103}$$

The meet of two blades can thus be evaluated from their join via

$$A_{\langle k\rangle} \vee B_{\langle l\rangle} = (A_{\langle k\rangle}\, J_{\langle m\rangle}^{-1}) \cdot B_{\langle l\rangle}. \tag{3.104}$$

This expression for the meet becomes even simpler if the join of the blades is the pseudoscalar, since in that case $A_{\langle k\rangle} \vee B_{\langle l\rangle} = A^*_{\langle k\rangle} \cdot B_{\langle l\rangle}$.

3.2.12.2 Example

As a simple example of a join and a meet, let $e_i := \overline{\mathbb{G}}^1_{p,q}[i]$, and let $A_{\langle 2\rangle}, B_{\langle 2\rangle} \in \mathbb{G}^{\varnothing}_{p,q}$, $p+q \geq 3$, be given by $A_{\langle 2\rangle} = e_1 e_2$ and $B_{\langle 2\rangle} = e_2 e_3$. Clearly, the join of $A_{\langle k\rangle}$ and $B_{\langle l\rangle}$ is $e_1 e_2 e_3$, which is not unique, since $e_1 e_3 e_2$ is as good a choice. However, if only the subspace represented by the join (i.e. the OPNS of the join) is of interest, then an overall scalar factor is of no importance.

Applying de Morgan's law directly to evaluate the meet of $A_{\langle k\rangle}$ and $B_{\langle l\rangle}$ gives

$$\begin{aligned} A_{\langle 2\rangle} \vee B_{\langle 2\rangle} &= \big((e_1 e_2\; e_3 e_2 e_1) \wedge (e_2 e_3\; e_3 e_2 e_1)\big)(e_1 e_2 e_3) \\ &= \big(e_3 \wedge e_1\big)(e_1 e_2 e_3) \\ &= -e_2, \end{aligned} \tag{3.105}$$

where $J_{\langle 3\rangle}^{-1} = e_3 e_2 e_1$. Hence, the result is the blade that both $A_{\langle 2\rangle}$ and $B_{\langle 2\rangle}$ have in common. Using (3.104) to evaluate the meet gives

$$\begin{aligned} A_{\langle 2\rangle} \vee B_{\langle 2\rangle} &= (e_1 e_2\; e_3 e_2 e_1) \cdot (e_2 e_3) \\ &= e_3 \cdot (e_2 e_3) \\ &= e_3\,(e_2 e_3) \\ &= -e_2. \end{aligned} \tag{3.106}$$

3.2.12.3 Properties

From the definition of the inner product, it is clear that the grade of $A_{\langle k\rangle} \cdot J_{\langle m\rangle}^{-1}$ is $m-k$ and the grade of $B_{\langle l\rangle} \cdot J_{\langle m\rangle}^{-1}$ is $m-l$. With the help of (3.51), it therefore follows that

$$(\boldsymbol{A}_{\langle k\rangle}\,\boldsymbol{J}^{-1}_{\langle m\rangle}) \wedge (\boldsymbol{B}_{\langle l\rangle}\,\boldsymbol{J}^{-1}_{\langle m\rangle}) = (-1)^{(m-k)(m-l)}\,(\boldsymbol{B}_{\langle l\rangle}\,\boldsymbol{J}^{-1}_{\langle m\rangle}) \wedge (\boldsymbol{A}_{\langle k\rangle}\,\boldsymbol{J}^{-1}_{\langle m\rangle}), \tag{3.107}$$

and thus

$$\boldsymbol{A}_{\langle k\rangle} \vee \boldsymbol{B}_{\langle l\rangle} = (-1)^{(m-k)(m-l)}\,\boldsymbol{B}_{\langle l\rangle} \vee \boldsymbol{A}_{\langle k\rangle}. \tag{3.108}$$

Let $\beta \in \mathbb{R}$ and $\boldsymbol{A}_{\langle k\rangle} \in \mathbb{G}^{\varnothing}_{p,q}$, $0 < k \leq n$; then their join is $\boldsymbol{A}_{\langle k\rangle} \dot{\wedge}\, \beta = \hat{\boldsymbol{A}}_{\langle k\rangle}$, where $\hat{\boldsymbol{A}}_{\langle k\rangle} := \boldsymbol{A}_{\langle k\rangle}/|\boldsymbol{A}_{\langle k\rangle}|$. Therefore, the meet is

$$\begin{aligned} \boldsymbol{A}_{\langle k\rangle} \vee \beta &= \big((\boldsymbol{A}_{\langle k\rangle}\hat{\boldsymbol{A}}^{-1}_{\langle k\rangle}) \wedge (\beta\hat{\boldsymbol{A}}^{-1}_{\langle k\rangle})\big)\,\hat{\boldsymbol{A}}_{\langle k\rangle} \\ &= \|\boldsymbol{A}_{\langle k\rangle}\|\,\beta\,\hat{\boldsymbol{A}}^{-1}_{\langle k\rangle}\,\hat{\boldsymbol{A}}_{\langle k\rangle} \\ &= \|\boldsymbol{A}_{\langle k\rangle}\|\,\beta. \end{aligned} \tag{3.109}$$

This shows that $\mathbb{NO}(\boldsymbol{A}_{\langle k\rangle} \vee \beta) = \emptyset$. Similarly, the meet of two scalars $\alpha, \beta \in \mathbb{R}$ is $\alpha \vee \beta = \alpha\beta$.

Let $\boldsymbol{A}_{\langle k\rangle} \in \mathbb{G}^{\varnothing}_{p,q}$, and let $\boldsymbol{I}$ denote the pseudoscalar of $\mathbb{G}_{p,q}$. Then the join of $\boldsymbol{A}_{\langle k\rangle}$ and $\boldsymbol{I}$ is $\boldsymbol{I}$, i.e. $\boldsymbol{A}_{\langle k\rangle} \dot{\wedge}\, \boldsymbol{I} = \boldsymbol{I}$. Because

$$\begin{aligned} \boldsymbol{A}_{\langle k\rangle} \vee \boldsymbol{I} &= (\boldsymbol{A}_{\langle k\rangle}\boldsymbol{I}^{-1})\,\boldsymbol{I} \\ &= \boldsymbol{A}_{\langle k\rangle}, \end{aligned} \tag{3.110}$$

the pseudoscalar is the identity element of the meet operation. Furthermore, the meet is associative; that is, for $\boldsymbol{A}_{\langle r\rangle}, \boldsymbol{B}_{\langle s\rangle}, \boldsymbol{C}_{\langle t\rangle} \in \mathbb{G}^{\varnothing}_{p,q}$,

$$(\boldsymbol{A}_{\langle r\rangle} \vee \boldsymbol{B}_{\langle s\rangle}) \vee \boldsymbol{C}_{\langle t\rangle} = \boldsymbol{A}_{\langle r\rangle} \vee (\boldsymbol{B}_{\langle s\rangle} \vee \boldsymbol{C}_{\langle t\rangle}). \tag{3.111}$$

3.2.12.4 Null Blades

The meet of general blades $\boldsymbol{A}_{\langle k\rangle} \in \mathbb{G}^{k}_{p,q}$ and $\boldsymbol{B}_{\langle l\rangle} \in \mathbb{G}^{l}_{p,q}$, which may be null-blades, can also be evaluated. The only aspect that has to be changed is the evaluation of the complements of the OPNS of the blades. Since the join $\boldsymbol{J}_{\langle m\rangle} = \boldsymbol{A}_{\langle k\rangle} \dot{\wedge}\, \boldsymbol{B}_{\langle l\rangle}$ may be a null-blade in this case, (3.87) can be used to define the complements of $\mathbb{A} := \mathbb{NO}(\boldsymbol{A}_{\langle k\rangle})$ and $\mathbb{B} := \mathbb{NO}(\boldsymbol{B}_{\langle l\rangle})$ as

$$\mathbb{A}^{c} := \mathbb{NO}\Big(\big(\boldsymbol{A}_{\langle k\rangle} \cdot \boldsymbol{J}^{+}_{\langle m\rangle}\big)^{\dagger}\Big) \quad \text{and} \quad \mathbb{B}^{c} := \mathbb{NO}\Big(\big(\boldsymbol{B}_{\langle l\rangle} \cdot \boldsymbol{J}^{+}_{\langle m\rangle}\big)^{\dagger}\Big), \tag{3.112}$$

if $1 \leq k, l \leq p+q$. Just as before, $\mathbb{A}^{c} \cap \mathbb{B}^{c} = \emptyset$ and

$$\mathbb{A}^{c} \oplus \mathbb{B}^{c} = \mathbb{NO}\Big(\big((\boldsymbol{A}_{\langle k\rangle} \cdot \boldsymbol{J}^{+}_{\langle m\rangle}) \wedge (\boldsymbol{B}_{\langle l\rangle} \cdot \boldsymbol{J}^{+}_{\langle m\rangle})\big)^{\dagger}\Big).$$

Therefore,

$$\boldsymbol{A}_{\langle k\rangle} \vee \boldsymbol{B}_{\langle l\rangle} = \big((\boldsymbol{A}_{\langle k\rangle} \cdot \boldsymbol{J}^{+}_{\langle m\rangle}) \wedge (\boldsymbol{B}_{\langle l\rangle} \cdot \boldsymbol{J}^{+}_{\langle m\rangle})\big) \cdot \boldsymbol{J}_{\langle m\rangle}. \tag{3.113}$$

If $1 \leq k, l \leq p+q$, $\boldsymbol{A}^2_{\langle k \rangle} \neq 0$, and $\boldsymbol{B}^2_{\langle l \rangle} \neq 0$, then this formula reduces to (3.103). Just as before, this equation can be simplified with the help of Lemma 3.12:

$$\begin{aligned}\boldsymbol{A}_{\langle k \rangle} \vee \boldsymbol{B}_{\langle l \rangle} &= (\boldsymbol{A}_{\langle k \rangle} \cdot \boldsymbol{J}^+_{\langle m \rangle}) \cdot \Big((\boldsymbol{B}_{\langle l \rangle} \cdot \boldsymbol{J}^+_{\langle m \rangle}) \cdot \boldsymbol{J}_{\langle m \rangle} \Big) \\ &= (\boldsymbol{A}_{\langle k \rangle} \cdot \boldsymbol{J}^+_{\langle m \rangle}) \cdot \boldsymbol{B}_{\langle l \rangle} .\end{aligned}$$

The second step follows since $(\boldsymbol{B}_{\langle l \rangle} \cdot \boldsymbol{J}^+_{\langle m \rangle}) \cdot \boldsymbol{J}_{\langle m \rangle}$ is the projection of $\boldsymbol{B}_{\langle l \rangle}$ onto $\boldsymbol{J}_{\langle m \rangle}$, as given in Definition 3.23. However, since $\mathbb{NO}(\boldsymbol{B}_{\langle l \rangle}) \subseteq \mathbb{NO}(\boldsymbol{J}_{\langle m \rangle})$, $\mathcal{P}_{\boldsymbol{J}_{\langle m \rangle}}(\boldsymbol{B}_{\langle l \rangle}) = \boldsymbol{B}_{\langle l \rangle}$. Therefore, the general expression for the meet of blades $\boldsymbol{A}_{\langle k \rangle}, \boldsymbol{B}_{\langle l \rangle} \in \mathbb{G}_{p,q}$, with $1 \leq k, l \leq p+q$, is

$$\boldsymbol{A}_{\langle k \rangle} \vee \boldsymbol{B}_{\langle l \rangle} = (\boldsymbol{A}_{\langle k \rangle} \cdot \boldsymbol{J}^+_{\langle m \rangle}) \cdot \boldsymbol{B}_{\langle l \rangle} . \tag{3.114}$$

This reduces to (3.104) if neither $\boldsymbol{A}_{\langle k \rangle}$ nor $\boldsymbol{B}_{\langle l \rangle}$ is a null blade. If $1 \leq k \leq p+q$ and $l = 0$, i.e. $\boldsymbol{B}_{\langle 0 \rangle} \in \mathbb{R}$ is a scalar, then $\boldsymbol{J}_{\langle m \rangle} = \boldsymbol{A}_{\langle k \rangle} / \|\boldsymbol{A}_{\langle k \rangle}\|$ and $\boldsymbol{A}_{\langle k \rangle} \cdot \boldsymbol{J}^+_{\langle m \rangle} = \|\boldsymbol{A}_{\langle k \rangle}\|$, and thus

$$\boldsymbol{A}_{\langle k \rangle} \vee \boldsymbol{B}_{\langle 0 \rangle} = \big((\boldsymbol{A}_{\langle k \rangle} \cdot \boldsymbol{J}^+_{\langle m \rangle}) \wedge (\boldsymbol{B}_{\langle 0 \rangle}\, \boldsymbol{J}^+_{\langle m \rangle}) \big) \cdot \boldsymbol{J}_{\langle m \rangle} = \|\boldsymbol{A}_{\langle k \rangle}\|\, \boldsymbol{B}_{\langle 0 \rangle} . \tag{3.115}$$

3.2.12.5 Summary

To summarize, the meet has the following properties. Let $\boldsymbol{A}_{\langle r \rangle}, \boldsymbol{B}_{\langle s \rangle}, \boldsymbol{C}_{\langle t \rangle} \in \mathbb{G}_{p,q}$, with $1 \leq r, s, t \leq p+q$, $\alpha \in \mathbb{R}$, and $\boldsymbol{J}_{\langle m \rangle} := \boldsymbol{A}_{\langle r \rangle} \dot{\wedge} \boldsymbol{B}_{\langle s \rangle}$, and let $\boldsymbol{I}$ be the pseudoscalar of $\mathbb{G}_{p,q}$; then the meet can in general be evaluated as

$$\boldsymbol{A}_{\langle k \rangle} \vee \boldsymbol{B}_{\langle l \rangle} = (\boldsymbol{A}_{\langle k \rangle} \cdot \boldsymbol{J}^+_{\langle m \rangle}) \cdot \boldsymbol{B}_{\langle l \rangle} ,$$

and its properties are the following:

1. $\boldsymbol{A}_{\langle r \rangle} \vee \boldsymbol{I} = \boldsymbol{I} \vee \boldsymbol{A}_{\langle r \rangle} = \boldsymbol{A}_{\langle r \rangle}$.
2. $\boldsymbol{A}_{\langle r \rangle} \vee 1 = \|\boldsymbol{A}_{\langle k \rangle}\|$.
3. $(\alpha \boldsymbol{A}_{\langle r \rangle}) \vee \boldsymbol{B}_{\langle s \rangle} = \boldsymbol{A}_{\langle r \rangle} \vee (\alpha \boldsymbol{B}_{\langle s \rangle}) = \alpha\, (\boldsymbol{A}_{\langle r \rangle} \vee \boldsymbol{B}_{\langle s \rangle})$.
4. $(\boldsymbol{A}_{\langle r \rangle} \vee \boldsymbol{B}_{\langle s \rangle}) \vee \boldsymbol{C}_{\langle t \rangle} = \boldsymbol{A}_{\langle r \rangle} \vee (\boldsymbol{B}_{\langle s \rangle} \vee \boldsymbol{C}_{\langle t \rangle})$.
5. $(\boldsymbol{A}_{\langle r \rangle} + \boldsymbol{B}_{\langle s \rangle}) \vee \boldsymbol{C}_{\langle t \rangle} = \boldsymbol{A}_{\langle r \rangle} \vee \boldsymbol{C}_{\langle t \rangle} + \boldsymbol{B}_{\langle s \rangle} \vee \boldsymbol{C}_{\langle t \rangle}$.
6. $\boldsymbol{A}_{\langle r \rangle} \vee \boldsymbol{B}_{\langle s \rangle} = (-1)^{(m-r)(m-s)}\, \boldsymbol{B}_{\langle s \rangle} \vee \boldsymbol{A}_{\langle r \rangle}$. (3.116)

3.2.13 Regressive Product

The regressive product is closely related to the meet. It is used to evaluate intersections in Grassmann–Cayley algebra (see Sect. 3.8.5), but is not as general as the meet. Basically, the regressive product is the meet where the join is replaced by the pseudoscalar. Owing to this replacement, the regressive product is defined for general multivectors and not just for blades.

Definition 3.28 (Regressive Product). The regressive product between two multivectors $\boldsymbol{A}, \boldsymbol{B} \in \mathbb{G}_{p,q}$ is denoted by $\triangledown$ and defined as

$$\boldsymbol{A} \triangledown \boldsymbol{B} := \left(\boldsymbol{A}^* \wedge \boldsymbol{B}^*\right) \boldsymbol{I}.$$

The regressive product can be regarded as the dual operation to the outer product, since for $\boldsymbol{A}, \boldsymbol{B} \in \mathbb{G}_{p,q}$,

$$\left(\boldsymbol{A}^* \triangledown \boldsymbol{B}^*\right)^* = \boldsymbol{A} \wedge \boldsymbol{B}. \tag{3.117}$$

Therefore, for blades $\boldsymbol{A}_{\langle k \rangle}, \boldsymbol{B}_{\langle l \rangle} \in \mathbb{G}_{p,q}$, the relation $\mathbb{NO}(\boldsymbol{A}_{\langle k \rangle} \wedge \boldsymbol{B}_{\langle l \rangle}) = \mathbb{NI}(\boldsymbol{A}^*_{\langle k \rangle} \triangledown \boldsymbol{B}^*_{\langle l \rangle})$ holds.

From the properties of the inner, outer, and geometric products, the following properties of the regressive product follow. Let $\boldsymbol{A}, \boldsymbol{B}, \boldsymbol{C}, \boldsymbol{X}_{\langle k \rangle}, \boldsymbol{Y}_{\langle l \rangle} \in \mathbb{G}_{p,q}$, let $\alpha \in \mathbb{R}$, and let $\boldsymbol{I} \in \mathbb{G}_{p,q}$ be the pseudoscalar of $\mathbb{G}_{p,q}$. We then have the following:

1. $\boldsymbol{A} \triangledown \boldsymbol{I} = \boldsymbol{I} \triangledown \boldsymbol{A} = \boldsymbol{A}\,.$
2. $\boldsymbol{A} \triangledown \alpha = \alpha \triangledown \boldsymbol{A} = 0\,.$
3. $(\alpha \boldsymbol{A}) \triangledown \boldsymbol{B} = \boldsymbol{A} \triangledown (\alpha \boldsymbol{B}) = \alpha\,(\boldsymbol{A} \triangledown \boldsymbol{B})\,.$
4. $(\boldsymbol{A} \triangledown \boldsymbol{B}) \triangledown \boldsymbol{C} = \boldsymbol{A} \triangledown (\boldsymbol{B} \triangledown \boldsymbol{C})\,.$
5. $(\boldsymbol{A} + \boldsymbol{B}) \triangledown \boldsymbol{C} = \boldsymbol{A} \triangledown \boldsymbol{C} + \boldsymbol{B} \triangledown \boldsymbol{C}\,.$
6. $\boldsymbol{X}_{\langle k \rangle} \triangledown \boldsymbol{Y}_{\langle l \rangle} = (-1)^{(n-k)(n-l)}\, \boldsymbol{Y}_{\langle l \rangle} \triangledown \boldsymbol{X}_{\langle k \rangle}\,.$ (3.118)

Owing to the associativity of the regressive product, $\boldsymbol{A} \triangledown (\boldsymbol{B} \triangledown \boldsymbol{C})$ can also be written as $\boldsymbol{A} \triangledown \boldsymbol{B} \triangledown \boldsymbol{C}$. This is dual to a blade, since

$$(\boldsymbol{A}^* \triangledown \boldsymbol{B}^* \triangledown \boldsymbol{C}^*)^* = \boldsymbol{A} \wedge \boldsymbol{B} \wedge \boldsymbol{C}.$$

3.3 Versors

Versors are a generalization of blades. They play an important role in the applications of geometric algebra because they allow the representation of transformations. In this section, the fundamental algebraic properties of versors are discussed. The term "versor" was coined by David Hestenes.

3.3.1 Definitions

Definition 3.29 (Versor). A *versor* is a multivector that can be expressed as the geometric product of a number of *non-null* 1-vectors. That is, a versor $\boldsymbol{V}$ can be written as $\boldsymbol{V} = \prod_{i=1}^{k} \boldsymbol{n}_i$, where $\{\boldsymbol{n}_1, \ldots, \boldsymbol{n}_k\} \subset \mathbb{G}_{p,q}^{\varnothing 1}$ with $k \in \mathbb{N}^+$, is a set of not necessarily linearly independent vectors.

Lemma 3.18. *For each versor $\boldsymbol{V} \in \mathbb{G}_{p,q}$, there exists an inverse $\boldsymbol{V}^{-1} \in \mathbb{G}_{p,q}$ such that $\boldsymbol{V}\,\boldsymbol{V}^{-1} = \boldsymbol{V}^{-1}\,\boldsymbol{V} = 1$.*

Proof. Let the versor $\boldsymbol{V}$ be defined as $\boldsymbol{V} = \prod_{i=1}^{k} \boldsymbol{n}_i$, where $\{\boldsymbol{n}_1, \ldots, \boldsymbol{n}_k\} \subset \mathbb{G}_{p,q}^{\varnothing 1}$, $k \in \mathbb{N}^+$. From Lemma 3.16, it follows that every non-null 1-vector $\boldsymbol{n}_i$ has an inverse. Hence, it can easily be seen by construction that $\boldsymbol{V}^{-1} = \prod_{i=k}^{1} \boldsymbol{n}_i^{-1}$.
□

Lemma 3.19 (Clifford Group). *The subset of versors of $\mathbb{G}_{p,q}$ together with the geometric product, forms a group, the* Clifford group, *denoted by $\mathfrak{G}_{p,q}$.*

Proof. From the properties of the geometric product, it follows immediately that associativity is satisfied and that a right and left identity element, the algebra identity element, exists. Furthermore, Lemma 3.18 shows that there exists an inverse for each element of $\mathfrak{G}_{p,q}$, which can be applied from both left and right.
□

Note that in the literature, the Clifford group is often denoted by Γ (see e.g. [148]). To keep the notation in this text consistent, a different font is used. It is important to realize that a Clifford group $\mathfrak{G}_{p,q}$ is in general not a *subalgebra* of $\mathbb{G}_{p,q}$. That is, the sum of two versors does not in general result in a versor.

Definition 3.30 (Unitary Versor). A versor $\boldsymbol{V} \in \mathfrak{G}_{p,q}$ is called *unitary* if $\boldsymbol{V}^{-1} = \widetilde{\boldsymbol{V}}$, i.e. $\boldsymbol{V}\,\widetilde{\boldsymbol{V}} = +1$.

Lemma 3.20 (Pin Group). *The set of unitary versors of $\mathfrak{G}_{p,q}$ forms a subgroup $\mathfrak{P}_{p,q}$ of the Clifford group $\mathfrak{G}_{p,q}$, called the* pin group.

Proof. Let $\boldsymbol{V}_1, \boldsymbol{V}_2 \in \mathfrak{G}_{p,q}$ denote two unitary versors of the Clifford group, that is $\boldsymbol{V}_1^{-1} = \widetilde{\boldsymbol{V}}_1$ and $\boldsymbol{V}_2^{-1} = \widetilde{\boldsymbol{V}}_2$. Then $\boldsymbol{U} := \boldsymbol{V}_1\,\boldsymbol{V}_2$ is a unitary versor, since $\widetilde{\boldsymbol{U}} = \widetilde{\boldsymbol{V}}_2\,\widetilde{\boldsymbol{V}}_1$ and

$$\boldsymbol{U}\,\widetilde{\boldsymbol{U}} = \boldsymbol{V}_1\,\boldsymbol{V}_2\,\widetilde{\boldsymbol{V}}_2\,\widetilde{\boldsymbol{V}}_1 = \boldsymbol{V}_1\,\boldsymbol{V}_2\,\boldsymbol{V}_2^{-1}\,\boldsymbol{V}_1^{-1} = 1\,.$$

Hence, $\widetilde{\boldsymbol{U}} = \boldsymbol{U}^{-1}$ is a unitary versor, which proves the assertion.
□

Definition 3.31 (Spinor). A versor $\boldsymbol{V} \in \mathfrak{G}_{p,q}$ is called a *spinor* if it is unitary ($\boldsymbol{V}\,\widetilde{\boldsymbol{V}} = 1$) and can be expressed as the geometric product of an even number of 1-vectors. This implies that a spinor is a linear combination of blades of even grade.

Lemma 3.21 (Spin Group). *The set of spinors of $\mathfrak{G}_{p,q}$ forms a subgroup of the pin group $\mathfrak{P}_{p,q}$, called the* spin group, *which is denoted by $\mathfrak{S}_{p,q}$.*

Proof. Let $\boldsymbol{V}_1, \boldsymbol{V}_2 \in \mathfrak{G}_{p,q}$ denote two spinors. Note that $\boldsymbol{V}_1$ and $\boldsymbol{V}_2$ are linear combinations of blades of even grade. It therefore follows from (3.39) that the geometric product $\boldsymbol{U} := \boldsymbol{V}_1\,\boldsymbol{V}_2$ has to result in a linear combination of blades of even grade. Since spinors are unitary versors and thus elements of the pin group $\mathfrak{P}_{p,q}$, $\boldsymbol{U}$ also has to be a unitary versor. Hence, $\boldsymbol{U}$ is a spinor, which proves the assertion.
□

To summarize, the set of versors of $\mathbb{G}_{p,q}$, together with the geometric product, forms a group, called the *Clifford group*, denoted by $\mathfrak{G}_{p,q}$. The set of unitary versors forms a subgroup of the Clifford group, called the *pin group*, $\mathfrak{P}_{p,q}$. Furthermore, the set of unitary versors that are generated by the geometric product of an even number of vectors forms a subgroup of the pin group, called the *spin group*, $\mathfrak{S}_{p,q}$. For a more detailed discussion, see for example [148].

Note also that a versor is a linear combination either of blades of even grade or of blades of odd grade. While the former generate rotations, the latter are combinations of rotations with a reflection, sometimes called *antirotations*.

Definition 3.32 (Null Versor). A *null-versor* is the geometric product of $k \in \mathbb{N}^+$ not necessarily linearly independent 1-vectors $\{\boldsymbol{n}_i\} \in \mathbb{G}^1_{p,q}$, i.e. $\boldsymbol{V} = \prod_{i=1}^k \boldsymbol{n}_i$, such that $\boldsymbol{V}\,\widetilde{\boldsymbol{V}} = 0$. Hence, at least one of the $\{\boldsymbol{n}_i\}$ is a null-vector.

It was shown in (3.49) in Sect. 3.2.2 that any blade may be represented by a geometric product of *perpendicular* vectors. Conversely, it can be said that a versor generated by the geometric product of a set of perpendicular vectors is a blade. Blades therefore form a subset of the set of versors. However, they do not form a group.

3.3.2 Properties

Consider the effect of the following operation between vectors $\boldsymbol{a}, \boldsymbol{n} \in \mathbb{G}^{\varnothing 1}_{p,q}$:

$$\boldsymbol{n}\,\boldsymbol{a}\,\boldsymbol{n}^{-1} = (\boldsymbol{n}\cdot\boldsymbol{a} + \boldsymbol{n}\wedge\boldsymbol{a})\,\boldsymbol{n}^{-1} = (\boldsymbol{n}\cdot\boldsymbol{a})\,\boldsymbol{n}^{-1} + (\boldsymbol{n}\wedge\boldsymbol{a})\cdot\boldsymbol{n}^{-1} + (\boldsymbol{n}\wedge\boldsymbol{a})\wedge\boldsymbol{n}^{-1}. \tag{3.119}$$

Since $\boldsymbol{n}^{-1} := \boldsymbol{n}/\|\boldsymbol{n}\|^2$, it is linearly dependent on $\boldsymbol{n}$ and thus $\boldsymbol{n} \wedge \boldsymbol{n}^{-1} = 0$. Also, just as in (3.73),

$$(\boldsymbol{n} \wedge \boldsymbol{a}) \cdot \boldsymbol{n}^{-1} = (\boldsymbol{a} \cdot \boldsymbol{n}^{-1})\, \boldsymbol{n} - (\boldsymbol{n} \cdot \boldsymbol{n}^{-1})\, \boldsymbol{a}.$$

Hence,

$$\boldsymbol{n}\, \boldsymbol{a}\, \boldsymbol{n}^{-1} = 2\,(\boldsymbol{a} \cdot \hat{\boldsymbol{n}})\, \hat{\boldsymbol{n}} - \boldsymbol{a} = 2\, \mathcal{P}_{\hat{\boldsymbol{n}}}(\boldsymbol{a}) - \boldsymbol{a}, \tag{3.120}$$

where $\hat{\boldsymbol{n}} := \boldsymbol{n}/\|\boldsymbol{n}\|$. Writing $\boldsymbol{a} = \mathcal{P}_{\hat{\boldsymbol{n}}}(\boldsymbol{a}) + \mathcal{P}^{\perp}_{\hat{\boldsymbol{n}}}(\boldsymbol{a})$, this results in

$$\boldsymbol{n}\, \boldsymbol{a}\, \boldsymbol{n}^{-1} = \mathcal{P}_{\hat{\boldsymbol{n}}}(\boldsymbol{a}) - \mathcal{P}^{\perp}_{\hat{\boldsymbol{n}}}(\boldsymbol{a})\,. \tag{3.121}$$

That is, $\boldsymbol{n}\, \boldsymbol{a}\, \boldsymbol{n}^{-1}$ results in a 1-vector, which has the same component as $\boldsymbol{a}$ parallel to $\boldsymbol{n}$ but whose component perpendicular to $\boldsymbol{n}$ is negated compared with $\boldsymbol{a}$. In other words, this operation results in the *reflection* of $\boldsymbol{a}$ in $\boldsymbol{n}$. Note that since a general versor $\boldsymbol{V}$ is the geometric product of a number of 1-vectors, the operation $\boldsymbol{V}\, \boldsymbol{a}\, \boldsymbol{V}^{-1}$ results in consecutive reflections of $\boldsymbol{a}$ in the constituent vectors of $\boldsymbol{V}$.

Lemma 3.22 (Grade Preservation). *Let $\boldsymbol{V} \in \mathbb{G}_{p,q}$ denote a versor and let $\boldsymbol{a} \in \mathbb{G}^1_{p,q}$; then $\boldsymbol{V}\, \boldsymbol{a}\, \boldsymbol{V}^{-1} \in \mathbb{G}^1_{p,q}$.*

Proof. Let the versor $\boldsymbol{V}$ be defined as $\boldsymbol{V} = \prod_{i=1}^{k} \boldsymbol{n}_i$, where $\{\, \boldsymbol{n}_1, \ldots, \boldsymbol{n}_k \,\} \subset \mathbb{G}^{\varnothing\, 1}_{p,q}$, $k \in \mathbb{N}^+$. Then,

$$\boldsymbol{V}\, \boldsymbol{a}\, \boldsymbol{V}^{-1} = \boldsymbol{n}_1 \cdots \boldsymbol{n}_{k-1}\, (\boldsymbol{n}_k\, \boldsymbol{a}\, \boldsymbol{n}_k^{-1})\, \boldsymbol{n}_{k-1}^{-1} \cdots \boldsymbol{n}_1^{-1}.$$

From (3.121), it follows that $\boldsymbol{n}_k\, \boldsymbol{a}\, \boldsymbol{n}_k^{-1} \in \mathbb{G}^1_{p,q}$. Through recursive application of this property and using the associativity of the geometric product, the proposition is proven.

□

Lemma 3.23 (Versor Outermorphism). *Let $\boldsymbol{V} \in \mathbb{G}_{p,q}$ denote a versor and let $\{\, \boldsymbol{a}_1, \ldots, \boldsymbol{a}_k \,\} \subset \mathbb{G}^1_{p,q}$ be a set of linearly independent vectors; then*

$$\boldsymbol{V} \left(\bigwedge_{i=1}^{k} \boldsymbol{a}_i \right) \boldsymbol{V}^{-1} = \bigwedge_{i=1}^{k} \left(\boldsymbol{V}\, \boldsymbol{a}_i\, \boldsymbol{V}^{-1} \right).$$

Proof. This may be shown by induction. Let $\boldsymbol{A}_{\langle k-j \rangle} := \bigwedge_{i=1}^{k-j} \boldsymbol{a}_i$, and note that

$$\boldsymbol{A}_{\langle k-1 \rangle} \wedge \boldsymbol{a}_k = \frac{1}{2}\,(\boldsymbol{A}_{\langle k-1 \rangle}\, \boldsymbol{a}_k + (-1)^{k-1}\, \boldsymbol{a}_k\, \boldsymbol{A}_{\langle k-1 \rangle}).$$

Hence,

$$
\begin{aligned}
&\boldsymbol{V}\,(\boldsymbol{A}_{\langle k-1\rangle} \wedge \boldsymbol{a}_k)\,\boldsymbol{V}^{-1} \\
&= \frac{1}{2}\,(\boldsymbol{V}\,\boldsymbol{A}_{\langle k-1\rangle}\,\boldsymbol{a}_k\,\boldsymbol{V}^{-1} + (-1)^{k-1}\,\boldsymbol{V}\,\boldsymbol{a}_k\,\boldsymbol{A}_{\langle k-1\rangle}\,\boldsymbol{V}^{-1}) \\
&= \frac{1}{2}\,(\boldsymbol{V}\,\boldsymbol{A}_{\langle k-1\rangle}\,\boldsymbol{V}^{-1}\,\boldsymbol{V}\,\boldsymbol{a}_k\,\boldsymbol{V}^{-1} \\
&\qquad\qquad + (-1)^{k-1}\,\boldsymbol{V}\,\boldsymbol{a}_k\,\boldsymbol{V}^{-1}\,\boldsymbol{V}\,\boldsymbol{A}_{\langle k-1\rangle}\,\boldsymbol{V}^{-1}) \\
&= (\boldsymbol{V}\,\boldsymbol{A}_{\langle k-1\rangle}\,\boldsymbol{V}^{-1}) \wedge (\boldsymbol{V}\,\boldsymbol{a}_k\,\boldsymbol{V}^{-1}).
\end{aligned}
$$

Through recursive application of this equation, the proposition is proven.
□

3.4 Linear Functions

While versors form a particularly important class of transformation operations in geometric algebra, they form a subgroup of general linear transformations. The goal of this section is to introduce linear functions on multivectors and to present some of their properties. It was noted in Chap. 1 that one of the advantageous properties of geometric algebra is the representation of particular transformation subgroups, such as the spin group. Nevertheless, general linear transformations can still be incorporated, albeit without an operator representation as for versors. That is, in Euclidean space there exist no grade-preserving transformation operators that represent general linear transformations.

Definition 3.33 (Linear Function). Let $\boldsymbol{A}, \boldsymbol{B} \in \mathbb{G}_n$ be two multivectors and let $a \in \mathbb{R}$. A linear function $\mathcal{F}$ on $\mathbb{G}_n$ is a map $\mathbb{G}_n \to \mathbb{G}_n$ which satisfies

$$
\begin{aligned}
\mathcal{F}(A + B) &= \mathcal{F}(A) + \mathcal{F}(B), \\
\mathcal{F}(a\,A) &= a\,\mathcal{F}(A).
\end{aligned}
$$

A type of linear function that is particularly interesting is *grade-preserving* linear functions, since they retain the algebraic and geometric identity of a multivector.

Definition 3.34 (Grade-Preserving Linear Function). A grade-preserving linear function $\mathcal{F}$ satisfies the following for all blades $\boldsymbol{A}_{\langle k\rangle} \in \mathbb{G}_n$ of arbitrary grade k:

$$
\mathrm{gr}(\mathcal{F}(\boldsymbol{A}_{\langle k\rangle})) = \mathrm{gr}(\boldsymbol{A}_{\langle k\rangle}) = k \qquad \text{if} \quad \mathcal{F}(\boldsymbol{A}_{\langle k\rangle}) \neq 0\,.
$$

Let $\mathcal{F}$ be a grade-preserving linear function defined on vectors. That is, $\mathcal{F}$ is a map $\mathbb{R}^n \to \mathbb{R}^n$. There then exist two vectors $\boldsymbol{a}, \boldsymbol{b} \in \mathbb{R}^n$ such that $\mathcal{F}(\boldsymbol{a}) = \boldsymbol{b}$. Any linear transformation of this type can be described by a matrix. Let $\{\boldsymbol{e}_i\} \equiv \overline{\mathbb{R}}^n$ denote the canonical basis of $\mathbb{R}^n$, such that vectors $\boldsymbol{a}$ and $\boldsymbol{b}$ can be defined as $\boldsymbol{a} := a^i\,\boldsymbol{e}_i$ and $\boldsymbol{b} := b^i\,\boldsymbol{e}_i$. There then exists a matrix $T^j{}_i$ such that

$$\mathcal{F}(\boldsymbol{a}) := a^i\, T^j{}_i\, \boldsymbol{e}_j = b^j\, \boldsymbol{e}_j ,$$

with $b^j = a^i\, T^j{}_i$ and an implicit summation over i. From this definition of $\mathcal{F}$ and the distributivity of linear functions, it follows that

$$\mathcal{F}(\boldsymbol{e}_i) = T^r{}_i\, \boldsymbol{e}_r \,.$$

Therefore, the bivector $\mathcal{F}(\boldsymbol{e}_i) \wedge \mathcal{F}(\boldsymbol{e}_j)$ becomes

$$\mathcal{F}(e_i) \wedge \mathcal{F}(e_j) = T^r{}_i\, T^s{}_j\, \boldsymbol{e}_r \wedge \boldsymbol{e}_s = U^{rs}{}_{ij}\, \boldsymbol{e}_r \wedge \boldsymbol{e}_s ,$$

where $U^{rs}{}_{ij} := T^r{}_i\, T^s{}_j$. The tensor $U^{rs}{}_{ij}$ now represents a grade-preserving linear function on bivectors. Since $U^{rs}{}_{ij}$ is directly related to the linear function $\mathcal{F}$, it is useful to extend the definition of $\mathcal{F}$ to blades of arbitrary grade via an outermorphism.

Definition 3.35 (Outermorphism of Linear Functions). Any grade-preserving linear function on grade 1 vectors in $\mathbb{G}_n$ is extended to blades of arbitrary grade via the outermorphism property. That is, if $\mathcal{F}$ is a grade-preserving linear function that maps $\mathbb{G}^1_n \to \mathbb{G}^1_n$, then for a set $\{\, \boldsymbol{a}_i \,\} \subset \mathbb{G}^1_n$ of k grade 1 vectors, the definition of $\mathcal{F}$ is extended as follows:

$$\mathcal{F}\Big(\bigwedge_{i=1}^{k} \boldsymbol{a}_k \Big) := \bigwedge_{i=1}^{k} \mathcal{F}(\boldsymbol{a}_k) \,.$$

3.4.1 Determinant

The concept of the determinant of a matrix is directly related to the evaluation of a linear function acting on the pseudoscalar. This can easily be demonstrated in $\mathbb{G}_2$. Let $\mathcal{F}$ be a grade-preserving linear function on $\mathbb{G}^1_2$, that is represented by the 2×2 matrix $T^j{}_i$. That is, given the canonical basis $\{\, \boldsymbol{e}_1, \boldsymbol{e}_2 \,\} \equiv \overline{\mathbb{R}}^2$, $\mathcal{F}(\boldsymbol{e}_i) = T^j{}_i\, \boldsymbol{e}_j$. The pseudoscalar $\boldsymbol{I} \in \mathbb{G}_2$ is given by $\boldsymbol{I} = \boldsymbol{e}_1 \wedge \boldsymbol{e}_2$. Therefore, using the extension of $\mathcal{F}$ via the outermorphism,

$$\mathcal{F}(I) = \mathcal{F}(\boldsymbol{e}_1 \wedge \boldsymbol{e}_2) = \mathcal{F}(\boldsymbol{e}_1) \wedge \mathcal{F}(\boldsymbol{e}_2) = T^j{}_1\, T^k{}_2\, \boldsymbol{e}_j \wedge \boldsymbol{e}_k ,$$

with an implicit summation over $j, k \in \{\, 1, 2 \,\}$. Expanding the implicit sums over j and k gives

$$\mathcal{F}(I) = \big(T^1{}_1\, T^1{}_2 - T^2{}_1\, T^1{}_2 \big)\, \boldsymbol{e}_1 \wedge \boldsymbol{e}_2 = \det(T^i{}_j)\, \boldsymbol{I} =: \det(\mathcal{F})\, \boldsymbol{I} ,$$

where $\det(\mathcal{F})$ is defined to denote the determinant of the associated transformation matrix. In fact, any grade-preserving linear function $\mathcal{F}$ on $\mathbb{G}^1_n$ satisfies $\mathcal{F}(\boldsymbol{I}) = \det(\mathcal{F})\, \boldsymbol{I}$, where $\boldsymbol{I}$ denotes the pseudoscalar of $\mathbb{G}_n$. This can be seen as follows.

Lemma 3.24. *Let $\mathcal{F}$ denote a grade-preserving linear function $\mathbb{G}^1_n \to \mathbb{G}^1_n$ that is defined as $\mathcal{F} : \boldsymbol{e}_i \mapsto T^j{}_i\, \boldsymbol{e}_j$, where $T^j{}_i \in \mathbb{R}^{n\times n}$. If $\boldsymbol{I} \in \mathbb{G}^n_n$ is the unit pseudoscalar, then*

$$\mathcal{F}(\boldsymbol{I}) = \det(T^j{}_i)\, \boldsymbol{I} .$$

Proof. By applying a singular-value decomposition (SVD), the matrix $\mathsf{T} \equiv T^j{}_i \in \mathbb{R}^{n\times n}$ can be written as

$$\mathsf{T} = \mathsf{U}\,\mathsf{D}\,\mathsf{V} ,$$

where $\mathsf{U}, \mathsf{V} \in \mathbb{R}^{n\times n}$ are unitary matrices and $\mathsf{D} \in \mathbb{R}^{n\times n}$ is a diagonal matrix (see e.g. [167, 168]). Define the linear functions $\mathcal{F}_U$, $\mathcal{F}_D$, and $\mathcal{F}_V$ as

$$\mathcal{F}_D : \boldsymbol{e}_i \mapsto D^j{}_i\, \boldsymbol{e}_j \;, \quad \mathcal{F}_U : \boldsymbol{e}_i \mapsto U^j{}_i\, \boldsymbol{e}_j \;, \quad \mathcal{F}_V : \boldsymbol{e}_i \mapsto V^j{}_i\, \boldsymbol{e}_j \;,$$

where $D^j{}_i \equiv \mathsf{D}$, $U^j{}_i \equiv \mathsf{U}$, and $V^j{}_i \equiv \mathsf{V}$. Define $\boldsymbol{u}_i := U^j{}_i\, \boldsymbol{e}_j$ as the column vectors of $U^j{}_i$. Since $U^j{}_i$ is unitary, the vectors $\{\boldsymbol{u}_i\}$ are orthogonal and have unit length, i.e. $\|\boldsymbol{u}_i\| = 1$. It follows that

$$\mathcal{F}_U(\boldsymbol{I}) = \bigwedge_{i=1}^{n} \mathcal{F}_U(\boldsymbol{e}_i) = \bigwedge_{i=1}^{n} \boldsymbol{u}_i = \prod_{i=1}^{n} \boldsymbol{u}_i \;,$$

where the last step follows since the $\{\boldsymbol{u}_i\}$ are orthogonal. The geometric product of n orthogonal grade 1 vectors in $\mathbb{G}_n$ has to be proportional to the pseudoscalar of $\mathbb{G}_n$, i.e.

$$\prod_{i=1}^{n} \boldsymbol{u}_i = \Big\| \prod_{i=1}^{n} \boldsymbol{u}_i \Big\|\, \boldsymbol{I} = \prod_{i=1}^{n} \|\boldsymbol{u}_i\|\, \boldsymbol{I} = \boldsymbol{I} \;,$$

since $\|\boldsymbol{u}_i\| = 1$ for all $i \in \{1, \ldots, n\}$ (cf. Sect. 3.2.6). Hence,

$$\mathcal{F}_U(\boldsymbol{I}) = \mathcal{F}_V(\boldsymbol{I}) = \boldsymbol{I} .$$

If the diagonal entries of $D^j{}_i$ are denoted by $d_i := D^i{}_i$, then $\mathcal{F}_D(\boldsymbol{e}_i) = d_i\, \boldsymbol{e}_i$, with no implicit summation over i. Therefore,

$$\mathcal{F}_D(\boldsymbol{I}) = \left(\prod_{i=1}^{n} d_i \right) \boldsymbol{I} .$$

It follows that

$$\mathcal{F}(\boldsymbol{I}) = \mathcal{F}_U\Big(\mathcal{F}_D\big(\mathcal{F}_V(\boldsymbol{I}) \big) \Big) = \left(\prod_{i=1}^{n} d_i \right) \boldsymbol{I} = \mathcal{F}_D(\boldsymbol{I}) .$$

From the properties of determinants, it also follows that

$$\det(\mathsf{T}) = \det(\mathsf{U}\,\mathsf{D}\,\mathsf{V}) = \det(\mathsf{U})\, \det(\mathsf{D})\, \det(\mathsf{V}) = \det(\mathsf{D}) \;,$$

since $\det(\mathsf{U}) = \det(\mathsf{V}) = 1$. Furthermore, the determinant of a diagonal matrix is the product of the diagonal entries, i.e. $\det(\mathsf{D}) = \prod_{i=1}^{n} d_i$. Thus,

$$\mathcal{F}(\boldsymbol{I}) = \det(\mathsf{D})\, \boldsymbol{I} = \det(\mathsf{T})\, \boldsymbol{I} .$$

Note that if T is not of full rank, D will have some zero entries on its diagonal, and thus $\det(\mathsf{T}) = 0$.

□

3.4.2 Determinant Product

Let $\mathcal{F}$ and $\mathcal{G}$ be grade-preserving linear functions on $\mathbb{G}[1]n$, with associated general $n \times n$ matrices $U^j{}_i$ and $V^j{}_i$, respectively.

Applying both functions consecutively to some vector $\boldsymbol{a} \in \mathbb{G}[1]n$, with $\boldsymbol{a} = a^i \boldsymbol{e}_i$, gives

$$(\mathcal{F} \circ \mathcal{G})(\boldsymbol{a}) = \mathcal{F}\Big(\mathcal{G}(\boldsymbol{a})\Big) = a^i \left(U^k{}_j\, V^j{}_i\right) \boldsymbol{e}_k = a^i\, W^k{}_i\, \boldsymbol{e}_k,$$

where $W^k{}_i := U^k{}_j\, V^j{}_i$ is the matrix product of $U^k{}_j$ and $V^j{}_i$. If we denote the linear function represented by $W^k{}_i$ as $\mathcal{H}$, then $\mathcal{F} \circ \mathcal{G} = \mathcal{H}$, i.e. $\mathcal{F}(\mathcal{G}(\boldsymbol{a})) = \mathcal{H}(\boldsymbol{a})$. Hence,

$$\mathcal{H}(\boldsymbol{I}) = \det(\mathcal{H})\, \boldsymbol{I}\,.$$

Expressing $\mathcal{H}$ as $\mathcal{F} \circ \mathcal{G}$ gives

$$\mathcal{H}(\boldsymbol{I}) = \mathcal{F}(\mathcal{G}(\boldsymbol{I})) = \mathcal{F}(\det(\mathcal{G})\, \boldsymbol{I}) = \det(\mathcal{G})\, \mathcal{F}(\boldsymbol{I}) = \det(\mathcal{G})\, \det(\mathcal{F})\, \boldsymbol{I}\,.$$

This demonstrates the determinant identity

$$\det(\mathsf{U}\,\mathsf{V}) = \det(\mathsf{U})\, \det(\mathsf{V}).$$

3.4.3 Inverse

Let $\mathcal{F}$ be an invertible, grade-preserving linear function on $\mathbb{G}^1_n$ and let $\mathcal{F}^{-1}$ denote its inverse. Then $\mathcal{F}^{-1} \circ \mathcal{F} = \mathcal{F} \circ \mathcal{F}^{-1} = 1$, where 1 denotes the identity function. If $\boldsymbol{I}$ denotes the pseudoscalar of $\mathbb{G}_n$, then

$$\boldsymbol{I} = \mathcal{F}^{-1}(\mathcal{F}(\boldsymbol{I})) = \mathcal{F}^{-1}(\det(\mathcal{F})\, \boldsymbol{I}) = \det(\mathcal{F})\, \det(\mathcal{F}^{-1})\, \boldsymbol{I}\,,$$

and thus

$$\det(\mathcal{F}^{-1}) = \frac{1}{\det(\mathcal{F})}.$$

3.4.4 Summary

In this section, the relations between grade-preserving linear functions in geometric algebra were introduced and their relation to determinants was shown. Linear functions are the only way to express general linear transformations of multivectors in geometric algebra. An additional feature of working with grade-preserving linear functions in geometric algebra, as compared with matrices, is that their definition can be extended to blades of arbitrary grade.

This means, for example, that geometric entities such as points, lines, and planes can be transformed with the same linear function, where otherwise a multilinear function would have to be defined in each case. That is, geometric algebra allows the representation of multilinear functions as linear functions.

3.5 Reciprocal Bases

Reciprocal bases play an important role in many aspects of geometric algebra. To introduce the idea behind reciprocal bases, consider first of all a geometric algebra with a Euclidean signature.

The orthonormal basis $\{\, \boldsymbol{e}_i \,\} := \overline{\mathbb{G}}^1_n$ has the nice property that $\boldsymbol{e}_i \cdot \boldsymbol{e}_j = \delta_{ij}$, where δ_{ij} is the Kronecker delta. That is, $\boldsymbol{e}_i \cdot \boldsymbol{e}_i = 1$ and $\boldsymbol{e}_i \cdot \boldsymbol{e}_j = 0$ if $i \neq j$. If a vector $\boldsymbol{x} \in \mathbb{G}^n$ is given in this basis as $\boldsymbol{x} = \mathsf{x}^i\, \boldsymbol{e}_i$, with $\{\, \mathsf{x}^i \,\} \subset \mathbb{R}$, then a particular component x^r can be extracted from $\boldsymbol{x}$ via $\boldsymbol{x} \cdot \boldsymbol{e}_r = \mathsf{x}^r$.

Clearly, $\boldsymbol{x}$ can be expressed in any basis of $\mathbb{G}^1_n$. Let $\{\, \boldsymbol{a}_i \,\} \subset \mathbb{G}^1_n$ be a basis of $\mathbb{G}^1_n$ and let $\{\, \mathsf{x}^i_a \,\} \subset \mathbb{R}$ be given such that $\boldsymbol{x} = \mathsf{x}^i_a \boldsymbol{a}_i$. The relation between $\{\, \mathsf{x}^i \,\}$ and $\{\, \mathsf{x}^i_a \,\}$ can be found by expressing the $\{\, \boldsymbol{a}_i \,\}$ in terms of the $\{\, \boldsymbol{e}_i \,\}$. Suppose that $\{\, \mathsf{A}^j{}_i \,\} \subset \mathbb{R}$ is such that $\boldsymbol{a}_i = \mathsf{A}^j{}_i\, \boldsymbol{e}_j$. It is then clear that

$$\boldsymbol{x} = \mathsf{x}^i_a\, \boldsymbol{a}_i = \mathsf{x}^i_a\, \mathsf{A}^j{}_i\, \boldsymbol{e}_j = \mathsf{x}^j\, \boldsymbol{e}_j. \tag{3.122}$$

The $\{\, \mathsf{A}^j{}_i \,\}$ can be regarded as the components of a matrix A, the $\{\, \mathsf{x}^i_a \,\}$ as a column vector x_a, and the $\{\, \mathsf{x}^i \,\}$ as a column vector x. Equation (3.122) can then be written as a matrix equation

$$\mathsf{x} = \mathsf{A}\, \mathsf{x}_a.$$

Since $\{\, \boldsymbol{a}_i \,\}$ and $\{\, \boldsymbol{e}_i \,\}$ are both bases of $\mathbb{G}^1_n$, the matrix A must be invertible, and thus $\mathsf{x}_a = \mathsf{A}^{-1}\mathsf{x}$. The components $\{\, \mathsf{x}^i_a \,\}$ could be found directly, given a basis $\{\, \boldsymbol{a}^i \,\}$ of $\mathbb{G}^1_n$ such that $\boldsymbol{a}_i \cdot \boldsymbol{a}^j = \delta_{ij}$. The basis $\{\, \boldsymbol{a}^i \,\}$ is called the reciprocal basis of $\{\, \boldsymbol{a}_i \,\}$. With its help,

$$\boldsymbol{x} \cdot \boldsymbol{a}^j = \mathsf{x}^i_a\, (\boldsymbol{a}_i \cdot \boldsymbol{a}^j) = \mathsf{x}^i_a\, \delta^j{}_i = \mathsf{x}^j_a.$$

On the other hand, $\boldsymbol{x} \cdot \boldsymbol{a}^j$ may be expanded as

$$\boldsymbol{x} \cdot \boldsymbol{a}^j = \mathsf{x}^i\, (\boldsymbol{e}_i \cdot \boldsymbol{a}^j) = \mathsf{x}^j_a.$$

Therefore, $\mathsf{A}^{-1} = \boldsymbol{e}_i \cdot \boldsymbol{a}^j$. The question that remains is how the reciprocal basis $\{\, \boldsymbol{a}^i \,\}$ of some basis $\{\, \boldsymbol{a}_i \,\}$ can be evaluated.

3.5.1 Definition

Definition 3.36 (Reciprocal Basis). Let $\{\, \boldsymbol{a}_i \,\} \subset \mathbb{G}^1_{p,q}$ be a basis of $\mathbb{G}^1_{p,q}$. By a *reciprocal basis* a set of vectors denoted by $\{\, \boldsymbol{a}^i \,\} \subset \mathbb{G}^1_{p,q}$ is meant, which also forms a basis of $\mathbb{G}^1_{p,q}$ and satisfies the equation $\boldsymbol{a}_i \cdot \boldsymbol{a}^j = \delta^j{}_i$, where $\delta^j{}_i$ is the Kronecker delta. Note that an orthonormal basis is its own reciprocal basis.

Lemma 3.25. *Let* $\{\, \boldsymbol{a}_i \,\} \subset \mathbb{G}^1_{p,q}$ *be a basis of* $\mathbb{G}^1_{p,q}$. *The reciprocal basis* $\{\, \boldsymbol{a}^i \,\} \subset \mathbb{G}^1_{p,q}$ *of* $\{\, \boldsymbol{a}_i \,\}$ *is then given by*

$$\boldsymbol{a}^j := (-1)^{j-1} \left[\Big(\bigwedge_{r=1}^{j-1} \boldsymbol{a}_r \Big) \wedge \Big(\bigwedge_{r=j+1}^{n} \boldsymbol{a}_r \Big) \right] \boldsymbol{I}_a^{-1},$$

where $\boldsymbol{I}_a := \bigwedge_{r=1}^{n} \boldsymbol{a}_i$. $\boldsymbol{I}_a$ *is therefore a pseudoscalar of* $\mathbb{G}_{p,q}$.

Proof. First of all, note that even though some or all of the $\{\, \boldsymbol{a}_i \,\}$ may be null-vectors, the pseudoscalar of $\mathbb{G}_{p,q}$ cannot be a null-blade. Initially, we show that $\boldsymbol{a}_i \cdot \boldsymbol{a}^i = 1$. It follows from the above definitions and Lemma 3.12 that

$$\begin{aligned}
\boldsymbol{a}_i \cdot \boldsymbol{a}^i &= (-1)^{i-1}\, \boldsymbol{a}_i \cdot \left[\Big(\bigwedge_{r=1}^{i-1} \boldsymbol{a}_r \Big) \wedge \Big(\bigwedge_{r=i+1}^{n} \boldsymbol{a}_r \Big) \right] \boldsymbol{I}_a^{-1} \\
&= (-1)^{i-1} \left[\boldsymbol{a}_i \wedge \Big(\bigwedge_{r=1}^{i-1} \boldsymbol{a}_r \Big) \wedge \Big(\bigwedge_{r=i+1}^{n} \boldsymbol{a}_r \Big) \right] \boldsymbol{I}_a^{-1} \\
&= (-1)^{i-1} (-1)^{i-1} \Big(\bigwedge_{r=1}^{n} \boldsymbol{a}_r \Big) \boldsymbol{I}_a^{-1} \\
&= \boldsymbol{I}_a\, \boldsymbol{I}_a^{-1} \\
&= 1.
\end{aligned} \tag{3.123}$$

Now we show that $\boldsymbol{a}_i \cdot \boldsymbol{a}^j = 0$ if $i \neq j$. Similarly to the argument above,

$$\begin{aligned}
\boldsymbol{a}_i \cdot \boldsymbol{a}^j &= (-1)^{j-1}\, \boldsymbol{a}_i \cdot \left[\Big(\bigwedge_{r=1}^{j-1} \boldsymbol{a}_r \Big) \wedge \Big(\bigwedge_{r=j+1}^{n} \boldsymbol{a}_r \Big) \right] \boldsymbol{I}_a^{-1} \\
&= (-1)^{j-1} \left[\boldsymbol{a}_i \wedge \Big(\bigwedge_{r=1}^{j-1} \boldsymbol{a}_r \Big) \wedge \Big(\bigwedge_{r=j+1}^{n} \boldsymbol{a}_r \Big) \right] \boldsymbol{I}_a^{-1} \\
&= 0,
\end{aligned} \tag{3.124}$$

since $\boldsymbol{a}_i$ also occurs in the outer product $\Big(\bigwedge_{r=1}^{j-1} \boldsymbol{a}_r \Big) \wedge \Big(\bigwedge_{r=j+1}^{n} \boldsymbol{a}_r \Big)$ if $i \neq j$.
□

Note that, given a basis of arbitrary k-blades, the corresponding reciprocal basis of k-blades can be found in much the same way as in the above lemma.

3.5.2 Example

Consider the algebra $\mathbb{G}_{1,1}$, let $\boldsymbol{e}_i := \overline{\mathbb{G}}^1_{1,1}[i]$, and define $\boldsymbol{n}_1 := \boldsymbol{e}_1 + \boldsymbol{e}_2$ and $\boldsymbol{n}_2 := \boldsymbol{e}_1 - \boldsymbol{e}_2$. Clearly, $(\boldsymbol{n}_1)^2 = (\boldsymbol{n}_2)^2 = 0$ and $\boldsymbol{n}_1 \cdot \boldsymbol{n}_2 = 2$, since $\boldsymbol{e}_1^2 = +1$

and $e_2^2 = -1$. Furthermore, $\boldsymbol{I}_n := \boldsymbol{n}_1 \wedge \boldsymbol{n}_2$, such that $\boldsymbol{I}_n^{-1} = \frac{1}{4}\boldsymbol{I}$, because

$$\boldsymbol{I}_n\,\boldsymbol{I}_n^{-1} = \frac{1}{4}(\boldsymbol{n}_1 \wedge \boldsymbol{n}_2) \cdot (\boldsymbol{n}_1 \wedge \boldsymbol{n}_2) = \frac{1}{4}\big(- (\boldsymbol{n}_1)^2\,(\boldsymbol{n}_2)^2 + (\boldsymbol{n}_1 \cdot \boldsymbol{n}_2)^2\big) = 1.$$

It also follows that

$$\boldsymbol{n}_1\,\boldsymbol{I}_n^{-1} = \frac{1}{4}\,\boldsymbol{n}_1 \cdot (\boldsymbol{n}_1 \wedge \boldsymbol{n}_2) = \frac{1}{4}\big((\boldsymbol{n}_1 \cdot \boldsymbol{n}_1)\,\boldsymbol{n}_2 - (\boldsymbol{n}_1 \cdot \boldsymbol{n}_2)\,\boldsymbol{n}_1\big) = -\frac{1}{2}\,\boldsymbol{n}_1$$

and, similarly, $\boldsymbol{n}_2\,\boldsymbol{I}_n^{-1} = \frac{1}{2}\,\boldsymbol{n}_2$. Lemma 3.25 defines the corresponding reciprocal basis as

$$\boldsymbol{n}^1 := \boldsymbol{n}_2\,\boldsymbol{I}_n^{-1} = \frac{1}{2}\,\boldsymbol{n}_2 \qquad \text{and} \qquad \boldsymbol{n}^2 := -\boldsymbol{n}_1\,\boldsymbol{I}_n^{-1} = \frac{1}{2}\,\boldsymbol{n}_1.$$

Thus

$$\boldsymbol{n}_1 \cdot \boldsymbol{n}^1 = \frac{1}{2}\,\boldsymbol{n}_1 \cdot \boldsymbol{n}_2 = 1 \qquad \text{and} \qquad \boldsymbol{n}_1 \cdot \boldsymbol{n}^2 = \frac{1}{2}\,\boldsymbol{n}_1 \cdot \boldsymbol{n}_1 = 0,$$

and similarly for $\boldsymbol{n}_2$.

3.6 Differentiation

Differentiation is clearly an important aspect of geometric algebra. This is particularly true from the point of view of applications, where the optimization of functions often necessitates using their derivative. In Chap. 5, the differentiation of multivector-valued functions is introduced by regarding multivectors as column vectors on the algebraic basis. The differentiation of a multivector then leads to a Jacobi matrix in general. While this is the most general form of a derivative on multivectors, the resultant Jacobi matrices cannot be represented directly in geometric-algebra terms. This is an advantageous approach when one is dealing with numerical optimization and error propagation. However, for analytical calculations, an approach where the differentiation operators are elements of geometric algebra is usually preferable.

3.6.1 Vector Derivative

The standard vector derivative in $\mathbb{R}^3$ is typically denoted by ∇ (see [126]) and defined as

$$\nabla := \boldsymbol{e}_1\,\frac{\partial}{\partial\,x} + \boldsymbol{e}_2\,\frac{\partial}{\partial\,y} + \boldsymbol{e}_3\,\frac{\partial}{\partial\,z}\,. \tag{3.125}$$

To emphasize the function of ∇ as a partial derivative in arbitrary dimensions, the partial differentiation operator with respect to $\boldsymbol{x} \in \mathbb{R}^{p,q}$, where $\boldsymbol{x} := x^i\, \boldsymbol{e}_i$ and $\{\boldsymbol{e}_i\} \equiv \overline{\mathbb{R}}^{p,q}$ is denoted by $\partial_{\boldsymbol{x}}$ and defined as

$$\partial_{\boldsymbol{x}} := \boldsymbol{e}^i\, \partial_{x^i} \,, \tag{3.126}$$

where $\boldsymbol{e}^i$ is the reciprocal vector of $\boldsymbol{e}_i$, ∂_{x^i} denotes partial differentiation with respect to x^i, and there is an implicit summation over i.

3.6.1.1 Gradient

Let $\mathcal{F} : \mathbb{G}^1_{p,q} \to \mathbb{R}$ denote a scalar-valued, not necessarily linear function; then

$$\partial_{\boldsymbol{x}}\, \mathcal{F}(\boldsymbol{x}) = \boldsymbol{e}^i\, \partial_{x^i}\, \mathcal{F}(\boldsymbol{x}) = \boldsymbol{e}^i\, \mathcal{F}_i(\boldsymbol{x}) \,, \tag{3.127}$$

where $\mathcal{F}_i(\boldsymbol{x}) := \partial_{x^i}\, \mathcal{F}(\boldsymbol{x})$ is the partial derivative of $\mathcal{F}$ with respect to x^i evaluated at $\boldsymbol{x}$. Equation (3.127) is called the *gradient* of $\mathcal{F}(\boldsymbol{x})$. The derivative of $\mathcal{F}$ in the direction of some unit vector $\hat{\boldsymbol{n}} \in \mathbb{G}^1_{p,q}$, with $\hat{\boldsymbol{n}} := n^i\, \boldsymbol{e}_i$, is then given by

$$(\hat{\boldsymbol{n}} \cdot \partial_{\boldsymbol{x}})\, \mathcal{F}(\boldsymbol{x}) = \hat{\boldsymbol{n}} \cdot\, (\partial_{\boldsymbol{x}}\, \mathcal{F}(\boldsymbol{x})) = \hat{\boldsymbol{n}} \cdot (\boldsymbol{e}^i\, \mathcal{F}_i(\boldsymbol{x})) = n^i\, \mathcal{F}_i(\boldsymbol{x}) \,. \tag{3.128}$$

The expression $\hat{\boldsymbol{n}} \cdot \partial_{\boldsymbol{x}}$ therefore denotes the directed differentiation operator.

3.6.1.2 Divergence and Curl

Let $\mathcal{F} : \mathbb{G}^1_{p,q} \to \mathbb{G}^1_{p,q}$ denote a vector-valued function, with

$$\mathcal{F} : \boldsymbol{x} \mapsto \mathcal{F}^i(\boldsymbol{x})\, \boldsymbol{e}_i \,,$$

where the $\mathcal{F}^i : \mathbb{G}^1_{p,q} \to \mathbb{R}$, for all i, are scalar-valued functions. The *divergence* of $\mathcal{F}$ is defined as

$$\partial_{\boldsymbol{x}} \cdot \mathcal{F}(\boldsymbol{x}) = \partial_{x^i}\, \mathcal{F}^i(\boldsymbol{x}) = \sum_i \mathcal{F}^i{}_i(\boldsymbol{x}) \,, \tag{3.129}$$

where $\mathcal{F}^i{}_i(\boldsymbol{x})$ denotes the derivative of $\mathcal{F}^i$ with respect to x^i evaluated at $\boldsymbol{x}$. Note that the set of components $\mathcal{F}^i{}_j$ forms the Jacobi matrix of $\mathcal{F}$ (cf. Sect. 5.1). The divergence of $\mathcal{F}$ is thus the trace of its Jacobi matrix.

The *curl* of $\mathcal{F}$ is defined only in 3D space. That is, if $\mathcal{F} : \mathbb{G}^1_3 \to \mathbb{G}^1_3$, the curl is given in terms of the vector cross product as

$$\partial_{\boldsymbol{x}} \times \mathcal{F}(\boldsymbol{x}) = \left(\partial_{\boldsymbol{x}} \wedge \mathcal{F}(\boldsymbol{x})\right)^* = \partial_{\boldsymbol{x}} \cdot \mathcal{F}^*(\boldsymbol{x}) = \partial^*_{\boldsymbol{x}} \cdot \mathcal{F}(\boldsymbol{x}) \,. \tag{3.130}$$

That is, the curl can be regarded as the divergence of the dual of $\mathcal{F}$ in geometric algebra. In geometric algebra, the definition of the curl can also be extended to any dimension, by considering the result of the curl to be a bivector. That is, instead of defining the curl via the vector cross product, it can be defined in terms of the outer product for arbitrary-dimensional functions $\mathcal{F} : \mathbb{G}^1_{p,q} \to \mathbb{G}^1_{p,q}$ as

$$\partial_{\boldsymbol{x}} \wedge \mathcal{F}(\boldsymbol{x}) \,. \tag{3.131}$$

In the special case of three dimensions, the standard curl can be recovered as the dual of the generalized curl.

3.6.2 Multivector Differentiation

A detailed introduction to multivector calculus is given in [91] and a short overview can be found, for example, in [110]. In the following, the fundamental ideas of multivector differentiation are presented. Note that this presentation differs somewhat from the presentation in [91] in that here direct reference is made to the canonical algebraic basis to simplify the definition of the differentiation operator. In [91], the initial definition is given without reference to any particular algebraic basis.

Let $\{\, \boldsymbol{E}_i \,\} := \overline{\mathbb{G}}_{p,q}$ denote the canonical algebraic basis of $\mathbb{G}_{p,q}$, as defined in Sect. 3.1.3. A general multivector $\boldsymbol{X} \in \mathbb{G}_{p,q}$ can then be defined as

$$\boldsymbol{X} := x^i\, \boldsymbol{E}_i \,, \tag{3.132}$$

with $\{x^i\} \subset \mathbb{R}$. The multivector differentiation operator with respect to $\boldsymbol{X}$ is defined in this context as

$$\partial_{\boldsymbol{X}} := \boldsymbol{E}^i\, \partial_{x^i} \,, \qquad \boldsymbol{E}^i := \boldsymbol{E}_i^\dagger \,. \tag{3.133}$$

Note that $\boldsymbol{E}^i := \boldsymbol{E}_i^\dagger$ denotes a reciprocal algebraic basis element *only* with respect to the scalar product. That is, $\boldsymbol{E}^i * \boldsymbol{E}_j = \delta^i{}_j$, as shown in Sect. 3.2.6.

3.6.2.1 Gradient

Let $\mathcal{F} : \mathbb{G}_{p,q} \to \mathbb{R}$ denote a scalar-valued function; then

$$\partial_{\boldsymbol{X}}\, \mathcal{F}(\boldsymbol{X}) = \boldsymbol{E}^i\, \partial_{x^i}\, \mathcal{F}(\boldsymbol{X}) = \boldsymbol{E}^i\, \mathcal{F}_i(\boldsymbol{X}) \,, \tag{3.134}$$

where $\mathcal{F}_i(\boldsymbol{X}) := \partial_{x^i}\, \mathcal{F}(\boldsymbol{X})$ is the partial derivative of $\mathcal{F}$ with respect to x^i, evaluated at $\boldsymbol{X}$. In analogy to the case of vector-valued functions, this is the gradient of $\mathcal{F}$. The derivative of $\mathcal{F}$ in the “direction” of a multivector

$\boldsymbol{A} \in \mathbb{G}_{p,q}$, with $\boldsymbol{A} := a^i\, \boldsymbol{E}_i$, is then given by

$$(\boldsymbol{A} * \partial_{\boldsymbol{X}})\, \mathcal{F}(\boldsymbol{X}) = \boldsymbol{A} * (\partial_{\boldsymbol{X}}\, \mathcal{F}(\boldsymbol{X})) = \boldsymbol{A} * (\boldsymbol{E}^i\, \mathcal{F}_i(\boldsymbol{X})) = a^i\, \mathcal{F}_i(\boldsymbol{X})\,. \quad (3.135)$$

The expression $\boldsymbol{A} * \partial_{\boldsymbol{X}}$ therefore represents the directed multivector derivative operator.

As an example of the gradient, consider the function

$$\mathcal{F} : \boldsymbol{X} \mapsto \boldsymbol{X} * \boldsymbol{A}\,,$$

where $\boldsymbol{A} \in \mathbb{G}_{p,q}$. Then

$$\partial_{\boldsymbol{X}}\, \mathcal{F}(\boldsymbol{X}) = \boldsymbol{E}^i\, \partial_{x^i}\, (x^j\, a^j\, \boldsymbol{E}_j * \boldsymbol{E}_j) = \sum_i \delta(x^i)\, a^i\, (\boldsymbol{E}_i * \boldsymbol{E}_i)\, \boldsymbol{E}^i\,,$$

where

$$\delta(x^i) := \begin{cases} 1, & x^i \neq 0\,, \\ 0, & x^i = 0\,. \end{cases}$$

Since $\boldsymbol{E}^i * \boldsymbol{E}_i = +1$, $\boldsymbol{E}^i$ can be written as

$$\boldsymbol{E}^i = \frac{\boldsymbol{E}_i}{\boldsymbol{E}_i * \boldsymbol{E}_i} \quad \Longleftrightarrow \quad \boldsymbol{E}_i = (\boldsymbol{E}_i * \boldsymbol{E}_i)\, \boldsymbol{E}^i\,.$$

Hence,

$$\partial_{\boldsymbol{X}}\, (\boldsymbol{X} * \boldsymbol{A}) = \delta(x^i)\, a^i\, \boldsymbol{E}_i\,.$$

That is, only those components i are considered that are also non-zero in $\boldsymbol{X} = x^i\, \boldsymbol{E}_i$. This can be regarded as the projection of $\boldsymbol{A}$ onto those components that are non-zero in $\boldsymbol{X}$.

3.6.2.2 General Derivative

For multivector-valued functions of the type $\mathcal{F} : \mathbb{G}_{p,q} \to \mathbb{G}_{p,q}$, care has to be taken when applying the multivector derivative. Consider first of all the trivial function

$$\mathcal{F} : \boldsymbol{X} \mapsto \boldsymbol{X}\,.$$

Then

$$\partial_{\boldsymbol{X}}\, \mathcal{F}(\boldsymbol{X}) = \boldsymbol{E}^i\, \partial_{x^i}\, x^j\, \boldsymbol{E}_j = \boldsymbol{E}^i\, \boldsymbol{E}_i = 2^n\,,$$

where $n = p + q$ is the dimension of the algebra. It follows right away from the associativity of the geometric product that

$$\partial_{\boldsymbol{X}}\, (\boldsymbol{X}\, \boldsymbol{A}) = 2^n\, \boldsymbol{A}\,,$$

where $\boldsymbol{A} \in \mathbb{G}_{p,q}$. However,

$$\partial_{\boldsymbol{X}} \, (\boldsymbol{A}\,\boldsymbol{X}) = \boldsymbol{E}^i \, \partial_{x^i} \, \boldsymbol{A} \, x^j \, \boldsymbol{E}_j = \boldsymbol{E}^i \, \boldsymbol{A} \, \boldsymbol{E}_i = \frac{\boldsymbol{E}_i \, \boldsymbol{A} \, \boldsymbol{E}_i}{\boldsymbol{E}_i \, \boldsymbol{E}_i} \,,$$

which has the form of a reflection of $\boldsymbol{A}$. Typically, it suffices to consider the scalar component of such a derivative, whence

$$\partial_{\boldsymbol{X}} * (\boldsymbol{A}\,\boldsymbol{X}) = \langle \partial_{\boldsymbol{X}} \, (\boldsymbol{A}\,\boldsymbol{X}) \rangle_0 = \left\langle \boldsymbol{E}^i \, \boldsymbol{A} \, \boldsymbol{E}_i \right\rangle_0 = \left\langle \boldsymbol{A} \, \boldsymbol{E}_i \, \boldsymbol{E}^i \right\rangle_0 = 2^n \, \langle \boldsymbol{A} \rangle_0 \,.$$

The third equality follows from Lemma 3.9 (p. 68).

3.6.3 Tensor Representation

This subsection gives a short introduction to the idea that geometric-algebra operations are essentially bilinear functions. This notion is discussed in detail later on in Sect. 5.1. If we make this identification of geometric-algebra operations with multilinear functions, differentiation and integration can be expressed in a straightforward manner.

As before, let $\{\, \boldsymbol{E}_i \,\} := \overline{\mathbb{G}}_{p,q}$ denote the canonical algebraic basis of $\mathbb{G}_{p,q}$ (see Sect. 3.1.3). A general multivector $\boldsymbol{X} \in \mathbb{G}_{p,q}$ can then be defined as

$$\boldsymbol{X} := x^i \, \boldsymbol{E}_i, \tag{3.136}$$

with $\{x^i\} \subset \mathbb{R}$. Since the $\{\boldsymbol{E}_i\}$ form an algebraic basis, it follows that the geometric product of two basis blades has to result in a basis blade, i.e.

$$\boldsymbol{E}_i \, \boldsymbol{E}_j = \Gamma^k{}_{ij} \, \boldsymbol{E}_k, \qquad \forall \, i, j \in \{1, \ldots, 2^2\}, \tag{3.137}$$

where $\Gamma^k{}_{ij} \in \mathbb{R}^{2^n \times 2^n \times 2^n}$ is a tensor encoding the geometric product.

If $\boldsymbol{A}, \boldsymbol{B}, \boldsymbol{C} \in \mathbb{G}_{p,q}$ are defined as $\boldsymbol{A} := a^i \, \boldsymbol{E}_i$, $\boldsymbol{B} := b^i \, \boldsymbol{E}_i$, and $\boldsymbol{C} := c^i \, \boldsymbol{E}_i$, then it follows from (3.137) that the components of $\boldsymbol{C}$ in the algebra equation $\boldsymbol{C} = \boldsymbol{A}\,\boldsymbol{B}$ can be evaluated via

$$c^k = a^i \, b^j \, \Gamma^k{}_{ij}, \tag{3.138}$$

where a summation over i and j is again implied. Such a summation of tensor indices is called *contraction*. Equation (3.138) shows that the geometric product is simply a bilinear function. In fact, all products in geometric algebra that are of interest in this text can be expressed in this form, as will be discussed in some detail in Sect. 5.1.

In the context of (3.138), the components c^k of $\boldsymbol{C}$ are functions of the components a^i and b^j. The differentiation of c^k with respect to a^i then results in the Jacobi matrix of c^k, i.e.

$$\frac{\partial \, c^k}{\partial \, a^i} = b^j \, \Gamma^k{}_{ij} \,, \tag{3.139}$$

which is a matrix with indices k and i. This result cannot be represented in geometric-algebra terms. Similarly, the Hesse tensor of c^k is

$$\frac{\partial^2 c^k}{\partial a^i \, \partial b^j} = \Gamma^k{}_{ij} \,. \tag{3.140}$$

Integration of c^k is equally simple:

$$\int c^k \, \mathrm{d}a^p = (1 - \frac{1}{2}\,\delta^{pi})\; a^p \, a^i \, b^j \, \Gamma^k{}_{ij} \,, \tag{3.141}$$

without summation over p.

While the above notions of differentiation and integration are very useful for solving geometric-algebra equations numerically, they are less well suited for analytic calculations. The main difference between the multivector differentiation previously discussed and differentiation in the the tensor representation is that in the former case the differentiation operator is a multivector itself, i.e. $\partial_{\boldsymbol{A}} = \partial_{a^i} \, \boldsymbol{E}_i^\dagger$. Therefore,

$$\partial_{\boldsymbol{A}} (\boldsymbol{A}\,\boldsymbol{B}) \quad \Longleftrightarrow \quad \partial_{a^p} \, C^{pq} \, \Gamma^r{}_{qk} \;\; a^i \, b^j \, \Gamma^k{}_{ij} = b^j \;\, C^{iq} \, \Gamma^r{}_{qk} \, \Gamma^k{}_{ij} \,,$$

where the tensor C^{pq} implements the conjugate operation. Similarly, defining $\mathrm{d}\boldsymbol{A} := \boldsymbol{E}_i \, \mathrm{d}a^i$ for integration gives

$$\int \boldsymbol{A} \, \mathrm{d}\boldsymbol{A} \quad \Longleftrightarrow \quad \int a^i \, \mathrm{d}a^j \, \Gamma^k{}_{ij} = (1 - \frac{1}{2}\,\delta^{ij})\; a^i \, a^j \, \Gamma^k{}_{ij} \,.$$

This result can be expressed in geometric-algebra terms as

$$(1 - \frac{1}{2}\,\delta^{ij})\; a^i \, a^j \, \Gamma^k{}_{ij} \quad \Longleftrightarrow \quad \boldsymbol{A}^2 - \frac{1}{2}\,(\boldsymbol{A} * \boldsymbol{A}) \,.$$

Hence,

$$\int \boldsymbol{A} \, \mathrm{d}\boldsymbol{A} = \boldsymbol{A}^2 - \frac{1}{2}\,(\boldsymbol{A} * \boldsymbol{A}) \,.$$

This shows that the tensor formalism can help to derive some results which are hard to obtain otherwise.

3.7 Algorithms

In this section, geometric-algebra algorithms are discussed, which use the basic algebraic operations introduced so far to perform more elaborate calculations. In particular, the factorization of blades plays an important role in many calculations. It is, for example, essential in the calculation of the join of two blades.

3.7.1 Basis Orthogonalization

The method presented here is basically the Gram–Schmidt method for the orthogonalization of a set of vectors in a vector space. Let $\{\, \boldsymbol{a}_i \,\}_{i=1}^{k} \subset \mathbb{G}^1_{p,q}$ be a set of k 1-vectors, not necessarily linearly independent. This set of vectors spans the subspace $\text{span}\{\, \boldsymbol{a}_i \,\} \subseteq \mathbb{G}^1_{p,q}$. The goal is to find an orthonormal set of vectors $\{\, \boldsymbol{n}_i \,\} \subseteq \mathbb{G}^1_{p,q}$ which spans the same subspace, i.e. $\text{span}\{\, \boldsymbol{a}_i \,\} = \text{span}\{\, \boldsymbol{n}_i \,\}$. Note that the set $\{\, \boldsymbol{a}_i \,\}$ may also contain null-vectors and hence $\{\, \boldsymbol{n}_i \,\}$ may contain null-vectors. The set $\{\, \boldsymbol{n}_i \,\}$ is therefore assumed to be orthonormal with respect to the Euclidean scalar product (see Definition 3.16).

The first vector $\boldsymbol{n}_1$ of the orthonormal set may be chosen to be

$$\boldsymbol{n}_1 := \frac{\boldsymbol{a}_1}{\|\boldsymbol{a}_1\|}. \tag{3.142}$$

The next vector has to lie in $\text{span}\{\, \boldsymbol{a}_i \,\}$, but also has to be perpendicular to $\boldsymbol{n}_1$. This can be achieved by evaluating the rejection of $\boldsymbol{a}_2$ on $\boldsymbol{n}_1$, for example:

$$\boldsymbol{a}'_2 := \mathcal{P}^{\perp}_{\boldsymbol{n}_1}(\boldsymbol{a}_2)\,. \tag{3.143}$$

Recall that the rejection is also defined for null-vectors. If $\boldsymbol{a}'_2 \neq 0$, i.e. if $\boldsymbol{a}_2$ is not parallel to $\boldsymbol{n}_1$, then $\boldsymbol{n}_2 = \boldsymbol{a}'_2/\|\boldsymbol{a}'_2\|$. The next vector of the orthonormal set now has to be perpendicular to $\boldsymbol{n}_1$ and $\boldsymbol{n}_2$. Therefore, it has to be perpendicular to $\boldsymbol{n}_1 \wedge \boldsymbol{n}_2$. Thus the rejection of $\boldsymbol{a}_3$ on $\boldsymbol{n}_1 \wedge \boldsymbol{n}_2$ is evaluated:

$$\boldsymbol{a}'_3 := \mathcal{P}^{\perp}_{\boldsymbol{n}_1 \wedge \boldsymbol{n}_2}(\boldsymbol{a}_3)\,. \tag{3.144}$$

If $\boldsymbol{a}'_3 \neq 0$, then $\boldsymbol{n}_2 = \boldsymbol{a}'_3/|\boldsymbol{a}'_3|$. The complete algorithm for the orthonormalization of a set of vectors is thus given by Algorithm 3.1.

3.7.2 Factorization of Blades

The factorization of a blade into constituent vectors is similar to the orthonormalization of a set of vectors. In both cases the goal is to find the basis of a subspace. Just as for the orthonormalization of a set of vectors, the factorization of a blade into vectors is not unique. For example, let $\boldsymbol{a}, \boldsymbol{b} \in \mathbb{G}^1_{p,q}$ and $\alpha \in \mathbb{R}$; then $(\alpha\boldsymbol{a}) \wedge \boldsymbol{b}$ could be factorized into $(\alpha\boldsymbol{a})$ and $\boldsymbol{b}$, but also into $\boldsymbol{a}$ and $\alpha\boldsymbol{b}$. Since the factorization of a blade is not unique, an arbitrary orthonormal basis of $\mathbb{G}^1_{p,q}$, with respect to which the factorization is performed, can be chosen.

The goal of the algorithm is to find, for a given blade $\boldsymbol{A}_{\langle k \rangle} \in \mathbb{G}_{p,q}$, a set of k 1-vectors $\{\, \boldsymbol{n}_i \,\} \subset \mathbb{G}^1_{p,q}$ such that $\mathbb{NO}(\boldsymbol{A}_{\langle k \rangle}) = \mathbb{NO}(\bigwedge_{i=1}^{k} \boldsymbol{n}_i)$. Since $\boldsymbol{A}_{\langle k \rangle}$ may be a null-blade, the set $\{\, \boldsymbol{n}_i \,\}$ may also contain null-vectors. The

Algorithm 3.1 OrthoNorm. Let $\{\boldsymbol{a}_i\}_{i=1}^{k} \subset \mathbb{G}^1_{p,q}$ be a set of k vectors not necessarily linearly independent. The algorithm OrthoNorm($\{\boldsymbol{a}_i\}$) returns a set of orthonormal vectors $\{\boldsymbol{n}_i\}$ such that $\text{span}\{\boldsymbol{n}_i\} = \text{span}\{\boldsymbol{a}_i\}$. When the algorithm is finished, the set $\{\boldsymbol{n}_i\}$ is an orthonormal basis of $\text{span}\{\boldsymbol{a}_i\}$, and $\boldsymbol{N}$ is the unit blade representing this subspace.

$\boldsymbol{n}_1 \leftarrow \boldsymbol{a}_1/\|\boldsymbol{a}_1\|$
$\boldsymbol{N} \leftarrow \boldsymbol{n}_1$
$j \leftarrow 1$
for all $i \in \{2, 3, \ldots, k\}$ **do**
  $\boldsymbol{a}'_i \leftarrow \mathcal{P}^{\perp}_{\boldsymbol{N}}(\boldsymbol{a}_i)$
  if $\|\boldsymbol{a}'_i\| \neq 0$ **then**
    $j \leftarrow j + 1$
    $\boldsymbol{n}_j \leftarrow \boldsymbol{a}'_i/\|\boldsymbol{a}'_i\|$
    $\boldsymbol{N} \leftarrow \boldsymbol{N} \wedge \boldsymbol{n}_j$
  end if
end for
return $\{\boldsymbol{n}_i\}$

method presented here for blade factorization projects the 1-vectors of an orthonormal basis of $\mathbb{G}^1_{p,q}$ onto the blade to find a set of constituent vectors.

We select an orthonormal basis $\boldsymbol{e}_i := \overline{\mathbb{G}}^1_{p,q}$. The first vector $\boldsymbol{n}_1$ of the factorization of $\boldsymbol{A}_{\langle k \rangle}$ is then given by

$$\boldsymbol{n}_1 = \mathcal{P}_{\boldsymbol{A}_{\langle k \rangle}}\left(\underset{\boldsymbol{e}_i}{\arg\max} \, \|\mathcal{P}_{\boldsymbol{A}_{\langle k \rangle}}(\boldsymbol{e}_i)\| \right). \tag{3.145}$$

The next vector of the factorization has to be one which lies in $\boldsymbol{A}_{\langle k \rangle}$ but is perpendicular to $\boldsymbol{n}_1$. This is obtained by first removing the linear dependence on $\boldsymbol{n}_1$ from $\boldsymbol{A}_{\langle k \rangle}$. Following Lemma 3.15, this is achieved using

$$\boldsymbol{A}_{\langle k-1 \rangle} := \hat{\boldsymbol{n}}_1^{\dagger} \cdot \boldsymbol{A}_{\langle k \rangle}, \tag{3.146}$$

where $\hat{\boldsymbol{n}}_1 = \boldsymbol{n}_1/\|\boldsymbol{n}_1\|$. Recall (3.89) in this context, where it is shown that $(\boldsymbol{A}_{\langle k \rangle} \cdot \boldsymbol{n}_i^{\dagger}) \wedge \boldsymbol{n}_i = \|\boldsymbol{n}_i\|^2 \, \boldsymbol{A}_{\langle k \rangle}$.

The next factorization vector $\boldsymbol{n}_2$ can now be found by applying (3.145) to $\boldsymbol{A}_{\langle k-1 \rangle}$. The whole factorization is obtained by repeated application of (3.145) and (3.146). The last vector of the factorization is given by $\boldsymbol{n}_k = \boldsymbol{A}_{\langle 1 \rangle}$. The whole algorithm is presented as Algorithm 3.2.

3.7.3 Evaluation of the Join

The evaluation of the join of two blades is closely related to the factorization of blades. Recall that if $\boldsymbol{A}_{\langle k \rangle} \in \mathbb{G}^k_{p,q}$ and $\boldsymbol{B}_{\langle l \rangle} \in \mathbb{G}^l_{p,q}$, then

---

Algorithm 3.2 FactBlade. Let $\boldsymbol{A}_{\langle k\rangle} \in \mathbb{G}^k_{p,q}$ be a blade which is to be factorized into a set of k orthonormal 1-vectors denoted by $\{\, \boldsymbol{n}_i \,\} \subset \mathbb{G}^1_{p,q}$. Furthermore, let $\boldsymbol{e}_i := \overline{\mathbb{G}}^1_{p,q}$ denote the canonical basis of $\mathbb{G}^1_{p,q}$. The algorithm FactBlade$(\boldsymbol{A}_{\langle k\rangle})$ returns the set of orthonormal vectors $\{\, \hat{\boldsymbol{n}}_i \,\}$ such that $\mathbb{NO}(\boldsymbol{A}_{\langle k\rangle}) = \mathbb{NO}(\bigwedge_{i=1}^{k} \hat{\boldsymbol{n}}_i)$.

$\boldsymbol{A} \leftarrow \boldsymbol{A}_{\langle k\rangle}$
for all $j \in \{\, 1, 2, \dots, k-1 \,\}$ **do**
  $\boldsymbol{n}_j \leftarrow \mathcal{P}_{\boldsymbol{A}}(\arg\max_{\boldsymbol{e}_i} \|\mathcal{P}_{\boldsymbol{A}}(\boldsymbol{e}_i)\|)$
  $\hat{\boldsymbol{n}}_j \leftarrow \boldsymbol{n}_j / \|\boldsymbol{n}_j\|$
  $\boldsymbol{A} \leftarrow \hat{\boldsymbol{n}}_j^{\dagger} \cdot \boldsymbol{A}$
end for
return $\{\, \hat{\boldsymbol{n}}_j \,\}$

$$\mathbb{NO}(\boldsymbol{A}_{\langle k\rangle} \dot{\wedge} \boldsymbol{B}_{\langle l\rangle}) = \mathbb{NO}(\boldsymbol{A}_{\langle k\rangle}) \oplus \mathbb{NO}(\boldsymbol{B}_{\langle l\rangle}).$$

Therefore, to evaluate the join of $\boldsymbol{A}_{\langle k\rangle}$ and $\boldsymbol{B}_{\langle l\rangle}$, a blade $\boldsymbol{J}_{\langle m\rangle}$ has to be found such that $\mathbb{NO}(\boldsymbol{J}_{\langle m\rangle}) = \mathbb{NO}(\boldsymbol{A}_{\langle k\rangle} \dot{\wedge} \boldsymbol{B}_{\langle l\rangle})$. This can be achieved by evaluating the set of $m-l$ orthonormal 1-vectors $\{\, \hat{\boldsymbol{n}}_i \,\}$ that lie in $\mathbb{NO}(\boldsymbol{A}_{\langle k\rangle})$ and are perpendicular to $\boldsymbol{B}_{\langle l\rangle}$. Note that the orthonormality of $\{\, \hat{\boldsymbol{n}}_i \,\}$ relates again to the Euclidean scalar product, since $\boldsymbol{A}_{\langle k\rangle}$ and/or $\boldsymbol{B}_{\langle l\rangle}$ may be null-blades.

The first step is to obtain the factorization of $\boldsymbol{A}_{\langle k\rangle}$ as $\{\, \hat{\boldsymbol{a}}_i \,\} =$ FactBlade $(\boldsymbol{A}_{\langle k\rangle})$, where the $\{\, \hat{\boldsymbol{a}}_i \,\}$ form an orthonormal set of vectors. The first vector $\boldsymbol{n}_1$ that lies in $\mathbb{NO}(\boldsymbol{A}_{\langle k\rangle})$ and is perpendicular to $\boldsymbol{B}_{\langle l\rangle}$ is evaluated via

$$\boldsymbol{n}_1 = \mathcal{P}^{\perp}_{\boldsymbol{B}_{\langle l\rangle}}\left(\arg\max_{\hat{\boldsymbol{a}}_i} \|\mathcal{P}^{\perp}_{\boldsymbol{B}_{\langle l\rangle}}(\hat{\boldsymbol{a}}_i)\| \right). \tag{3.147}$$

If $\|\boldsymbol{n}_1\| = 0$, then $\mathbb{NO}(\boldsymbol{A}_{\langle k\rangle}) \subseteq \mathbb{NO}(\boldsymbol{B}_{\langle l\rangle})$ and the join is $\boldsymbol{B}_{\langle l\rangle}$. Otherwise, let $\hat{\boldsymbol{n}}_1 = \boldsymbol{n}_1/\|\boldsymbol{n}_1\|$. Now, $\boldsymbol{B}_{\langle l\rangle}$ is updated to $\boldsymbol{B}_{\langle l+1\rangle} := \hat{\boldsymbol{n}}_1 \wedge \boldsymbol{B}_{\langle l\rangle}$ and the vector $\boldsymbol{n}_2$ is evaluated from (3.147) using $\boldsymbol{B}_{\langle l+1\rangle}$ instead of $\boldsymbol{B}_{\langle l\rangle}$. The final result $\boldsymbol{B}_{\langle m\rangle}$ is normalized to give the join in accordance with its definition. Note that $\boldsymbol{B}_{\langle m\rangle}$ may be a null-blade.

3.7.4 Versor Factorization

The goal of the algorithm for versor factorization is, given a versor $\boldsymbol{V} \in \mathbb{G}_{p,q}$, to find a set of k 1-vectors $\{\, \boldsymbol{n}_i \,\} \subset \mathbb{G}^{\varnothing 1}_{p,q}$ such that $\boldsymbol{V} = \prod_{i=1}^{k} \boldsymbol{n}_i$. In contrast to a blade, a versor cannot in general be written as a geometric product of orthogonal vectors, which complicates the factorization.

It was mentioned in Sect. 3.3, on versors, that a versor is either a linear combination of blades of even grade or a linear combination of blades of odd

Algorithm 3.3 Join. Let $\boldsymbol{A}_{\langle k \rangle} \in \mathbb{G}^k_{p,q}$ and $\boldsymbol{B}_{\langle l \rangle} \in \mathbb{G}^l_{p,q}$ be two blades whose join is to be evaluated. The algorithm returns a blade $\boldsymbol{J}_{\langle m \rangle}$ such that $\mathbb{NO}(\boldsymbol{J}_{\langle m \rangle}) = \mathbb{NO}(\boldsymbol{A}_{\langle k \rangle}) \oplus \mathbb{NO}(\boldsymbol{B}_{\langle l \rangle})$ and $\|\boldsymbol{J}_{\langle m \rangle}\| = 1$.

$\{\, \hat{\boldsymbol{a}}_i \,\} \leftarrow$ FactBlade($\boldsymbol{A}_{\langle k \rangle}$)
$\boldsymbol{J} \leftarrow \boldsymbol{B}_{\langle l \rangle}$
$\boldsymbol{n} \leftarrow 1$
repeat
  $\boldsymbol{J} \leftarrow \boldsymbol{J} \wedge (\boldsymbol{n}/\|\boldsymbol{n}\|)$
  $\boldsymbol{n} \leftarrow \mathcal{P}^{\perp}_{\boldsymbol{J}}(\arg\max_{\hat{\boldsymbol{a}}_i} \|\mathcal{P}^{\perp}_{\boldsymbol{J}}(\hat{\boldsymbol{a}}_i)\|)$
until $\|\boldsymbol{n}\| = 0$
$\boldsymbol{J} \leftarrow \boldsymbol{J}/\|\boldsymbol{J}\|$
return $\boldsymbol{J}$

Algorithm 3.4 FactVersor. Let $\boldsymbol{V} \in \mathbb{G}_{p,q}$ denote a versor. The algorithm returns a set of k normalized 1-vectors $\{\, \hat{\boldsymbol{n}}_i \,\} \subset \mathbb{G}^{\varnothing 1}_{p,q}$ and a scalar factor $\nu \in \mathbb{R}$ such that $\boldsymbol{V} = \nu \prod_{i=1}^{k} \hat{\boldsymbol{n}}_i$.

$j \leftarrow 1$
while $\mathrm{gr}(\langle \boldsymbol{V} \rangle_{\max}) > 0$ **do**
  $\boldsymbol{A}_{\langle l \rangle} \leftarrow \langle \boldsymbol{V} \rangle_{\max}$
  $\{\, \hat{\boldsymbol{a}}_i \,\} \leftarrow$ FactBlade($\boldsymbol{A}_{\langle l \rangle}$)
  $i \leftarrow 1$
  while $\hat{\boldsymbol{a}}_i \cdot \hat{\boldsymbol{a}}_i = 0$ **do**
    $i \leftarrow i + 1$
  end while
  $\hat{\boldsymbol{n}}_j \leftarrow \hat{\boldsymbol{a}}_i$
  $\boldsymbol{V} \leftarrow \boldsymbol{V}\, \hat{\boldsymbol{n}}_j$
  $j \leftarrow j + 1$
end while
$\nu \leftarrow \boldsymbol{V}$
return ν, $\{\, \hat{\boldsymbol{n}}_i \,\}$

grade. Clearly, the grade of the blade of maximum grade in a versor gives the minimum number of 1-vectors that are needed to generate the versor.

Let $\langle \cdot \rangle_{\max}$ denote the grade projection bracket that returns the component blades of maximum grade. For example, if $\boldsymbol{V} := \boldsymbol{A}_{\langle 0 \rangle} + \boldsymbol{A}_{\langle 2 \rangle} + \boldsymbol{A}_{\langle 4 \rangle}$, then $\langle \boldsymbol{V} \rangle_{\max} = \boldsymbol{A}_{\langle 4 \rangle}$. In the following, let $\boldsymbol{A}_{\langle k \rangle} := \langle \boldsymbol{V} \rangle_{\max}$, such that k is the maximum grade in a versor.

Recall that a vector $\boldsymbol{n} \in \mathbb{G}^{\varnothing 1}_{p,q}$ with $\mathbb{NO}(\boldsymbol{n}) \subset \mathbb{NO}(\boldsymbol{A}_{\langle k \rangle})$ satisfies $\boldsymbol{n}\, \boldsymbol{A}_{\langle k \rangle} = \boldsymbol{n} \cdot \boldsymbol{A}_{\langle k \rangle}$. Hence, $\boldsymbol{V}' := \boldsymbol{V}\, \boldsymbol{n}^{-1}$ reduces the grade of the blade of maximum grade in $\boldsymbol{V}$ by one, i.e. $\mathrm{gr}\,(\langle \boldsymbol{V}' \rangle_{\max}) = k-1$. Furthermore, since the geometric product is invertible, $\boldsymbol{V}'\boldsymbol{n} = \boldsymbol{V}$. Through repeated application of this grade reduction process, a set of vectors can be evaluated whose geometric product results in the initial versor. The complete algorithm is given as Algorithm 3.4. This algorithm does not work for null-versors, i.e. if $\boldsymbol{V}\widetilde{\boldsymbol{V}} = 0$.

Note, however, that some of the constituent blades of a non-null versor may be null-blades. For example, let $\boldsymbol{n} \in \mathbb{G}^{\circ 1}_{p,q}$ be a null-vector and let $\hat{\boldsymbol{a}} \in \mathbb{G}^{\varnothing 1}_{p,q}$ satisfy $\hat{\boldsymbol{a}}^2 = 1$, with $\hat{\boldsymbol{a}} \cdot \boldsymbol{n} = 0$. Then, $\boldsymbol{b} := \hat{\boldsymbol{a}} + \boldsymbol{n}$ is not a null-vector, since $\boldsymbol{b}^2 = \hat{\boldsymbol{a}}^2 = 1$ and thus $\boldsymbol{V} := \hat{\boldsymbol{a}}\,\boldsymbol{b}$ is a versor. However,

$$\boldsymbol{V} = \hat{\boldsymbol{a}}\,\boldsymbol{b} = \hat{\boldsymbol{a}}^2 + \hat{\boldsymbol{a}}\,\boldsymbol{n} = 1 + \hat{\boldsymbol{a}} \wedge \boldsymbol{n}.$$

Hence, $\langle \boldsymbol{V} \rangle_{\max} = \hat{\boldsymbol{a}} \wedge \boldsymbol{n}$, which is a null-blade since

$$(\hat{\boldsymbol{a}} \wedge \boldsymbol{n})^2 = \hat{\boldsymbol{a}}\,\boldsymbol{n}\,\hat{\boldsymbol{a}}\,\boldsymbol{n} = -\hat{\boldsymbol{a}}^2\,\boldsymbol{n}^2 = 0.$$

Nevertheless, not all constituent vectors of $\langle \boldsymbol{V} \rangle_{\max}$ can be null-vectors, since otherwise $\boldsymbol{V}$ would be a null-blade.

3.8 Related Algebras

Clearly, geometric algebra is not the only algebra that describes geometry. In this section, other algebras that relate to geometry and their relation to geometric algebra are discussed.

3.8.1 Gibbs's Vector Algebra

Basically, the inner product between vectors in geometric algebra is equivalent to the scalar product of vectors in Gibbs's vector algebra. Furthermore, since the dual of the outer product of two vectors $\boldsymbol{a}, \boldsymbol{b} \in \mathbb{R}^3$ gives a vector perpendicular to the plane spanned by $\boldsymbol{a}$ and $\boldsymbol{b}$, it should be no surprise that the outer product is related to the cross product in the following way.

$$\boldsymbol{a} \times \boldsymbol{b} = (\boldsymbol{a} \wedge \boldsymbol{b})^*. \tag{3.148}$$

Using this relation, the identities of Gibbs's vector algebra can be translated into geometric algebra. For example, the triple scalar product of three vectors $\boldsymbol{a}, \boldsymbol{b}, \boldsymbol{c} \in \mathbb{R}^3$ can be translated as follows:

$$\begin{aligned} \boldsymbol{a} \cdot (\boldsymbol{b} \times \boldsymbol{c}) &= \boldsymbol{a} \cdot (\boldsymbol{b} \wedge \boldsymbol{c})^* \\ &= \boldsymbol{a} \cdot \big((\boldsymbol{b} \wedge \boldsymbol{c}) \cdot \boldsymbol{I}^{-1}\big) \\ &= (\boldsymbol{a} \wedge \boldsymbol{b} \wedge \boldsymbol{c}) \cdot \boldsymbol{I}^{-1} \\ &= (\boldsymbol{a} \wedge \boldsymbol{b} \wedge \boldsymbol{c})^* \\ &= \det\big([\boldsymbol{a}, \boldsymbol{b}, \boldsymbol{c}]\big). \end{aligned} \tag{3.149}$$

Note that $\det([\boldsymbol{a}, \boldsymbol{b}, \boldsymbol{c}])$ is the volume of the parallelepiped spanned by $\boldsymbol{a}$, $\boldsymbol{b}$, and $\boldsymbol{c}$. This shows again that the outer product of three vectors spans a volume element. Another often used identity is the triple vector product $\boldsymbol{a} \times (\boldsymbol{b} \times \boldsymbol{c})$. This is usually expanded as

$$\boldsymbol{a} \times (\boldsymbol{b} \times \boldsymbol{c}) = \boldsymbol{b}\,(\boldsymbol{a} \cdot \boldsymbol{b}) - \boldsymbol{c}\,(\boldsymbol{a} \cdot \boldsymbol{b}).$$

Translating this expression into geometric algebra gives

$$\begin{aligned} \boldsymbol{a} \times (\boldsymbol{b} \times \boldsymbol{c}) &= \Big(\boldsymbol{a} \wedge \big((\boldsymbol{b} \wedge \boldsymbol{c})\boldsymbol{I}^{-1}\big)\Big)\boldsymbol{I}^{-1} \\ &= \boldsymbol{a} \cdot \Big(\big((\boldsymbol{b} \wedge \boldsymbol{c})\boldsymbol{I}^{-1}\big)\boldsymbol{I}^{-1}\Big) \\ &= -\boldsymbol{a} \cdot (\boldsymbol{b} \wedge \boldsymbol{c}) \\ &= \boldsymbol{b}\,(\boldsymbol{a} \cdot \boldsymbol{c}) - \boldsymbol{c}\,(\boldsymbol{a} \cdot \boldsymbol{b}). \end{aligned} \tag{3.150}$$

The expansion in geometric algebra is valid in any dimension, whereas the vector cross product is defined only in a 3D-vector space.

3.8.2 Complex Numbers

The algebra of complex numbers may also be regarded as a geometric algebra, where the real and imaginary parts of a complex number are interpreted as the two coordinates of a point in a 2D space. A complex number $z \in \mathbb{C}$ can be expressed in two equivalent ways:

$$z = \mathsf{a} + \mathsf{i}\mathsf{b} = \mathsf{r}\,\exp(\mathsf{i}\,\theta),$$

where $\mathsf{i} = \sqrt{-1}$ denotes the imaginary unit, and $\mathsf{a}, \mathsf{b}, \mathsf{r}, \theta \in \mathbb{R}$. The relation between a, b and r and θ is $\mathsf{r} = \sqrt{\mathsf{a}^2 + \mathsf{b}^2}$ and $\theta = \tan^{-1}(\mathsf{b}/\mathsf{a})$. Recall that a unit bivector in $\mathbb{G}_n$ squares to minus one and thus may replace the imaginary unit i. Accordingly, the definition of the exponential function may be extended to multivectors, and thus rotors can also be written in exponential form.

The exponential function can indeed be used to write any multivector $\boldsymbol{A} \in \mathbb{G}_n$ defined as $\boldsymbol{A} = \mathsf{a} + \boldsymbol{U}_{\langle 2\rangle}\,\mathsf{b}$, as

$$\boldsymbol{A} = \mathsf{r}\,\exp(\boldsymbol{U}_{\langle 2\rangle}\,\theta),$$

where $\boldsymbol{U}_{\langle 2\rangle} \in \mathbb{G}_n^2$ is a unit bivector (cf. equation (4.10), p. 132). Note that $\boldsymbol{A}$ is an element of a subalgebra $\mathbb{G}_2 \subseteq \mathbb{G}_n$, $n \geq 2$. More precisely, it is an element of the even subalgebra $\mathbb{G}_2^+ \subset \mathbb{G}_2$, which consists of the linear combinations of the even-grade elements of $\mathbb{G}_2$. The even subalgebra $\mathbb{G}_2^+$ of $\mathbb{G}_2$ has basis $\{1, \boldsymbol{U}_{\langle 2\rangle}\}$, where $\boldsymbol{U}_{\langle 2\rangle}$ is also the pseudoscalar of $\mathbb{G}_2$ and $\boldsymbol{U}_{\langle 2\rangle}^2 = -1$.

Hence, the complex numbers $\mathbb{C}$ and the geometric algebra $\mathbb{G}_2^+$ are isomorphic, such that the product between complex numbers becomes the geometric product. Note that the complex conjugate becomes the reverse, since the reverse of $\boldsymbol{A}$ is

$$\widetilde{\boldsymbol{A}} = \mathsf{r}\,\exp(\widetilde{\boldsymbol{U}}_{\langle 2\rangle}\,\theta) = \mathsf{r}\,\exp(-\boldsymbol{U}_{\langle 2\rangle}\,\theta),$$

which is equivalent to

$$z^* = \mathsf{r}\,\exp(-\mathsf{i}\,\theta).$$

3.8.3 Quaternions

The interesting aspect of the isomorphism between $\mathbb{C}$ and $\mathbb{G}_2^+$ is that $\mathbb{G}_n$ has $\binom{n}{k}$ bivectors and thus the same number of different even subalgebras $\mathbb{G}_2^+$. That is, different complex algebras can be combined in geometric algebra. One effect of this is that there is an isomorphism between quaternions ($\mathbb{H}$) and $\mathbb{G}_3^+$. Before this isomorphism is presented, the basic properties of quaternions will be recapitulated.

The name "quaternion" literally means a combination of four parts. The quaternions discussed here consist of a scalar component and three imaginary components. The imaginary components are typically denoted by $\mathsf{i}, \mathsf{j}, \mathsf{k}$, and they satisfy the following relations:

$$\begin{aligned}
&\mathsf{i}^2 = \mathsf{j}^2 = \mathsf{k}^2 = -1,\\
&\mathsf{ij} = \mathsf{k},\ \mathsf{jk} = \mathsf{i},\ \mathsf{ki} = \mathsf{j},\\
&\mathsf{ij} = -\mathsf{ji},\ \mathsf{jk} = -\mathsf{kj},\ \mathsf{ki} = -\mathsf{ik},\\
&\mathsf{ijk} = -1.
\end{aligned} \tag{3.151}$$

A general quaternion is then given by

$$a = \mathsf{a}_0 + \mathsf{a}_1\mathsf{i} + \mathsf{a}_2\mathsf{j} + \mathsf{a}_3\mathsf{k},$$

with $\{\,\mathsf{a}_i\,\} \subset \mathbb{R}$. A *pure* quaternion is one with no scalar component, i.e. $\bar{a} = \mathsf{a}_1\mathsf{i} + \mathsf{a}_2\mathsf{j} + \mathsf{a}_3\mathsf{k}$ is a pure quaternion. The square of a pure quaternion gives

$$\bar{a}^2 = (\mathsf{a}_1\mathsf{i} + \mathsf{a}_2\mathsf{j} + \mathsf{a}_3\mathsf{k})^2 = -\big((\mathsf{a}_1)^2 + (\mathsf{a}_2)^2 + (\mathsf{a}_3)^2\big).$$

The complex conjugate of a quaternion a is denoted by a^*. Taking the conjugate negates all imaginary components. Therefore,

$$\begin{aligned}
aa^* &= (\mathsf{a}_0 + \mathsf{a}_1\mathsf{i} + \mathsf{a}_2\mathsf{j} + \mathsf{a}_3\mathsf{k})\,(\mathsf{a}_0 - \mathsf{a}_1\mathsf{i} - \mathsf{a}_2\mathsf{j} - \mathsf{a}_3\mathsf{k})\\
&= (\mathsf{a}_0)^2 + (\mathsf{a}_1)^2 + (\mathsf{a}_2)^2 + (\mathsf{a}_3)^2.
\end{aligned}$$

A unit pure quaternion $\hat{a}$ satisfies $\hat{a}\hat{a}^* = 1$ and thus $\hat{a}\hat{a} = -1$. A quaternion a may therefore also be written as

$$
\begin{aligned}
a &= (\mathsf{a}_0 + \mathsf{a}_1\mathsf{i} + \mathsf{a}_2\mathsf{j} + \mathsf{a}_3\mathsf{k}) \\
&= \mathsf{r}\,(\cos\theta + \hat{\bar{a}}\,\sin\theta),
\end{aligned}
$$

where $\mathsf{r} = \sqrt{aa^*}$, $\theta = \tan^{-1}(\bar{a}\bar{a}^*/\mathsf{a}_0)$, $\bar{a} = \mathsf{a}_1\mathsf{i} + \mathsf{a}_2\mathsf{j} + \mathsf{a}_3\mathsf{k}$, and $\hat{\bar{a}} = \bar{a}/\sqrt{\bar{a}\bar{a}^*}$. Since $\hat{\bar{a}}$ squares to minus one, there is again an isomorphism between the complex numbers $\mathbb{C}$ and a subalgebra of $\mathbb{H}$. The definition of the exponential function can also be extended to quaternions to give

$$
a = (\mathsf{a}_0 + \mathsf{a}_1\mathsf{i} + \mathsf{a}_2\mathsf{j} + \mathsf{a}_3\mathsf{k}) = \mathsf{r}\,\exp(\theta\,\hat{\bar{a}}),
$$

where r, θ, and $\hat{\bar{a}}$ are given as before. It can be shown that the operation $\hat{r}\bar{a}\hat{r}^*$ between a unit quaternion $\hat{r} = \exp(\frac{1}{2}\theta\hat{\bar{r}})$ and a pure quaternion $\bar{a}$ represents a rotation of $\bar{a}$. That is, if we regard $\bar{a} = \mathsf{a}_1\mathsf{i}+\mathsf{a}_2\mathsf{j}+\mathsf{a}_3\mathsf{k}$ as a vector $(\mathsf{a}_1, \mathsf{a}_2, \mathsf{a}_3)$, then $\hat{r}\bar{a}\hat{r}^*$ rotates this vector by an angle θ about the vector represented by $\hat{\bar{r}}$.

Consider two simple examples of this. Assuming $(\mathsf{i}, \mathsf{j}, \mathsf{k})$ to form the basis of a right-handed coordinate system, the pure quaternion k can be written in exponential form as $\mathsf{k} = \exp(\frac{1}{2}\pi\,\mathsf{k})$. Therefore, it should rotate the pure quaternion i about 180 degrees, if applied as kik^*:

$$
\mathsf{kik}^* = -\mathsf{ikk}^* = -\mathsf{i}.
$$

This example also shows that operators and the elements operated on can be of the same type.

Here is a more complex example. Consider the rotation operator for a rotation about the k-axis, $r = \exp(\frac{1}{2}\theta\mathsf{k})$. Expand r to read $r = \cos\frac{1}{2}\theta + \mathsf{k}\,\sin\frac{1}{2}\theta$. If r is applied to i, it should rotate i in the ij plane by an angle θ:

$$
\begin{aligned}
r\,\mathsf{i}\,r^* &= (\cos\tfrac{1}{2}\theta + \mathsf{k}\,\sin\tfrac{1}{2}\theta)\,\mathsf{i}\,(\cos\tfrac{1}{2}\theta - \mathsf{k}\,\sin\tfrac{1}{2}\theta) \\
&= \cos^2\tfrac{1}{2}\theta\,\mathsf{i} - \cos\tfrac{1}{2}\theta\,\sin\tfrac{1}{2}\theta\,\mathsf{ik} + \cos\tfrac{1}{2}\theta\,\sin\tfrac{1}{2}\theta\,\mathsf{ki} - \sin^2\tfrac{1}{2}\theta\,\mathsf{kik} \\
&= (\cos^2\tfrac{1}{2}\theta - \sin^2\tfrac{1}{2}\theta)\mathsf{i} + 2\cos\tfrac{1}{2}\theta\,\sin\tfrac{1}{2}\theta\,\mathsf{j} \\
&= \cos\theta\,\mathsf{i} + \sin\theta\,\mathsf{j}.
\end{aligned}
$$

This shows that $r = \exp(\frac{1}{2}\theta\,\mathsf{k})$ is indeed an operator for a rotation by a mathematically positive angle θ. If this is compared with rotors in geometric algebra, it can be seen that there is a difference in sign. Recall that a rotor for a rotation by a mathematically positive angle θ is given by $\exp(-\frac{1}{2}\theta\boldsymbol{U}_{\langle 2\rangle})$. This difference in sign stems from the way in which bivectors are interpreted.

This will become clear once the isomorphism between quaternions and a geometric algebra is given.

What has been discussed so far about quaternions shows already how similar they are to rotors, which were discussed earlier. This also gives a hint for how to find an isomorphism. Basically, a set of multivectors has to be found which have the same properties as i, j, and k, and together with the unit scalar, form the basis of an even subalgebra. A possible isomorphism is given by the following identifications:

$$\mathsf{i} \to \boldsymbol{e}_2\boldsymbol{e}_3, \quad \mathsf{j} \to \boldsymbol{e}_1\boldsymbol{e}_2, \quad \mathsf{k} \to \boldsymbol{e}_3\boldsymbol{e}_1, \tag{3.152}$$

where the $\{\boldsymbol{e}_1, \boldsymbol{e}_2, \boldsymbol{e}_3\} \subset \mathbb{R}^3$ are an orthonormal basis of $\mathbb{R}^3$. Therefore, the geometric algebra $\mathbb{G}_3^+$ with basis $\{1, \boldsymbol{e}_2\boldsymbol{e}_3, \boldsymbol{e}_1\boldsymbol{e}_2, \boldsymbol{e}_3\boldsymbol{e}_1\}$ is isomorphic to the quaternions $\mathbb{H}$ if the above identifications are made. Note that this is only one possible isomorphism. We have

$$\begin{aligned}
\mathsf{ij} &\to \boldsymbol{e}_2\boldsymbol{e}_3\,\boldsymbol{e}_1\boldsymbol{e}_2 = \boldsymbol{e}_3\boldsymbol{e}_1 \to \mathsf{k},\\
\mathsf{jk} &\to \boldsymbol{e}_1\boldsymbol{e}_2\,\boldsymbol{e}_3\boldsymbol{e}_1 = \boldsymbol{e}_2\boldsymbol{e}_3 \to \mathsf{i},\\
\mathsf{ki} &\to \boldsymbol{e}_3\boldsymbol{e}_1\,\boldsymbol{e}_2\boldsymbol{e}_3 = \boldsymbol{e}_1\boldsymbol{e}_2 \to \mathsf{j}.
\end{aligned} \tag{3.153}$$

It can now be seen where the sign difference in the rotation operators comes from. When one is working with vectors, it is usually assumed that the vectors are defined in a right-handed system and that the coordinates are given in the order of the x-, y-, and z-axes, for example $(\mathsf{a}_1, \mathsf{a}_2, \mathsf{a}_3)$. When one is using quaternions, the imaginary units i, j, and k are defined with the three coordinate axes in this order. In geometric algebra, on the other hand, the three axes are denoted by $\boldsymbol{e}_1$, $\boldsymbol{e}_2$, and $\boldsymbol{e}_3$. Now, recall that the plane of rotation is given by a unit bivector, for example $\boldsymbol{U}_{\langle 2\rangle} \in \mathbb{G}_3^+$, and the corresponding rotation axis is given by $\boldsymbol{U}^*_{\langle 2\rangle}$. Since

$$(\boldsymbol{e}_2\boldsymbol{e}_3)^* = \boldsymbol{e}_1,\ (\boldsymbol{e}_1\boldsymbol{e}_2)^* = \boldsymbol{e}_3,\ (\boldsymbol{e}_3\boldsymbol{e}_1)^* = \boldsymbol{e}_2,$$

the rotation axis $(\mathsf{a}_1\mathsf{i} + \mathsf{a}_2\mathsf{j} + \mathsf{a}_3\mathsf{k})$ corresponds to the rotation axis $(\mathsf{a}_1\boldsymbol{e}_1 + \mathsf{a}_2\boldsymbol{e}_3 + \mathsf{a}_3\boldsymbol{e}_2)$ in geometric algebra using the above identification for i, j, and k. That is, the y- and z-axes are exchanged. Therefore, when quaternions are embedded into geometric algebra, they cannot be applied directly to vectors, but only to other embedded quaternions. When quaternions are translated into rotors, the appropriate exchange of axes has to be made, which also introduces the minus sign into the rotor.

It has been shown that quaternions are isomorphic to the space of rotors in $\mathbb{G}_3$, which is the even subalgebra $\mathbb{G}_3^+ \subset \mathbb{G}_3$. The main advantages of rotors in geometric algebra over quaternions are that rotors may be defined in any dimension and that a rotor can rotate blades of any grade. That is, not only vectors but also lines, planes, and any other geometric objects that can be represented by a blade may be rotated with a rotor.

3.8.4 Grassmann Algebra

Today, Grassmann algebra is usually taken as a synonym for *exterior algebra.* Although Hermann Grassmann also developed exterior algebra, he looked at the whole subject from a much more general point of view. In fact, he developed some fundamental results of what is today known as *universal algebra.* In his book *Die lineale Ausdehnungslehre, ein neuer Zweig der Mathematik, dargestellt und durch Anwendungen auf die übrigen Zweige der Mathematik, wie auch auf die Statik, Mechanik, die Lehre vom Magnetismus und die Krystallonomie erläutert von Hermann Grassmann* [81], Grassmann basically developed linear algebra, with a theory of bases and dimensions for finite-dimensional linear spaces. He called vectors *extensive quantities* and a basis $\{\, \boldsymbol{e}_1, \boldsymbol{e}_2, \ldots, \boldsymbol{e}_n \,\}$ a *system of units.* The vector space spanned by a basis was called a *region.* He then introduced a very general product on the extensive quantities (vectors). Given two vectors $\boldsymbol{a} = \mathsf{a}^i \boldsymbol{e}_i$ and $\boldsymbol{b} = \mathsf{b}^i \boldsymbol{e}_i$, a general product of the two is written as

$$\boldsymbol{ab} = \mathsf{a}^i \mathsf{b}^j \, (\boldsymbol{e}_i \boldsymbol{e}_j),$$

with an implicit summation over i and j. Grassmann made no additional assumptions at first about the elements $(\boldsymbol{e}_i \boldsymbol{e}_j)$, apart from noting that they were extensive quantities themselves. The set of products that can be formed with extensive quantities was called a *product structure.* For example, for a vector basis $\{\, \boldsymbol{e}_1, \boldsymbol{e}_2 \,\}$, the set of products is

$$\{\, \boldsymbol{e}_1, \boldsymbol{e}_2, (\boldsymbol{e}_1\boldsymbol{e}_1), (\boldsymbol{e}_1\boldsymbol{e}_2), (\boldsymbol{e}_2\boldsymbol{e}_1), (\boldsymbol{e}_2\boldsymbol{e}_2), \boldsymbol{e}_1(\boldsymbol{e}_1\boldsymbol{e}_1), \boldsymbol{e}_1(\boldsymbol{e}_1\boldsymbol{e}_2), \ldots \}.$$

This product structure may then be constrained by a *determining equation.* That is, if the elements of the product structure are denoted by $\{\, E_i \,\}$, a determining equation is of the form $\mathsf{a}^i E_i = 0$, $\mathsf{a}^i \in \mathbb{R}$. For example, the determining equation could be $(\boldsymbol{e}_1\boldsymbol{e}_2) + (\boldsymbol{e}_2\boldsymbol{e}_1) = 0$. In this case $(\boldsymbol{e}_1\boldsymbol{e}_2)$ is linearly dependent on $(\boldsymbol{e}_2\boldsymbol{e}_1)$. Or, more generally, $(\boldsymbol{e}_i\boldsymbol{e}_j) + (\boldsymbol{e}_j\boldsymbol{e}_i) = 0$ for all i and j. This also implies that $\boldsymbol{e}_i\boldsymbol{e}_i = 0$. Assuming that the product is associative, the basis for the algebra generated by $\{\, \boldsymbol{e}_1, \boldsymbol{e}_2 \,\}$ becomes

$$\{\, \boldsymbol{e}_1, \boldsymbol{e}_2, (\boldsymbol{e}_1\boldsymbol{e}_2) \,\}.$$

Grassmann found that the only determining equations that stay invariant under a change of basis were, for two vectors $\boldsymbol{a}$ and $\boldsymbol{b}$, $\boldsymbol{ab} = 0$, $\boldsymbol{ab} - \boldsymbol{ba} = 0$, and $\boldsymbol{ab} + \boldsymbol{ba} = 0$. He then considered at some length the algebra generated by the determining equation $\boldsymbol{ab} + \boldsymbol{ba} = 0$. This algebra is today called exterior algebra, and the product which satisfies this determining equation is called the exterior product. In the following, the exterior product is denoted by $\wedge$, just like the outer product. In fact, "outer product" is just another name for "exterior product".

Today, exterior algebra is described in much the same way, albeit more generally and rigorously. The general product that Grassmann introduced is replaced by the *tensor product.*

Grassmann also introduced an inner product between extensive quantities of the same grade. He did this in a very interesting way, by first defining what is essentially the dual. For an extensive quantity E, the dual is denoted by E^* and is defined such that $E^* \wedge E$ is an extensive quantity of the highest grade, i.e. a pseudoscalar. Since the pseudoscalars span a one-dimensional subspace, he equated the extensive quantity $\boldsymbol{e}_1 \wedge \boldsymbol{e}_2 \wedge \ldots \wedge \boldsymbol{e}_n$ with the scalar 1. With this definition, $E^* \wedge E$ is indeed a scalar. The inner product of two extensive quantities E, F of same grade is then defined as

$$< E, F >:= E^* \wedge F.$$

3.8.5 Grassmann–Cayley Algebra

The main difference between Grassmann and Grassmann–Cayley algebra is that there is also a grade-reducing inner product defined between blades of different grade in Grassmann–Cayley algebra. This product may also be called the shuffle product or the regressive product. Sometimes this product is also called the meet, and the exterior product is called the join. These should not be confused with the meet and join defined previously in this text. Another source of confusion is the meaning of the symbols $\wedge$ and $\vee$, which is exactly the opposite of what they mean in geometric algebra. The symbol $\wedge$ usually stands for the meet (inner product) and the $\vee$ stands for the join (outer product). This is actually somewhat more logical than the usage in geometric algebra, since it is comparable to the use of the symbols for union ($\cup$) and intersection ($\cap$). Unfortunately, not all authors who use Grassmann–Cayley algebra follow this convention. Sometimes Grassmann algebra is also taken to mean Grassmann–Cayley algebra. At times, even completely different symbols ($\triangledown$ and $\triangle$) are used for the meet and join.

Despite these notational differences, Grassmann–Cayley algebra and geometric algebra are equivalent in the sense that anything expressed in one of them can also be expressed in the other.

The shuffle product is defined with respect to the bracket operator $[]$. The bracket operator is defined for elements of the highest grade in an algebra (pseudoscalars), for which it evaluates their magnitude. In the following, the geometric-algebra notation is used. If $\boldsymbol{A}_{\langle k \rangle}, \boldsymbol{B}_{\langle l \rangle} \in \mathbb{G}_n$ are given by $\boldsymbol{A}_{\langle k \rangle} = \bigwedge_{i=1}^{k} \boldsymbol{a}_i$ and $\boldsymbol{B}_{\langle l \rangle} = \bigwedge_{i=1}^{l} \boldsymbol{b}_i$, with $k + l \geq n$ and $k \geq l$, then the shuffle product of $\boldsymbol{A}_{\langle k \rangle}$ and $\boldsymbol{B}_{\langle l \rangle}$, which is temporarily denoted here by $\odot$, is defined as

$$\boldsymbol{A}_{\langle k \rangle} \odot \boldsymbol{B}_{\langle l \rangle} := \sum_{\sigma} \operatorname{sgn}(\sigma) \left[\boldsymbol{a}_{\sigma(1)}\boldsymbol{a}_{\sigma(2)} \ldots \boldsymbol{a}_{\sigma(n-l)}\; \boldsymbol{b}_1 \wedge \ldots \boldsymbol{b}_l\right] \\ \times \boldsymbol{a}_{\sigma(n-l+1)} \wedge \ldots \boldsymbol{a}_{\sigma(k)}. \tag{3.154}$$

The sum is taken over all permutations σ of $\{1, \ldots, k\}$, such that $\sigma(1) < \sigma(2) < \ldots \sigma(n-l)$ and $\sigma(n-l+1) < \sigma(n-l+2) < \ldots \sigma(n)$. These types of permutations are called *shuffles* of the $(n-l,\, k-(n-l))$ split of $\boldsymbol{A}_{\langle k \rangle}$. If σ is an even permutation of $\{1, \ldots, k\}$, then $\operatorname{sgn}(\sigma) = +1$; otherwise, $\operatorname{sgn}(\sigma) = -1$. For example, the shuffles of a $(2,1)$ split of $\{1,2,3\}$ are

$$(\{1,2\}, \{3\}),\; (\{1,3\}, \{2\}),\; (\{2,3\}, \{1\}),$$

where

$$\operatorname{sgn}(\{1,2,3\}) = +1,\; \operatorname{sgn}(\{1,3,2\}) = -1,\; \operatorname{sgn}(\{2,3,1\}) = +1.$$

Therefore, for $\{\boldsymbol{a}_1, \boldsymbol{a}_2, \boldsymbol{a}_3\} \subset \mathbb{R}^3$ and $\boldsymbol{b} \in \mathbb{R}^3$, it follows that

$$(\boldsymbol{a}_1 \wedge \boldsymbol{a}_2 \wedge \boldsymbol{a}_3) \odot \boldsymbol{b} = [\boldsymbol{a}_1\boldsymbol{a}_2\boldsymbol{b}]\, \boldsymbol{a}_3 - [\boldsymbol{a}_1\boldsymbol{a}_3\boldsymbol{b}]\, \boldsymbol{a}_2 + [\boldsymbol{a}_2\boldsymbol{a}_3\boldsymbol{b}]\, \boldsymbol{a}_1.$$

If $\{\boldsymbol{e}_1, \boldsymbol{e}_2, \boldsymbol{e}_3\}$ is an orthonormal basis of $\mathbb{R}^3$, and $\boldsymbol{b} = \mathsf{b}^i \boldsymbol{e}_i$, then

$$\begin{aligned}(\boldsymbol{e}_1 \wedge \boldsymbol{e}_2 \wedge \boldsymbol{e}_3) \odot \boldsymbol{b} &= [\boldsymbol{e}_1\boldsymbol{e}_2\boldsymbol{b}]\, \boldsymbol{e}_3 - [\boldsymbol{e}_1\boldsymbol{e}_3\boldsymbol{b}]\, \boldsymbol{e}_2 + [\boldsymbol{e}_2\boldsymbol{e}_3\boldsymbol{b}]\, \boldsymbol{e}_1 \\ &= [\boldsymbol{e}_1\boldsymbol{e}_2\, \mathsf{b}^3\boldsymbol{e}_3]\, \boldsymbol{e}_3 - [\boldsymbol{e}_1\boldsymbol{e}_3\, \mathsf{b}^2\boldsymbol{e}_2]\, \boldsymbol{e}_2 + [\boldsymbol{e}_2\boldsymbol{e}_3\, \mathsf{b}^1\boldsymbol{e}_1]\, \boldsymbol{e}_1 \\ &= \boldsymbol{b},\end{aligned}$$

since $[\boldsymbol{e}_1\boldsymbol{e}_2\boldsymbol{e}_3] = 1$. This shows that the pseudoscalar is the unit element with respect to the shuffle product. This property appeared earlier in this book when the regressive product was introduced (see Sect. 3.2.13). In fact, it can be shown that the regressive product as defined earlier *is* the shuffle product. That is,

$$\boldsymbol{A}_{\langle k \rangle} \odot \boldsymbol{B}_{\langle l \rangle} \equiv \boldsymbol{A}_{\langle k \rangle} \triangledown \boldsymbol{B}_{\langle l \rangle} = \left(\boldsymbol{A}^*_{\langle k \rangle} \wedge \boldsymbol{B}^*_{\langle l \rangle}\right) \boldsymbol{I}.$$

The shuffle product is usually used to evaluate the intersection of subspaces. As was shown in the discussion of the meet and join, this can be done only in the case where the join of the two subspaces is the whole space. The shuffle product also cannot fully replace the inner product of geometric algebra, since it is defined to be zero for two blades $\boldsymbol{A}_{\langle k \rangle}, \boldsymbol{B}_{\langle l \rangle} \in \mathbb{G}_n$ if $k+l < n$. It is nonetheless possible to recover the inner product from the shuffle product through the definition of the Hodge dual. This is basically the same as the dual defined in geometric algebra. The only difference is that the Hodge dual of the Hodge dual of a blade is again that blade in any space. The dual of the dual of a blade in geometric algebra is either the blade or the negated blade. The geometric-algebra inner product may then be expressed in terms of the shuffle product as

$$\boldsymbol{A}_{\langle k \rangle} \cdot \boldsymbol{B}_{\langle l \rangle} \iff \boldsymbol{A}^*_{\langle k \rangle} \odot \boldsymbol{B}_{\langle l \rangle}.$$

This follows immediately from the definition of the regressive product. Translating the Hodge dual of $\boldsymbol{A}_{\langle k\rangle}$ as $\boldsymbol{A}_{\langle k\rangle}\boldsymbol{I}$, we then obtain

$$\boldsymbol{A}^*_{\langle k\rangle} \odot \boldsymbol{B}_{\langle l\rangle} \Rightarrow (\boldsymbol{A}_{\langle k\rangle}\boldsymbol{I}) \triangledown \boldsymbol{B}_{\langle l\rangle} = (\boldsymbol{A}_{\langle k\rangle} \wedge \boldsymbol{B}^*_{\langle l\rangle})\boldsymbol{I} = \boldsymbol{A}_{\langle k\rangle} \cdot \boldsymbol{B}_{\langle l\rangle}.$$

Grassmann–Cayley algebra is probably used most widely in the areas of computer vision [68, 69] and robotics [174, 175]. There is still a lively, ongoing discussion within the research community about whether Grassmann–Cayley or geometric algebra is better suited for these fields. This is probably a matter of personal preference to a large extent and therefore this decision is left to the reader's intuition.

Chapter 4
Geometries

The goal of this chapter is to show how geometric algebra can be used to represent geometry. This includes the representation of geometric entities, incidence relations, and transformations. The algebraic entities by themselves have no a priori geometric interpretation. They are simply mathematical objects that have particular algebraic properties. In order to associate the algebra with geometry, a particular representation has to be defined. The goal is, of course, to define a representation where the standard algebraic operations relate to geometrically meaningful and useful operations. Given such a representation, geometric problems can be translated into algebraic expressions, whence they can be solved using algebraic operations. In this sense, the algebra is then an algebra of geometry, a *geometric algebra*.

There are many examples of these types of algebras, some of which we described in Sect. 3.8. The geometric algebra discussed in this text encompasses all the algebras described in Sect. 3.8, which is the justification for calling it *the* geometric algebra.

Note that the field of *algebraic geometry* is closely related to this method of representing geometry. In algebraic geometry, geometric entities are represented through the null space of a set of polynomials, an *affine variety* (see e.g. [38]). The representations of geometric entities discussed in this chapter can all be regarded as affine varieties or as intersections of affine varieties. In this context, geometric algebra offers a convenient way to work with certain types of affine varieties.

The representation of geometry by algebraic entities chosen for geometric algebra can be formulated quite generally and is closely related to the outer- and inner-product null spaces of blades. Recall the definition of the OPNS of a blade $\boldsymbol{A}_{\langle k \rangle} \in \mathbb{G}^k_{p,q}$:

$$\mathbb{NO}(\boldsymbol{A}_{\langle k \rangle}) := \{\, \boldsymbol{x} \in \mathbb{G}^1_{p,q} \;:\; \boldsymbol{x} \wedge \boldsymbol{A}_{\langle k \rangle} = 0 \,\}.$$

This definition implies that if $\boldsymbol{A}_{\langle k \rangle} := \bigwedge_{i=1}^{k} \boldsymbol{a}_i$ with $\{\, \boldsymbol{a}_i \,\} \subset \mathbb{G}^1_{p,q}$, then $\mathbb{NO}(\boldsymbol{A}_{\langle k \rangle}) = \operatorname{span}\{\, \boldsymbol{a}_i \,\}$. That is, $\mathbb{NO}(\boldsymbol{A}_{\langle k \rangle}) \subset \mathbb{R}^{p,q}$ is a linear subspace of

C. Perwass, *Geometric Algebra with Applications in Engineering.* Geometry and Computing.

$\mathbb{R}^{p,q}$, which may be interpreted as an infinitely extended geometric entity. This is indeed the simplest choice of representation that can be made. A point is an element of a vector space, and thus a vector space represents the geometric entity that consists of all points that have a representation in it. A blade can then be taken to represent the geometric entity that is represented by its OPNS or IPNS.

However, this representation is somewhat limiting, because only linear subspaces that include the origin can be represented. For example, using this representation, a vector $\boldsymbol{a} \in \mathbb{G}^1_{p,q}$ represents a line through the origin in the direction of $\boldsymbol{a}$, since

$$\mathbb{NO}(\boldsymbol{a}) := \{\, \boldsymbol{x} \in \mathbb{G}^1_{p,q} \;:\; \boldsymbol{x} \wedge \boldsymbol{a} = 0 \,\} = \{\, \alpha\, \boldsymbol{a} \;:\; \alpha \in \mathbb{R} \,\}.$$

More general representations can be achieved if the initial space in which geometry is to be represented is embedded in a higher-dimensional space in a possibly non-linear manner. A linear subspace in this embedding space may then represent a non-linear subspace in the initial space. This may be formulated mathematically as follows. Let $\mathbb{R}^{r,s}$ denote the initial vector space in which geometric entities are to be represented. A point is thus represented by a vector in $\mathbb{R}^{r,s}$. Now define a possibly non-linear embedding $\mathcal{X}$ of $\mathbb{R}^{r,s}$ into a higher-dimensional vector space $\mathbb{R}^{p,q}$ with $r + s \leq p + q$, such that $\mathcal{X}(\mathbb{R}^{r,s}) \subset \mathbb{R}^{p,q}$. That is, $\mathcal{X}(\mathbb{R}^{r,s})$ need not be a vector space anymore. For example, $\mathcal{X}$ could map the 1D Euclidean space $\mathbb{R}^1$ onto the unit circle in $\mathbb{R}^2$. In this case, not every vector in $\mathbb{R}^2$ can be mapped back to $\mathbb{R}^1$.

A blade in $\mathbb{G}_{p,q}$ may now represent the geometric entity that consists of those points in $\mathbb{R}^{r,s}$ whose embedding by use of $\mathcal{X}$ results in a vector in the OPNS or IPNS of the blade. This will be called the *geometric* OPNS or IPNS.

Definition 4.1. Let $\mathbb{R}^{r,s}$ denote the vector space in which geometric entities are to be represented, and denote by $\mathcal{X}$ a bijective mapping $\mathbb{G}^1_{r,s} \to \mathbb{X}$, where $\mathbb{X} \subset \mathbb{G}^1_{p,q}$, with $r + s \leq p + q$. Then the *geometric* outer- and inner-product null spaces (GOPNS and GIPNS, respectively) of a blade $\boldsymbol{A}_{\langle k \rangle} \in \mathbb{G}^k_{p,q}$ are defined as

$$\mathbb{NO}_G(\boldsymbol{A}_{\langle k \rangle}) := \{\, \boldsymbol{x} \in \mathbb{G}^1_{r,s} \;:\; \mathcal{X}(\boldsymbol{x}) \wedge \boldsymbol{A}_{\langle k \rangle} = 0 \,\},$$
$$\mathbb{NI}_G(\boldsymbol{A}_{\langle k \rangle}) := \{\, \boldsymbol{x} \in \mathbb{G}^1_{r,s} \;:\; \mathcal{X}(\boldsymbol{x}) \cdot \boldsymbol{A}_{\langle k \rangle} = 0 \,\}.$$

If $\mathcal{X}^{-1} : \mathbb{X} \to \mathbb{R}^{r,q}$ denotes the inverse of $\mathcal{X}$, then these geometric null spaces can also be written as

$$\mathbb{NO}_G(\boldsymbol{A}_{\langle k \rangle}) = \mathcal{X}^{-1}\big(\mathbb{X} \cap \mathbb{NO}(\boldsymbol{A}_{\langle k \rangle})\big) \quad \text{and} \quad \mathbb{NI}_G(\boldsymbol{A}_{\langle k \rangle}) = \mathcal{X}^{-1}\big(\mathbb{X} \cap \mathbb{NI}(\boldsymbol{A}_{\langle k \rangle})\big).$$

If $\mathcal{X}$ is the identity, then $\mathbb{NO}_G(\boldsymbol{A}_{\langle k \rangle}) = \mathbb{NO}(\boldsymbol{A}_{\langle k \rangle})$ and $\mathbb{NI}_G(\boldsymbol{A}_{\langle k \rangle}) = \mathbb{NI}(\boldsymbol{A}_{\langle k \rangle})$.

These geometric null spaces form the basis for the representation of geometry with geometric algebra. In the literature, many authors call the representation of a geometric entity through the GOPNS the "standard" representation and, accordingly, they call the representation through the GIPNS the "dual" representation. This stems from the fact that $\mathbb{NO}_G(\boldsymbol{A}_{\langle k \rangle}) = \mathbb{NI}_G(\boldsymbol{A}^*_{\langle k \rangle})$. The terms "standard" and "direct" can also be found in use for the two types of representation. The problem with these terminologies is that the names do not describe what they relate to. While the representations in $\mathbb{NO}_G$ and $\mathbb{NI}_G$ are dual *to each other*, there is no a priori preferred "standard" representation, since $\mathbb{NI}_G(\boldsymbol{A}_{\langle k \rangle}) = \mathbb{NO}_G(\boldsymbol{A}^*_{\langle k \rangle})$ also. Different authors do indeed define what is "standard" differently.

On the other hand, the terms "GOPNS" and "GIPNS" are somewhat cumbersome when used in spoken language. In this text, the two different representations are also referred to as the *outer-product* or OP representation, and the *inner-product* or IP representation. For example, if the GOPNS of a blade $\boldsymbol{A}_{\langle k \rangle}$ represents a line, then this is described as "$\boldsymbol{A}_{\langle k \rangle}$ is an outer-product representation of a line", or, even shorter, "$\boldsymbol{A}_{\langle k \rangle}$ is an outer-product line".

In the remainder of this chapter, this representation of geometry by geometric algebra is discussed in a number of settings, the most important of which is *conformal space*. The geometric algebra over this conformal space is used extensively in the application chapters later on. Note that the geometric algebras over conic space and conformal conic space were developed by the author and are published here for the first time in full detail.

4.1 Euclidean Space

In this section, the representation of geometry in Euclidean space is discussed. In terms of Definition 4.1, the space in which geometric entities are to be represented is $\mathbb{R}^n$ and the transformation $\mathcal{X}$ is the identity. Hence, the geometric inner- and outer-product null spaces of blades are equivalent to the algebraic inner- and outer-product null spaces.

In the following, the 3D Euclidean space $\mathbb{R}^3$ and the corresponding geometric algebra $\mathbb{G}_3$ are of particular interest. The algebraic basis of $\mathbb{G}_3$ is given in Table 4.1. Throughout this section, the symbols $\{ \boldsymbol{e}_i \}$ are used to denote the algebraic basis of $\mathbb{G}_3$, as shown in this table.

Table 4.1 Algebra basis of $\mathbb{G}_3$, where the geometric product of basis vectors $\boldsymbol{e}_i := \overline{\mathbb{G}}^1_3[i]$ is denoted by combining their indices, i.e. $\boldsymbol{e}_1 \boldsymbol{e}_2 \equiv \boldsymbol{e}_{12}$

Type	No.	Basis Elements
Scalar	1	1
1-Vector	3	$\boldsymbol{e}_1, \boldsymbol{e}_2, \boldsymbol{e}_3$
2-Vector	3	$\boldsymbol{e}_{12}, \boldsymbol{e}_{13}, \boldsymbol{e}_{23}$
3-Vector	1	$\boldsymbol{e}_{123}$

The remainder of this section is structured as follows. First, the geometric entities that can be represented in $\mathbb{G}_3$ are discussed. Then the effect of algebra operations on the geometric representations is looked at in more detail. This is followed by a presentation of the transformation operators that are available in $\mathbb{G}_3$.

4.1.1 Outer-Product Representations

The geometric entities of $\mathbb{R}^3$ that can be represented in $\mathbb{G}_3$ are only lines and planes through the origin. Consider first the GOPNSs of blades. Clearly, the GOPNS of a 1-vector $\boldsymbol{a} \in \mathbb{G}_3^1$ is

$$\mathbb{NO}_G(\boldsymbol{a}) = \{\, \boldsymbol{x} \in \mathbb{G}_3^1 \,:\, \boldsymbol{x} \wedge \boldsymbol{a} = 0 \,\} = \{\, \alpha\, \boldsymbol{a} \,:\, \alpha \in \mathbb{R} \,\}, \tag{4.1}$$

which is a line through the origin with its orientation given by $\boldsymbol{a}$. Similarly, given a second vector $\boldsymbol{b} \in \mathbb{G}_3^1$ with $\boldsymbol{a} \wedge \boldsymbol{b} \neq 0$,

$$\mathbb{NO}_G(\boldsymbol{a} \wedge \boldsymbol{b}) = \{\, \boldsymbol{x} \in \mathbb{G}_3^1 \,:\, \boldsymbol{x} \wedge \boldsymbol{a} \wedge \boldsymbol{b} = 0 \,\} = \{\, \alpha\, \boldsymbol{a} + \beta\, \boldsymbol{b} \,:\, (\alpha, \beta) \in \mathbb{R}^2 \,\}, \tag{4.2}$$

which is the plane spanned by $\boldsymbol{a}$ and $\boldsymbol{b}$. Note that points cannot be represented through null spaces in $\mathbb{G}_3$. This is only possible in projective spaces, which are discussed later on. The GOPNS of the outer product of three linearly independent vectors $\boldsymbol{a}, \boldsymbol{b}, \boldsymbol{c} \in \mathbb{G}_3^1$ is the whole space:

$$\begin{aligned} \mathbb{NO}_G(\boldsymbol{a} \wedge \boldsymbol{b} \wedge \boldsymbol{c}) &= \{\, \boldsymbol{x} \in \mathbb{G}_3^1 \,:\, \boldsymbol{x} \wedge \boldsymbol{a} \wedge \boldsymbol{b} \wedge \boldsymbol{c} = 0 \,\} \\ &= \{\, \alpha\, \boldsymbol{a} + \beta\, \boldsymbol{b} + \gamma\, \boldsymbol{c} \,:\, (\alpha, \beta, \gamma) \in \mathbb{R}^3 \,\}. \end{aligned} \tag{4.3}$$

Since blades of grade 3 are blades of the highest grade in $\mathbb{G}_3$, $\boldsymbol{a} \wedge \boldsymbol{b} \wedge \boldsymbol{c} \propto \boldsymbol{I}$, where $\boldsymbol{I} := \overline{\mathbb{G}}_3[8]$ is the unit pseudoscalar of $\mathbb{G}_3$. In Sect. 3.2.6, it was shown that the magnitude of $\boldsymbol{a} \wedge \boldsymbol{b} \wedge \boldsymbol{c}$ is the product of the magnitudes of a set of orthogonal vectors $\{\, \boldsymbol{a}', \boldsymbol{b}', \boldsymbol{c}' \,\}$, say, which is also equivalent to the determinant of a matrix where the $\{\, \boldsymbol{a}', \boldsymbol{b}', \boldsymbol{c}' \,\}$ form the rows or columns. Geometrically, this means that the magnitude of the blade $\boldsymbol{a} \wedge \boldsymbol{b} \wedge \boldsymbol{c}$ is the volume of the parallelepiped spanned by $\boldsymbol{a}$, $\boldsymbol{b}$, and $\boldsymbol{c}$. The pseudoscalar $\boldsymbol{I}$ may thus also be understood as the unit volume element.

Similarly, the magnitude of the blade $\boldsymbol{a} \wedge \boldsymbol{b}$ relates to the area of the parallelogram spanned by $\boldsymbol{a}$ and $\boldsymbol{b}$. To see this, we write $\boldsymbol{b} = \boldsymbol{b}^{\parallel} + \boldsymbol{b}^{\perp}$, where $\boldsymbol{b}^{\parallel} := \mathcal{P}_{\boldsymbol{a}}(\boldsymbol{b})$ and $\boldsymbol{b}^{\perp} := \mathcal{P}_{\boldsymbol{a}}^{\perp}(\boldsymbol{b})$. See Sects. 3.2.10 and 3.2.11 for the definitions of $\mathcal{P}_{\boldsymbol{a}}(\boldsymbol{b})$ and $\mathcal{P}_{\boldsymbol{a}}^{\perp}(\boldsymbol{b})$. Then $\boldsymbol{a} \wedge \boldsymbol{b} = \boldsymbol{a} \wedge \boldsymbol{b}^{\perp}$, because the outer product of two parallel vectors is zero. As discussed in Sect. 3.2.6, the norm of the blade is given by

$$\|\boldsymbol{a} \wedge \boldsymbol{b}\|^2 = (\boldsymbol{a} \wedge \boldsymbol{b})\,(\boldsymbol{b} \wedge \boldsymbol{a}) = \boldsymbol{a}\, \boldsymbol{b}^{\perp}\, \boldsymbol{b}^{\perp}\, \boldsymbol{a} = \boldsymbol{a}^2\, (\boldsymbol{b}^{\perp})^2.$$

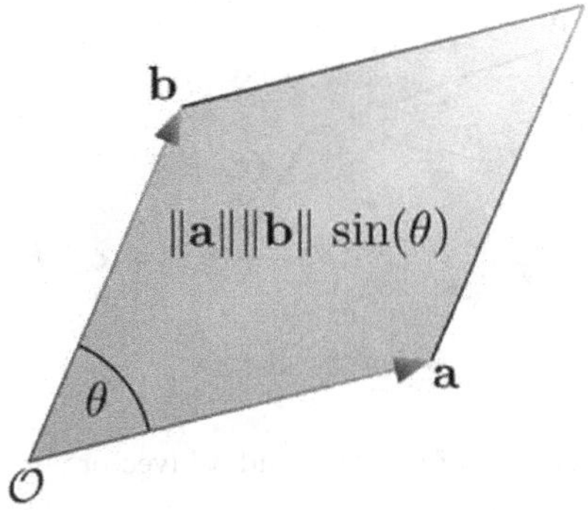

Fig. 4.1 Magnitude of blade $\boldsymbol{a} \wedge \boldsymbol{b}$ is area of parallelogram spanned by $\boldsymbol{a}$ and $\boldsymbol{b}$

If $\theta := \angle(\boldsymbol{a}, \boldsymbol{b})$ is the angle between $\boldsymbol{a}$ and $\boldsymbol{b}$, then $\|\boldsymbol{b}^{\perp}\| = \|\boldsymbol{b}\| \sin\theta$. Hence,

$$\|\boldsymbol{a} \wedge \boldsymbol{b}\| = \|\boldsymbol{a}\| \, \|\boldsymbol{b}\| \, \sin\theta,$$

which is the area of the parallelogram spanned by $\boldsymbol{a}$ and $\boldsymbol{b}$ (Fig. 4.1).

4.1.2 Geometric Interpretation of the Inner Product

The geometric interpretation of the effect of the inner product on blades is directly related to the effect of the inner product on the null spaces of the blades. Recall (3.85), which states that the OPNS of the inner product of two blades is the direct difference between the blades' OPNSs. That is, if $\boldsymbol{A}_{\langle k \rangle} \in \mathbb{G}^{\varnothing\, k}_{p,q}$ and $\boldsymbol{B}_{\langle l \rangle} \in \mathbb{G}^{\varnothing\, l}_{p,q}$ with $k \leq l$, then

$$\mathbb{NO}(\boldsymbol{A}_{\langle k \rangle} \cdot \boldsymbol{B}_{\langle l \rangle}) = \mathbb{NO}(\boldsymbol{B}_{\langle l \rangle}) \ominus \mathbb{NO}(\boldsymbol{A}_{\langle k \rangle}).$$

In particular, consider $\boldsymbol{x}, \boldsymbol{y}, \boldsymbol{a}, \boldsymbol{b} \in \mathbb{G}^1_3$ and let

$$\boldsymbol{y} := \boldsymbol{x} \cdot (\boldsymbol{a} \wedge \boldsymbol{b}) = (\boldsymbol{x} \cdot \boldsymbol{a})\, \boldsymbol{b} - (\boldsymbol{x} \cdot \boldsymbol{b})\, \boldsymbol{a}.$$

It follows that

$$\begin{aligned} \boldsymbol{x} \cdot \boldsymbol{y} &= \boldsymbol{x} \cdot \big[(\boldsymbol{x} \cdot \boldsymbol{a})\, \boldsymbol{b} - (\boldsymbol{x} \cdot \boldsymbol{b})\, \boldsymbol{a}\big] \\ &= (\boldsymbol{x} \cdot \boldsymbol{a})\,(\boldsymbol{x} \cdot \boldsymbol{b}) - (\boldsymbol{x} \cdot \boldsymbol{b})\,(\boldsymbol{x} \cdot \boldsymbol{a}) \\ &= 0. \end{aligned}$$

That is, $\boldsymbol{x}$ is perpendicular to $\boldsymbol{y}$, as expected, since the inner product $\boldsymbol{x} \cdot (\boldsymbol{a} \wedge \boldsymbol{b})$ has "subtracted" the OPNS of $\boldsymbol{x}$ from the OPNS of $\boldsymbol{a} \wedge \boldsymbol{b}$.

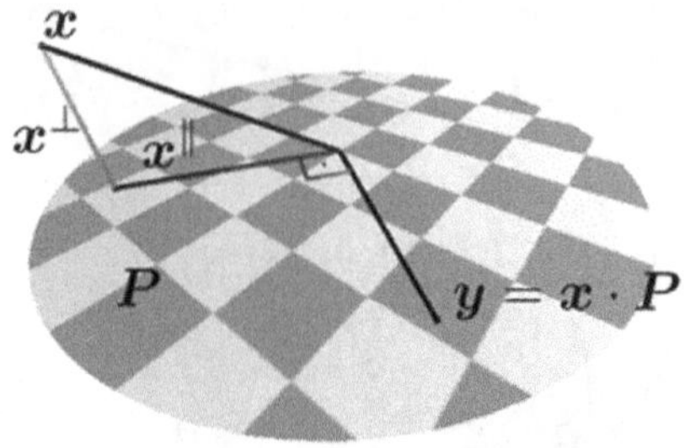

Fig. 4.2 Inner product of a vector and a bivector

This subspace subtraction is illustrated in Fig. 4.2. Here $\boldsymbol{P}$ denotes the bivector $\boldsymbol{a} \wedge \boldsymbol{b} \in \mathbb{G}_3^2$, which represents a plane through the origin in $\mathbb{R}^3$. A vector $\boldsymbol{x} \in \mathbb{R}^3$ will in general have a component parallel to $\boldsymbol{P}$, $\boldsymbol{x}^{\|}$, and a component perpendicular to $\boldsymbol{P}$, $\boldsymbol{x}^{\perp}$, such that $\boldsymbol{x} = \boldsymbol{x}^{\|} + \boldsymbol{x}^{\perp}$. Therefore,

$$\boldsymbol{y} := \boldsymbol{x} \cdot \boldsymbol{P} = (\boldsymbol{x}^{\|} + \boldsymbol{x}^{\perp}) \cdot \boldsymbol{P} = \boldsymbol{x}^{\|} \cdot \boldsymbol{P}.$$

The inner product $\boldsymbol{x}^{\|} \cdot \boldsymbol{P}$ now "subtracts" the subspace represented by $\boldsymbol{x}^{\|}$ from the subspace represented by $\boldsymbol{P}$, which results in a vector that lies in $\boldsymbol{P}$ and is perpendicular to $\boldsymbol{x}$, as shown in Fig. 4.2.

4.1.3 Inner Product Representation

In Sect. 3.2.8, it was shown that the OPNS and IPNS of a blade are related by the dual. Given vectors $\boldsymbol{a}, \boldsymbol{b} \in \mathbb{G}_3^1$, it is thus clear that

$$\mathbb{NO}_G(\boldsymbol{a}) = \mathbb{NI}_G(\boldsymbol{a}^*) \qquad \text{and} \qquad \mathbb{NO}_G(\boldsymbol{a} \wedge \boldsymbol{b}) = \mathbb{NI}_G\big((\boldsymbol{a} \wedge \boldsymbol{b})^*\big). \tag{4.4}$$

4.1.3.1 Planes

Consider first of all $\mathbb{NI}_G\big((\boldsymbol{a} \wedge \boldsymbol{b})^*\big)$. Equation (3.97) in Sect. 3.2.8 (p. 80) states that $\boldsymbol{a} \times \boldsymbol{b} = (\boldsymbol{a} \wedge \boldsymbol{b})^*$, where $\times$ denotes the standard vector cross product. Therefore, $\boldsymbol{n} := (\boldsymbol{a} \wedge \boldsymbol{b})^*$ is the normal of the plane represented by the OPNS of $\boldsymbol{a} \wedge \boldsymbol{b}$. This also follows from the fact that the OPNSs of a blade and its dual are complementary spaces.

A plane can thus be represented either by the OPNS of the outer product of two linearly independent vectors that lie in the plane or by the IPNS of a vector that is normal (i.e. perpendicular) to the plane. Representation of a plane by its normal is possible only in 3D Euclidean space $\mathbb{R}^3$, though. In $\mathbb{R}^2$ or $\mathbb{R}^n$ with $n > 3$, only the outer-product representation can be used.

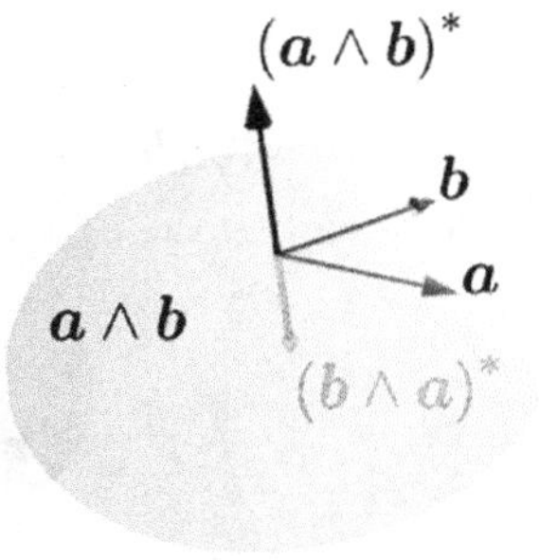

Fig. 4.3 Dual of a plane represented by a bivector $\boldsymbol{a} \wedge \boldsymbol{b}$

It may be shown by a straightforward calculation that $\boldsymbol{a}$, $\boldsymbol{b}$, and $\boldsymbol{n} := (\boldsymbol{a} \wedge \boldsymbol{b})^*$ form a right-handed coordinate system. Conversely, this construction may be used to define what is meant by a right-handed coordinate system in $\mathbb{R}^3$. The simplest way to explain this is to consider two vectors $\boldsymbol{a}$ and $\boldsymbol{b}$ that lie in the plane of this piece of paper. If the rotation that rotates $\boldsymbol{a}$ into $\boldsymbol{b}$ along the shortest path is anticlockwise, then the vector $(\boldsymbol{a} \wedge \boldsymbol{b})^*$ points out of the paper towards the reader. This is illustrated in Fig. 4.3.

Because $\boldsymbol{a} \wedge \boldsymbol{b} = -\boldsymbol{b} \wedge \boldsymbol{a}$, it is clear that $(\boldsymbol{b} \wedge \boldsymbol{a})^* = -(\boldsymbol{a} \wedge \boldsymbol{b})^*$, which means that the normal of the plane $\boldsymbol{b} \wedge \boldsymbol{a}$ points in the direction opposite to the normal of the plane $\boldsymbol{a} \wedge \boldsymbol{b}$. A bivector thus represents an *oriented plane*.

4.1.3.2 Lines

As shown in (4.4), an inner-product line is represented by the dual of a vector in $\mathbb{G}_3$,

$$\mathbb{NO}_G(\boldsymbol{a}) = \mathbb{NI}_G(\boldsymbol{a}^*).$$

The dual of a vector in $\mathbb{G}_3$ is clearly a bivector. Consider the bivector $\boldsymbol{n} \wedge \boldsymbol{m} \in \mathbb{G}_3^2$, with GIPNS

$$\mathbb{NI}_G(\boldsymbol{n} \wedge \boldsymbol{m}) = \left\{ \boldsymbol{x} \in \mathbb{G}_3^1 \; : \; \boldsymbol{x} \cdot (\boldsymbol{n} \wedge \boldsymbol{m}) = 0 \right\}.$$

With the help of (3.73), it follows that

$$\boldsymbol{x} \cdot (\boldsymbol{n} \wedge \boldsymbol{m}) = (\boldsymbol{x} \cdot \boldsymbol{n})\, \boldsymbol{m} - (\boldsymbol{x} \cdot \boldsymbol{m})\, \boldsymbol{n}.$$

Since $\boldsymbol{n} \wedge \boldsymbol{m} \neq 0$ by definition, $\boldsymbol{n}$ and $\boldsymbol{m}$ have to be linearly independent. Therefore, the above expression can become zero if and only if

$$\boldsymbol{x} \cdot \boldsymbol{n} = 0 \quad \text{and} \quad \boldsymbol{x} \cdot \boldsymbol{m} = 0.$$

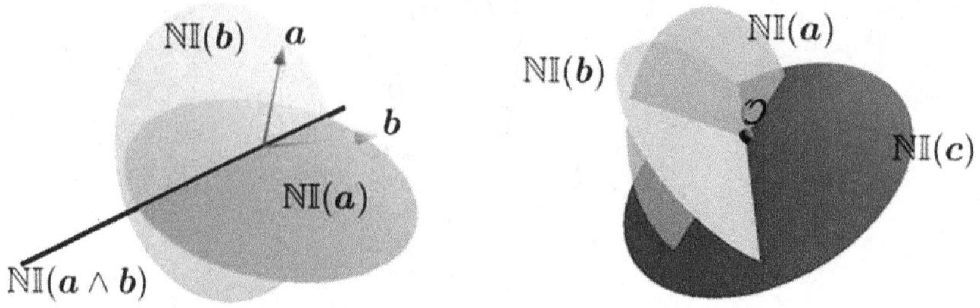

(a) Intersection of two planes (b) Intersection of three planes

Fig. 4.4 Intersection of inner-product planes

Geometrically, this means that $\boldsymbol{x}$ has to lie on the inner-product plane $\boldsymbol{n}$ *and* the inner product plane $\boldsymbol{m}$. Hence, $\boldsymbol{x}$ lies on the intersection of the two planes represented by $\boldsymbol{n}$ and $\boldsymbol{m}$. This shows that with respect to the GIPNS, the outer product of two vectors represents the intersection of their separately represented geometric entities. This is illustrated in Fig. 4.4(a).

In terms of sets, this reads

$$\mathbb{NI}_G(\boldsymbol{n} \wedge \boldsymbol{m}) = \mathbb{NI}_G(\boldsymbol{n}) \cap \mathbb{NI}_G(\boldsymbol{m}). \tag{4.5}$$

The same intersection may also be evaluated using the meet operation in the GOPNS representation. We define $\boldsymbol{n} := (\boldsymbol{a}_1 \wedge \boldsymbol{a}_2)^*$ and $\boldsymbol{m} := (\boldsymbol{b}_1 \wedge \boldsymbol{b}_2)^*$; then

$$\mathbb{NO}_G(\boldsymbol{a}_1 \wedge \boldsymbol{a}_2) = \mathbb{NI}_G(\boldsymbol{n}) \quad \text{and} \quad \mathbb{NO}_G(\boldsymbol{b}_1 \wedge \boldsymbol{b}_2) = \mathbb{NI}_G(\boldsymbol{m}).$$

Note that because in $\mathbb{G}_3$ the pseudoscalar squares to -1, $-\boldsymbol{n}^* = \boldsymbol{a}_1 \wedge \boldsymbol{a}_2$ and $-\boldsymbol{m}^* = \boldsymbol{b}_1 \wedge \boldsymbol{b}_2$. Furthermore, the join of $\boldsymbol{a}_1 \wedge \boldsymbol{a}_2$ and $\boldsymbol{b}_1 \wedge \boldsymbol{b}_2$ is the whole space, i.e. the pseudoscalar $\boldsymbol{I}$, if the blades are linearly independent. Therefore, using (3.104),

$$\begin{aligned}(\boldsymbol{a}_1 \wedge \boldsymbol{a}_2) \vee (\boldsymbol{b}_1 \wedge \boldsymbol{b}_2) &= (\boldsymbol{a}_1 \wedge \boldsymbol{a}_2)^* \cdot (\boldsymbol{b}_1 \wedge \boldsymbol{b}_2) \\ &= -\boldsymbol{n} \cdot \boldsymbol{m}^* \\ &= -(\boldsymbol{n} \wedge \boldsymbol{m})^*,\end{aligned} \tag{4.6}$$

where the last equality follows as in (3.79). Thus,

$$\mathbb{NO}_G\big((\boldsymbol{a}_1 \wedge \boldsymbol{a}_2) \vee (\boldsymbol{b}_1 \wedge \boldsymbol{b}_2)\big) = \mathbb{NI}_G(\boldsymbol{n} \wedge \boldsymbol{m}).$$

The outer product can only be used in this way for the intersection of geometric entities represented by their GIPNS if the join of the geometric entities is the whole space. In this case the meet becomes the regressive product, which shows again that

$$\mathbb{NO}_G(\boldsymbol{n}^* \triangledown \boldsymbol{m}^*) = \mathbb{NI}_G(\boldsymbol{n} \wedge \boldsymbol{m}).$$

The intersection line also has an orientation, which in this case is given by the vector $-(\boldsymbol{n} \wedge \boldsymbol{m})^*$.

Note that in $\mathbb{R}^3$, two parallel but not identical planes cannot be represented by null spaces, since all planes that can be represented in this way pass through the origin.

4.1.3.3 Points

The last type of blade whose GIPNS representation can be discussed in $\mathbb{R}^3$ is a 3-blade, or *trivector*. It has been shown earlier that a trivector $\boldsymbol{A}_{\langle 3 \rangle} \in \mathbb{G}_3$ is a pseudoscalar and thus $\boldsymbol{A}_{\langle 3 \rangle} = \|\boldsymbol{A}_{\langle 3 \rangle}\| \, \boldsymbol{I}$, where $\boldsymbol{I}$ is the unit pseudoscalar of $\mathbb{G}_3$. Suppose $\boldsymbol{A}_{\langle 3 \rangle}$ is defined as

$$\boldsymbol{A}_{\langle 3 \rangle} := \boldsymbol{a} \wedge \boldsymbol{b} \wedge \boldsymbol{c}.$$

If $\boldsymbol{A}_{\langle 3 \rangle} \neq 0$, then $\boldsymbol{a}$, $\boldsymbol{b}$, and $\boldsymbol{c}$ are linearly independent. In order to find the GIPNS of $\boldsymbol{A}_{\langle 3 \rangle}$, the set of vectors $\{\, \boldsymbol{x} \,\}$ that satisfies $\boldsymbol{x} \cdot \boldsymbol{A}_{\langle 3 \rangle} = 0$ has to be found. Using (3.73) again, it follows that

$$\begin{aligned} \boldsymbol{x} \cdot \boldsymbol{A}_{\langle 3 \rangle} = \quad & (\boldsymbol{x} \cdot \boldsymbol{a})\, (\boldsymbol{b} \wedge \boldsymbol{c}) \\ & - (\boldsymbol{x} \cdot \boldsymbol{b})\, (\boldsymbol{a} \wedge \boldsymbol{c}) \\ & + (\boldsymbol{x} \cdot \boldsymbol{c})\, (\boldsymbol{a} \wedge \boldsymbol{b}). \end{aligned}$$

The bivectors $(\boldsymbol{b} \wedge \boldsymbol{c})$, $(\boldsymbol{a} \wedge \boldsymbol{c})$, and $(\boldsymbol{a} \wedge \boldsymbol{b})$ are linearly independent, and thus $\boldsymbol{x} \cdot \boldsymbol{A}_{\langle 3 \rangle} = 0$ if and only if

$$\boldsymbol{x} \cdot \boldsymbol{a} = 0 \quad \text{and} \quad \boldsymbol{x} \cdot \boldsymbol{b} = 0 \quad \text{and} \quad \boldsymbol{x} \cdot \boldsymbol{c} = 0.$$

Geometrically, this means that $\boldsymbol{x} \cdot \boldsymbol{A}_{\langle 3 \rangle} = 0$ if and only if $\boldsymbol{x}$ lies on the intersection of the three inner-product planes represented by $\boldsymbol{a}$, $\boldsymbol{b}$, and $\boldsymbol{c}$. Since all planes represented through the GIPNSs of vectors pass through the origin, the only point that all three planes can meet in is the origin. Hence, the only solution for $\boldsymbol{x}$ that satisfies $\boldsymbol{x} \cdot \boldsymbol{A}_{\langle 3 \rangle} = 0$ is the trivial solution $\boldsymbol{x} = \boldsymbol{0} \in \mathbb{R}^3$, as shown in Fig. 4.4(b).

4.1.4 Reflections

So far, it has been shown how to construct linear subspaces using the outer product and to subtract linear subspaces from one another using the inner product. It is also clear how to intersect linear subspaces using the meet and

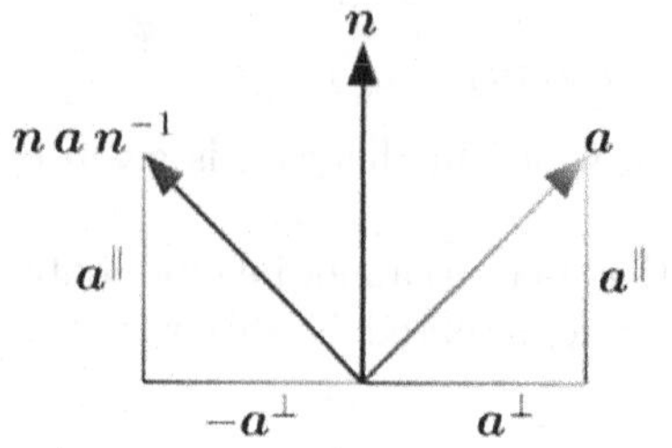

(a) Reflection of vector $\boldsymbol{a}$ on vector $\boldsymbol{n}$.

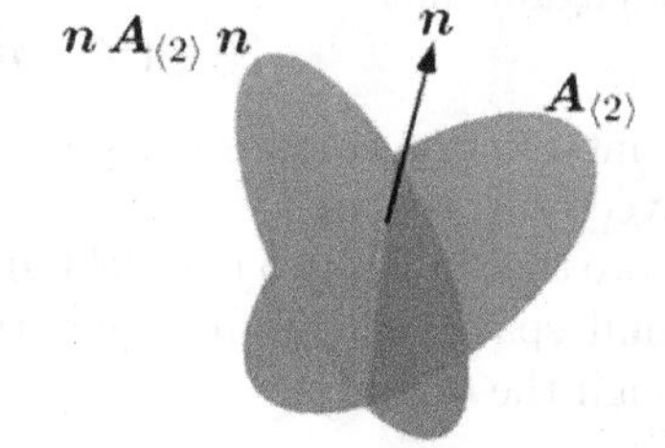

(b) Reflection of bivector $\boldsymbol{A}_{\langle 2\rangle}$ on vector $\boldsymbol{n}$.

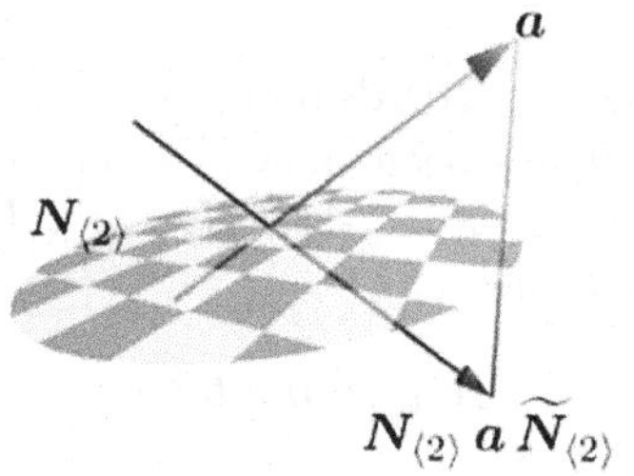

(c) Reflection of vector $\boldsymbol{a}$ on bivector $\boldsymbol{N}_{\langle 2\rangle}$.

Fig. 4.5 Examples of reflection operations

how to form their union with the join. Now, operations on subspaces which keep the dimensionality unchanged, will be discussed. For example, rotating a line results in another line, not a point or a plane. Recall that an operation on a blade that does not change its grade is called *grade-preserving.* In Lemma 3.22 it was shown that versors are grade-preserving operators.

In Sect. 3.3, (3.121) presented the fundamental versor operation as

$$\boldsymbol{n}\,\boldsymbol{a}\,\boldsymbol{n}^{-1} = \mathcal{P}_{\hat{\boldsymbol{n}}}(\boldsymbol{a}) - \mathcal{P}^{\perp}_{\hat{\boldsymbol{n}}}(\boldsymbol{a})\,,$$

where $\boldsymbol{n}, \boldsymbol{a} \in \mathbb{G}^{\varnothing\,1}_{p,q}$ and $\hat{\boldsymbol{n}} := \boldsymbol{n}/\|\boldsymbol{n}\|$. Since the component of $\boldsymbol{a}$ perpendicular to $\boldsymbol{n}$ is negated while the component of $\boldsymbol{a}$ parallel to $\boldsymbol{n}$ is left unchanged, this operation represents a reflection of $\boldsymbol{a}$ in the outer-product line represented by $\boldsymbol{n}$. Figure 4.5(a) shows an illustration of this type of reflection.

This type of reflection operation is in fact the only transformation that will be used in this text. However, depending on the embedding of the initial space, a reflection operation in the embedding space may represent a completely different transformation in the initial space. This will become apparent in particular in the discussion of conformal space in Sect. 4.3.

Because of the outermorphism property of versors, such a reflection may be applied to any blade. For example, if a plane is given as a bivector $\boldsymbol{A}_{\langle 2\rangle} \in \mathbb{G}_3$, it can be reflected in a unit vector $\hat{\boldsymbol{n}} \in \mathbb{G}^1_3$ simply by evaluating $\hat{\boldsymbol{n}}\,\boldsymbol{A}_{\langle 2\rangle}\,\hat{\boldsymbol{n}}$.

That is, if $\boldsymbol{A}_{\langle 2\rangle} = \boldsymbol{a}_1 \wedge \boldsymbol{a}_2$ with $\boldsymbol{a}_1, \boldsymbol{a}_2 \in \mathbb{G}^1_3$, then

$$\hat{\boldsymbol{n}}\, \boldsymbol{A}_{\langle 2\rangle}\, \hat{\boldsymbol{n}} = \big(\hat{\boldsymbol{n}}\, \boldsymbol{a}_1\, \hat{\boldsymbol{n}}\big) \;\wedge\; \big(\hat{\boldsymbol{n}}\, \boldsymbol{a}_2\, \hat{\boldsymbol{n}}\big).$$

This type of reflection is visualized in Fig. 4.5(b).

A blade may also be reflected in another blade, which follows directly from the fact that a blade is a special case of a versor (see Sect. 3.3.1). Fig. 4.5(c) shows the reflection of a vector $\boldsymbol{a} \in \mathbb{G}^1_3$ in a unit bivector $\boldsymbol{N}_{\langle 2\rangle} \in \mathbb{G}^2_3$ obtained by evaluating $-\boldsymbol{N}_{\langle 2\rangle}\boldsymbol{a}\widetilde{\boldsymbol{N}}_{\langle 2\rangle}$. This operation again results in

$$-\boldsymbol{N}_{\langle 2\rangle}\, \boldsymbol{a}\, \widetilde{\boldsymbol{N}}_{\langle 2\rangle} = \boldsymbol{a}^{\parallel} - \boldsymbol{a}^{\perp},$$

where $\boldsymbol{a}^{\parallel} := \mathcal{P}_{\boldsymbol{N}_{\langle 2\rangle}}(\boldsymbol{a})$ and $\boldsymbol{a}^{\perp} := \mathcal{P}^{\perp}_{\boldsymbol{N}_{\langle 2\rangle}}(\boldsymbol{a})$ are, respectively, this time, the parallel and perpendicular components of $\boldsymbol{a}$ with respect to $\boldsymbol{N}_{\langle 2\rangle}$. The general formula for the reflection of a vector $\boldsymbol{a} \in \mathbb{G}^1_n$ in a blade $\boldsymbol{N}_{\langle k\rangle} \in \mathbb{G}^k_n$ is given by

$$(-1)^{k+1}\, \boldsymbol{N}_{\langle k\rangle}\, \boldsymbol{a}\, \boldsymbol{N}^{-1}_{\langle k\rangle} = \mathcal{P}_{\boldsymbol{N}_{\langle k\rangle}}(\boldsymbol{a}) - \mathcal{P}^{\perp}_{\boldsymbol{N}_{\langle k\rangle}}(\boldsymbol{a})\,. \tag{4.7}$$

Later on, it will be useful to have an explicit notation to represent the reflection operator that reflects one vector into another, including a possible scaling. This idea is captured in the following lemma.

Lemma 4.1. *Let $\boldsymbol{x}, \boldsymbol{y} \in \mathbb{G}^1_n$ with $n \geq 2$; then the vector that reflects $\boldsymbol{x}$ into $\boldsymbol{y}$, which we denote by $\mathsf{ref}(\boldsymbol{x}, \boldsymbol{y})$, is given by*

$$\mathsf{ref}(\boldsymbol{x}, \boldsymbol{y}) := \sqrt{\frac{\|\boldsymbol{y}\|}{\|\boldsymbol{x}\|}}\, \frac{\hat{\boldsymbol{x}} + \hat{\boldsymbol{y}}}{\|\hat{\boldsymbol{x}} + \hat{\boldsymbol{y}}\|}\,,$$

where $\hat{\boldsymbol{x}} := \boldsymbol{x}/\|\boldsymbol{x}\|$ and $\hat{\boldsymbol{y}} := \boldsymbol{y}/\|\boldsymbol{y}\|$.

Proof. First of all, it is clear that $(\hat{\boldsymbol{x}} + \hat{\boldsymbol{y}})/\|\hat{\boldsymbol{x}} + \hat{\boldsymbol{y}}\|$ is the unit vector bisecting $\boldsymbol{x}$ and $\boldsymbol{y}$. This implies right away that the proposition should be true. However, this can also be shown algebraically. It has to be shown that

$$\begin{aligned} \boldsymbol{y} &= \mathsf{ref}(\boldsymbol{x}, \boldsymbol{y})\, \boldsymbol{x}\, \mathsf{ref}(\boldsymbol{x}, \boldsymbol{y}) \\ &= \frac{\|\boldsymbol{y}\|}{\|\boldsymbol{x}\|}\, \frac{\|\boldsymbol{x}\|}{\|\hat{\boldsymbol{x}} + \hat{\boldsymbol{y}}\|^2}\, (\hat{\boldsymbol{x}} + \hat{\boldsymbol{y}})\, \hat{\boldsymbol{x}}\, (\hat{\boldsymbol{x}} + \hat{\boldsymbol{y}}). \end{aligned} \tag{4.8}$$

Consider first the second part:

$$\begin{aligned} (\hat{\boldsymbol{x}} + \hat{\boldsymbol{y}})\, \hat{\boldsymbol{x}}\, (\hat{\boldsymbol{x}} + \hat{\boldsymbol{y}}) &= \hat{\boldsymbol{x}} + \hat{\boldsymbol{y}}\, \hat{\boldsymbol{x}}\, \hat{\boldsymbol{y}} + 2\, \hat{\boldsymbol{y}} \\ &= \hat{\boldsymbol{x}} + \hat{\boldsymbol{x}}^{\parallel} - \hat{\boldsymbol{x}}^{\perp} + 2\hat{\boldsymbol{y}} \\ &= 2\, (\hat{\boldsymbol{x}}^{\parallel} + \hat{\boldsymbol{y}}) \\ &= 2\, (\hat{\boldsymbol{x}} \cdot \hat{\boldsymbol{y}} + 1)\, \hat{\boldsymbol{y}}, \end{aligned}$$

where $\hat{\boldsymbol{x}}^{\parallel} := \mathcal{P}_{\hat{\boldsymbol{y}}}(\hat{\boldsymbol{x}})$ and $\hat{\boldsymbol{x}}^{\perp} := \mathcal{P}^{\perp}_{\hat{\boldsymbol{y}}}(\hat{\boldsymbol{x}})$. The second step follows from the properties of the reflection operation, that is, $\hat{\boldsymbol{y}}\, \hat{\boldsymbol{x}}\, \hat{\boldsymbol{y}} = \hat{\boldsymbol{x}}^{\parallel} - \hat{\boldsymbol{x}}^{\perp}$. Furthermore,

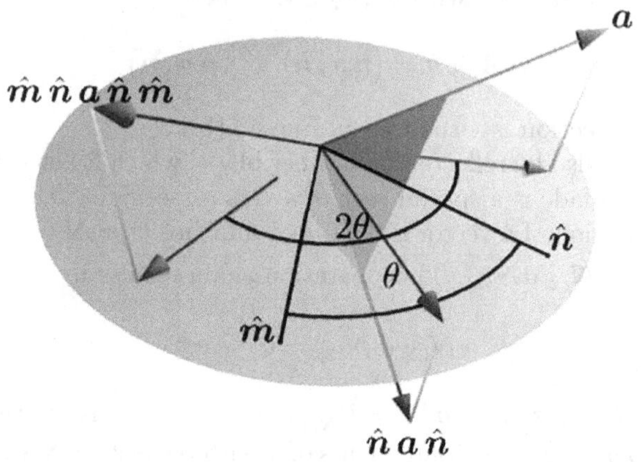

Fig. 4.6 Rotation of vector $\boldsymbol{a}$ by consecutive reflections of $\boldsymbol{a}$ in $\boldsymbol{n}$ and $\boldsymbol{m}$

$\hat{\boldsymbol{x}} \cdot \hat{\boldsymbol{y}} = \|\hat{\boldsymbol{x}}^{\|}\|$ by the definition of the inner product, and thus $\hat{\boldsymbol{x}}^{\|} = (\hat{\boldsymbol{x}} \cdot \hat{\boldsymbol{y}})\,\hat{\boldsymbol{y}}$. It is also not difficult to show that $\|\hat{\boldsymbol{x}} + \hat{\boldsymbol{y}}\|^2 = 2\,(1 + \hat{\boldsymbol{x}} \cdot \hat{\boldsymbol{y}})$. Substituting these results into (4.8) then proves the proposition.
□

As mentioned before, reflection is the most fundamental transformation in Geometric algebra. Most other transformations are obtained by combining a number of different reflections. In Euclidean space, this confines the set of transformations to reflections and rotations about axes that pass through the origin, as is shown in Sect. 4.1.5. To extend the set of available transformations, Euclidean space has to be embedded in other spaces, which is discussed later on.

4.1.5 Rotations

Reflections with respect to a normalized vector $\hat{\boldsymbol{n}} \in \mathbb{G}^1_n$ are always reflections in a line with direction $\hat{\boldsymbol{n}}$, passing through the origin. It may be shown that two consecutive reflections in different normalized vectors $\hat{\boldsymbol{n}}$ and $\hat{\boldsymbol{m}} \in \mathbb{G}^1_n$ are equivalent to a rotation by twice the angle between $\hat{\boldsymbol{n}}$ and $\hat{\boldsymbol{m}}$.

Figure 4.6 shows such a setup in 3D Euclidean space $\mathbb{R}^3$. The normalized vectors $\hat{\boldsymbol{n}}, \hat{\boldsymbol{m}} \in \mathbb{G}^1_3$ enclose an angle $\angle(\hat{\boldsymbol{n}}, \hat{\boldsymbol{m}}) = \theta$ and define a rotation plane through their outer product $\hat{\boldsymbol{n}} \wedge \hat{\boldsymbol{m}}$. Reflecting a vector $\boldsymbol{a} \in \mathbb{G}^1_3$ first in $\hat{\boldsymbol{n}}$ and then in $\hat{\boldsymbol{m}}$ rotates the component of $\boldsymbol{a}$ that lies in the rotation plane by 2θ. The component of $\boldsymbol{a}$ perpendicular to the rotation plane remains unchanged.

The rotation of the vector $\boldsymbol{a}$ in the plane $\hat{\boldsymbol{n}} \wedge \hat{\boldsymbol{m}}$ by an angle 2θ may then be written as

$$b := \hat{\boldsymbol{m}}\,\hat{\boldsymbol{n}}\,\boldsymbol{a}\,\hat{\boldsymbol{n}}\,\hat{\boldsymbol{m}}. \tag{4.9}$$

Defining $\boldsymbol{R} := \hat{\boldsymbol{m}}\,\hat{\boldsymbol{n}}$, and noting that $\widetilde{\boldsymbol{R}} = \hat{\boldsymbol{n}}\,\hat{\boldsymbol{m}}$ by definition, this equation becomes $\boldsymbol{b} = \boldsymbol{R}\,\boldsymbol{a}\,\widetilde{\boldsymbol{R}}$. The multivector $\boldsymbol{R}$ is usually called a *rotor*, because it has the effect of rotating blades. In order for $\boldsymbol{R}$ to be a proper rotor, it must not scale the entities that it rotates. Hence, it is demanded that $\boldsymbol{R}\,\widetilde{\boldsymbol{R}} = 1$. In the current example, this is clearly the case, since $\boldsymbol{R}\,\widetilde{\boldsymbol{R}} = \hat{\boldsymbol{m}}\,\hat{\boldsymbol{n}}\,\hat{\boldsymbol{n}}\,\hat{\boldsymbol{m}} = \hat{\boldsymbol{n}}^2\,\hat{\boldsymbol{m}}^2$. We now provide a formal proof that a rotor such as $\boldsymbol{R}$ does indeed rotate entities.

Lemma 4.2 (Rotor). *A rotation operator in geometric algebra is called a* rotor. *A rotor that rotates by an angle θ in a mathematically positive sense, in the plane spanned by unit vectors $\hat{\boldsymbol{n}}, \hat{\boldsymbol{m}} \in \mathbb{G}^1_n$, is denoted by* $\mathsf{rot}(\theta,\, \hat{\boldsymbol{n}} \wedge \hat{\boldsymbol{m}})$ *and is given by*

$$\mathsf{rot}(\theta,\, \hat{\boldsymbol{n}} \wedge \hat{\boldsymbol{m}}) = \cos\theta/2 - \sin\theta/2\, \frac{\hat{\boldsymbol{n}} \wedge \hat{\boldsymbol{m}}}{\|\hat{\boldsymbol{n}} \wedge \hat{\boldsymbol{m}}\|}.$$

Proof. Let the angle between $\hat{\boldsymbol{n}}$ and $\hat{\boldsymbol{m}}$ be half the rotation angle θ, i.e. $\angle(\hat{\boldsymbol{n}}, \hat{\boldsymbol{m}}) = \theta/2$, and let $\boldsymbol{N}_{\langle 2\rangle} := (\hat{\boldsymbol{n}} \wedge \hat{\boldsymbol{m}})/\|\hat{\boldsymbol{n}} \wedge \hat{\boldsymbol{m}}\|$ be the unit bivector representing the rotation plane. Also, let $\boldsymbol{R} := \hat{\boldsymbol{m}}\,\hat{\boldsymbol{n}}$; then

$$\boldsymbol{R} = \hat{\boldsymbol{m}} \cdot \hat{\boldsymbol{n}} + \hat{\boldsymbol{m}} \wedge \hat{\boldsymbol{n}} = \cos\theta/2 - \sin\theta/2\, \boldsymbol{N}_{\langle 2\rangle},$$

and, accordingly, $\widetilde{\boldsymbol{R}} = \cos\theta/2 + \sin\theta/2\, \boldsymbol{N}_{\langle 2\rangle}$. Now consider a vector $\boldsymbol{x} \in \mathbb{G}^1_n$ and define $\boldsymbol{x}^{\parallel} := \mathcal{P}_{\boldsymbol{N}_{\langle 2\rangle}}(\boldsymbol{x})$ and $\boldsymbol{x}^{\perp} := \mathcal{P}^{\perp}_{\boldsymbol{N}_{\langle 2\rangle}}(\boldsymbol{x})$, such that $\boldsymbol{x} \cdot \boldsymbol{N}_{\langle 2\rangle} = \boldsymbol{x}^{\parallel} \cdot \boldsymbol{N}_{\langle 2\rangle}$ and $\boldsymbol{x} \wedge \boldsymbol{N}_{\langle 2\rangle} = \boldsymbol{x}^{\perp} \wedge \boldsymbol{N}_{\langle 2\rangle}$. From the properties of the inner and the outer product, it follows that $\boldsymbol{N}_{\langle 2\rangle}\, \boldsymbol{x}^{\parallel} = -\boldsymbol{x}^{\parallel}\, \boldsymbol{N}_{\langle 2\rangle}$ and $\boldsymbol{N}_{\langle 2\rangle}\, \boldsymbol{x}^{\perp} = \boldsymbol{x}^{\perp}\, \boldsymbol{N}_{\langle 2\rangle}$, and thus

$$\boldsymbol{N}_{\langle 2\rangle}\, \boldsymbol{x} = (-\boldsymbol{x}^{\parallel} + \boldsymbol{x}^{\perp})\, \boldsymbol{N}_{\langle 2\rangle}.$$

Therefore,

$$\begin{aligned} \boldsymbol{R}\boldsymbol{x} &= (\cos\theta/2 - \sin\theta/2\, \boldsymbol{N}_{\langle 2\rangle})\, (\boldsymbol{x}^{\parallel} + \boldsymbol{x}^{\perp}) \\ &= \boldsymbol{x}^{\parallel}\, (\cos\theta/2 + \sin\theta/2\, \boldsymbol{N}_{\langle 2\rangle}) + \boldsymbol{x}^{\perp}\, (\cos\theta/2 - \sin\theta/2\, \boldsymbol{N}_{\langle 2\rangle}) \\ &= \boldsymbol{x}^{\parallel}\, \widetilde{\boldsymbol{R}} + \boldsymbol{x}^{\perp}\, \boldsymbol{R}. \end{aligned}$$

Hence,

$$\boldsymbol{R}\boldsymbol{x}\,\widetilde{\boldsymbol{R}} = \boldsymbol{x}^{\parallel}\, \widetilde{\boldsymbol{R}}^2 + \boldsymbol{x}^{\perp},$$

which shows, as expected, that the component of $\boldsymbol{x}$ that is perpendicular to the rotation plane is not changed. Now,

$$\widetilde{\boldsymbol{R}}^2 = \cos^2\theta/2 - \sin^2\theta/2 + 2\,\cos\theta/2\,\sin\theta/2\, \boldsymbol{N}_{\langle 2\rangle} = \cos\theta + \sin\theta\, \boldsymbol{N}_{\langle 2\rangle},$$

where $\boldsymbol{N}^2_{\langle 2\rangle} = -1$ and standard trigonometric identities have been used. Therefore,

$$\boldsymbol{R}\boldsymbol{x}\,\widetilde{\boldsymbol{R}} = \cos\theta\, \boldsymbol{x}^{\parallel} + \sin\theta\, \boldsymbol{x}^{\parallel} \cdot \boldsymbol{N}_{\langle 2\rangle} + \boldsymbol{x}^{\perp}.$$

In $\mathbb{G}_3$, the normal $\hat{\boldsymbol{r}}$ of the rotation plane $\boldsymbol{N}_{\langle 2\rangle}$, i.e. the rotation axis, is given by $\hat{\boldsymbol{r}} = \boldsymbol{N}^*_{\langle 2\rangle}$. Note that it makes sense to talk about a rotation axis in $\mathbb{R}^3$ only. In higher-dimensional spaces, rotation planes have to be used directly. If

$\boldsymbol{I}$ denotes the pseudoscalar of $\mathbb{G}_3$, then $\boldsymbol{N}_{\langle 2\rangle} = \hat{\boldsymbol{r}}\,\boldsymbol{I}$. Furthermore,

$$\boldsymbol{x}^{\parallel} \cdot \boldsymbol{N}_{\langle 2\rangle} = \boldsymbol{x}^{\parallel} \cdot (\hat{\boldsymbol{r}}\,\boldsymbol{I}) = (\boldsymbol{x}^{\parallel} \wedge \hat{\boldsymbol{r}})\,\boldsymbol{I} = -(\boldsymbol{x}^{\parallel} \wedge \hat{\boldsymbol{r}})^{*} = -\boldsymbol{x}^{\parallel} \times \hat{\boldsymbol{r}}.$$

It therefore follows that in $\mathbb{G}_3$,

$$\boldsymbol{R}\,\boldsymbol{x}\,\widetilde{\boldsymbol{R}} = \cos\theta\,\boldsymbol{x}^{\parallel} - \sin\theta\,(\boldsymbol{x}^{\parallel} \times \hat{\boldsymbol{r}}) + \boldsymbol{x}^{\perp}.$$

Since the vectors $\boldsymbol{x}^{\parallel}$, $\hat{\boldsymbol{r}}$, and $\boldsymbol{x}^{\parallel} \times \hat{\boldsymbol{r}}$ form an orthogonal coordinate system of $\mathbb{R}^3$ in a right-handed sense, $\boldsymbol{R}\,\boldsymbol{x}\,\widetilde{\boldsymbol{R}}$ rotates $\boldsymbol{x}$ about $\hat{\boldsymbol{r}}$ anticlockwise by an angle θ.

□

Note that in $\mathbb{G}_3$, a rotor may also be defined using the rotation axis instead of the rotation plane. If $\hat{\boldsymbol{r}}$ denotes the rotation axis, then the rotor that rotates about $\hat{\boldsymbol{r}}$ by an angle θ is written as $\mathsf{rot}(\theta, \hat{\boldsymbol{r}})$.

In the above proof, the fact that $\boldsymbol{N}^2_{\langle 2\rangle} = -1$ was used. This stems from the definition of $\boldsymbol{N}_{\langle 2\rangle}$ as a unit blade; that is, $\boldsymbol{N}_{\langle 2\rangle} \star \boldsymbol{N}_{\langle 2\rangle} = 1$. However, since $\boldsymbol{N}_{\langle 2\rangle} \in \mathbb{G}^2_n$, it cannot be a null-blade, and thus $\boldsymbol{N}_{\langle 2\rangle} \star \boldsymbol{N}_{\langle 2\rangle} = \boldsymbol{N}_{\langle 2\rangle} \cdot \widetilde{\boldsymbol{N}}_{\langle 2\rangle}$. From the definition of the reverse, it follows that $\widetilde{\boldsymbol{N}}_{\langle 2\rangle} = -\boldsymbol{N}_{\langle 2\rangle}$ and thus $\boldsymbol{N}^2_{\langle 2\rangle} = -\boldsymbol{N}_{\langle 2\rangle}\,\widetilde{\boldsymbol{N}}_{\langle 2\rangle} = -1$.

Since $\boldsymbol{N}_{\langle 2\rangle}$ squares to -1, the expression for a rotor in Lemma 4.2 is similar to that of a complex number z in the polar representation

$$z = r\,(\cos\theta + \mathrm{i}\,\sin\theta),$$

where $\mathrm{i} := \sqrt{-1}$ represents the imaginary unit and $r \in \mathbb{R}$ is the radius. For complex numbers, it is well known that the above expression can also be written as

$$z = r\,\exp(\mathrm{i}\,\theta).$$

The definition of the exponential function can be extended to geometric algebra, and it can be shown that the Taylor series of $\exp(\theta\,\boldsymbol{N}_{\langle 2\rangle})$ does indeed converge to

$$\exp\big(\theta\,\boldsymbol{N}_{\langle 2\rangle}\big) = \cos\theta + \sin\theta\,\boldsymbol{N}_{\langle 2\rangle}. \tag{4.10}$$

Therefore, a rotor that rotates in an outer-product plane $\boldsymbol{N}_{\langle 2\rangle} \in \mathbb{G}^2_n$ by an angle θ can be expressed in polar form as

$$\mathsf{rot}(\theta, \boldsymbol{N}_{\langle 2\rangle}) = \exp\big(-\theta/2\,\boldsymbol{N}_{\langle 2\rangle}\big). \tag{4.11}$$

Note that this offers a very concise and clear form for defining a rotor in any dimension $n \geq 2$.

Since a rotor is a versor and versors satisfy an outermorphism, a rotor can rotate any blade. That is, with the same rotor it is possible to rotate vectors, bivectors, etc., and even other versors. For blades, this means that given a blade $\boldsymbol{A}_{\langle k\rangle} = \bigwedge_{i=1}^{k} \boldsymbol{a}_i \in \mathbb{G}^k_n$, with $\{\boldsymbol{a}_i\} \subset \mathbb{G}^1_n$, and a rotor $\boldsymbol{R}$, the expression

$\boldsymbol{R} \boldsymbol{A}_{\langle k \rangle} \widetilde{\boldsymbol{R}}$ can be expanded as

$$\boldsymbol{R} \, \boldsymbol{A}_{\langle k \rangle} \, \widetilde{\boldsymbol{R}} = \big(\boldsymbol{R} \, \boldsymbol{a}_1 \, \widetilde{\boldsymbol{R}}\big) \wedge \big(\boldsymbol{R} \, \boldsymbol{a}_2 \, \widetilde{\boldsymbol{R}}\big) \wedge \ldots \wedge \big(\boldsymbol{R} \, \boldsymbol{a}_k \, \widetilde{\boldsymbol{R}}\big). \tag{4.12}$$

Hence, the rotation of the outer product of a number of vectors is the same as the outer product of a number of rotated vectors.

4.1.6 Mean Rotor

A standard problem is to evaluate the mean of a set of rotors. While it is straight forward to evaluate the mean of a set of blades in $\mathbb{G}_3$, simply by summing them and dividing by the number of blades, it is not obvious that this is also the correct approach for evaluating the mean of a set of rotors in $\mathbb{G}_3$. The problem here is that the space of rotors in $\mathbb{G}_3$, i.e. $\mathbb{G}_3^+$, is isomorphic to $\mathbb{S}^4$. Hence, the sum of two rotors does not in general result in a rotor. The appropriate operation for evaluating the "sum" of rotors is to take their geometric product. Similarly, since rotors are unitary versors, the appropriate distance measure between rotors $\boldsymbol{R}_1$ and $\boldsymbol{R}_2$ is $\boldsymbol{R}_{12} = \widetilde{\boldsymbol{R}}_1 \, \boldsymbol{R}_2$, such that $\boldsymbol{R}_1 \, \boldsymbol{R}_{12} = \boldsymbol{R}_2$. The "difference" between a rotor and itself therefore results in unity, i.e. $\widetilde{\boldsymbol{R}}_1 \, \boldsymbol{R}_1 = 1$.

An appropriate evaluation procedure for the mean of a set of rotors has to account for this particular metric. Note that this problem has been treated more generally for Clifford groups by Buchholz and Sommer in [30].

Let $\{\boldsymbol{R}_i\} \subset \mathbb{G}_3^+$ denote a set of N known rotors, and let $\boldsymbol{R}_M \in \mathbb{G}_3^+$ denote the unknown mean rotor. If we define $\boldsymbol{D}_i := \widetilde{\boldsymbol{R}}_i \, \boldsymbol{R}_M - 1$, one approach to evaluating the mean rotor is to minimize the measure $\sum_i \widetilde{\boldsymbol{D}}_i \, \boldsymbol{D}_i$. Let $\boldsymbol{R}_{\Delta_i} := \widetilde{\boldsymbol{R}}_i \, \boldsymbol{R}_M$ and $\boldsymbol{R}_{\Delta_i} = \cos\theta_i - \sin\theta_i \, \boldsymbol{U}_i$, where $\boldsymbol{U}_i$ is a unit bivector that represents the respective rotation plane. Then

$$\begin{aligned} \widetilde{\boldsymbol{D}}_i \, \boldsymbol{D}_i &= (\widetilde{\boldsymbol{R}}_{\Delta_i} - 1)\,(\boldsymbol{R}_{\Delta_i} - 1) \\ &= 2 - \boldsymbol{R}_{\Delta_i} - \widetilde{\boldsymbol{R}}_{\Delta_i} \\ &= 2\,(1 - \cos\theta_i) \\ &= 4\,\sin^2(\theta_i/2). \end{aligned} \tag{4.13}$$

Minimizing $\sum_i \widetilde{\boldsymbol{D}}_i \, \boldsymbol{D}_i$ therefore minimizes $\sum_i 4\,\sin^2(\theta_i/2)$. If $\theta_i \ll 1$ for all i, then

$$\sum_i 4\,\sin^2(\theta_i/2) \;\approx\; \sum_i \theta_i^2\,,$$

which is the measure that should ideally be minimized. This shows that the rotor $\boldsymbol{R}_M$ that minimizes $\sum_i \widetilde{\boldsymbol{D}}_i \, \boldsymbol{D}_i$ approximates the "true" mean rotor quite well if the $\{\boldsymbol{R}_i\}$ are "fairly" similar.

It remains to be shown how the $\boldsymbol{R}_M$ that minimizes $\sum_i \widetilde{\boldsymbol{D}}_i \boldsymbol{D}_i$ can be evaluated. From (4.13), it follows that

$$\boldsymbol{\Delta} := \frac{1}{N} \sum_{i=1}^{N} (\widetilde{\boldsymbol{D}}_i \boldsymbol{D}) = 2 - \Big(\frac{1}{N} \sum_{i=1}^{N} \widetilde{\boldsymbol{R}}_i\Big) \boldsymbol{R}_M - \widetilde{\boldsymbol{R}}_M \Big(\frac{1}{N} \sum_{i=1}^{N} \boldsymbol{R}_i\Big). \quad (4.14)$$

Let us denote the arithmetic mean of $\{\boldsymbol{R}_i\}$ by $\bar{\boldsymbol{R}}_i$, i.e. $\bar{\boldsymbol{R}}_i = (1/N) \sum_{i=1}^{N} \boldsymbol{R}_i$. Since $\sum_i \widetilde{\boldsymbol{R}}_i = (\sum_i \boldsymbol{R}_i)^{\sim}$, it follows that $\boldsymbol{\Delta} = 2 - \widetilde{\bar{\boldsymbol{R}}} \boldsymbol{R}_M - \widetilde{\boldsymbol{R}}_M \bar{\boldsymbol{R}}$. In general $\bar{\boldsymbol{R}}$ is not a unitary versor, but $\widetilde{\bar{\boldsymbol{R}}} \bar{\boldsymbol{R}} = \rho^2$, $\rho \in \mathbb{R}$. Replacing $\boldsymbol{R}_M$ by $\rho^{-1} \bar{\boldsymbol{R}}$ in the expression for $\boldsymbol{\Delta}$ results in $\boldsymbol{\Delta} = 0$, i.e. it minimizes $\boldsymbol{\Delta}$. However, $\rho^{-2} \bar{\boldsymbol{R}}$ is not a unitary versor. Hence, the rotor that minimizes $\boldsymbol{\Delta}$ has to be the unitary versor closest to $\rho^{-2} \bar{\boldsymbol{R}}$, that is $\boldsymbol{R}_M = \rho^{-1} \bar{\boldsymbol{R}}$. Substituting this expression for $\boldsymbol{R}_M$ into (4.14) gives $\boldsymbol{\Delta} = 2(1 - \rho)$.

In summary, a good approximation to the mean rotor of a set of N rotors $\{\boldsymbol{R}_i\}$ is given by

$$\boldsymbol{R}_M = \frac{\sum_{i=1}^{N} \boldsymbol{R}_i}{\sqrt{(\sum_{i=1}^{N} \widetilde{\boldsymbol{R}}_i)(\sum_{i=1}^{N} \boldsymbol{R}_i)}}. \quad (4.15)$$

4.2 Projective Space

Projective space is given its name by the fact that it consists of equivalence classes of points which form projection rays. The idea of projective space is derived from a camera obscura or pinhole camera, where one light ray generates exactly one image point. This means that an image point may be generated by any point along the path of a light ray, or *projective ray*. Projective spaces can be defined in a rather abstract manner without reference to coordinates (see e.g. [25]). In this text, however, the more applied view in terms of coordinates will be introduced right away.

4.2.1 Definition

4.2.1.1 Basics

In the above spirit, a projective space, also called a *homogeneous space*, is generated from a vector space $\mathbb{R}^n$ by regarding the elements of $\mathbb{R}^n$ as representatives of equivalence classes. In particular, a vector $\boldsymbol{a} \in \mathbb{R}^n$ represents an equivalence class often denoted by $[\boldsymbol{a}]$, which is defined as

$$[\boldsymbol{a}] := \{\, \alpha\, \boldsymbol{a} \;:\; \alpha \in \mathbb{R}\backslash 0 \,\}. \quad (4.16)$$

The equivalence class of $\boldsymbol{a}$ is thus the set of points that lie on the line through the origin of $\mathbb{R}^n$ and through $\boldsymbol{a}$, without the origin itself. The set of these equivalence classes of all vectors of $\mathbb{R}^n$ form the projective space of $\mathbb{R}^n$, often denoted by $\mathbb{RP}^{n-1}$:

$$\mathbb{RP}^{n-1} := \{ \, [\boldsymbol{a}] \; : \; \boldsymbol{a} \in \mathbb{R}^n \backslash \boldsymbol{0} \, \}. \tag{4.17}$$

Note that $\mathbb{RP}^{n-1}$ is not a vector space anymore. The transformation from an element of $\mathbb{RP}^{n-1}$ back to a Euclidean space is basically the selection of appropriate representatives for the equivalence classes. This can be achieved by choosing a (hyper)plane in $\mathbb{R}^n$ that does not pass through the origin and using the intersections of the equivalence classes (or projective rays) with this (hyper)plane as the representatives. This set of representatives then forms again a vector space, of dimension $n-1$. However, the equivalence classes that are parallel to the chosen (hyper)plane have no representatives on that plane; instead, they map to infinity. This shows one of the advantages of working in projective space: points at infinity in $\mathbb{R}^{n-1}$ can be represented by equivalence classes in $\mathbb{R}^n$.

4.2.1.2 Geometric Algebra

From the discussion of the geometric algebra of Euclidean space in Sect. 4.1 and in particular the definitions of the IPNS and OPNS, it is clear that the null space representations in geometric algebra are ideally suited to representing equivalence classes in a projective space. The equivalence class of a vector $\boldsymbol{a} \in \mathbb{R}^n$ as defined in (4.16) is the same as the OPNS of $\boldsymbol{a}$ apart from the exclusion of the origin.

To generate representatives of equivalence classes of projective space from vectors in Euclidean space, Euclidean vectors are first embedded in an *affine space*. For a detailed discussion of affine space, see for example [25, 75]. Here it should suffice to say that the affine space of $\mathbb{R}^n$ is represented by the following embedding of a vector $\boldsymbol{a} \in \mathbb{R}^n$ in $\mathbb{R}^{n+1}$:

$$\mathcal{H} \; : \; \boldsymbol{a} \in \mathbb{R}^n \; \mapsto \; \boldsymbol{a} + \boldsymbol{e}_{n+1} \in \mathbb{R}^{n+1}, \tag{4.18}$$

where the $\boldsymbol{e}_i := \overline{\mathbb{R}}^{n+1}[i]$ are the canonical basis of $\mathbb{R}^{n+1}$. This embedding is also called *homogenization* of $\boldsymbol{a}$, and hence 1-vectors in $\mathbb{R}^{n+1}$ are called *homogeneous vectors*. If $\boldsymbol{a} \in \mathbb{R}^n$, then the corresponding homogenized vector $\mathcal{H}(\boldsymbol{a})$ is denoted by capitalizing the vector symbol, i.e. $\boldsymbol{A} := \mathcal{H}(\boldsymbol{a})$.

The OPNS of $\boldsymbol{A} = \mathcal{H}(\boldsymbol{a})$ without the origin is then the equivalence class of $\mathcal{H}(\boldsymbol{a})$. The geometric algebra of $\mathbb{R}^{n+1}$ can be used in this way to represent entities in $\mathbb{R}^n$.

In terms of the discussion at the beginning of this chapter, $\mathcal{H}$ is the bijective embedding operator that maps $\mathbb{R}^n$ to a subspace of $\mathbb{R}^{n+1}$. In this case, it is

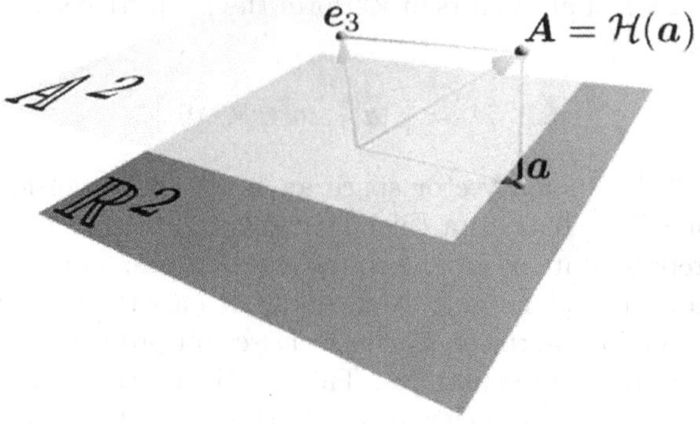

Fig. 4.7 Embedding of a Euclidean vector $\boldsymbol{a} \in \mathbb{G}_2^1$ in the affine plane $\mathbb{A}^2$ as $\boldsymbol{A} = \mathcal{H}(\boldsymbol{a})$

in fact a linear mapping, as illustrated in Fig. 4.7 for $\mathbb{R}^2$. That is, Euclidean space $\mathbb{R}^n$ is embedded as a particular hyperplane $\mathcal{H}(\mathbb{R}^n)$ in $\mathbb{R}^{n+1}$. The origin of $\mathbb{R}^n$ maps to $\boldsymbol{e}_{n+1}$, which means that the origin of Euclidean space, as represented in $\mathbb{R}^{n+1}$, is not a special point anymore. For example, whereas the scalar product of a vector with the origin in Euclidean space is always identically zero, this is not necessarily the case in projective space.

The hyperplane $\mathcal{H}(\mathbb{R}^n)$ represents the affine space of $\mathbb{R}^n$. Since this hyperplane plays an important role, it is given its own symbol $\mathbb{A}^n := \mathcal{H}(\mathbb{R}^n)$. Note that in this representation $\mathbb{A}^n$ does not form a vector space, since the sum of two vectors on the hyperplane $\mathbb{A}^n$ does not lie on the hyperplane anymore.

As an aside, note that *affine transformations* are in fact those *linear* transformations that map $\mathbb{A}^n$ onto itself. Transformations that map $\mathbb{A}^n$ into $\mathbb{R}^{n+1}$ are called *projective transformations.*

Given an arbitrary vector $\boldsymbol{A} \in \mathbb{G}_{n+1}^1$, the corresponding Euclidean vector $\boldsymbol{a} \in \mathbb{G}_n^1$ cannot immediately be evaluated, since $\mathcal{H}^{-1}$, the inverse of $\mathcal{H}$, is defined only for 1-vectors in the subspace $\mathbb{A}^n := \mathcal{H}(\mathbb{G}_n^1) \subset \mathbb{G}_{n+1}^1$. However, $\boldsymbol{A}$ represents the equivalence class $\mathbb{NO}(\boldsymbol{A})$ and thus any vector in $\mathbb{NO}(\boldsymbol{A})$ can be used to represent the same point in $\mathcal{H}(\mathbb{G}_n^1)$. In particular, the point in $\mathbb{NO}(\boldsymbol{A})$ that also lies in $\mathbb{A}^n$ can be used. Therefore, $\boldsymbol{A}$ is projected back to $\mathbb{G}_n^1$ via

$$\boldsymbol{a} = \mathcal{H}^{-1}\big(\mathbb{NO}(\boldsymbol{A}) \cap \mathbb{A}^n\big) \tag{4.19}$$

if $\mathbb{NO}(\boldsymbol{A}) \cap \mathbb{A}^n$ exists, which is not the case if $\boldsymbol{A}$ has no $\boldsymbol{e}_{n+1}$ component. This intersection can also be evaluated as $\boldsymbol{A}/(\boldsymbol{A} \cdot \boldsymbol{e}_{n+1})$. Hence,

$$\boldsymbol{a} = \frac{\boldsymbol{A}}{\boldsymbol{A} \cdot \boldsymbol{e}_{n+1}} - \boldsymbol{e}_{n+1}. \tag{4.20}$$

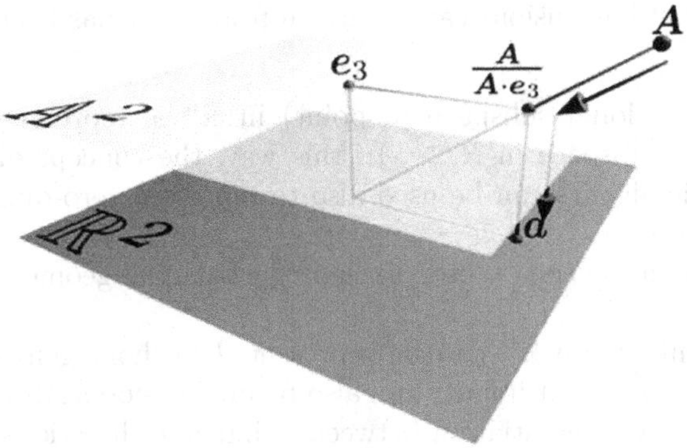

Fig. 4.8 Projection of a homogeneous vector $\boldsymbol{A} \in \mathbb{G}_3^1$ into the corresponding Euclidean space $\mathbb{G}_2^1$ as $\boldsymbol{a} = \mathbb{NO}_G(\boldsymbol{A})$

For convenience, we define the operator $\mathcal{A} : \mathbb{R}^{n+1} \to \mathbb{A}^n$ that evaluates the intersection $\mathbb{NO}(\boldsymbol{A}) \cap \mathbb{A}^n$ as

$$\mathcal{A} : \boldsymbol{A} \in \mathbb{R}^{n+1} \mapsto \frac{\boldsymbol{A}}{\boldsymbol{A} \cdot \boldsymbol{e}_{n+1}}. \tag{4.21}$$

This operator is useful because homogeneous vectors on $\mathbb{A}^n$ can be immediately identified with their corresponding Euclidean vectors in $\mathbb{R}^n$.

The projection of a vector $\boldsymbol{A} \in \mathbb{G}_{n+1}^1$ to $\mathbb{G}_n^1$ using (4.20) is equivalent to evaluating the geometric OPNS of $\boldsymbol{A}$ as given in Definition 4.1. Recall that

$$\mathbb{NO}_G(\boldsymbol{A}) = \left\{\, \boldsymbol{a} \in \mathbb{G}_n^1 \,:\, \mathcal{H}(\boldsymbol{a}) \wedge \boldsymbol{A} = 0 \,\right\} = \mathcal{H}^{-1}\big(\mathbb{NO}(\boldsymbol{A}) \cap \mathbb{A}^n\big),$$

which is equivalent to (4.19). This projection is shown in Fig. 4.8.

As noted before, this transformation is valid only for homogeneous vectors that have a non-zero $\boldsymbol{e}_{n+1}$ component, also called the *homogeneous component*. Those homogeneous vectors that do have a zero homogeneous component would map to infinity, and are accordingly called points at infinity or *direction* vectors.

A very useful property of homogeneous vectors is that their overall scale is of no importance. For example, given a vector $\boldsymbol{a} \in \mathbb{G}_n^1$ and a scale $\alpha \in \mathbb{R} \setminus 0$, then

$$\mathbb{NO}_G\Big(\alpha\, \mathcal{H}(\boldsymbol{a})\Big) = \boldsymbol{a}.$$

4.2.1.3 Conclusions

Three important conclusions can be drawn from what has been presented so far:

1. A zero-dimensional subspace (a point) in $\mathbb{R}^n$ is represented by a one-dimensional subspace in $\mathbb{R}^{n+1}$. In this way, the concept of a null space in geometric algebra can be used also to represent zero-dimensional entities, i.e points.
2. Scaling a homogeneous vector has no effect on the geometric entity that it represents.
3. Points at infinity in $\mathbb{R}^n$ can be represented by homogeneous vectors in $\mathbb{R}^{n+1}$. These points at infinity can also be understood as direction vectors, which allows a differentiation between points and directions.

4.2.2 Outer-Product Representations

In the previous subsection, it was shown how the concept of the geometric inner- and outer-product null spaces can be used to represent a point in Euclidean space $\mathbb{R}^n$ by the GOPNS of a vector in $\mathbb{R}^{n+1}$. In this subsection, the other geometric entities that can be represented in this way are presented. Of particular interest later on is the space $\mathbb{R}^3$ and its homogeneous embedding. However, for better visualization, initially the space $\mathbb{R}^2$ and its homogeneous embedding in $\mathbb{R}^3$ are treated.

4.2.2.1 Lines

Using the concept of the geometric OPNS, a bivector $\boldsymbol{A}_{\langle 2\rangle} \in \mathbb{G}_3^2$ represents the set of those points in $\mathbb{G}_2^1$ whose embedding in $\mathbb{G}_3^1$ lies in $\mathbb{NO}(\boldsymbol{A}_{\langle 2\rangle})$, i.e.

$$\mathbb{NO}_G(\boldsymbol{A}_{\langle 2\rangle}) := \{\, \boldsymbol{x} \in \mathbb{G}_2^1 \,:\, \mathcal{H}(\boldsymbol{x}) \wedge \boldsymbol{A}_{\langle 2\rangle} = 0 \,\}. \tag{4.22}$$

The set of points that lie in $\mathbb{NO}_G(\boldsymbol{A}_{\langle 2\rangle})$ can be expressed in parametric form. To show this, let $\boldsymbol{A}_{\langle 2\rangle} = \boldsymbol{A} \wedge \boldsymbol{B}$, with $\boldsymbol{A}, \boldsymbol{B} \in \mathbb{G}_3^1$ being homogeneous embeddings of $\boldsymbol{a}, \boldsymbol{b} \in \mathbb{G}_2^1$, respectively, i.e $\boldsymbol{A} := \mathcal{H}(\boldsymbol{a})$ and $\boldsymbol{B} := \mathcal{H}(\boldsymbol{b})$. In this case

$$\mathbb{NO}(\boldsymbol{A}_{\langle 2\rangle}) = \{\, \alpha\,\boldsymbol{A} + \beta\,\boldsymbol{B} \,:\, (\alpha, \beta) \in \mathbb{R}^2 \,\}.$$

First of all, we evaluate the intersection of $\mathbb{NO}(\boldsymbol{A}_{\langle 2\rangle})$ with $\mathbb{A}^2$. That is,

$$\boldsymbol{X} := \mathbb{NO}(\boldsymbol{A}_{\langle 2\rangle}) \,\cap\, \mathbb{A}^2 = \frac{\alpha\,\boldsymbol{A} + \beta\,\boldsymbol{B}}{(\alpha\,\boldsymbol{A} + \beta\,\boldsymbol{B}) \cdot \boldsymbol{e}_3}.$$

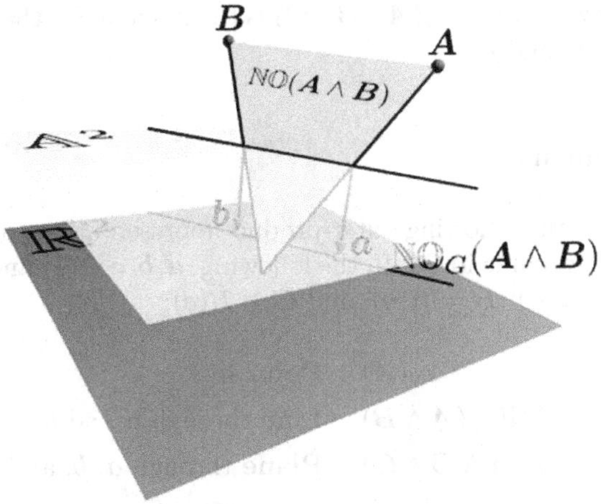

Fig. 4.9 Representation of a line in $\mathbb{R}^2$ through a bivector in $\mathbb{G}_3^2$

From the definition of $\boldsymbol{A}$ and $\boldsymbol{B}$, it follows that $\boldsymbol{A} \cdot \boldsymbol{e}_3 = \boldsymbol{B} \cdot \boldsymbol{e}_3 = 1$. Adding and subtracting $\alpha\,\boldsymbol{B}$ gives

$$\boldsymbol{X} = \frac{\alpha}{\alpha+\beta}\,(\boldsymbol{a}-\boldsymbol{b}) + \boldsymbol{b} + \boldsymbol{e}_3.$$

Hence,

$$\boldsymbol{x} = \mathcal{H}^{-1}(\boldsymbol{X}) = \lambda\,(\boldsymbol{a}-\boldsymbol{b}) + \boldsymbol{b},$$

where $\lambda = \alpha/(\alpha+\beta)$ is a free parameter. This shows that the Euclidean vectors that lie in $\mathbb{NO}(\boldsymbol{A}_{\langle 2\rangle})$ after homogenization are those vectors that lie on the line through $\boldsymbol{a}$ and $\boldsymbol{b}$. This construction is illustrated in Fig. 4.9. Conversely, the line passing through the Euclidean vectors $\boldsymbol{a}$ and $\boldsymbol{b}$ is represented by $\mathcal{H}(\boldsymbol{a}) \wedge \mathcal{H}(\boldsymbol{b})$. This is a very direct, clear, and easy-to-use representation of lines.

Note that arbitrary lines can be represented in this way, which is another advantage of working with the homogeneous embeddings of Euclidean vectors. In Euclidean space, only lines through the origin can be represented by null spaces of blades.

4.2.2.2 Planes

Without going into detail, it may be shown that the Euclidean OPNS of the outer product of three homogeneous vectors in $\mathbb{G}_3$ represents a plane in $\mathbb{R}^3$. That is, given vectors $\boldsymbol{a}, \boldsymbol{b}, \boldsymbol{c} \in \mathbb{G}_3^1$ and $\boldsymbol{A}, \boldsymbol{B}, \boldsymbol{C} \in \mathbb{G}_4^1$ with

$$\boldsymbol{A} = \mathcal{H}(\boldsymbol{a}), \quad \boldsymbol{B} = \mathcal{H}(\boldsymbol{b}), \quad \text{and} \quad \boldsymbol{C} = \mathcal{H}(\boldsymbol{c}),$$

it may be shown that $\mathbb{NO}_G(\boldsymbol{A}\wedge\boldsymbol{B}\wedge\boldsymbol{C})$ is the plane in $\mathbb{R}^3$ that passes through the points $\boldsymbol{a}$, $\boldsymbol{b}$, and $\boldsymbol{c}$.

4.2.2.3 Summary

To summarize, the following outer-product representations of geometric entities of $\mathbb{R}^3$ are available in $\mathbb{G}_4$. In the following, $\boldsymbol{a},\boldsymbol{b},\boldsymbol{c}\in\mathbb{G}_3^1$ and $\boldsymbol{A},\boldsymbol{B},\boldsymbol{C}\in\mathbb{G}_4^1$, whereby $\boldsymbol{A}=\mathcal{H}(\boldsymbol{a})$, $\boldsymbol{B}=\mathcal{H}(\boldsymbol{b})$, and $\boldsymbol{C}=\mathcal{H}(\boldsymbol{c})$:

$$\begin{aligned}
\mathbb{NO}_G(\boldsymbol{A}) \quad & \textbf{Point } \boldsymbol{a}\,. \\
\mathbb{NO}_G(\boldsymbol{A}\wedge\boldsymbol{B}) \quad & \textbf{Line} \text{ through } \boldsymbol{a} \text{ and } \boldsymbol{b}\,. \\
\mathbb{NO}_G(\boldsymbol{A}\wedge\boldsymbol{B}\wedge\boldsymbol{C}) \quad & \textbf{Plane} \text{ through } \boldsymbol{a},\ \boldsymbol{b}, \text{ and } \boldsymbol{c}\,.
\end{aligned} \tag{4.23}$$

4.2.3 Inner-Product Representations

Just as for Euclidean space, the duality of the GOPNS and the GIPNS also holds for projective space. In the following, the GIPNS of blades in $\mathbb{G}_4$ is investigated.

4.2.3.1 Planes

It was noted in the previous subsection that a plane in $\mathbb{R}^3$ can be represented by the GOPNS of the outer product of three 1-vectors in $\mathbb{G}_4$. Since the dual of a 3-blade is a 1-blade, i.e. a 1-vector, the GIPNS of a 1-vector has to represent a plane. Let $\boldsymbol{A}\in\mathbb{G}_4^1$ be given by

$$\boldsymbol{A}=\hat{\boldsymbol{a}}-\alpha\,\boldsymbol{e}_4,$$

where $\hat{\boldsymbol{a}}\in\mathbb{G}_3^1$ and $\|\hat{\boldsymbol{a}}\|=1$, and, furthermore, $\alpha\in\mathbb{R}$ and $\boldsymbol{e}_4$ denotes the homogeneous dimension. Recall that the GIPNS of $\boldsymbol{A}$ is given by

$$\mathbb{NI}_G(\boldsymbol{A}) := \{\ \boldsymbol{x}\in\mathbb{G}_3^1\ :\ \mathcal{H}(\boldsymbol{x})\cdot\boldsymbol{A}=0\ \}.$$

Let $\boldsymbol{X}\in\mathbb{G}_4^1$ be defined by $\boldsymbol{X}:=\mathcal{H}(\boldsymbol{x})=\boldsymbol{x}+\boldsymbol{e}_4$ with $\boldsymbol{x}\in\mathbb{G}_3^1$; then

$$\begin{aligned}
\boldsymbol{A}\cdot\boldsymbol{X}=0 \iff & (\hat{\boldsymbol{a}}-\alpha\,\boldsymbol{e}_4)\cdot(\boldsymbol{x}+\boldsymbol{e}_4)=0 \\
\iff & \hat{\boldsymbol{a}}\cdot\boldsymbol{x}-\alpha=0 \\
\iff & \hat{\boldsymbol{a}}\cdot\boldsymbol{x}^{\|}-\alpha=0 \\
\iff & \boldsymbol{x}^{\|}=\alpha\,\hat{\boldsymbol{a}},
\end{aligned}$$

where $\boldsymbol{x}^{\parallel} := \mathcal{P}_{\hat{\boldsymbol{a}}}(\boldsymbol{x})$ is the component of $\boldsymbol{x}$ parallel to $\hat{\boldsymbol{a}}$. If we define $\boldsymbol{x}^{\perp} := \mathcal{P}^{\perp}_{\hat{\boldsymbol{a}}}(\boldsymbol{x})$ such that $\boldsymbol{x} = \boldsymbol{x}^{\parallel} + \boldsymbol{x}^{\perp}$, then it follows that for all $\alpha \in \mathbb{R}$,

$$\boldsymbol{x} = \alpha\, \hat{\boldsymbol{a}} + \boldsymbol{x}^{\perp}$$

lies in the GIPNS of $\boldsymbol{A}$. Hence, $\boldsymbol{A}$ represents a plane with a normal $\hat{\boldsymbol{a}}$ and a distance α from the origin in $\mathbb{R}^3$.

4.2.3.2 Lines

A line in $\mathbb{R}^3$ may be represented by the GIPNS of a bivector. In contrast to the GOPNS of a bivector, which also represents a line, the GIPNS of a bivector can be interpreted as the intersection of two planes. Recall that we have encountered the same construction in Euclidean space.

If $\boldsymbol{A}, \boldsymbol{B} \in \mathbb{G}^1_4$ are defined as $\boldsymbol{A} := \mathcal{H}(\boldsymbol{a})$ and $\boldsymbol{B} := \mathcal{H}(\boldsymbol{b})$ with $\boldsymbol{a}, \boldsymbol{b} \in \mathbb{G}^1_3$, then the GIPNS of $\boldsymbol{A} \wedge \boldsymbol{B}$ is given by

$$\mathbb{NI}_G(\boldsymbol{A} \wedge \boldsymbol{B}) := \left\{ \boldsymbol{x} \in \mathbb{G}^1_3 \;:\; \mathcal{H}(\boldsymbol{x}) \cdot (\boldsymbol{A} \wedge \boldsymbol{B}) = 0 \right\}.$$

Let $\boldsymbol{X} := \mathcal{H}(\boldsymbol{x})$, and assume that $\boldsymbol{A} \wedge \boldsymbol{B} \neq 0$; then

$$\boldsymbol{X} \cdot (\boldsymbol{A} \wedge \boldsymbol{B}) = (\boldsymbol{X} \cdot \boldsymbol{A})\, \boldsymbol{B} - (\boldsymbol{X}\, \boldsymbol{B})\, \boldsymbol{A},$$

which can be zero only if $\boldsymbol{X} \cdot \boldsymbol{A} = 0$ *and* $\boldsymbol{X} \cdot \boldsymbol{B} = 0$. Therefore, $\boldsymbol{X} \cdot (\boldsymbol{A} \wedge \boldsymbol{B}) = 0$ if and only if $\boldsymbol{x}$ lies on both of the planes represented by the GIPNSs of $\boldsymbol{A}$ and $\boldsymbol{B}$, which means that it lies on the intersection line of those planes. This shows that in terms of the GIPNS, the outer product represents the intersection of geometric entities.

The parametric representation of the line $\mathbb{NI}_G(\boldsymbol{A} \wedge \boldsymbol{B})$ is more easily evaluated from the equivalent representation $\mathbb{NO}_G\big((\boldsymbol{A} \wedge \boldsymbol{B})^*\big)$.

4.2.3.3 Points

Just as a line is represented by a 2-blade, a point can be represented by the GIPNS of a 3-blade. This corresponds to the intersection point of three planes.

4.2.3.4 Summary

Given homogeneous vectors $\boldsymbol{A}, \boldsymbol{B}, \boldsymbol{C} \in \mathbb{G}^1_4$ defined as $\boldsymbol{A} = \mathcal{H}(\boldsymbol{a})$, $\boldsymbol{B} = \mathcal{H}(\boldsymbol{b})$, and $\boldsymbol{C} = \mathcal{H}(\boldsymbol{c})$, where $\boldsymbol{a}, \boldsymbol{b}, \boldsymbol{c} \in \mathbb{G}^1_3$, the following geometric entities in $\mathbb{R}^3$ can be represented:

$$\begin{array}{rl} \mathbb{NI}_G(\boldsymbol{A}) & \textbf{Plane} \\ \mathbb{NI}_G(\boldsymbol{A}\wedge\boldsymbol{B}) & \textbf{Line}\ \mathbb{NI}_G(\boldsymbol{A}) \cap \mathbb{NI}_G(\boldsymbol{B}) \\ \mathbb{NI}_G(\boldsymbol{A}\wedge\boldsymbol{B}\wedge\boldsymbol{C}) & \textbf{Point}\ \mathbb{NI}_G(\boldsymbol{A}) \cap \mathbb{NI}_G(\boldsymbol{B}) \cap \mathbb{NI}_G(\boldsymbol{C}) \end{array}$$

4.2.4 Reflections in Projective Space

By going from Euclidean to projective space, an additional dimension, the homogeneous dimension, is introduced. The question is what effect this has when the reflection operator introduced earlier is used.

First of all, consider a vector $\boldsymbol{a} \in \mathbb{G}_2^1$ and its homogeneous representation

$$\boldsymbol{A} := \mathcal{H}(\boldsymbol{a}) = \boldsymbol{a} + \boldsymbol{e}_3,$$

where $\boldsymbol{e}_3$ is the homogeneous dimension. A reflection in $\boldsymbol{e}_3$ gives

$$\begin{aligned} \boldsymbol{e}_3\,\boldsymbol{A}\,\boldsymbol{e}_3 &= \boldsymbol{e}_3\,\boldsymbol{a}\,\boldsymbol{e}_3 + \boldsymbol{e}_3\,\boldsymbol{e}_3\,\boldsymbol{e}_3 \\ &= -\boldsymbol{a}\,\boldsymbol{e}_3\,\boldsymbol{e}_3 + \boldsymbol{e}_3 \\ &= -\boldsymbol{a} + \boldsymbol{e}_3, \end{aligned}$$

where the fact that $\boldsymbol{e}_3$ is perpendicular to all vectors in $\mathbb{G}_2^1$ has been used. Therefore,

$$\boldsymbol{e}_3\,\boldsymbol{a} = \boldsymbol{e}_3 \wedge \boldsymbol{a} = -\boldsymbol{a} \wedge \boldsymbol{e}_3 = -\boldsymbol{a}\boldsymbol{e}_3.$$

It therefore follows that

$$\mathbb{NO}_G\Big(\boldsymbol{e}_3\,\mathcal{H}(\boldsymbol{a})\,\boldsymbol{e}_3\Big) = -\boldsymbol{a},$$

which shows that a reflection of $\boldsymbol{A}$ in $\boldsymbol{e}_3$ represents a reflection of $\boldsymbol{a}$ in the origin in $\mathbb{R}^2$.

Next, consider a vector $\boldsymbol{N} \in \mathbb{G}_3^1$, with $\|\boldsymbol{N}\| = 1$ and $\boldsymbol{N} \cdot \boldsymbol{e}_3 = 0$, i.e. $\boldsymbol{N}$ is a point at infinity, or direction vector. Let $\boldsymbol{A} \in \mathbb{G}_3^1$ be a homogeneous vector as before; then

$$\begin{aligned} \boldsymbol{N}\,\boldsymbol{A}\,\boldsymbol{N} &= \boldsymbol{N}\,(\boldsymbol{a} + \boldsymbol{e}_3)\,\boldsymbol{N} \\ &= \boldsymbol{N}\,\boldsymbol{a}\,\boldsymbol{N} + \boldsymbol{N}\,\boldsymbol{e}_3\,\boldsymbol{N} \\ &= \boldsymbol{N}\,\boldsymbol{a}\,\boldsymbol{N} - \boldsymbol{e}_3\,\boldsymbol{N}^2 \\ &= \boldsymbol{N}\,\boldsymbol{a}\,\boldsymbol{N} - \boldsymbol{e}_3. \end{aligned}$$

Using the projection operator $\mathcal{A}$ that maps a homogeneous vector to the affine hyperplane, it follows that

$$\begin{aligned}\mathcal{A}(\boldsymbol{N}\,\boldsymbol{A}\,\boldsymbol{N}) &= -\boldsymbol{N}\,\boldsymbol{a}\,\boldsymbol{N} + e_3\\ &= -(\boldsymbol{a}^{\|} - \boldsymbol{a}^{\perp}) + e_3\\ &= \boldsymbol{a}^{\perp} - \boldsymbol{a}^{\|} + e_3,\end{aligned}$$

where $\boldsymbol{a}^{\|} := \mathcal{P}_{\boldsymbol{N}}(\boldsymbol{a})$ and $\boldsymbol{a}^{\perp} := \mathcal{P}^{\perp}_{\boldsymbol{N}}(\boldsymbol{a})$ are the orthogonal and parallel components, respectively, of $\boldsymbol{a}$ with respect to $\boldsymbol{N}$. This shows that the component of the homogeneous vector $\boldsymbol{A}$ that is *parallel* to the reflection direction $\boldsymbol{N}$ is reflected, and not the part perpendicular to it. Figure 4.10 shows this setup.

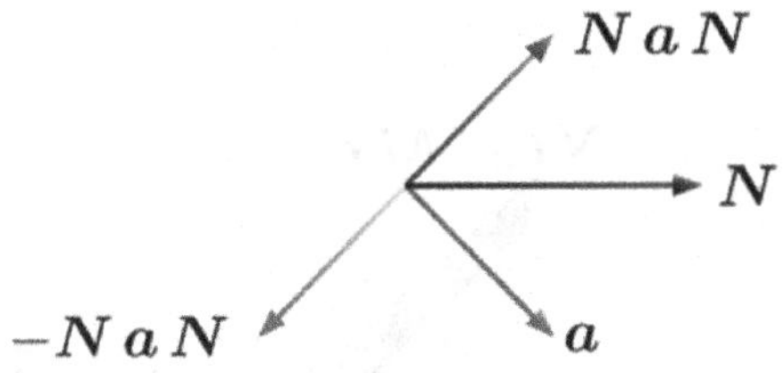

Fig. 4.10 Effect in $\mathbb{R}^2$ of reflection of a homogeneous vector in a direction vector $\boldsymbol{N} \in \mathbb{R}^3$

This is not the kind of reflection that is usually desired. This situation can be remedied by reflecting $\boldsymbol{N}\,\boldsymbol{A}\,\boldsymbol{N}$ again in the origin. That is, in order to reflect a homogeneous vector in a line with direction $\boldsymbol{N}$, the operator that has to be used is $(\boldsymbol{N}\,e_3)$ instead of $\boldsymbol{N}$:

$$\begin{aligned}(\boldsymbol{N}\,e_3)\,\boldsymbol{A}\,(e_3\,\boldsymbol{N}) &= \boldsymbol{N}\,(-\boldsymbol{a} + e_3)\,\boldsymbol{N}\\ &= -\boldsymbol{N}\,\boldsymbol{a}\,\boldsymbol{N} + \boldsymbol{N}\,e_3\,\boldsymbol{N}\\ &= -\boldsymbol{N}\,\boldsymbol{a}\,\boldsymbol{N} - e_3,\end{aligned}$$

and thus

$$\mathcal{A}\Big((\boldsymbol{N}\,e_3)\,\boldsymbol{A}\,(e_3\,\boldsymbol{N})\Big) = \boldsymbol{N}\,\boldsymbol{a}\,\boldsymbol{N} + e_3.$$

Note that this analysis is true in any dimension. Only the homogeneous-dimension basis vector has to be adapted.

4.2.5 Rotations in Projective Space

In the previous subsection, it was shown how a reflection in $\mathbb{R}^2$ has to be expressed in terms of homogeneous vectors in $\mathbb{G}_3$. Since a rotation expressed by a rotor is nothing more than two consecutive reflections, a rotor may also take on a different form in projective space.

Consider the rotation of a vector $\boldsymbol{a} \in \mathbb{G}_2^1$ represented by $\boldsymbol{A} := \mathcal{H}(\boldsymbol{a})$ by reflecting it first in a unit direction vector $\boldsymbol{N} \in \mathbb{G}_3^1$ and then in a unit direction vector $\boldsymbol{M} \in \mathbb{G}_3^1$. A reflection in $\boldsymbol{N}$ has to be expressed as $(\boldsymbol{N}\, \boldsymbol{e}_3)$ and a reflection in $\boldsymbol{M}$ as $(\boldsymbol{M}\, \boldsymbol{e}_3)$. Hence, the rotation of $\boldsymbol{A}$ becomes

$$(\boldsymbol{M}\, \boldsymbol{e}_3)\,(\boldsymbol{N}\, \boldsymbol{e}_3)\,\boldsymbol{A}\,(\boldsymbol{e}_3\, \boldsymbol{N})\,(\boldsymbol{e}_3\, \boldsymbol{M}) = \boldsymbol{R}\,\boldsymbol{A}\,\widetilde{\boldsymbol{R}}, \quad \boldsymbol{R} := (\boldsymbol{M}\, \boldsymbol{e}_3)\,(\boldsymbol{N}\, \boldsymbol{e}_3).$$

Such a double reflection is illustrated in Fig. 4.11. In this figure, the vector $\boldsymbol{a} \in \mathbb{G}_2^1$ is represented by $\boldsymbol{A} := \mathcal{H}(\boldsymbol{a})$. The first reflection, of $\boldsymbol{A}$ in $\boldsymbol{N}\, \boldsymbol{e}_3$, gives $\boldsymbol{B}$, and the next reflection, of $\boldsymbol{B}$ in $\boldsymbol{M}\, \boldsymbol{e}_3$, gives $\boldsymbol{C}$. Note that $\mathbb{NO}(\boldsymbol{M}\, \boldsymbol{e}_3)$ and $\mathbb{NO}(\boldsymbol{N}\, \boldsymbol{e}_3)$ are planes, as shown in the figure.

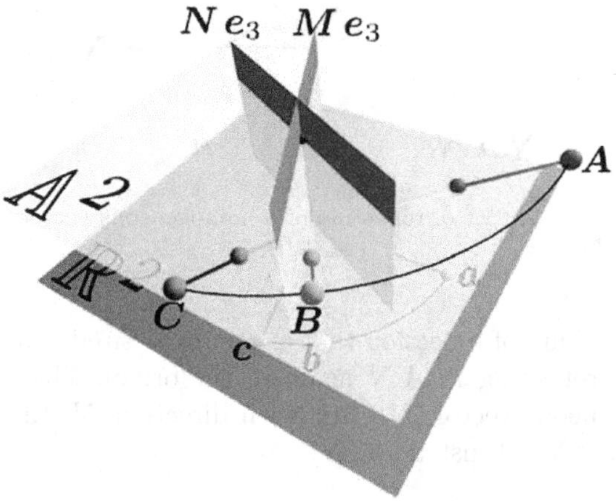

Fig. 4.11 Double reflection of a homogenous vector $\boldsymbol{A}$ in reflection planes $\boldsymbol{N}\, \boldsymbol{e}_3$ and $\boldsymbol{M}\, \boldsymbol{e}_3$ in $\mathbb{G}_3$

However, the expression for $\boldsymbol{R}$ can be simplified:

$$\begin{aligned} \boldsymbol{R} &= (\boldsymbol{M}\, \boldsymbol{e}_3)\,(\boldsymbol{N}\, \boldsymbol{e}_3) \\ &= -\boldsymbol{M}\,\boldsymbol{N}\,\boldsymbol{e}_3\,\boldsymbol{e}_3 \\ &= -\boldsymbol{M}\,\boldsymbol{N}. \end{aligned}$$

That is, compared with the expression for the rotor in Euclidean space, a minus sign is introduced. This, however, cancels out when the rotor is applied:

$$\boldsymbol{R}\,\boldsymbol{A}\,\widetilde{\boldsymbol{R}} = (-\boldsymbol{M}\,\boldsymbol{N})\,\boldsymbol{A}\,(-\boldsymbol{N}\,\boldsymbol{M}) = (\boldsymbol{M}\,\boldsymbol{N})\,\boldsymbol{A}\,(\boldsymbol{N}\,\boldsymbol{M}).$$

It may also be argued that since an overall scalar factor is of no importance for homogeneous vectors with respect to their projection into Euclidean space, the minus sign of the rotor in projective space may be neglected. Hence, because direction vectors can also be regarded as vectors in the corresponding

Euclidean space, *the same representation of a rotor can be used in both Euclidean and projective space.*

4.3 Conformal Space

In this introduction to conformal space, many of the concepts introduced in the discussion of projective space will reappear. Again, Euclidean space is embedded in a higher-dimensional space, where the extra dimensions have particular properties such that linear subspaces in conformal space represent geometric entities in Euclidean space that are of interest. In projective space, the addition of one dimension allows the representation of null-dimensional spaces, i.e. points, in Euclidean space by one-dimensional subspaces in projective space. This allows points to be distinguished from directions.

Conformal space is introduced here in two steps where, each time, we extend the dimensionality of the space by one. In the first step, Euclidean space is embedded in a non-linear way in a higher-dimensional space. The actual conformal space that is used later on is a special homogenization of the initial non-linear embedding of Euclidean space.

What is meant by conformal space is therefore the projective space of a conformal embedding of Euclidean space. The geometric algebra over this conformal space is called *conformal geometric algebra* (CGA), even though this terminology is not completely exact.

Before the conformal embedding of Euclidean space is discussed, it is helpful to first know what "conformal" actually means. A conformal transformation is one that preserves angles locally. For example, a conformal transformation of two intersecting straight lines in Euclidean space may result in two intersecting circles on a sphere. However, the angle at which the circles intersect is the same as the intersection angle of the lines.

It turns out that all conformal transformations can be expressed by means of combinations of inversions. In a 1D Euclidean space $\mathbb{R}^1 \equiv \mathbb{R}$, the inversion of a vector $\boldsymbol{x} \in \mathbb{R}^1$ in the unit one-dimensional sphere centered on the origin is simply $\boldsymbol{x}^{-1}$. In $\mathbb{R}^3$, the inversion of a plane in the unit sphere centered on the origin is a sphere, as shown in Fig. 4.12.

Note that inversions are closely related to reflections, in that a reflection is a special case of an inversion. In fact, an inversion in a sphere with an infinite radius, i.e. a plane, is a reflection. All Euclidean transformations can be represented by combinations of reflections. This has already been seen for rotations, which are combinations of two reflections. A translation may be represented by the reflections in two parallel reflection planes. Since all Euclidean transformations can be represented by combinations of reflections and all conformal transformations by combinations of inversions, it follows that the Euclidean transformations form a subset of the conformal transformations.

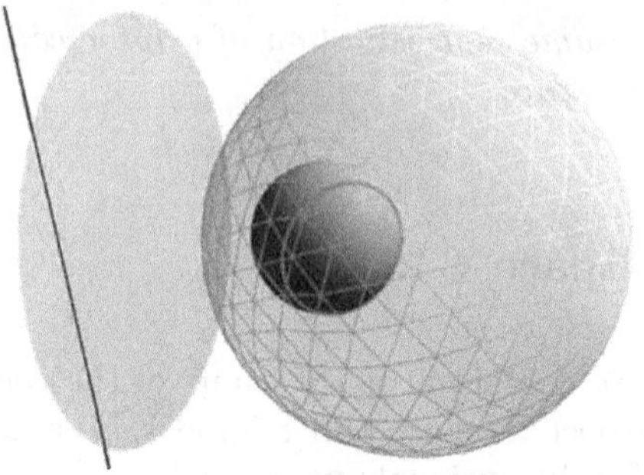

Fig. 4.12 Inversion of a plane and a line in a sphere in $\mathbb{R}^3$

The reason why the embedding of Euclidean space in conformal space is particularly useful is that a reflection in the conformal embedding space represents an inversion in the corresponding Euclidean space. The reflection of a blade in any other blade is, however, the fundamental transformation operation of geometric algebra.

The remainder of this section is structured as follows. First, the embedding in conformal space is introduced. This is followed by a discussion of the various geometric entities that can be represented in conformal space. Then, the available transformation versors are discussed.

4.3.1 Stereographic Embedding of Euclidean Space

The first step in embedding Euclidean space in conformal space is a stereographic embedding. This is illustrated in Fig. 4.13(a) for the embedding of $\mathbb{R}^1$ into $\mathbb{R}^2$. The 1D Euclidean space is mapped onto the unit circle in $\mathbb{R}^2$ by the following geometric construction. Given a vector $\boldsymbol{x} \in \mathbb{R}^1$, the line from $\boldsymbol{x}$ to $\boldsymbol{e}_+$ is drawn. The point where this line intersects the unit circle is the corresponding embedded point. The inverse procedure is called a *stereographic projection*. Clearly, $\pm\infty$ maps to $\boldsymbol{e}_+$ and the origin maps to $-\boldsymbol{e}_+$.

In general, an n-dimensional Euclidean space $\mathbb{R}^n$ is embedded in $\mathbb{R}^{n+1}$. The embedding function is $\mathcal{S} : \mathbb{R}^n \to \mathbb{S}^n \subset \mathbb{R}^{n+1}$ and is defined as

$$\mathcal{S} : \boldsymbol{x} \mapsto \frac{2}{\boldsymbol{x}^2+1}\,\boldsymbol{x} + \frac{\boldsymbol{x}^2-1}{\boldsymbol{x}^2+1}\,\boldsymbol{e}_+, \tag{4.24}$$

where $\boldsymbol{e}_+ := \boldsymbol{e}_{n+1}$ has been defined for brevity, and $\mathbb{S}^n \subset \mathbb{R}^{n+1}$ denotes the unit sphere in $\mathbb{R}^{n+1}$ such that

$$\|\mathcal{S}(\boldsymbol{x})\| = 1. \tag{4.25}$$

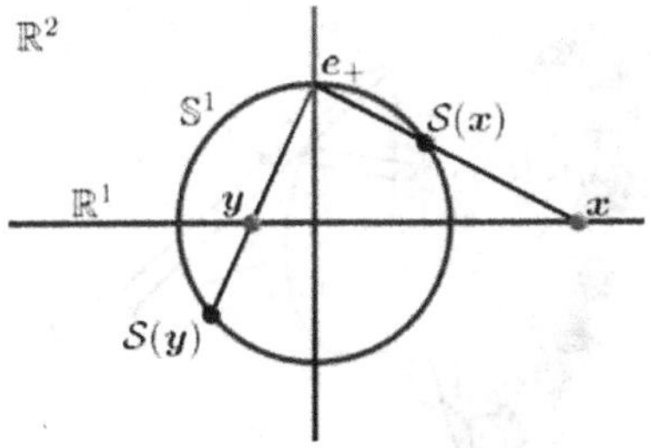

(a) Stereographic embedding of points $x, y \in \mathbb{R}^1$ onto the unit circle $\mathbb{S}^1$

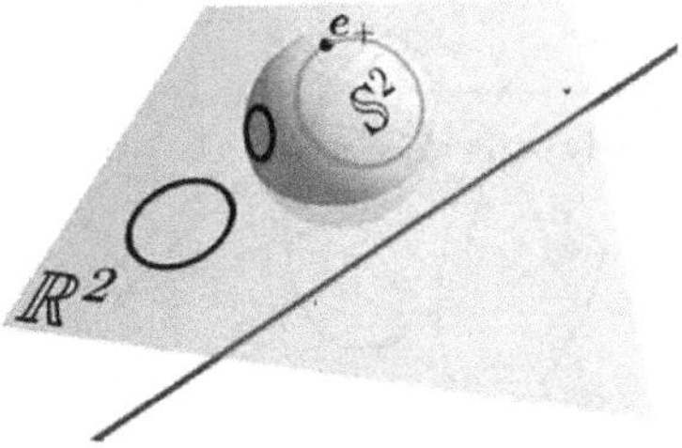

(b) Stereographic embedding of a line and a circle in $\mathbb{R}^2$ onto $\mathbb{S}^2$

Fig. 4.13 Examples of stereographic embedding

Figure 4.13(b) shows the stereographic embedding of $\mathbb{R}^2$ in $\mathbb{R}^3$. A line maps to a circle on $\mathbb{S}^2$ that passes through e_+, and a circle in $\mathbb{R}^2$ maps to a circle on $\mathbb{S}^2$ that does not pass through e_+.

Because the stereographic embedding operator $\mathcal{S}$ maps $\mathbb{R}^n$ to $\mathbb{S}^2 \subset \mathbb{R}^{n+1}$, the inverse transformation $\mathcal{S}^{-1}$ is defined only for elements of $\mathbb{S}^n$. Recall that for projective space, a similar restriction applied: the plane of homogeneous vectors with a zero e_{n+1} component could not be transformed back to Euclidean space.

The stereographic projection operator $\mathcal{S}^{-1} : \mathbb{S}^n \to \mathbb{R}^n$ is given by

$$\mathcal{S}^{-1} : X \mapsto \frac{1}{1 - X \cdot e_+} \, \mathcal{P}^{\perp}_{e_+}(X) . \tag{4.26}$$

A vector $X \in \mathbb{R}^{n+1}$ is an element of $\mathbb{S}^n$ if and only if $\|X\| = 1$.

4.3.2 Homogenization of Stereographic Embedding

The stereographic embedding of Euclidean space is now homogenized similarly to the homogenization of Euclidean space described in Sect. 4.2. Specifically, $\mathbb{R}^{n+1} \supset \mathbb{S}^n$ is embedded in a projective space represented in $\mathbb{R}^{n+1,1}$. The space $\mathbb{R}^{n+1,1}$ is of dimension $n+2$, where its orthonormal basis contains $n+1$ basis vectors that square to $+1$ and one basis vector that squares to -1. This type of space is called *Minkowski space*. The effect of using a negatively squaring homogeneous dimension is quite substantial, as we shall see throughout the rest of this text.

The homogeneous embedding of some $\mathbb{R}^n$ in the Minkowski space $\mathbb{R}^{n,1}$ is denoted by $\mathcal{H}_M : \mathbb{R}^n \to \mathbb{A}^n_M \subset \mathbb{R}^{n,1}$ and defined as

$$\mathcal{H}_M : x \in \mathbb{R}^n \mapsto X = x + e_- , \tag{4.27}$$

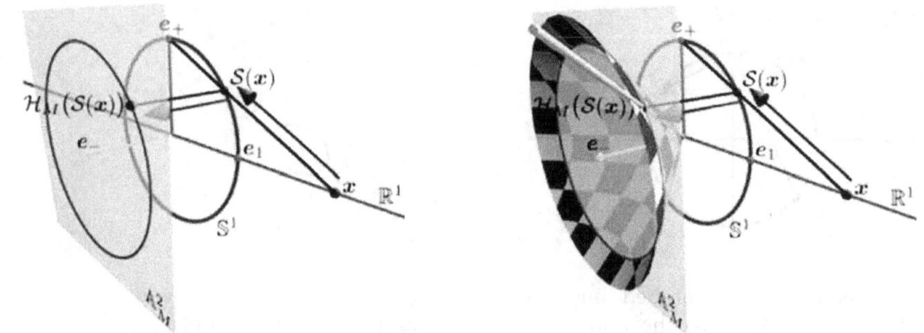

(a) Embedding of a vector $\boldsymbol{x} \in \mathbb{R}^1$ first into $\mathbb{S}^1$ and then into $\mathbb{R}^{2,1}$

(b) The same embedding as in (a) with visualization of the null-cone

Fig. 4.14 Homogenization of stereographic embedding

where the homogeneous dimension is $\boldsymbol{e}_- := \boldsymbol{e}_{n+1}$ with $\boldsymbol{e}_- \cdot \boldsymbol{e}_- = -1$, and $\mathbb{A}^n_M$ denotes the Minkowski affine space. Figure 4.14(a) illustrates the embedding of a vector from $\mathbb{R}^1$ into $\mathbb{S}^1 \subset \mathbb{R}^2$ and then into $\mathbb{R}^{2,1}$. The Minkowski affine plane $\mathbb{A}^2_M$ is also drawn.

One immediate result that follows from the use of a homogeneous dimension with a negative signature is that for $\boldsymbol{x} \in \mathbb{R}^n$,

$$
\begin{aligned}
\Big(\alpha\, \mathcal{H}_M\big(\mathcal{S}(\boldsymbol{x})\big)\Big)^2 &= \alpha^2\, \big(\mathcal{S}(\boldsymbol{x}) + \boldsymbol{e}_-\big)^2 \\
&= \alpha^2\, \Big(\big(\mathcal{S}(\boldsymbol{x})\big)^2 + \big(\boldsymbol{e}_-\big)^2\Big) \\
&= \alpha^2\, (1 - 1) \\
&= 0,
\end{aligned}
$$

where $\alpha \in \mathbb{R} \setminus 0$ is some scalar. That is, all vectors in $\mathbb{R}^{n+1,1}$ that result from an embedding of a Euclidean vector from $\mathbb{R}^n$ square to zero. For $\mathbb{R}^1$, the set of points in $\mathbb{R}^{2,1}$ that satisfy this condition lie on a cone. Hence, all null-vectors in $\mathbb{R}^{n+1,1}$ are said to lie on the *null cone*, as shown in Fig. 4.14(b). It also follows that, just as in projective space, all scaled versions of a null-vector represent the same point in Euclidean space.

The set of null-vectors, i.e. the null-cone, is denoted by $\mathbb{K}^{n+1} \subset \mathbb{R}^{n+1,1}$ and is defined as

$$
\mathbb{K}^{n+1} := \big\{\, \boldsymbol{X} \in \mathbb{R}^{n+1,1} \;:\; \boldsymbol{X}^2 = 0 \,\big\}. \tag{4.28}
$$

From the previous considerations, it follows that

$$
\mathbb{K}^{n+1} = \big\{\, \alpha\, \mathcal{H}_M\big(\mathcal{S}(\mathbb{R}^n)\big) \;:\; \alpha \in \mathbb{R} \setminus 0 \,\big\}.
$$

Let $\mathbb{SA}^n_M \subset \mathbb{A}^{n+1}_M$ denote the embedding of $\mathbb{S}^n$ in the affine space $\mathbb{A}^{n+1}_M$, i.e. $\mathbb{SA}^n_M = \mathbb{A}^{n+1}_M \cap \mathbb{K}^{n+1} = \mathcal{H}_M(\mathcal{S}(\mathbb{R}^n))$. The inverse transformation $\mathcal{H}_M^{-1} : \mathbb{SA}^n_M \to \mathbb{S}^n$ is given by

$$\mathcal{H}_M^{-1} : \boldsymbol{X} \mapsto \boldsymbol{X} - \boldsymbol{e}_- . \tag{4.29}$$

This implies that the only elements of $\mathbb{R}^{n+1,1}$ that can be mapped back to $\mathbb{R}^n$ are those in $\mathbb{SA}^n_M$. All other vectors are given a geometric meaning through their geometric OPNS or IPNS. That is, for $\boldsymbol{A} \in \mathbb{R}^{n+1,1}$,

$$\begin{aligned} \mathbb{NO}_G(\boldsymbol{A}) &:= \{\, \boldsymbol{x} \in \mathbb{R}^n \; : \; \mathcal{H}_M(\mathcal{S}(\boldsymbol{x})) \wedge \boldsymbol{A} = 0 \,\}, \\ \mathbb{NI}_G(\boldsymbol{A}) &:= \{\, \boldsymbol{x} \in \mathbb{R}^n \; : \; \mathcal{H}_M(\mathcal{S}(\boldsymbol{x})) \cdot \boldsymbol{A} = 0 \,\}. \end{aligned} \tag{4.30}$$

The geometric OPNS of a vector $\boldsymbol{X} \in \mathbb{R}^{n+1,1}$ is thus independent of the vector's scale. This is indicated in Fig. 4.14(b) by the line passing through the origin and $\mathcal{H}_M(\mathcal{S}(\boldsymbol{x}))$.

To summarize, the embedding of a Euclidean vector $\boldsymbol{x} \in \mathbb{R}^n$ in the (homogeneous) conformal space $\mathbb{R}^{n+1,1}$ is given by

$$\mathcal{H}_M\big(\mathcal{K}(\boldsymbol{x})\big) = \frac{2}{\boldsymbol{x}^2+1}\,\boldsymbol{x} \;+\; \frac{\boldsymbol{x}^2-1}{\boldsymbol{x}^2+1}\,\boldsymbol{e}_+ \;+\; \boldsymbol{e}_- . \tag{4.31}$$

Since this is an element of a projective space, an overall scale does not influence the representation in the corresponding Euclidean space. The vector may therefore be scaled without changing its representation in Euclidean space. A convenient scaling is a multiplication by $\frac{1}{2}(\boldsymbol{x}^2+1)$, which cannot be zero:

$$\begin{aligned} \frac{1}{2}(\boldsymbol{x}^2+1)\,\mathcal{H}_M\big(\mathcal{S}(\boldsymbol{x})\big) &= \boldsymbol{x} \;+\; \frac{1}{2}(\boldsymbol{x}^2-1)\,\boldsymbol{e}_+ \;+\; \frac{1}{2}(\boldsymbol{x}^2+1)\,\boldsymbol{e}_- \\ &= \boldsymbol{x} \;+\; \frac{1}{2}\,\boldsymbol{x}^2\,(\boldsymbol{e}_- + \boldsymbol{e}_+) \;+\; \frac{1}{2}\,(\boldsymbol{e}_- - \boldsymbol{e}_+) \\ &= \boldsymbol{x} \;+\; \frac{1}{2}\,\boldsymbol{x}^2\,\boldsymbol{e}_\infty \;+\; \boldsymbol{e}_o, \end{aligned} \tag{4.32}$$

where

$$\boldsymbol{e}_\infty := \boldsymbol{e}_- + \boldsymbol{e}_+ \quad \text{and} \quad \boldsymbol{e}_o := \frac{1}{2}\,(\boldsymbol{e}_- - \boldsymbol{e}_+). \tag{4.33}$$

The properties of $\boldsymbol{e}_\infty$ and $\boldsymbol{e}_o$ are easily derived from the properties of $\boldsymbol{e}_+$ and $\boldsymbol{e}_-$:

$$\boldsymbol{e}_\infty^2 = \boldsymbol{e}_o^2 = 0 \quad \text{and} \quad \boldsymbol{e}_\infty \cdot \boldsymbol{e}_o = -1. \tag{4.34}$$

Even though such a basis is rather uncommon, using $\boldsymbol{e}_\infty$ and $\boldsymbol{e}_o$ instead of $\boldsymbol{e}_-$ and $\boldsymbol{e}_+$ is simply a basis transformation. In the *null* basis formed by $\boldsymbol{e}_o$ and $\boldsymbol{e}_\infty$, $\boldsymbol{e}_o$ is now regarded as the homogeneous dimension. Therefore, the following embedding of Euclidean vectors into conformal space is defined, which is also the embedding that will be used throughout the rest of this

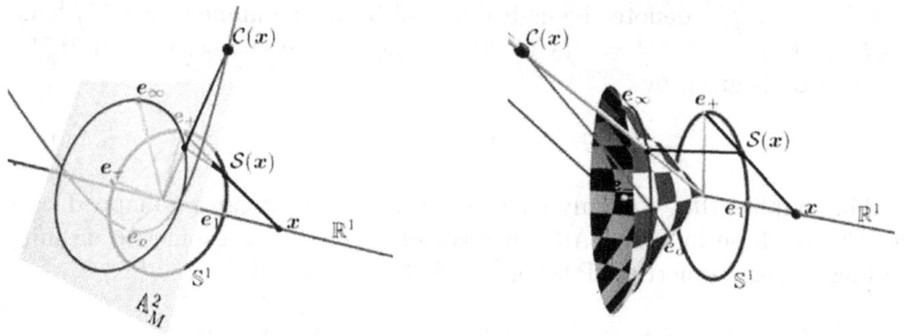

(a) Embedding of a vector $\boldsymbol{x} \in \mathbb{R}^1$ as $\mathcal{C}(\boldsymbol{x})$

(b) A different view of the same embedding with the null-cone drawn

Fig. 4.15 Embedding using $\mathcal{C}$

text. For this embedding, the operator $\mathcal{C} : \mathbb{R}^n \rightarrow \mathbb{K}^{n+1} \subset \mathbb{R}^{n+1,1}$ is defined by

$$\mathcal{C} : \boldsymbol{x} \mapsto \frac{1}{2}(\boldsymbol{x}^2 + 1)\, \mathcal{H}_M\big(\mathcal{S}(\boldsymbol{x})\big), \tag{4.35}$$

such that

$$\mathcal{C}(\boldsymbol{x}) = \boldsymbol{x} + \frac{1}{2}\boldsymbol{x}^2\, \boldsymbol{e}_\infty + \boldsymbol{e}_o. \tag{4.36}$$

Figure 4.15 illustrates this type of embedding. The vector $\boldsymbol{x} \in \mathbb{R}^1$ is embedded in $\mathbb{R}^{2,1}$ just as in Fig. 4.14 and is scaled such that its $\boldsymbol{e}_o$ component is unity. It then lies on a parabola, which is called a *horosphere*. That is, the 1D Euclidean space $\mathbb{R}^1$ is mapped onto a particular parabola on the null cone $\mathbb{K}^2$.

The inverse operator $\mathcal{C}^{-1}$ is defined only for vectors on the null-cone $\mathbb{K}^{n+1}$, i.e. $\mathcal{C}^{-1} : \mathbb{K}^{n+1} \rightarrow \mathbb{R}^n$,

$$\mathcal{C}^{-1} : \boldsymbol{X} \in \mathbb{K}^{n+1} \mapsto \mathcal{P}^{\perp}_{\boldsymbol{e}_\infty \wedge \boldsymbol{e}_o}\left(\frac{\boldsymbol{X}}{-\boldsymbol{X} \cdot \boldsymbol{e}_\infty}\right). \tag{4.37}$$

Dividing by $-\boldsymbol{X} \cdot \boldsymbol{e}_\infty$ has the effect of orthographically projecting $\boldsymbol{X}$ along $\boldsymbol{e}_o$ onto the parabola that passes through $\boldsymbol{e}_o$. This is basically the same as the projection onto the affine plane in projective space.

4.3.3 Geometric Algebra on $\mathbb{R}^{n+1,1}$

Just as for projective space, a geometric algebra can be constructed over $\mathbb{R}^{n+1,1}$, and is denoted by $\mathbb{G}_{n+1,1}$. Blades in $\mathbb{G}_{n+1,1}$ still represent linear subspaces through their IPNS and OPNS with respect to $\mathbb{R}^{n+1,1}$ itself. However,

Fig. 4.16 OPNS of the outer product of vectors $\boldsymbol{X}, \boldsymbol{Y} \in \mathbb{G}^1_{2,1}$

the *geometric* IPNS and OPNS, which are given in much the same way as for vectors, are the main things of interest here. Let $\boldsymbol{A}_{\langle k \rangle} \in \mathbb{G}^k_{n+1,1}$; then

$$\begin{aligned} \mathbb{NO}_G(\boldsymbol{A}_{\langle k \rangle}) &:= \{\, \boldsymbol{x} \in \mathbb{R}^n \,:\, \mathcal{C}(\boldsymbol{x}) \wedge \boldsymbol{A}_{\langle k \rangle} = 0 \,\}, \\ \mathbb{NI}_G(\boldsymbol{A}_{\langle k \rangle}) &:= \{\, \boldsymbol{x} \in \mathbb{R}^n \,:\, \mathcal{C}(\boldsymbol{x}) \cdot \boldsymbol{A}_{\langle k \rangle} = 0 \,\}. \end{aligned} \tag{4.38}$$

Since all vectors on the null cone $\mathbb{K}^{n+1}$ can be projected into $\mathbb{R}^n$, the geometric null spaces can also be expressed as

$$\begin{aligned} \mathbb{NO}_G(\boldsymbol{A}_{\langle k \rangle}) &= \mathcal{C}^{-1}\Big(\{\, \boldsymbol{X} \in \mathbb{K}^{n+1} \,:\, \boldsymbol{X} \wedge \boldsymbol{A}_{\langle k \rangle} = 0 \,\}\Big), \\ \mathbb{NI}_G(\boldsymbol{A}_{\langle k \rangle}) &= \mathcal{C}^{-1}\Big(\{\, \boldsymbol{X} \in \mathbb{K}^{n+1} \,:\, \boldsymbol{X} \cdot \boldsymbol{A}_{\langle k \rangle} = 0 \,\}\Big). \end{aligned}$$

This can also be interpreted as the inverse conformal embedding of those vectors in $\mathbb{R}^{n+1,1}$ that lie in the intersection of the null space represented by $\boldsymbol{A}_{\langle k \rangle}$ and the null cone $\mathbb{K}^{n+1}$. That is,

$$\begin{aligned} \mathbb{NO}_G(\boldsymbol{A}_{\langle k \rangle}) &= \mathcal{C}^{-1}\Big(\mathbb{NO}(\boldsymbol{A}_{\langle k \rangle}) \cap \mathbb{K}^{n+1}\Big), \\ \mathbb{NI}_G(\boldsymbol{A}_{\langle k \rangle}) &= \mathcal{C}^{-1}\Big(\mathbb{NI}(\boldsymbol{A}_{\langle k \rangle}) \cap \mathbb{K}^{n+1}\Big). \end{aligned}$$

An example of the OPNS of a bivector in $\mathbb{G}_{2,1}$ is shown in Fig. 4.16. The vectors $\boldsymbol{X}, \boldsymbol{Y} \in \mathbb{K}^2$ span a 2D subspace in $\mathbb{R}^{2,1}$, the plane $\mathbb{NO}(\boldsymbol{X} \wedge \boldsymbol{Y})$. However, the geometric OPNS of $\boldsymbol{X} \wedge \boldsymbol{Y}$ is the set of points in $\mathbb{R}^1$ whose conformal embeddings in $\mathbb{R}^{n+1,1}$ lie in $\mathbb{NO}(\boldsymbol{X} \wedge \boldsymbol{Y})$; these are the points $\mathcal{C}^{-1}(\boldsymbol{X})$ and $\mathcal{C}^{-1}(\boldsymbol{Y})$. Hence, $\mathbb{NO}_G(\boldsymbol{X} \wedge \boldsymbol{Y})$ represents a *point pair*.

4.3.4 Inner-Product Representations in $\mathbb{G}_{4,1}$

In contrast to the cases of Euclidean and projective space, it is initially easier to look at the geometric IPNS of blades in $\mathbb{G}_{4,1}$. Before specific geometric entities are treated, some general properties will be presented.

Consider Euclidean vectors $\boldsymbol{a}, \boldsymbol{b} \in \mathbb{R}^3$ with conformal embeddings $\boldsymbol{A} := \mathcal{C}(\boldsymbol{a})$ and $\boldsymbol{B} := \mathcal{C}(\boldsymbol{b})$; that is,

$$\boldsymbol{A} = \boldsymbol{a} + \frac{1}{2}\boldsymbol{a}^2\, \boldsymbol{e}_\infty + \boldsymbol{e}_o \quad \text{and} \quad \boldsymbol{B} = \boldsymbol{b} + \frac{1}{2}\boldsymbol{b}^2\, \boldsymbol{e}_\infty + \boldsymbol{e}_o.$$

Using the properties of $\boldsymbol{e}_\infty$ and $\boldsymbol{e}_o$, it follows that

$$\begin{aligned} \boldsymbol{A} \cdot \boldsymbol{B} &= (\boldsymbol{a} + \frac{1}{2}\boldsymbol{a}^2\, \boldsymbol{e}_\infty + \boldsymbol{e}_o) \cdot (\boldsymbol{b} + \frac{1}{2}\boldsymbol{b}^2\, \boldsymbol{e}_\infty + \boldsymbol{e}_o) \\ &= \boldsymbol{a} \cdot \boldsymbol{b} - \frac{1}{2}\boldsymbol{a}^2 - \frac{1}{2}\boldsymbol{b}^2 \\ &= -\frac{1}{2}(\boldsymbol{a} - \boldsymbol{b})^2 \\ &= -\frac{1}{2}\|\boldsymbol{a} - \boldsymbol{b}\|^2. \end{aligned} \tag{4.39}$$

Thus, the inner product of two conformal vectors gives a measure of the Euclidean distance between their corresponding Euclidean vectors. This fundamental feature of conformal space plays an important role in the following.

4.3.4.1 Points

The geometric IPNS of a vector $\boldsymbol{A} \in \mathbb{K}^4 \subset \mathbb{G}^1_{4,1}$ on the null-cone is given by

$$\mathbb{NI}_G(\boldsymbol{A}) = \mathcal{C}^{-1}\Big(\mathbb{NI}(\boldsymbol{A})\Big).$$

Because vectors on the null cone are null-vectors,

$$\mathbb{NI}(\boldsymbol{A}) = \{\, \alpha\, \boldsymbol{A} \;:\; \alpha \in \mathbb{R} \,\},$$

and thus

$$\mathbb{NI}_G(\boldsymbol{A}) = \mathcal{C}^{-1}(\boldsymbol{A}).$$

Just as for projective space, a null-dimensional entity in Euclidean space (a point) can be represented by a one-dimensional subspace in (homogeneous) conformal space.

4.3.4.2 Spheres

While vectors on the null-cone represent points in Euclidean space in terms of their geometric IPNS, this is only a subset of all vectors $\mathbb{G}^1_{4,1}$, and it is not immediately clear what vectors off the null-cone represent. Consider the vector $\boldsymbol{a} \in \mathbb{R}^3$ and define $\boldsymbol{A} := \mathcal{C}(\boldsymbol{a})$, which lies on the null cone. Furthermore, define a vector $\boldsymbol{S} \in \mathbb{G}^1_{4,1}$ off the null-cone by

$$\boldsymbol{S} := \boldsymbol{A} - \frac{1}{2}\rho^2\, \boldsymbol{e}_\infty, \tag{4.40}$$

where $\rho \in \mathbb{R}$. To evaluate the geometric IPNS of $\boldsymbol{S}$, consider the inner product of a vector $\boldsymbol{X} \in \mathbb{K}^4$ with $\boldsymbol{S}$:

$$\begin{aligned} \boldsymbol{S} \cdot \boldsymbol{X} &= \boldsymbol{A} \cdot \boldsymbol{X} - \frac{1}{2}\rho^2\, \boldsymbol{e}_\infty \cdot \boldsymbol{X} \\ &= -\frac{1}{2}(\boldsymbol{a} - \boldsymbol{x})^2 + \frac{1}{2}\rho^2. \end{aligned} \tag{4.41}$$

Hence,

$$\boldsymbol{S} \cdot \boldsymbol{X} = 0 \iff (\boldsymbol{a} - \boldsymbol{x})^2 = \rho^2.$$

That is, the inner product of $\boldsymbol{S}$ and $\boldsymbol{X}$ is zero if and only if $\boldsymbol{x} = \mathcal{C}^{-1}(\boldsymbol{X})$ lies on a sphere centered on $\boldsymbol{a} = \mathcal{C}^{-1}(\boldsymbol{A})$ with radius ρ. Therefore, the geometric IPNS of $\boldsymbol{S}$ is a sphere:

$$\mathbb{NI}_G(\boldsymbol{S}) = \{\, \boldsymbol{x} \in \mathbb{R}^3 \;:\; \|\boldsymbol{x} - \boldsymbol{a}\|^2 = \rho^2 \,\}. \tag{4.42}$$

Note that, as for points, every scaled version of $\boldsymbol{S}$ represents the same sphere. The "normalized" form of $\boldsymbol{S}$ as given in (4.40) has the added advantage that the sphere's radius can be evaluated via

$$\boldsymbol{S}^2 = \boldsymbol{A}^2 - \rho^2\, \boldsymbol{A} \cdot \boldsymbol{e}_\infty = \rho^2. \tag{4.43}$$

Given an arbitrarily scaled version of $\boldsymbol{S}$, the radius can be evaluated via

$$\left(\frac{\boldsymbol{S}}{-\boldsymbol{S} \cdot \boldsymbol{e}_\infty}\right)^2 = \rho^2. \tag{4.44}$$

Whether a point $\boldsymbol{X} := \mathcal{C}(\boldsymbol{x})$, with $\boldsymbol{x} \in \mathbb{R}^3$, lies inside, on, or outside a sphere represented by $\boldsymbol{S}$ can be evaluated quite easily. From (4.41), it follows that

$$\frac{\boldsymbol{S} \cdot \boldsymbol{X}}{(\boldsymbol{S} \cdot \boldsymbol{e}_\infty)\,(\boldsymbol{X} \cdot \boldsymbol{e}_\infty)} \begin{cases} > 0 : & \boldsymbol{x} \text{ inside sphere,} \\ = 0 : & \boldsymbol{x} \text{ on sphere,} \\ < 0 : & \boldsymbol{x} \text{ outside sphere.} \end{cases} \tag{4.45}$$

This feature also forms the basic idea behind the hypersphere neuron [14, 15]. This may be represented as a perceptron with two "bias" components, and allows the separation of the input space of a multilayer perceptron in terms of hyperspheres, not hyperplanes.

4.3.4.3 Imaginary Spheres

Consider now a vector $\boldsymbol{S} \in \mathbb{G}^1_{4,1}$ defined by

$$\boldsymbol{S} := \boldsymbol{A} + \frac{1}{2}\rho^2\, \boldsymbol{e}_\infty. \tag{4.46}$$

The inner product of $\boldsymbol{S}$ with $\boldsymbol{X} := \mathcal{C}(\boldsymbol{x})$, $\boldsymbol{x} \in \mathbb{R}^3$, gives

$$\boldsymbol{S} \cdot \boldsymbol{X} = -\frac{1}{2}(\boldsymbol{a} - \boldsymbol{x})^2 - \frac{1}{2}\rho^2,$$

such that

$$\boldsymbol{S} \cdot \boldsymbol{X} = 0 \iff (\boldsymbol{a} - \boldsymbol{x})^2 = -\rho^2.$$

Since $\boldsymbol{x} \in \mathbb{R}^3$, this condition is never satisfied for $\rho \neq 0$. However, if $\boldsymbol{x} \in \mathbb{C}^3$ were an element of a complex vector space, then, together with an appropriate definition of the norm, the solution would be

$$\|\boldsymbol{a} - \boldsymbol{x}\| = \mathsf{i}\,\rho,$$

where $\mathsf{i} = \sqrt{-1}$ is the imaginary unit. In this context, $\boldsymbol{S}$ as defined in (4.46) represents a sphere with an *imaginary* radius ρ.

A nice way to construct an imaginary sphere is by the addition of two null vectors. For example, given $\boldsymbol{a}, \boldsymbol{b} \in \mathbb{R}^3$, and defining $\boldsymbol{A} := \mathcal{C}(\boldsymbol{a})$ and $\boldsymbol{B} := \mathcal{C}(\boldsymbol{b})$, then $\boldsymbol{S} = \boldsymbol{A} + \boldsymbol{B}$ represents an imaginary sphere, whose center lies at the midpoint between $\boldsymbol{a}$ and $\boldsymbol{b}$, and whose radius is $\rho = \mathsf{i}\,\|\boldsymbol{a} - \boldsymbol{b}\|/2$.

Note that any vector in $\mathbb{G}^1_{4,1}$ may be brought into the form

$$\boldsymbol{S} = \boldsymbol{A} \pm \frac{1}{2}\rho^2\, \boldsymbol{e}_\infty,$$

where $\boldsymbol{A} := \mathcal{C}(\boldsymbol{a})$ and $\boldsymbol{a} \in \mathbb{R}^3$. From a visual point of view, it can be said that vectors of the type $\boldsymbol{S} = \boldsymbol{A} - \frac{1}{2}\rho^2\, \boldsymbol{e}_\infty$ lie outside the null cone and vectors of the type $\boldsymbol{S} = \boldsymbol{A} + \frac{1}{2}\rho^2\, \boldsymbol{e}_\infty$ lie inside the null cone. It follows that any vector in $\mathbb{G}^1_{4,1}$ represents a sphere with either a positive, zero, or imaginary radius. In terms of the geometric IPNS, the basic "building blocks" of homogeneous conformal space are therefore spheres.

4.3.4.4 Planes

It was mentioned earlier that a plane can be regarded as a sphere with an infinite radius. Since we are working in a homogeneous space, infinity can be represented by setting the homogeneous component of a vector to zero. And this is also all that it takes to make a sphere into a plane, which becomes clear from (4.44). Consider a vector $\boldsymbol{P} \in \mathbb{G}^1_{4,1}$ defined by

$$\boldsymbol{P} := \boldsymbol{A} - \boldsymbol{e}_o - \frac{1}{2}\rho^2\,\boldsymbol{e}_\infty = \boldsymbol{a} + \frac{1}{2}\boldsymbol{a}^2\,\boldsymbol{e}_\infty - \frac{1}{2}\rho^2\,\boldsymbol{e}_\infty$$

that has no $\boldsymbol{e}_o$ component, i.e. it is zero in the homogeneous dimension. The inner product of $\boldsymbol{P}$ with some vector $\boldsymbol{X} := \mathcal{C}(\boldsymbol{x})$, $\boldsymbol{x} \in \mathbb{R}^3$, gives

$$\begin{aligned}\boldsymbol{P}\cdot\boldsymbol{X} &= \boldsymbol{a}\cdot\boldsymbol{x} - \frac{1}{2}\boldsymbol{a}^2 + \frac{1}{2}\rho^2\\ &= \|\boldsymbol{a}\|\,\|\boldsymbol{x}^{\|}\| - \frac{1}{2}(\boldsymbol{a}^2 - \rho^2),\end{aligned}$$

where $\boldsymbol{x}^{\|} := \mathcal{P}_{\boldsymbol{a}}(\boldsymbol{x})$ is the component of $\boldsymbol{x}$ parallel to $\boldsymbol{a}$. Therefore,

$$\boldsymbol{P}\cdot\boldsymbol{X} = 0 \iff \|\boldsymbol{x}^{\|}\| = \frac{\boldsymbol{a}^2 - \rho^2}{2\,\|\boldsymbol{a}\|}.$$

Hence, all vectors $\boldsymbol{x}$ whose component parallel to $\boldsymbol{a}$ has a fixed length lie in the geometric IPNS of $\boldsymbol{P}$, which thus represents a plane with an orthogonal distance $(\boldsymbol{a}^2 - \rho^2)/(2\|\boldsymbol{a}\|)$ from the origin and a normal $\boldsymbol{a}$.

In general, a plane with a normal $\hat{\boldsymbol{a}} := \boldsymbol{a}/\|\boldsymbol{a}\|$ and an orthogonal distance $\alpha \in \mathbb{R}$ from the origin is represented by

$$\boldsymbol{P} = \hat{\boldsymbol{a}} + \alpha\,\boldsymbol{e}_\infty\,. \tag{4.47}$$

Another nice representation of planes is provided by the difference of two vectors from the null cone. That is, for $\boldsymbol{A}, \boldsymbol{B} \in \mathbb{K}^4$, we define $\boldsymbol{P} := \boldsymbol{A} - \boldsymbol{B}$. The inner product of $\boldsymbol{P}$ with a vector $\boldsymbol{X}$ then gives

$$\begin{aligned}\boldsymbol{P}\cdot\boldsymbol{X} &= \boldsymbol{A}\cdot\boldsymbol{X} - \boldsymbol{B}\cdot\boldsymbol{X}\\ &= -\frac{1}{2}(\boldsymbol{a} - \boldsymbol{x})^2 + \frac{1}{2}(\boldsymbol{b} - \boldsymbol{x})^2.\end{aligned}$$

It follows that

$$\boldsymbol{P}\cdot\boldsymbol{X} = 0 \iff \frac{1}{2}(\boldsymbol{a} - \boldsymbol{x})^2 = \frac{1}{2}(\boldsymbol{b} - \boldsymbol{x})^2.$$

This is the case if $\boldsymbol{x}$ lies on the plane halfway between $\boldsymbol{a}$ and $\boldsymbol{b}$, with normal $\boldsymbol{a} - \boldsymbol{b}$.

4.3.4.5 Circles

In the inner-product representation, circles are constructed by intersecting two spheres. Let $\boldsymbol{S}_1, \boldsymbol{S}_2 \in \mathbb{G}^1_{4,1}$ represent two spheres as described before; then $\boldsymbol{C} := \boldsymbol{S}_1 \wedge \boldsymbol{S}_2$ represents the spheres' intersection circle, because

$$\mathbb{NI}_G(\boldsymbol{C}) = \{\, \boldsymbol{x} \in \mathbb{R}^3 \;:\; \mathcal{C}(\boldsymbol{x}) \cdot \boldsymbol{C} = 0 \,\} \tag{4.48}$$

and

$$\boldsymbol{X} \cdot (\boldsymbol{S}_1 \wedge \boldsymbol{S}_2) = (\boldsymbol{X} \cdot \boldsymbol{S}_1)\, \boldsymbol{S}_2 - (\boldsymbol{X} \cdot \boldsymbol{S}_2)\, \boldsymbol{S}_1,$$

where $\boldsymbol{X} := \mathcal{C}(\boldsymbol{x})$. If $\boldsymbol{S}_1$ and $\boldsymbol{S}_2$ are linearly independent, i.e. they represent different spheres, then $\boldsymbol{X} \cdot (\boldsymbol{S}_1 \wedge \boldsymbol{S}_2)$ is zero if and only if $\boldsymbol{X} \cdot \boldsymbol{S}_1 = 0$ and $\boldsymbol{X} \cdot \boldsymbol{S}_2 = 0$. Thus, $\mathbb{NI}_G(\boldsymbol{C})$ is the set of those points that lie on both spheres, which is the spheres' intersection circle.

If the two spheres intersect only in a single point, then $\boldsymbol{C}$ represents that point. If the spheres do not intersect, then $\mathbb{NI}_G(\boldsymbol{C}) = \emptyset$. However, in all cases the algebraic element $\boldsymbol{C}$ is of the same grade, and contains geometric information about the intersection. In fact, if $\boldsymbol{S}_1$ and $\boldsymbol{S}_2$ intersect in a single point, then $\boldsymbol{C}$ represents their intersection point, but the algebraic entity also contains the normal to the plane that is tangential to the spheres at the intersection point. Furthermore, if the spheres do not intersect, then $\boldsymbol{C}$ represents an imaginary circle.

It is quite instructive to consider the case of two spheres intersecting in a point, in some more detail. Note, that a sphere centered on some vector $\boldsymbol{a} \in \mathbb{R}^3$, that passes through the origin, must have radius $\|\boldsymbol{a}\|$. Hence, the representation of such a sphere in $\mathbb{G}_{4,1}$ is given by

$$\boldsymbol{S} = \boldsymbol{a} + \frac{1}{2}\,\boldsymbol{a}^2\, \boldsymbol{e}_\infty + \boldsymbol{e}_o - \frac{1}{2}\,\boldsymbol{a}^2\, \boldsymbol{e}_\infty = \boldsymbol{a} + \boldsymbol{e}_o\,.$$

Without loss of generality, we can consider the case of two spheres intersecting only in the origin. We define

$$\boldsymbol{S}_1 := \alpha\, \boldsymbol{a} + \boldsymbol{e}_o\,, \quad \boldsymbol{S}_2 := \beta\, \boldsymbol{a} + \boldsymbol{e}_o\,, \quad \alpha, \beta \in \mathbb{R} \setminus \{0\}\,, \quad \alpha \neq \beta\,.$$

The intersection circle $\boldsymbol{C}$ is then given by

$$\boldsymbol{C} = \boldsymbol{S}_1 \wedge \boldsymbol{S}_2 = \alpha\, \boldsymbol{a} \wedge \boldsymbol{e}_o + \beta\, \boldsymbol{e}_o \wedge \boldsymbol{a} = (\alpha - \beta)\, \boldsymbol{a} \wedge \boldsymbol{e}_o\,.$$

Clearly, $\boldsymbol{C} \cdot \boldsymbol{e}_o = 0$, and, thus, $\boldsymbol{C}$ represents the origin. However, $\boldsymbol{C}$ also contains the vector $\boldsymbol{a}$, which is normal to the plane that is tangential to the spheres at the intersection point. In fact, $\boldsymbol{a}$ is the inner-product representation of just that plane (see (4.47)).

An entity of the type $\boldsymbol{n} \wedge \mathcal{C}(\boldsymbol{x})$, with $\boldsymbol{n}, \boldsymbol{x} \in \mathbb{R}^3$, can be regarded as representing a tangential plane with normal $\boldsymbol{n}$ at the point $\boldsymbol{x}$.

4.3.4.6 Lines

The inner-product representation of lines is constructed in much the same way as the inner-product representation of circles: a line is the intersection of two planes. Let $\boldsymbol{P}_1, \boldsymbol{P}_2 \in \mathbb{G}^1_{4,1}$ represent two inner-product planes; then $\boldsymbol{L} := \boldsymbol{P}_1 \wedge \boldsymbol{P}_2$ represents the planes' intersection line. This follows, as before, because

$$\mathbb{NI}_G(\boldsymbol{L}) = \{\, \boldsymbol{x} \in \mathbb{R}^3 \;:\; \mathcal{C}(\boldsymbol{x}) \cdot \boldsymbol{L} = 0 \,\} \tag{4.49}$$

and

$$\boldsymbol{X} \cdot (\boldsymbol{P}_1 \wedge \boldsymbol{P}_2) = (\boldsymbol{X} \cdot \boldsymbol{P}_1)\, \boldsymbol{P}_2 - (\boldsymbol{X} \cdot \boldsymbol{P}_2)\, \boldsymbol{P}_1,$$

where $\boldsymbol{X} := \mathcal{C}(\boldsymbol{x})$. If $\boldsymbol{P}_1$ and $\boldsymbol{P}_2$ are linearly independent vectors, i.e. they represent different planes, then $\boldsymbol{X} \cdot \boldsymbol{L} = 0$ if and only if $\boldsymbol{X} \cdot \boldsymbol{P}_1 = 0$ and $\boldsymbol{X} \cdot \boldsymbol{P}_2 = 0$, which implies that $\mathbb{NI}_G(\boldsymbol{L})$ is the planes' intersection line.

As an example, consider the case of two parallel planes, with $\boldsymbol{P}_1 := \hat{\boldsymbol{a}} + \alpha\, \boldsymbol{e}_\infty$ and $\boldsymbol{P}_2 := \hat{\boldsymbol{a}} + \beta\, \boldsymbol{e}_\infty$, where $\hat{\boldsymbol{a}} \in \mathbb{R}^3$ is a unit vector that gives the normal of the planes, and $\alpha, \beta \in \mathbb{R}$. Then

$$\boldsymbol{P}_1 \wedge \boldsymbol{P}_2 = (\hat{\boldsymbol{a}} + \alpha\, \boldsymbol{e}_\infty) \wedge (\hat{\boldsymbol{a}} + \beta\, \boldsymbol{e}_\infty) = (\beta - \alpha)\, \hat{\boldsymbol{a}} \wedge \boldsymbol{e}_\infty,$$

which should represent the planes' intersection line at infinity. Using $\boldsymbol{X} := \mathcal{C}(\boldsymbol{x})$ again as before, then

$$\boldsymbol{X} \cdot (\hat{\boldsymbol{a}} \wedge \boldsymbol{e}_\infty) = (\boldsymbol{x} \cdot \hat{\boldsymbol{a}})\, \boldsymbol{e}_\infty + \hat{\boldsymbol{a}}.$$

This is zero only if $\hat{\boldsymbol{a}} = 0$, and thus $\mathbb{NI}_G(\boldsymbol{P}_1 \wedge \boldsymbol{P}_2) = \emptyset$; that is, the parallel planes have no intersection that is representable in $\mathbb{R}^3$, as expected. However, $\hat{\boldsymbol{a}} \wedge \boldsymbol{e}_\infty$ can be regarded as representing a line at infinity, which is the intersection of two planes with a normal $\hat{\boldsymbol{a}}$.

4.3.4.7 Point Pairs

An algebraic entity that represents a point pair has to exist in $\mathbb{G}_{4,1}$ because all intersections of a circle and a plane or of three spheres have to be representable. Consider the case of three spheres represented by $\boldsymbol{S}_1, \boldsymbol{S}_2, \boldsymbol{S}_3 \in \mathbb{G}^1_{4,1}$ that intersect in two points; then $\boldsymbol{A} := \boldsymbol{S}_1 \wedge \boldsymbol{S}_2 \wedge \boldsymbol{S}_3$ is an inner-product representation of this point pair. This follows immediately, using the same method as before. First of all,

$$\mathbb{NI}_G(\boldsymbol{A}) = \{\, \boldsymbol{x} \in \mathbb{R}^3 \;:\; \mathcal{C}(\boldsymbol{x}) \cdot \boldsymbol{A} = 0 \,\} \tag{4.50}$$

and

$$\boldsymbol{X} \cdot \boldsymbol{A} = (\boldsymbol{X} \cdot \boldsymbol{S}_1)\, (\boldsymbol{S}_2 \wedge \boldsymbol{S}_3) - (\boldsymbol{X} \cdot \boldsymbol{S}_2)\, (\boldsymbol{S}_1 \wedge \boldsymbol{S}_3) + (\boldsymbol{X} \cdot \boldsymbol{S}_3)\, (\boldsymbol{S}_1 \wedge \boldsymbol{S}_2),$$

where $\boldsymbol{X} := \mathcal{C}(\boldsymbol{x})$. Because $\boldsymbol{S}_1$, $\boldsymbol{S}_2$, and $\boldsymbol{S}_3$ represent different spheres they are linearly independent vectors, and thus $\boldsymbol{X} \cdot \boldsymbol{A} = 0$ if and only if $\boldsymbol{X} \cdot \boldsymbol{S}_i = 0$ for all $i \in \{1, 2, 3\}$. Hence, $\mathbb{NI}_G(\boldsymbol{A})$ is the intersection of all three spheres, which in this case is a point pair, by definition. Note that a point pair is more easily represented in the GOPNS, where it is simply the outer product of the two points.

4.3.5 Outer-Product Representations in $\mathbb{G}_{4,1}$

The outer-product representation is dual to the inner-product representation, as has been shown in Sect. 3.2.8. This property is used below to verify some of the representations.

4.3.5.1 Points

It was shown earlier that a point $\boldsymbol{a} \in \mathbb{R}^3$ is represented in $\mathbb{G}_{4,1}$ by $\boldsymbol{A} := \mathcal{C}(\boldsymbol{a})$. Since $\boldsymbol{A}^2 = 0$, $\boldsymbol{A}$ is an inner-product representation of the point $\boldsymbol{a}$. It is also an outer-product representation of $\boldsymbol{a}$, since $\boldsymbol{A} \wedge \boldsymbol{A} = 0$ by the definition of the outer product. That is,

$$\mathbb{NO}_G(\boldsymbol{A}) = \{\, \boldsymbol{x} \in \mathbb{R}^3 \;:\; \mathcal{C}(\boldsymbol{x}) \wedge \boldsymbol{A} = 0 \,\} = \{\, \boldsymbol{a} \,\}. \tag{4.51}$$

The main difference between the GIPNS and the GOPNS is that the GIPNS depends on the metric of the space, whereas the GOPNS depends only on the algebraic properties.

4.3.5.2 Point Pairs

It has been mentioned before that the outer product of two vectors $\boldsymbol{A}, \boldsymbol{B} \in \mathbb{G}^1_{4,1}$, defined as $\boldsymbol{A} := \mathcal{C}(\boldsymbol{a})$ and $\boldsymbol{B} := \mathcal{C}(\boldsymbol{b})$ with $\boldsymbol{a}, \boldsymbol{b} \in \mathbb{R}^3$, represents the point pair $\boldsymbol{a}$, $\boldsymbol{b}$. This can be seen as follows. First of all,

$$\mathbb{NO}(\boldsymbol{A}\wedge\boldsymbol{B}) = \{\, \boldsymbol{X} \in \mathbb{G}^1_{4,1} \;:\; \boldsymbol{X}\wedge\boldsymbol{A}\wedge\boldsymbol{B} = 0 \,\} = \{\, \alpha\, \boldsymbol{A} + \beta\, \boldsymbol{B} \;:\; (\alpha, \beta) \in \mathbb{R}^2 \,\}.$$

The GOPNS of $\boldsymbol{A} \wedge \boldsymbol{B}$ is the subset of $\mathbb{NO}(\boldsymbol{A} \wedge \boldsymbol{B})$ whose constituent vectors square to zero, because these are the vectors that lie on the null cone and thus represent points in $\mathbb{R}^3$. Let $\boldsymbol{X} := \alpha\, \boldsymbol{A} + \beta\, \boldsymbol{B}$; then

$$\boldsymbol{X}^2 = 2\, \alpha\, \beta\; \boldsymbol{A} \cdot \boldsymbol{B},$$

because $\boldsymbol{A}^2 = \boldsymbol{B}^2 = 0$. Hence, $\boldsymbol{X}^2 = 0$ has the non-trivial solutions that either $\alpha = 0$ or $\beta = 0$. Thus,

$$\mathbb{NO}_G(\boldsymbol{A} \wedge \boldsymbol{B}) = \left\{\, \boldsymbol{x} \in \mathbb{R}^3 \;:\; \mathcal{C}(\boldsymbol{x}) \wedge \boldsymbol{A} \wedge \boldsymbol{B} = 0 \,\right\} = \{\, \boldsymbol{a},\, \boldsymbol{b} \,\}. \tag{4.52}$$

4.3.5.3 Homogeneous Points

A *homogeneous point* is a point pair consisting of a Euclidean point and the point at infinity. That is, if $\boldsymbol{a} \in \mathbb{R}^3$ and $\boldsymbol{A} := \mathcal{C}(\boldsymbol{a})$, then $\boldsymbol{H} := \boldsymbol{A} \wedge \boldsymbol{e}_\infty$ represents a homogeneous point. Clearly,

$$\mathbb{NO}_G(\boldsymbol{H}) = \left\{\, \boldsymbol{x} \in \mathbb{R}^3 \;:\; \mathcal{C}(\boldsymbol{x}) \wedge \boldsymbol{H} = 0 \,\right\} = \{\, \boldsymbol{a} \,\}, \tag{4.53}$$

since the point at infinity has no representation in $\mathbb{R}^3$. That is, $\boldsymbol{H}$ and $\boldsymbol{A}$ are both outer-product representations of $\boldsymbol{a}$. However,

$$\boldsymbol{A} = \boldsymbol{a} + \frac{1}{2}\, \boldsymbol{a}^2\, \boldsymbol{e}_\infty + \boldsymbol{e}_o \qquad \text{and} \qquad \boldsymbol{H} = \boldsymbol{a} \wedge \boldsymbol{e}_\infty + \boldsymbol{e}_o \wedge \boldsymbol{e}_\infty,$$

so that the latter can be regarded as a homogeneous point, whereby $\boldsymbol{e}_o \wedge \boldsymbol{e}_\infty$ takes on the function of the homogeneous dimension. This is discussed in more detail in Sect. 4.3.7.

4.3.5.4 Lines

The outer-product representation of lines in $\mathbb{G}_{4,1}$ is very similar to that in projective space. Let $\boldsymbol{a}, \boldsymbol{b} \in \mathbb{R}^3$ and $\boldsymbol{A} := \mathcal{C}(\boldsymbol{a})$, $\boldsymbol{B} := \mathcal{C}(\boldsymbol{b})$; then the line that passes through $\boldsymbol{a}$ and $\boldsymbol{b}$ is represented by $\boldsymbol{L} := \boldsymbol{A} \wedge \boldsymbol{B} \wedge \boldsymbol{e}_\infty$. This can be seen as follows:

$$\mathbb{NO}_G(\boldsymbol{L}) = \left\{\, \boldsymbol{x} \in \mathbb{R}^3 \;:\; \mathcal{C}(\boldsymbol{x}) \wedge \boldsymbol{L} = 0 \,\right\}. \tag{4.54}$$

Define $\boldsymbol{X} := \mathcal{C}(\boldsymbol{x})$; then

$$\begin{aligned} &\boldsymbol{X} \wedge \boldsymbol{L} = 0 \\ \iff\; &\boldsymbol{x} \wedge \boldsymbol{a} \wedge \boldsymbol{b} \wedge \boldsymbol{e}_\infty + \Big(\boldsymbol{x} \wedge (\boldsymbol{b} - \boldsymbol{a}) - \boldsymbol{a} \wedge \boldsymbol{b}\Big) \wedge \boldsymbol{e}_\infty \wedge \boldsymbol{e}_o = 0. \end{aligned}$$

The two terms in the sum have to be separately zero, because they are linearly independent. The first term, $\boldsymbol{x} \wedge \boldsymbol{a} \wedge \boldsymbol{b} \wedge \boldsymbol{e}_\infty = 0$, implies that $\boldsymbol{x} = \alpha\, \boldsymbol{a} + \beta\, \boldsymbol{b}$, with $\alpha, \beta \in \mathbb{R}$. The second term is zero if and only if

$$\begin{aligned} &\boldsymbol{x} \wedge (\boldsymbol{b} - \boldsymbol{a}) = \boldsymbol{a} \wedge \boldsymbol{b} \\ \iff\; &(\alpha\, \boldsymbol{a} + \beta\, \boldsymbol{b}) \wedge (\boldsymbol{b} - \boldsymbol{a}) = \boldsymbol{a} \wedge \boldsymbol{b} \\ \iff\; &(\alpha + \beta)\, \boldsymbol{a} \wedge \boldsymbol{b} = \boldsymbol{a} \wedge \boldsymbol{b} \\ \iff\; &\alpha + \beta = 1. \end{aligned}$$

Therefore,

$$\boldsymbol{x} = \alpha\,\boldsymbol{a} + \beta\,\boldsymbol{b} + \alpha\,\boldsymbol{b} - \alpha\,\boldsymbol{b} = \alpha\,(\boldsymbol{a} - \boldsymbol{b}) + \boldsymbol{b},$$

which is a parametric representation of the line passing through $\boldsymbol{a}$ and $\boldsymbol{b}$. Thus,

$$\mathbb{NO}_G(\boldsymbol{A} \wedge \boldsymbol{B} \wedge \boldsymbol{e}_\infty) = \{\, \alpha\,(\boldsymbol{a} - \boldsymbol{b}) + \boldsymbol{b} \;:\; \alpha \in \mathbb{R} \,\}. \tag{4.55}$$

4.3.5.5 Planes

The outer-product representation of planes is very similar to that of lines. Let $\boldsymbol{a}, \boldsymbol{b}, \boldsymbol{c} \in \mathbb{R}^3$ and $\boldsymbol{A} := \mathcal{C}(\boldsymbol{a})$, $\boldsymbol{B} := \mathcal{C}(\boldsymbol{b})$, and $\boldsymbol{C} := \mathcal{C}(\boldsymbol{c})$; then the outer-product representation of the plane passing through $\boldsymbol{a}$, $\boldsymbol{b}$, and $\boldsymbol{c}$ is given by $\boldsymbol{P} := \boldsymbol{A} \wedge \boldsymbol{B} \wedge \boldsymbol{C} \wedge \boldsymbol{e}_\infty$. In much the same way as for lines, it may be shown that

$$\begin{aligned}\mathbb{NO}_G(\boldsymbol{P}) &= \{\, \boldsymbol{x} \in \mathbb{R}^3 \;:\; \mathcal{C}(\boldsymbol{x}) \wedge \boldsymbol{P} = 0 \,\} \\ &= \{\, \alpha\,(\boldsymbol{a} - \boldsymbol{c}) + \beta\,(\boldsymbol{b} - \boldsymbol{c}) + \boldsymbol{c} \;:\; (\alpha, \beta) \in \mathbb{R}^2 \,\}.\end{aligned} \tag{4.56}$$

4.3.5.6 Circles

The outer-product representation of a circle that passes through points $\boldsymbol{a}, \boldsymbol{b}, \boldsymbol{c} \in \mathbb{R}^3$ is given by $\boldsymbol{K} := \boldsymbol{A} \wedge \boldsymbol{B} \wedge \boldsymbol{C}$, where $\boldsymbol{A} := \mathcal{C}(\boldsymbol{a})$, $\boldsymbol{B} := \mathcal{C}(\boldsymbol{b})$, and $\boldsymbol{C} := \mathcal{C}(\boldsymbol{c})$. It is not easy to show this directly, but this result can be argued indirectly as follows. First of all, because

$$\mathbb{NO}_G(\boldsymbol{K}) = \{\, \boldsymbol{x} \in \mathbb{R}^3 \;:\; \mathcal{C}(\boldsymbol{x}) \wedge \boldsymbol{K} = 0 \,\}, \tag{4.57}$$

the points $\boldsymbol{a}$, $\boldsymbol{b}$, and $\boldsymbol{c}$ lie on the entity represented by $\boldsymbol{K}$. Secondly, $\mathbb{NO}_G(\boldsymbol{K}) = \mathbb{NI}_G(\boldsymbol{K}^*)$, and because $\boldsymbol{K} \in \mathbb{G}^3_{4,1}$ is a 3-blade it follows that $\boldsymbol{K}^* \in \mathbb{G}^2_{4,1}$ is a 2-blade. As was shown in Sect. 4.3.4, the GIPNS of a 2-blade is a line, a line at infinity, a circle, or an imaginary circle. However, because the points $\boldsymbol{a}$, $\boldsymbol{b}$, and $\boldsymbol{c}$ lie on the entity, it has to be a finite entity, i.e. a line or a circle. Thus $\boldsymbol{K}$ represents either a line or a circle through the points $\boldsymbol{a}$, $\boldsymbol{b}$, and $\boldsymbol{c}$. In fact, $\boldsymbol{K}$ represents a line only if the three Euclidean vectors are collinear.

4.3.5.7 Spheres

Just as the outer-product representation of a circle is the outer product of three vectors representing points, the outer-product representation of a sphere is given by the outer product of four vectors representing points. That is, if $\boldsymbol{a}, \boldsymbol{b}, \boldsymbol{c}, \boldsymbol{d} \in \mathbb{R}^3$ and $\boldsymbol{A} := \mathcal{C}(\boldsymbol{a})$, $\boldsymbol{B} := \mathcal{C}(\boldsymbol{b})$, $\boldsymbol{C} := \mathcal{C}(\boldsymbol{c})$, and $\boldsymbol{D} := \mathcal{C}(\boldsymbol{d})$, the outer-product representation of the sphere passing through $\boldsymbol{a}$, $\boldsymbol{b}$, $\boldsymbol{c}$, and $\boldsymbol{d}$ is given by $\boldsymbol{S} := \boldsymbol{A} \wedge \boldsymbol{B} \wedge \boldsymbol{C} \wedge \boldsymbol{D}$.

As in the case of the representation of a circle, this is most easily shown indirectly. First of all, because

$$\mathbb{NO}_G(\boldsymbol{S}) = \{\, \boldsymbol{x} \in \mathbb{R}^3 \ : \ \mathcal{C}(\boldsymbol{x}) \wedge \boldsymbol{S} = 0 \,\}, \tag{4.58}$$

the points $\boldsymbol{a}$, $\boldsymbol{b}$, $\boldsymbol{c}$, and $\boldsymbol{d}$ lie on the entity represented by $\boldsymbol{S}$. Secondly, $\mathbb{NO}_G(\boldsymbol{S}) = \mathbb{NI}_G(\boldsymbol{S}^*)$, where $\boldsymbol{S}^* \in \mathbb{G}^1_{4,1}$ is a 1-blade, since $\boldsymbol{S} \in \mathbb{G}^4_{4,1}$ is a 4-blade. The inner-product representation of a 1-blade is either a point, a sphere, or an imaginary sphere. It follows that because the points $\boldsymbol{a}$, $\boldsymbol{b}$, $\boldsymbol{c}$, and $\boldsymbol{d}$, which are different points, have to lie on $\boldsymbol{S}$, $\boldsymbol{S}$ represents a sphere through the four Euclidean points.

4.3.6 Summary of Representations

In this subsection, the various geometric entities and their representations in the CGA $\mathbb{G}_{4,1}$ are summarized. These representations are also related to the algebra basis elements of the subspace that they lie in. For this purpose, the algebra basis of $\mathbb{G}_{4,1}$ is given in Table 4.2. This table assumes that $\boldsymbol{e}_i := \overline{\mathbb{G}}^1_{4,1}[i]$ and $\boldsymbol{e}_{ijk} := \boldsymbol{e}_i\, \boldsymbol{e}_j\, \boldsymbol{e}_k$. Instead of $\boldsymbol{e}_+$ and $\boldsymbol{e}_-$, the null-basis vectors $\boldsymbol{e}_\infty := \boldsymbol{e}_+ + \boldsymbol{e}_-$ and $\boldsymbol{e}_o := \frac{1}{2}(\boldsymbol{e}_- - \boldsymbol{e}_+)$ are used. The order of the elements of the algebra basis is not the same as that of the canonical algebra basis. Instead, the algebra basis-blades are ordered in an easily readable fashion.

Table 4.3 summarizes the geometric entities that can be represented in $\mathbb{G}_{4,1}$, their algebraic outer-product representation, and their algebra basis. In this table, $\boldsymbol{A}, \boldsymbol{B}, \boldsymbol{C}, \boldsymbol{D} \in \mathbb{G}^1_{4,1}$ represent points in $\mathbb{R}^3$.

Table 4.4 does the same for the inner-product representation of geometric entities. In this table, the $\{\, \boldsymbol{S}_i \,\} \subset \mathbb{G}^1_{4,1}$ are inner-product representations of spheres and the $\{\, \boldsymbol{P}_i \,\} \subset \mathbb{G}^1_{4,1}$ are inner-product representations of planes. Furthermore, the spheres represented by $\boldsymbol{S}_1$ and $\boldsymbol{S}_2$ intersect in a circle, the spheres represented by $\boldsymbol{S}_1$, $\boldsymbol{S}_2$, and $\boldsymbol{S}_3$ intersect in a point pair, and the spheres represented by $\boldsymbol{S}_1$, $\boldsymbol{S}_2$, $\boldsymbol{S}_3$, and $\boldsymbol{S}_4$ intersect in a point. Similarly, the planes represented by $\boldsymbol{P}_1$ and $\boldsymbol{P}_2$ intersect in a line, and the planes represented by $\boldsymbol{P}_1$, $\boldsymbol{P}_2$, and $\boldsymbol{P}_3$ intersect in a point.

4.3.7 Stratification of Spaces

With respect to Table 4.3, it is interesting to see that those geometric entities that can also be represented in projective space are represented by the outer

Table 4.2 Algebra basis of the CGA $\mathbb{G}_{4,1}$

Type	No.	Basis elements
Scalar	1	1
Vector	5	$e_1, e_2, e_3, e_\infty, e_o$
2-Vector	10	$e_{23}, e_{31}, e_{12}, e_{1o}, e_{2o}, e_{3o}, e_{1\infty}, e_{2\infty}, e_{3\infty}, e_{o\infty}$
3-Vector	10	$e_{23\infty}, e_{31\infty}, e_{12\infty}, e_{23o}, e_{31o}, e_{12o}, e_{1\infty o}, e_{2\infty o}, e_{3\infty o}, e_{123}$
4-Vector	5	$e_{123\infty}, e_{123o}, e_{23\infty o}, e_{31\infty o}, e_{12\infty o}$
5-Vector	1	$e_{123\infty o}$

Table 4.3 Geometric entities, their GOPNS representation in $\mathbb{G}_{4,1}$, and their algebra basis

Entity	Grade	No.	Basis elements
Point $\boldsymbol{A}$	1	5	$e_1, e_2, e_3, e_\infty, e_o$
Homogeneous point $\boldsymbol{A} \wedge e_\infty$	2	4	$e_{1\infty}, e_{2\infty}, e_{3\infty}, e_{o\infty}$
Point Pair $\boldsymbol{A} \wedge \boldsymbol{B}$	2	10	$e_{23}, e_{31}, e_{12}, e_{1o}, e_{2o}, e_{3o}$, $e_{1\infty}, e_{2\infty}, e_{3\infty}, e_{o\infty}$
Line $\boldsymbol{A} \wedge \boldsymbol{B} \wedge e_\infty$	3	6	$e_{23\infty}, e_{31\infty}, e_{12\infty}$, $e_{1o\infty}, e_{2o\infty}, e_{3o\infty}$
Circle $\boldsymbol{A} \wedge \boldsymbol{B} \wedge \boldsymbol{C}$	3	10	$e_{23\infty}, e_{31\infty}, e_{12\infty}, e_{23o}, e_{31o}, e_{12o}$, $e_{1o\infty}, e_{2o\infty}, e_{3o\infty}, e_{123}$
Plane $\boldsymbol{A} \wedge \boldsymbol{B} \wedge \boldsymbol{C} \wedge e_\infty$	4	4	$e_{123\infty}, e_{23o\infty}, e_{31o\infty}, e_{12o\infty}$
Sphere $\boldsymbol{A} \wedge \boldsymbol{B} \wedge \boldsymbol{C} \wedge \boldsymbol{D}$	4	5	$e_{123\infty}, e_{123o}, e_{23o\infty}, e_{31o\infty}, e_{12o\infty}$

product of a blade of null vectors and e_∞. Consider, for example, a vector $\boldsymbol{a} \in \mathbb{R}^3$ and define

$$\boldsymbol{A} := \mathcal{C}(\boldsymbol{a}) = \boldsymbol{a} + \frac{1}{2}\boldsymbol{a}^2\, e_\infty + e_o.$$

Taking the outer product of $\boldsymbol{A}$ with e_∞ results in

$$\boldsymbol{A} \wedge e_\infty = \boldsymbol{a} \wedge e_\infty + e_o \wedge e_\infty.$$

If we identify $e_o \wedge e_\infty$ with the homogeneous dimension and the bivectors $\{e_i \wedge e_\infty\}$ for $i \in \{1, 2, 3\}$ with the orthonormal basis vectors of a vector space, then $\boldsymbol{A} \wedge e_\infty$ can be regarded as the element of the projective space of $\mathbb{R}^3$. This also carries over to blades of the type $\boldsymbol{A}_{\langle k \rangle} \wedge e_\infty$, where $\boldsymbol{A}_{\langle k \rangle}$ is a blade of null vectors excluding e_∞.

In a similar way, the geometric entities that are representable in the geometric algebra of Euclidean space can be expressed in $\mathbb{G}_{4,1}$. This time, the outer product of a vector $\boldsymbol{A}$ with $e_\infty \wedge e_o$ is taken to give

$$\boldsymbol{A} \wedge e_\infty \wedge e_o = \boldsymbol{a} \wedge e_\infty \wedge e_o.$$

Table 4.4 Geometric entities, their GIPNS representation in $\mathbb{G}_{4,1}$, and their algebra basis

Entity	**Grade**	**No.**	**Basis Elements**
Sphere, Point $\boldsymbol{S}_1$	1	5	$\boldsymbol{e}_1, \boldsymbol{e}_2, \boldsymbol{e}_3, \boldsymbol{e}_\infty, \boldsymbol{e}_o$
Plane $\boldsymbol{P}_1$	1	4	$\boldsymbol{e}_1, \boldsymbol{e}_2, \boldsymbol{e}_3, \boldsymbol{e}_\infty$
Line $\boldsymbol{P}_1 \wedge \boldsymbol{P}_2$	2	6	$\boldsymbol{e}_{23}, \boldsymbol{e}_{31}, \boldsymbol{e}_{12}, \boldsymbol{e}_{1\infty}, \boldsymbol{e}_{2\infty}, \boldsymbol{e}_{3\infty}$
Circle $\boldsymbol{S}_1 \wedge \boldsymbol{S}_2$	2	10	$\boldsymbol{e}_{23}, \boldsymbol{e}_{31}, \boldsymbol{e}_{12}, \boldsymbol{e}_{1o}, \boldsymbol{e}_{2o}, \boldsymbol{e}_{3o},$ $\boldsymbol{e}_{1\infty}, \boldsymbol{e}_{2\infty}, \boldsymbol{e}_{3\infty}, \boldsymbol{e}_{o\infty}$
Point Pair $\boldsymbol{S}_1 \wedge \boldsymbol{S}_2 \wedge \boldsymbol{S}_3$	3	10	$\boldsymbol{e}_{23\infty}, \boldsymbol{e}_{31\infty}, \boldsymbol{e}_{12\infty}, \boldsymbol{e}_{23o}, \boldsymbol{e}_{31o}, \boldsymbol{e}_{12o},$ $\boldsymbol{e}_{1o\infty}, \boldsymbol{e}_{2o\infty}, \boldsymbol{e}_{3o\infty}, \boldsymbol{e}_{123}$
Homogeneous Point $\boldsymbol{P}_1 \wedge \boldsymbol{P}_2 \wedge \boldsymbol{P}_3$	3	4	$\boldsymbol{e}_{23\infty}, \boldsymbol{e}_{31\infty}, \boldsymbol{e}_{12\infty}, \boldsymbol{e}_{123}$
Point $\boldsymbol{S}_1 \wedge \boldsymbol{S}_2 \wedge \boldsymbol{S}_3 \wedge \boldsymbol{S}_4$	4	5	$\boldsymbol{e}_{123\infty}, \boldsymbol{e}_{123o}, \boldsymbol{e}_{23o\infty}, \boldsymbol{e}_{31o\infty}, \boldsymbol{e}_{12o\infty}$

The $\{\boldsymbol{e}_i \wedge \boldsymbol{e}_\infty \wedge \boldsymbol{e}_o\}$ for $i \in \{1, 2, 3\}$ may now be identified with the orthonormal basis of a Euclidean space $\mathbb{R}^3$. In fact, $\mathbb{NO}_G(\boldsymbol{a} \wedge \boldsymbol{e}_\infty \wedge \boldsymbol{e}_o)$ is a line through the origin in the direction of $\boldsymbol{a}$, that is, exactly the same as $\mathbb{NO}_G(\boldsymbol{a})$. Similarly, $\mathbb{NO}_G(\boldsymbol{a} \wedge \boldsymbol{b} \wedge \boldsymbol{e}_\infty \wedge \boldsymbol{e}_o)$ is the same plane through the origin as $\mathbb{NO}_G(\boldsymbol{a} \wedge \boldsymbol{b})$, where $\boldsymbol{b} \in \mathbb{R}^3$.

This analysis shows that conformal space combines the features of Euclidean and projective space. This also carries over to the operators, as will be seen in the following subsections. This embedding of Euclidean and projective space in a single framework immediately offers the possibility to implement the ideas of Faugeras regarding the stratification of three-dimensional vision [67], without changing spaces or representations. This has been used quite successfully in, for example, [150, 158].

4.3.8 Reflections in $\mathbb{G}_{n+1,1}$

Reflections are represented by planes in conformal space. This is most easily shown with inner-product planes. Let $\boldsymbol{a} \in \mathbb{R}^n$, define $\hat{\boldsymbol{a}} := \boldsymbol{a}/\|\boldsymbol{a}\|$, and let $\alpha \in \mathbb{R}$. A plane with a normal $\hat{\boldsymbol{a}}$ and an orthogonal separation α from the origin can then be represented in $\mathbb{G}_{n+1,1}$ through the GIPNS of $\boldsymbol{P} := \hat{\boldsymbol{a}} + \alpha\, \boldsymbol{e}_\infty \in \mathbb{G}^1_{n+1,1}$. Given a vector $\boldsymbol{x} \in \mathbb{R}^n$ and its representation $\boldsymbol{X} := \mathcal{C}(\boldsymbol{x})$ in $\mathbb{G}_{4,1}$, it may be shown that

$$\begin{aligned} \boldsymbol{P}\,\boldsymbol{X}\,\boldsymbol{P} &= (\hat{\boldsymbol{a}} + \alpha\,\boldsymbol{e}_\infty)\,(\boldsymbol{x} + \frac{1}{2}\,\boldsymbol{x}^2\,\boldsymbol{e}_\infty + \boldsymbol{e}_o)\,(\hat{\boldsymbol{a}} + \alpha\,\boldsymbol{e}_\infty) \\ &\simeq \boldsymbol{x} + 2\,(\alpha - \hat{\boldsymbol{a}} \cdot \boldsymbol{x})\,\hat{\boldsymbol{a}} + \frac{1}{2}\,(\boldsymbol{x}^2 + 4\,\alpha^2 - 4\,\alpha\,\hat{\boldsymbol{a}} \cdot \boldsymbol{x})\,\boldsymbol{e}_\infty + \boldsymbol{e}_o, \end{aligned} \tag{4.59}$$

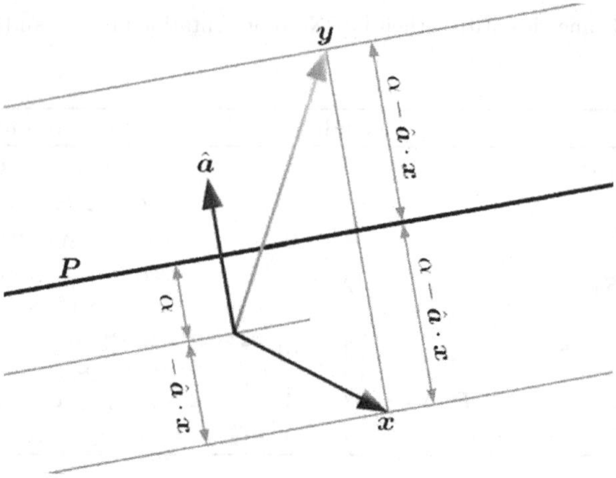

Fig. 4.17 Reflection of a vector $\boldsymbol{x}$ in a plane $\boldsymbol{P}$ with a normal $\boldsymbol{a}$ and an orthogonal separation α from the origin, which results in $\boldsymbol{y}$

where "$\simeq$" denotes equality up to a scalar factor, which does not influence the geometric representation. The result has to represent a point because it is an element of grade 1 and it lies on the null-cone:

$$(\boldsymbol{P}\,\boldsymbol{X}\,\boldsymbol{P})^2 = \boldsymbol{P}\,\boldsymbol{X}\,\boldsymbol{P}\,\boldsymbol{P}\,\boldsymbol{X}\,\boldsymbol{P} = \boldsymbol{P}^4\,\boldsymbol{X}^2 = 0,$$

which follows since $\boldsymbol{P}^2 \in \mathbb{R}$ and $\boldsymbol{X}^2 = 0$. Projecting the result of (4.59) into $\mathbb{R}^n$ gives

$$\begin{aligned} \boldsymbol{y} &:= \mathcal{C}^{-1}\Big(\boldsymbol{x} + 2\,(\alpha - \hat{\boldsymbol{a}}\cdot\boldsymbol{x})\,\hat{\boldsymbol{a}} + \frac{1}{2}\,(\boldsymbol{x}^2 + 4\,\alpha^2 - 4\,\alpha\,\hat{\boldsymbol{a}}\cdot\boldsymbol{x})\,\boldsymbol{e}_\infty + \boldsymbol{e}_o\Big) \\ &= \boldsymbol{x} + 2\,(\alpha - \hat{\boldsymbol{a}}\cdot\boldsymbol{x})\,\hat{\boldsymbol{a}}. \end{aligned} \tag{4.60}$$

Figure 4.17 illustrates the result of (4.59) in $\mathbb{R}^2$, which shows that $\boldsymbol{P}\,\boldsymbol{X}\,\boldsymbol{P}$ represents the reflection of $\boldsymbol{X}$ in the plane $\boldsymbol{P}$. In contrast to reflections in $\mathbb{G}_n$, the reflection plane need not pass through the origin. As will be seen later on, this feature allows the representation of translations and rotations about arbitrary axes.

4.3.9 Inversions in $\mathbb{G}_{n+1,1}$

It was noted in the introduction to Sect. 4.3 that conformal space takes its name from the conformal mappings that are possible within it. Furthermore, it was noted that a conformal transformation can be expressed by a combination of inversions. However, so far it has not been shown how an inversion may be expressed in $\mathbb{G}_{n+1,1}$.

Before general inversions in $\mathbb{G}_{n+1,1}$ are discussed, it is helpful to take a look at the initial stereographic embedding (see Sect. 4.3.1), which was defined for a Euclidean vector $\boldsymbol{x} \in \mathbb{R}^n$ by

$$\mathcal{S}(\boldsymbol{x}) = \frac{2}{\boldsymbol{x}^2+1}\,\boldsymbol{x} + \frac{\boldsymbol{x}^2-1}{\boldsymbol{x}^2+1}\,\boldsymbol{e}_+ .$$

The inverse of $\boldsymbol{x}$ can be written as

$$\boldsymbol{x}^{-1} = \frac{\boldsymbol{x}}{\boldsymbol{x}^2},$$

which is the same as the inversion of $\boldsymbol{x}$ in the unit sphere centered at the origin. The embedding of $\boldsymbol{x}^{-1}$ in $\mathbb{R}^{n+1,1}$ gives

$$\begin{aligned}
\mathcal{S}(\boldsymbol{x}^{-1}) &= \frac{2}{\dfrac{\boldsymbol{x}^2}{\boldsymbol{x}^4}+1}\,\frac{\boldsymbol{x}}{\boldsymbol{x}^2} + \frac{\dfrac{\boldsymbol{x}^2}{\boldsymbol{x}^4}-1}{\dfrac{\boldsymbol{x}^2}{\boldsymbol{x}^4}+1}\,\boldsymbol{e}_+ \\
&= \frac{2}{\dfrac{1}{\boldsymbol{x}^2}+1}\,\frac{\boldsymbol{x}}{\boldsymbol{x}^2} + \frac{\dfrac{1}{\boldsymbol{x}^2}-1}{\dfrac{1}{\boldsymbol{x}^2}+1}\,\boldsymbol{e}_+ \\
&= \boldsymbol{x}^2\,\frac{2}{1+\boldsymbol{x}^2}\,\frac{\boldsymbol{x}}{\boldsymbol{x}^2} + \frac{\dfrac{1}{\boldsymbol{x}^2}}{\dfrac{1}{\boldsymbol{x}^2}}\,\frac{1-\boldsymbol{x}^2}{1+\boldsymbol{x}^2}\,\boldsymbol{e}_+ \\
&= \frac{2}{\boldsymbol{x}^2+1}\,\boldsymbol{x} - \frac{\boldsymbol{x}^2-1}{\boldsymbol{x}^2+1}\,\boldsymbol{e}_+ .
\end{aligned}$$

This shows that in order to invert a vector in $\mathbb{R}^n$, only its $\boldsymbol{e}_+$ component in its stereographic embedding has to be negated, which is equivalent to a reflection in the Euclidean subspace. Consider, for example, the stereographic embedding of $\mathbb{R}^1$ in $\mathbb{R}^2$. A vector $\boldsymbol{x} \in \mathbb{R}^1$ defined as $\boldsymbol{x} := \alpha\,\boldsymbol{e}_1$, with $\alpha \in \mathbb{R}$, becomes

$$\mathcal{S}(\boldsymbol{x}) = \frac{2\alpha}{\alpha^2+1}\,\boldsymbol{e}_1 + \frac{\alpha^2-1}{\alpha^2+1}\,\boldsymbol{e}_+ .$$

The inverse of $\boldsymbol{x}$ is then given by

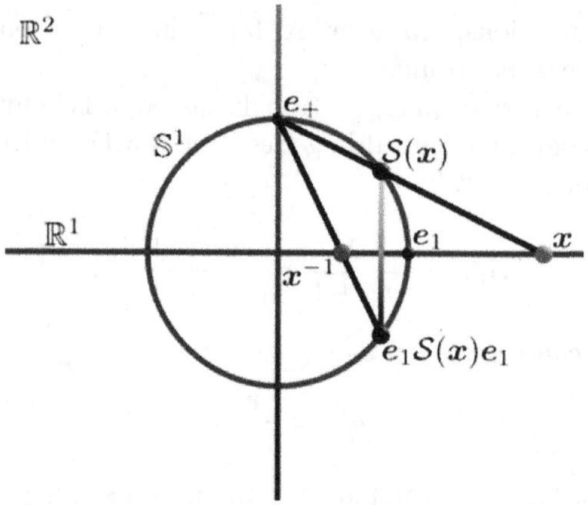

Fig. 4.18 Inversion of a vector $\boldsymbol{x}$

$$\begin{aligned}\boldsymbol{x}^{-1} &= \mathcal{S}^{-1}\big(e_1\,\mathcal{S}(\boldsymbol{x})\,e_1\big)\\ &= \mathcal{S}^{-1}\Big(\frac{2\alpha}{\alpha^2+1}\,e_1 - \frac{\alpha^2-1}{\alpha^2+1}\,e_+\Big)\\ &= \frac{1}{1+\dfrac{\alpha^2-1}{\alpha^2+1}}\,\frac{2\alpha}{\alpha^2+1}\,e_1\\ &= \frac{2\alpha}{(\alpha^2+1)+(\alpha^2-1)}\,e_1\\ &= \alpha^{-1}\,e_1,\end{aligned}$$

with $e_1\,e_1\,e_1 = e_1$ and $e_1\,e_+\,e_1 = -e_+\,e_1\,e_1 = -e_+$; (4.26) has been used to evaluate $\mathcal{S}^{-1}$. Figure 4.18 illustrates this example.

In $\mathbb{G}_{4,1}$, it turns out that an inversion of a vector in the unit sphere centered at the origin is given by a reflection in e_+. Given a vector $\boldsymbol{x} \in \mathbb{R}^n$, and if we define $\boldsymbol{X} := \mathcal{C}(\boldsymbol{x})$,

$$\begin{aligned}e_+\,\boldsymbol{X}\,e_+ &= e_+\,(\boldsymbol{x} + \frac{1}{2}\boldsymbol{x}^2\,e_\infty + e_o)\,e_+\\ &= -1\,(\boldsymbol{x} + \boldsymbol{x}^2\,e_o + \frac{1}{2}\,e_\infty)\\ &= -\boldsymbol{x}^2\,(\boldsymbol{x}^{-1} + \frac{1}{2}\boldsymbol{x}^{-2}\,e_\infty + e_o).\end{aligned}$$

Projecting this vector back into $\mathbb{R}^n$ then gives

$$\mathcal{C}^{-1}(e_+\,\mathcal{C}(\boldsymbol{x})\,e_+) = \boldsymbol{x}^{-1}.$$

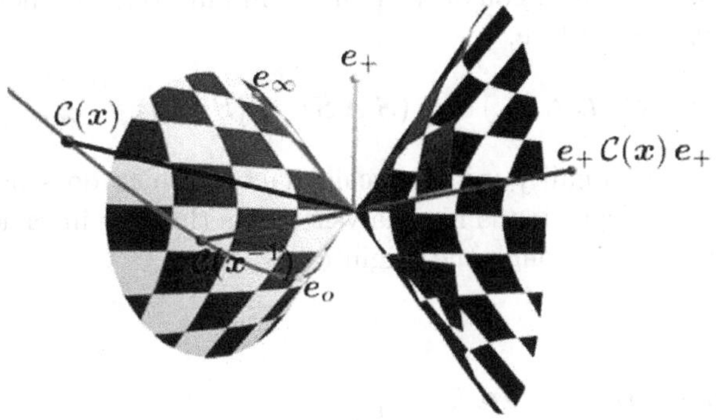

Fig. 4.19 Inversion of vector $\boldsymbol{x} \in \mathbb{R}^1$ via reflection of $\mathcal{C}(\boldsymbol{x})$ on $\boldsymbol{e}_+$

This is visualized in Fig. 4.19. The vector $\boldsymbol{X} \in \mathbb{K}^2$ on the null cone is reflected in $\boldsymbol{e}_+$, which gives $\boldsymbol{e}_+ \boldsymbol{X} \boldsymbol{e}_+$. Scaling the reflected vector such that its $\boldsymbol{e}_o$ component is unity results in the vector $\boldsymbol{Y}$. Projecting $\boldsymbol{Y}$ back into $\mathbb{R}^1$ then gives the inverse of $\mathcal{C}^{-1}(\boldsymbol{X})$.

Note that using the definitions $\boldsymbol{e}_\infty = \boldsymbol{e}_- + \boldsymbol{e}_+$ and $\boldsymbol{e}_o = \frac{1}{2}(\boldsymbol{e}_- - \boldsymbol{e}_+)$, the unit sphere $\boldsymbol{S}$ centered at the origin can be expressed as

$$\boldsymbol{S} = \underbrace{\boldsymbol{e}_o}_{\text{origin}} - \underbrace{\frac{1}{2}\boldsymbol{e}_\infty}_{\text{radius 1}} = \frac{1}{2}(\boldsymbol{e}_- - \boldsymbol{e}_+) - \frac{1}{2}(\boldsymbol{e}_- + \boldsymbol{e}_+) = -\boldsymbol{e}_+ .$$

This shows that it is possible to use vectors in $\mathbb{G}^1_{n+1,1}$ that represent spheres in their GIPNS, i.e. inner-product spheres, to invert representations of vectors in $\mathbb{G}^1_{n+1,1}$ in them. That is, the inversion of $\boldsymbol{x} \in \mathbb{R}^n$ in the unit sphere is represented by $\boldsymbol{S}\,\boldsymbol{X}\,\boldsymbol{S}$, where $\boldsymbol{X} := \mathcal{C}(\boldsymbol{x})$. Since 1-vectors are versors, the representation of any blade in $\mathbb{G}_{n+1,1}$ can be inverted in this way, owing to the versor's outermorphism. For example, for $\boldsymbol{A}, \boldsymbol{B} \in \mathbb{G}^1_{n+1,1}$,

$$\boldsymbol{S}\,(\boldsymbol{A} \wedge \boldsymbol{B})\,\boldsymbol{S} = (\boldsymbol{S}\,\boldsymbol{A}\,\boldsymbol{S}) \wedge (\boldsymbol{S}\,\boldsymbol{B}\,\boldsymbol{S}).$$

4.3.9.1 Example

As another example, consider the inversion of the line $\boldsymbol{L} := \boldsymbol{A} \wedge \boldsymbol{B} \wedge \boldsymbol{e}_\infty$ in $\boldsymbol{S}$:

$$\boldsymbol{S}\,(\boldsymbol{A} \wedge \boldsymbol{B} \wedge \boldsymbol{e}_\infty)\,\boldsymbol{S} = (\boldsymbol{S}\,\boldsymbol{A}\,\boldsymbol{S}) \wedge (\boldsymbol{S}\,\boldsymbol{B}\,\boldsymbol{S}) \wedge (\boldsymbol{S}\,\boldsymbol{e}_\infty\,\boldsymbol{S}).$$

From the definitions of $\boldsymbol{S}$, $\boldsymbol{e}_\infty$, and $\boldsymbol{e}_o$, it follows that

$$\boldsymbol{S}\,\boldsymbol{e}_\infty\,\boldsymbol{S} = \boldsymbol{e}_+\,(\boldsymbol{e}_+ + \boldsymbol{e}_-)\,\boldsymbol{e}_+ = \boldsymbol{e}_+ - \boldsymbol{e}_- = -2\,\boldsymbol{e}_o,$$

which means that the inversion of the point at infinity ($\boldsymbol{e}_\infty$) in the unit sphere is the origin ($\boldsymbol{e}_o$). Therefore,

$$\boldsymbol{S}\,(\boldsymbol{A} \wedge \boldsymbol{B} \wedge \boldsymbol{e}_\infty)\,\boldsymbol{S} \simeq (\boldsymbol{S}\,\boldsymbol{A}\,\boldsymbol{S}) \;\wedge\; (\boldsymbol{S}\,\boldsymbol{B}\,\boldsymbol{S}) \;\wedge\; \boldsymbol{e}_o,$$

where "$\simeq$" denotes equality up to a scalar factor, which does not influence the geometric representation. This analysis shows that the inversion of a line is a circle that passes through the origin $\boldsymbol{e}_o$.

4.3.10 Translations in $\mathbb{G}_{n+1,1}$

At the end of Sect. 4.3.8 it was mentioned that translations, as well as rotations, can be constructed by use of two consecutive reflections. In the following, it is shown that a translation can be represented by two consecutive reflections in parallel planes.

Let $\boldsymbol{P}_A, \boldsymbol{P}_B \in \mathbb{G}^1_{n+1,1}$ represent two parallel planes with respect to their GIPNS. Given a normalized vector $\hat{\boldsymbol{a}} \in \mathbb{R}^3$ and $\alpha, \beta \in \mathbb{R}$, the two planes may be defined by $\boldsymbol{P}_A := \hat{\boldsymbol{a}} + \alpha\,\boldsymbol{e}_\infty$ and $\boldsymbol{P}_B := \hat{\boldsymbol{a}} + \beta\,\boldsymbol{e}_\infty$, such that $\hat{\boldsymbol{a}}$ is the normal of the planes and α and β give the orthogonal distances of the planes from the origin. A point $\boldsymbol{x} \in \mathbb{R}^3$ represented by $\boldsymbol{X} := \mathcal{C}(\boldsymbol{x})$ behaves as follows under consecutive reflections in $\boldsymbol{P}_A$ and $\boldsymbol{P}_B$:

$$\boldsymbol{y} := \mathcal{C}^{-1}(\boldsymbol{P}_A\,\boldsymbol{X}\,\boldsymbol{P}_A) = \boldsymbol{x} + 2\,(\alpha - \hat{\boldsymbol{a}} \cdot \boldsymbol{x})\,\hat{\boldsymbol{a}}, \tag{4.61}$$

as shown in (4.60). The second reflection then results in

$$\begin{aligned} \boldsymbol{z} &:= \mathcal{C}^{-1}(\boldsymbol{P}_B\,\boldsymbol{P}_A\;\boldsymbol{X}\;\boldsymbol{P}_A\,\boldsymbol{P}_B) \\ &= \boldsymbol{y} + 2\,(\beta - \hat{\boldsymbol{a}} \cdot \boldsymbol{y})\,\hat{\boldsymbol{a}} \\ &= \boldsymbol{x} + 2\,(\beta - \alpha)\,\hat{\boldsymbol{a}}. \end{aligned} \tag{4.62}$$

The transformation of $\boldsymbol{x}$ due to both reflections therefore depends only on the planes' normal $\hat{\boldsymbol{a}}$ and their separation $\beta - \alpha$. In fact, the double reflection results in a translation of $\boldsymbol{x}$ in the direction of the planes' normal by an amount equal to twice the separation of the planes.

The multivector that results from the product $\boldsymbol{P}_B\,\boldsymbol{P}_A$ may thus be regarded as a translation operator, a *translator*:

$$\begin{aligned} \boldsymbol{P}_B\,\boldsymbol{P}_A &= (\hat{\boldsymbol{a}} + \beta\,\boldsymbol{e}_\infty)\,(\hat{\boldsymbol{a}} + \alpha\,\boldsymbol{e}_\infty) \\ &= 1 - (\beta - \alpha)\,\hat{\boldsymbol{a}}\,\boldsymbol{e}_\infty \\ &= 1 - \frac{1}{2}\,\boldsymbol{t}\,\boldsymbol{e}_\infty, \end{aligned} \tag{4.63}$$

where $\boldsymbol{t} := 2\,(\beta - \alpha)\,\hat{\boldsymbol{a}}$ is the translation vector. That is, the translator $\boldsymbol{T}$ for a translation by $\boldsymbol{t}$ is given by $\boldsymbol{T} = 1 - \frac{1}{2}\,\boldsymbol{t}\,\boldsymbol{e}_\infty$.

A translator has a scalar and a bivector part, just like a rotor. In fact, in the space $\mathbb{R}^{n+1,1}$ itself, a translator expresses a rotation. However, the rotation plane lies not in the Euclidean subspace but in a mixed subspace, and the effect of the rotation in $\mathbb{R}^{n+1,1}$ is a translation in $\mathbb{R}^n$.

4.3.10.1 Properties

A translator may also be given in exponential form as

$$T = \exp\big(-\frac{1}{2}\boldsymbol{t}\,\boldsymbol{e}_\infty\big), \tag{4.64}$$

which follows from the Taylor expansion of the exponential function and the fact that

$$(\boldsymbol{t}\,\boldsymbol{e}_\infty)^2 = \boldsymbol{t}\,\boldsymbol{e}_\infty\,\boldsymbol{t}\,\boldsymbol{e}_\infty = -\boldsymbol{t}\,\boldsymbol{t}\,\boldsymbol{e}_\infty\,\boldsymbol{e}_\infty = 0.$$

A translator is a unitary versor, since $T\widetilde{T} = 1$, which also follows from the above equation. Applying $\boldsymbol{T}$ to the origin $\boldsymbol{e}_o$ results in

$$T\,\boldsymbol{e}_o\widetilde{T} = \boldsymbol{t} + \frac{1}{2}\boldsymbol{t}^2\,\boldsymbol{e}_\infty + \boldsymbol{e}_o,$$

and the translation of the point at infinity, $\boldsymbol{e}_\infty$, gives

$$T\,\boldsymbol{e}_\infty\,\widetilde{T} = \boldsymbol{e}_\infty.$$

4.3.10.2 Example

With the help of a translator, it becomes sufficient to prove many properties at the origin. By applying the translation operator, it is then possible to show that the property in question holds everywhere in space. A simple example may elucidate this. It was shown earlier that a sphere of radius ρ centered at the origin can be expressed by

$$\boldsymbol{S} = \boldsymbol{e}_0 - \frac{1}{2}\rho^2\,\boldsymbol{e}_\infty,$$

and thus

$$\boldsymbol{S}\cdot\boldsymbol{S} = \rho^2.$$

This raises the question of whether this true for any sphere. Suppose now that $\boldsymbol{S}'$ is a sphere with center $\boldsymbol{t}$ and radius ρ, and let $\boldsymbol{T}$ denote a translator representing a translation by $\boldsymbol{t}$; then $\boldsymbol{S}' = \widetilde{T}\,\boldsymbol{S}\,T$. It follows that

$$S' S' = T S \widetilde{T} T S \widetilde{T} = T S S \widetilde{T} = \rho^2 T \widetilde{T} = \rho^2 .$$

Thus, the property that $S^2 = \rho^2$, which was shown first to hold for a sphere centered at the origin, can easily be shown to hold for any sphere.

4.3.11 Rotations in $\mathbb{G}_{n+1,1}$

It was stated previously that the group of Euclidean transformations is a subgroup of the conformal group. Whereas the conformal transformation group is generated by combinations of inversions, the Euclidean transformation group is generated by combinations of reflections. In previous sections, it was shown that reflections are represented by planes and inversions by spheres. Therefore, it is possible to represent all conformal and Euclidean transformations in $\mathbb{G}_{n+1,1}$.

It is interesting to note that just as planes are special cases of spheres (with infinite radius), Euclidean transformations are special cases of conformal transformations.

In Sect. 4.1.5, it was shown that a rotation in $\mathbb{R}^3$ about an axis that passes through the origin is given by two consecutive reflections in planes that pass through the origin. In conformal space, a plane that passes through the origin can be represented in terms of the GIPNS as the plane's normal. That is, if $\hat{a}, \hat{b} \in \mathbb{R}^3$ are two normalized vectors, then the planes with normals $\hat{a}$ and $\hat{b}$ can be represented by vectors $P_A, P_B \in \mathbb{G}^1_{4,1}$ as $P_A = \hat{a}$ and $P_B = \hat{b}$, respectively. Hence,

$$P_B \, P_A = \hat{b} \, \hat{a} = \hat{b} \cdot \hat{a} + \hat{b} \wedge \hat{a} = \cos\theta - \sin\theta \, U_{\langle 2\rangle} = \mathsf{rot}(2\theta, U_{\langle 2\rangle}), \quad (4.65)$$

where $U_{\langle 2\rangle} := (\hat{a} \wedge \hat{b})/\|\hat{a} \wedge \hat{b}\|$ is the unit bivector representing the rotation plane in 3D Euclidean space and $\theta := \angle(\hat{a}, \hat{b})$ is the angle between $\hat{a}$ and $\hat{b}$. The rotation axis is $U^*_{\langle 2\rangle}$, which is the direction of the intersection line of the planes P_A and P_B. Thus, the intersection line of the reflection planes gives the rotation axis.

A rotor for a rotation about an axis through the origin is therefore the same in conformal and Euclidean space. However, in conformal space a rotor may be translated with a translator. Such a general rotation operator may be given simply by

$$G = T R \widetilde{T}, \quad (4.66)$$

where T is a translator and R a rotor. The effect of applying G to a vector $X \in \mathbb{G}^1_{4,1}$ representing a point $x \in \mathbb{R}^3$, i.e. $X = \mathcal{C}(x)$, is

$$\boldsymbol{G}\boldsymbol{X}\boldsymbol{G} = \underbrace{\boldsymbol{T}\,\overbrace{\boldsymbol{R}\quad \underbrace{\widetilde{\boldsymbol{T}}\boldsymbol{X}\boldsymbol{T}}_{\text{translation by } -\boldsymbol{t}}\quad \widetilde{\boldsymbol{R}}}^{\text{rotation}}\,\widetilde{\boldsymbol{T}}}_{\text{translation by } \boldsymbol{t}}.$$

If the rotor $\boldsymbol{R} := \boldsymbol{P}_B\,\boldsymbol{P}_A$, as above, then

$$\boldsymbol{T}\,\boldsymbol{R}\widetilde{\boldsymbol{T}} = (\boldsymbol{T}\,\boldsymbol{P}_B\,\widetilde{\boldsymbol{T}})\,(\boldsymbol{T}\,\boldsymbol{P}_A\,\widetilde{\boldsymbol{T}}).$$

It therefore follows that two consecutive reflections in arbitrary planes result in a rotation about the intersection line of the two planes by twice the angle between the planes. In the special case where the planes are parallel, the consecutive reflections in the planes result in a rotation about an axis at infinity, i.e. a translation. This shows very clearly that a translation is just a special case of a rotation.

The transformations that can be represented by operators of the form $\boldsymbol{T}\,\boldsymbol{R}\widetilde{\boldsymbol{T}}$ are not the most general transformations, because a translation along the rotation axis cannot be represented in this way. A general Euclidean transformation can be represented by

$$\boldsymbol{M} := \boldsymbol{T}_2\,\boldsymbol{T}_1\,\boldsymbol{R}\widetilde{\boldsymbol{T}}_1, \tag{4.67}$$

where $\boldsymbol{T}_1$ translates the rotor $\boldsymbol{R}$ in the rotation plane and $\boldsymbol{T}_2$ is a translation along the rotation axis. Such a general transformation operator is called a *motor* or a *screw*. Note that $\boldsymbol{M}$ can also be parameterized as

$$\boldsymbol{M} := \boldsymbol{T}'\,\boldsymbol{R}', \tag{4.68}$$

where $\boldsymbol{T}'$ is an appropriate translator and $\boldsymbol{R}'$ an appropriate rotor.

The multivector $\boldsymbol{M}$ is a unitary versor and thus has the same effect for all blades. That is, $\boldsymbol{M}$ can be used to transform points, lines, planes, circles, spheres, and any operators.

4.3.12 Dilations in $\mathbb{G}_{n+1,1}$

A *dilation* is an isotropic scaling; that is, an equal scaling in all dimensions. A dilation can be achieved by two consecutive inversions in concentric spheres of different radii, which is how a dilation is constructed in CGA.

Let $\boldsymbol{S}_1, \boldsymbol{S}_2 \in \mathbb{G}^1_{4,1}$ be outer-product representations of two spheres centered at the origin, with radii $r_1, r_2 \in \mathbb{R}$; that is, $\boldsymbol{S}_i := \boldsymbol{e}_o - \frac{1}{2}\,r_i^2\,\boldsymbol{e}_\infty$. If $\boldsymbol{x} \in \mathbb{R}^3$ and $\boldsymbol{X} := \mathcal{C}(\boldsymbol{x})$, then

$$\boldsymbol{S}_1\,\boldsymbol{X}\,\boldsymbol{S}_1 \simeq \frac{r_1^2}{\boldsymbol{x}^2}\,\boldsymbol{x} + \frac{1}{2}\,\frac{r_1^4}{\boldsymbol{x}^2}\,\boldsymbol{e}_\infty + \boldsymbol{e}_o,$$

where "$\simeq$" denotes equality up to a scalar factor. Therefore,

$$\boldsymbol{S}_2\,\boldsymbol{S}_1\;\boldsymbol{X}\;\boldsymbol{S}_1\,\boldsymbol{S}_2 \simeq \frac{r_2^2}{r_1^2}\,\boldsymbol{x} + \frac{1}{2}\,\frac{r_2^4}{r_1^4}\,\boldsymbol{x}^2\,\boldsymbol{e}_\infty + \boldsymbol{e}_o.$$

Hence, the operator $\boldsymbol{D} := \boldsymbol{S}_2\,\boldsymbol{S}_1$ scales the vector $\boldsymbol{x}$ by a factor r_2^2/r_1^2. The dilation operator, or *dilator*, $\boldsymbol{D}$ is therefore given by

$$\begin{aligned}\boldsymbol{D} &= \boldsymbol{S}_2\,\boldsymbol{S}_1\\ &= \frac{1}{2}\,(r_1^2 + r_2^2) + \frac{1}{2}\,(r_1^2 - r_2^2)\;\boldsymbol{e}_\infty \wedge \boldsymbol{e}_o\\ &\simeq 1 + \frac{r_1^2 - r_2^2}{r_1^2 + r_2^2}\;\boldsymbol{e}_\infty \wedge \boldsymbol{e}_o\\ &\simeq 1 + \frac{1-d}{1+d}\;\boldsymbol{e}_\infty \wedge \boldsymbol{e}_o,\end{aligned}$$

where $d := r_2^2/r_1^2$ is the dilation factor. That is, a dilation operator that represents a dilation by a factor d about the origin is defined as

$$\boldsymbol{D} := 1 + \frac{1-d}{1+d}\;\boldsymbol{e}_\infty \wedge \boldsymbol{e}_o. \tag{4.69}$$

A dilation centered at a point $\boldsymbol{t} \in \mathbb{R}^3$ can be constructed from the above dilator and an appropriate translator. We define $\boldsymbol{T} := 1 - \frac{1}{2}\,\boldsymbol{t}\,\boldsymbol{e}_\infty$, which is the translation operator for a translation by $\boldsymbol{t}$. If $\boldsymbol{x} \in \mathbb{R}^3$ and $\boldsymbol{X} := \mathcal{C}(\boldsymbol{x})$, then

$$\boldsymbol{T}\,\boldsymbol{D}\,\widetilde{\boldsymbol{T}}\;\boldsymbol{X}\;\boldsymbol{T}\,\boldsymbol{D}\,\widetilde{\boldsymbol{T}}$$

is a dilation of $\boldsymbol{x}$ centered on $\boldsymbol{t}$. Thus,

$$\boldsymbol{D}_t := \boldsymbol{T}\,\boldsymbol{D}\,\widetilde{\boldsymbol{T}}, \tag{4.70}$$

is the dilation operator for a dilation about $\boldsymbol{t}$.

4.3.13 Summary of Operator Representations

Table 4.5 summarizes the operators available in $\mathbb{G}_{4,1}$ and lists their algebraic bases. The basis blades relate again to the algebra basis of $\mathbb{G}_{4,1}$ listed in Table 4.2.

Table 4.5 Operators in $\mathbb{G}_{4,1}$ and their algebraic basis. Note that the operators are mostly multivectors of mixed grade

Entity	Grades	No.	Basis Elements
Reflection	1	4	$\boldsymbol{e}_1, \boldsymbol{e}_2, \boldsymbol{e}_3, \boldsymbol{e}_\infty$
Inversion	1	5	$\boldsymbol{e}_1, \boldsymbol{e}_2, \boldsymbol{e}_3, \boldsymbol{e}_\infty, \boldsymbol{e}_o$
Rotor $\boldsymbol{R}$	0,2	4	$1, \boldsymbol{e}_{23}, \boldsymbol{e}_{31}, \boldsymbol{e}_{12}$
Translator $\boldsymbol{T}$	0,2	4	$1, \boldsymbol{e}_{1\infty}, \boldsymbol{e}_{2\infty}, \boldsymbol{e}_{3\infty}$
Dilator $\boldsymbol{D}$	0,2	2	$1, \boldsymbol{e}_{\infty o}$
General Dilator $\boldsymbol{T}\boldsymbol{D}\widetilde{\boldsymbol{T}}$	0,2	2	$1, \boldsymbol{e}_{1\infty}, \boldsymbol{e}_{2\infty}, \boldsymbol{e}_{3\infty}, \boldsymbol{e}_{\infty o}$
Motor $\boldsymbol{RT}$	0,2,4	8	$1, \boldsymbol{e}_{23}, \boldsymbol{e}_{31}, \boldsymbol{e}_{12}, \boldsymbol{e}_{1\infty}, \boldsymbol{e}_{2\infty}, \boldsymbol{e}_{3\infty}, \boldsymbol{e}_{123\infty}$
General Rotor $\boldsymbol{TR}\widetilde{\boldsymbol{T}}$	0,2	7	$1, \boldsymbol{e}_{23}, \boldsymbol{e}_{31}, \boldsymbol{e}_{12}, \boldsymbol{e}_{1\infty}, \boldsymbol{e}_{2\infty}, \boldsymbol{e}_{3\infty}$

4.3.14 Incidence Relations

Various types of incidence relations exist. In this section, algebraic operations between blades are presented that result in zero if a particular incidence relation is satisfied. The most common such relation between a geometric entity with an outer-product representation $\boldsymbol{A}_{\langle k \rangle}$ and a point with a representation $\boldsymbol{X}$ is that $\boldsymbol{A}_{\langle k \rangle} \wedge \boldsymbol{X} = 0$ if the point represented by $\boldsymbol{X}$ lies on the geometric entity represented by $\boldsymbol{A}_{\langle k \rangle}$. This follows immediately from the definition of the GOPNS.

4.3.14.1 Containment Relations

Table 4.6 lists the *containment relations* between geometric entities in the outer-product representation. For example, two outer-product line representations $\boldsymbol{L}$ and $\boldsymbol{K}$ represent lines that are contained in one another, i.e. they represent the same line, if and only if $\boldsymbol{L} \underline{\times} \boldsymbol{K} = 0$, where $\underline{\times}$ denotes the commutator product.

Table 4.6 Constraints between outer-product representations of geometric entities that are zero if the corresponding geometric entities are contained in one another

	Point $\boldsymbol{X}$	Line $\boldsymbol{L}$	Plane $\boldsymbol{P}$	Circle $\boldsymbol{C}$	Sphere $\boldsymbol{S}$
Point $\boldsymbol{Y}$	$\boldsymbol{X} \wedge \boldsymbol{Y}$	$\boldsymbol{L} \wedge \boldsymbol{Y}$	$\boldsymbol{P} \wedge \boldsymbol{Y}$	$\boldsymbol{C} \wedge \boldsymbol{Y}$	$\boldsymbol{S} \wedge \boldsymbol{Y}$
Line $\boldsymbol{K}$		$\boldsymbol{L} \underline{\times} \boldsymbol{K}$	$\boldsymbol{P} \overline{\times} \boldsymbol{K}$		
Plane $\boldsymbol{O}$			$\boldsymbol{P} \underline{\times} \boldsymbol{O}$		
Circle $\boldsymbol{B}$				$\boldsymbol{C} \underline{\times} \boldsymbol{B}$	$\boldsymbol{S} \overline{\times} \boldsymbol{B}$
Sphere $\boldsymbol{R}$					$\boldsymbol{S} \underline{\times} \boldsymbol{R}$

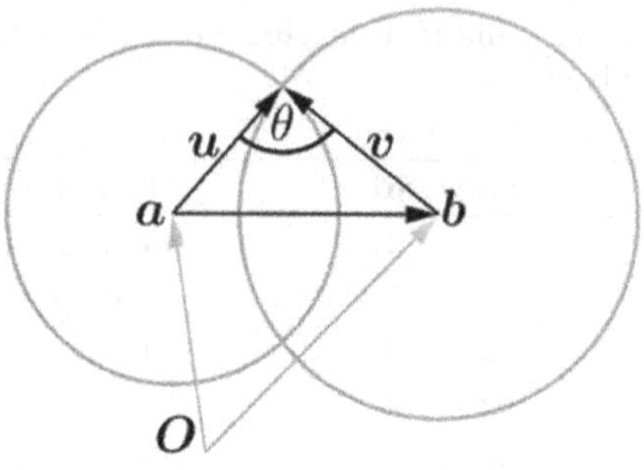

Fig. 4.20 The angle θ of intersection of two spheres is the angle between the surface normals of the spheres at the points of intersection

4.3.14.2 Inner Product of Spheres

It is interesting to look at the inner-product of two inner product representations of spheres. Let $\boldsymbol{a}, \boldsymbol{b} \in \mathbb{R}^3$, $\boldsymbol{A} := \mathcal{C}(\boldsymbol{a})$, and $\boldsymbol{B} := \mathcal{C}(\boldsymbol{b})$, such that two spheres can be defined by $\boldsymbol{S}_1 := \boldsymbol{A} - \frac{1}{2}\,\rho_a^2\,\boldsymbol{e}_\infty$ and $\boldsymbol{S}_2 := \boldsymbol{B} - \frac{1}{2}\,\rho_b^2\,\boldsymbol{e}_\infty$ where $\rho_a, \rho_b \in \mathbb{R}$ are the radii of the spheres. Then

$$\boldsymbol{S}_1 \cdot \boldsymbol{S}_2 = \boldsymbol{A} \cdot \boldsymbol{B} + \frac{1}{2}\,\rho_b^2 + \frac{1}{2}\,\rho_a^2 = \frac{1}{2}\,(\rho_a^2 + \rho_b^2) - \frac{1}{2}\,(\boldsymbol{a} - \boldsymbol{b})^2. \qquad (4.71)$$

Therefore,

$$\boldsymbol{S}_1 \cdot \boldsymbol{S}_2 = 0 \qquad \iff \qquad \rho_a^2 + \rho_b^2 = (\boldsymbol{a} - \boldsymbol{b})^2,$$

which is simply the Pythagorean theorem, where the vector between the centers of the spheres is the hypotenuse and the radii ρ_a and ρ_b are the legs of the right-angled triangle. This implies that the inner product of $\boldsymbol{S}_1$ and $\boldsymbol{S}_2$ is zero if and only if the corresponding spheres intersect at a right angle.

This result motivates the question in what way the inner product of two spheres is related to their angle of intersection, if the spheres do indeed intersect. The intersection angle of two spheres is defined as the angle between the surface normals at the points of intersection, as shown in Fig. 4.20. Note that this angle is the same at all points on the intersection circle. From Fig. 4.20, it can be seen that $\boldsymbol{u} - \boldsymbol{v} = \boldsymbol{b} - \boldsymbol{a}$, $\boldsymbol{u}^2 = \rho_a^2$, $\boldsymbol{v}^2 = \rho_b^2$, and $\boldsymbol{u} \cdot \boldsymbol{v} = \rho_a\,\rho_b\,\cos\theta$. The inner product of the spheres $\boldsymbol{S}_1$ and $\boldsymbol{S}_2$ can thus be brought into a relation with the intersection angle θ by substituting the expression $(\boldsymbol{u} - \boldsymbol{v})^2$ for $(\boldsymbol{a} - \boldsymbol{b})^2$ in (4.71), where

$$(\boldsymbol{a} - \boldsymbol{b})^2 = (\boldsymbol{u} - \boldsymbol{v})^2 = \rho_a^2 + \rho_b^2 - 2\,\rho_a\,\rho_b\,\cos\theta\,.$$

Upon substitution, the inner product of $\boldsymbol{S}_1$ and $\boldsymbol{S}_2$ results in

$$\boldsymbol{S}_1 \cdot \boldsymbol{S}_2 = \rho_a\,\rho_b\,\cos\theta\,. \qquad (4.72)$$

This result is valid only if $\boldsymbol{S}_1$ and $\boldsymbol{S}_2$ do indeed intersect.

Using this result, it can be easily demonstrated that the inversion operation is locally angle-preserving, as is to be expected, since it is a conformal mapping. For this purpose, consider two planes $\boldsymbol{P}_a$ and $\boldsymbol{P}_b$ with inner-product representations

$$\boldsymbol{P}_a := \hat{\boldsymbol{a}} + \alpha\, e_\infty\,, \quad \boldsymbol{P}_b := \hat{\boldsymbol{b}} + \beta\, e_\infty\,,$$

where $\hat{\boldsymbol{a}}$ and $\hat{\boldsymbol{b}}$ are unit vectors giving the normals of the planes, and $\alpha, \beta \in \mathbb{R}$ are the orthogonal separations of the planes from the origin. The angle ϕ between the two planes is the angle between their normals, that is, $\hat{\boldsymbol{a}} \cdot \hat{\boldsymbol{b}} = \cos\phi$. Clearly,

$$\boldsymbol{P}_a \cdot \boldsymbol{P}_b = \hat{\boldsymbol{a}} \cdot \hat{\boldsymbol{b}} = \cos\phi\,.$$

Now we define two spheres $\boldsymbol{S}_a$ and $\boldsymbol{S}_b$ as the inversions of $\boldsymbol{P}_a$ and $\boldsymbol{P}_b$, respectively. That is,

$$\boldsymbol{S}_a := \boldsymbol{S}\,\boldsymbol{P}_a\,\boldsymbol{S}\,, \quad \boldsymbol{S}_b := \boldsymbol{S}\,\boldsymbol{P}_b\,\boldsymbol{S}\,,$$

where $\boldsymbol{S} := \boldsymbol{e}_o - \frac{1}{2}\, e_\infty$ is the unit sphere centered at the origin. If a plane passes through the origin, than an inversion of this plane in $\boldsymbol{S}$ results again in a plane. To avoid this case, we assume in the following that $\alpha, \beta \neq 0$; that is, the planes $\boldsymbol{P}_a$ and $\boldsymbol{P}_b$ do not pass through the origin. By a straightforward calculation, it may be shown that

$$\boldsymbol{S}_a = \boldsymbol{S}\,\boldsymbol{P}_a\,\boldsymbol{S} = -\hat{\boldsymbol{a}} - 2\,\alpha\,\boldsymbol{e}_o = -\frac{1}{\rho_a}\left(\boldsymbol{a} + \frac{1}{2}\,\boldsymbol{a}^2\,e_\infty + \boldsymbol{e}_o - \frac{1}{2}\,\rho_a^2\,e_\infty\right),$$

where $\rho_a := 1/(2\alpha)$ and $\boldsymbol{a} := \rho_a\,\hat{\boldsymbol{a}}$. If the planes $\boldsymbol{P}_a$ and $\boldsymbol{P}_b$ are not parallel, they intersect in a line. The inversion of this line in the unit sphere gives a real circle, which is the intersection of the inversions of the planes. If the planes are parallel, they intersect in the line at infinity, which maps under inversion to the origin of the inversion sphere. Therefore, the spheres $\boldsymbol{S}_a$ and $\boldsymbol{S}_b$ always intersect, which implies that (4.71) is applicable. Hence,

$$\boldsymbol{S}_a \cdot \boldsymbol{S}_b = \frac{1}{\rho_a\,\rho_b}\,\rho_a\,\rho_b\,\cos\theta = \cos\theta,$$

where θ is the intersection angle of the spheres $\boldsymbol{S}_a$ and $\boldsymbol{S}_b$. The relation between the the inner product of the spheres $\boldsymbol{S}_a$, $\boldsymbol{S}_b$ and the planes $\boldsymbol{P}_a$, $\boldsymbol{P}_b$ can be found as follows:

$$\begin{aligned}\boldsymbol{S}_a \cdot \boldsymbol{S}_b &= (\boldsymbol{S}\,\boldsymbol{P}_a\,\boldsymbol{S}) \cdot (\boldsymbol{S}\,\boldsymbol{P}_b\,\boldsymbol{S})\\ &= \frac{1}{2}\,(\boldsymbol{S}\,\boldsymbol{P}_a\,\boldsymbol{S}\,\boldsymbol{S}\,\boldsymbol{P}_b\,\boldsymbol{S} + \boldsymbol{S}\,\boldsymbol{P}_b\,\boldsymbol{S}\,\boldsymbol{S}\,\boldsymbol{P}_a\,\boldsymbol{S})\\ &= \boldsymbol{S}\,(\boldsymbol{P}_a \cdot \boldsymbol{P}_b)\,\boldsymbol{S}\\ &= \boldsymbol{P}_a \cdot \boldsymbol{P}_b\,,\end{aligned} \tag{4.73}$$

where $\boldsymbol{S}\,\boldsymbol{S} = 1$. Hence,

$$\boldsymbol{S}_a \cdot \boldsymbol{S}_b = \boldsymbol{P}_a \cdot \boldsymbol{P}_b \quad \Longleftrightarrow \quad \cos\theta = \cos\phi\,,$$

which shows that the intersection angle of two planes is preserved under inversion.

4.3.15 Analysis of Blades

The goal of this subsection is to show how, given a blade that represents some geometric entity, the parameters describing that entity can be extracted. For example, given an outer-product representation of a circle, the question is how the center, normal, and radius of the circle can be extracted. The analysis methods presented here are user, for example, in the software CLUCalc to automatically analyze multivectors for their geometric content.

4.3.15.1 Planes

Let $\boldsymbol{P} \in \mathbb{G}^4_{4,1}$ be an outer-product representation of a plane; then $\boldsymbol{P}^*$ is of the form

$$\boldsymbol{P}^* = \alpha\,(\hat{\boldsymbol{a}} + d\,\boldsymbol{e}_\infty),$$

where $\hat{\boldsymbol{a}} \in \mathbb{R}^3$ is the normal of the plane, $d \in \mathbb{R}$ is the orthogonal separation of the plane from the origin, and $\alpha \in \mathbb{R}$ is some general scale. The parameters of the plane can be extracted via

$$\boldsymbol{a} = \mathcal{P}_{\mathbf{e}_{123}}(\boldsymbol{P}^*)\,, \qquad \alpha = \|\boldsymbol{a}\|, \qquad d = -\frac{\boldsymbol{P}^* \cdot \boldsymbol{e}_o}{\alpha}. \tag{4.74}$$

4.3.15.2 Spheres

Let $\boldsymbol{S} \in \mathbb{G}^4_{4,1}$ be an outer-product representation of a sphere; then $\boldsymbol{S}^*$ is of the form

$$\boldsymbol{S}^* = \alpha\,(\boldsymbol{A} - \frac{1}{2}\,r^2\,\boldsymbol{e}_\infty),$$

where $\boldsymbol{A} := \mathcal{C}(\boldsymbol{a})$, $\boldsymbol{a} \in \mathbb{R}^3$ is the center of the sphere, $r \in \mathbb{R}$ is its radius, and $\alpha \in \mathbb{R}$ is some scale. The parameters of the sphere can be extracted via

$$r^2 = \frac{(\boldsymbol{S}^*)^2}{(\boldsymbol{S}^* \cdot \boldsymbol{e}_\infty)^2}, \qquad \boldsymbol{a} = \frac{\mathcal{P}_{\mathbf{e}_{123}}(\boldsymbol{S}^*)}{-\boldsymbol{S}^* \cdot \boldsymbol{e}_\infty}. \tag{4.75}$$

4.3.15.3 Lines

The outer-product representation of a line that passes through $\boldsymbol{a}, \boldsymbol{b} \in \mathbb{R}^3$ is given by

$$\boldsymbol{L} := \boldsymbol{A} \wedge \boldsymbol{B} \wedge \boldsymbol{e}_\infty = \boldsymbol{a} \wedge \boldsymbol{b} \wedge \boldsymbol{e}_\infty - (\boldsymbol{b} - \boldsymbol{a}) \wedge \boldsymbol{e}_\infty \wedge \boldsymbol{e}_o,$$

where $\boldsymbol{A} := \mathcal{C}(\boldsymbol{a})$ and $\boldsymbol{B} := \mathcal{C}(\boldsymbol{b})$. Since $(\boldsymbol{e}_\infty \wedge \boldsymbol{e}_o)^2 = 1$,

$$\boldsymbol{L} \cdot (\boldsymbol{e}_\infty \wedge \boldsymbol{e}_o) = \boldsymbol{b} - \boldsymbol{a},$$

which is the direction of the line. On the other hand, $\boldsymbol{d} := \boldsymbol{b} - \boldsymbol{a}$ can be interpreted as the inner-product representation of a plane that passes through the origin with normal $\boldsymbol{d}$. This plane intersects the line $\boldsymbol{L}$ at a right angle in the point on the line that is closest to the origin. This point can be found by calculating the intersection of the line and the plane.

If we define $\boldsymbol{P}^* := \boldsymbol{d}$, then $\boldsymbol{P}$ is the outer-product representation of the plane, and the intersection point $\boldsymbol{X}$ of $\boldsymbol{L}$ and $\boldsymbol{P}$ is given by

$$\boldsymbol{X} = \boldsymbol{P} \vee \boldsymbol{L} = \boldsymbol{P}^* \cdot \boldsymbol{L},$$

since the join of $\boldsymbol{P}$ and $\boldsymbol{L}$ is the whole space, represented by the pseudoscalar. Estimating $\boldsymbol{X}$ gives

$$\begin{aligned} \boldsymbol{X} &= \boldsymbol{P}^* \cdot \boldsymbol{L} \\ &= (\boldsymbol{d} \cdot \boldsymbol{a})\, \boldsymbol{b} \wedge \boldsymbol{e}_\infty - (\boldsymbol{d} \cdot \boldsymbol{b})\, \boldsymbol{a} \wedge \boldsymbol{e}_\infty + \boldsymbol{d}^2\, \boldsymbol{e}_\infty \wedge \boldsymbol{e}_o \\ &\simeq (\boldsymbol{b} \cdot \boldsymbol{d}^{-1})\, \boldsymbol{a} \wedge \boldsymbol{e}_\infty - (\boldsymbol{a} \cdot \boldsymbol{d}^{-1})\, \boldsymbol{b} \wedge \boldsymbol{e}_\infty + \boldsymbol{e}_o \wedge \boldsymbol{e}_\infty, \end{aligned}$$

where "$\simeq$" denotes equality up to a scalar factor, and $\boldsymbol{d}^{-1} = \boldsymbol{d}/\boldsymbol{d}^2$. Hence, $\boldsymbol{X}$ is a homogeneous point.

To summarize, the direction $\boldsymbol{d}$ of the line and the point $\boldsymbol{X}$ on the line that is closest to the origin can be estimated via

$$\boldsymbol{d} = \boldsymbol{L} \cdot (\boldsymbol{e}_\infty \wedge \boldsymbol{e}_o) \qquad \text{and} \qquad \boldsymbol{X} = \boldsymbol{d} \cdot \boldsymbol{L}. \tag{4.76}$$

Note that $\boldsymbol{X}$ can also be written as

$$\boldsymbol{X} = \boldsymbol{d} \cdot \boldsymbol{L} \simeq \Big((\boldsymbol{e}_\infty \wedge \boldsymbol{e}_o) \cdot \boldsymbol{L} \Big) \cdot \boldsymbol{L} \simeq \mathcal{P}_{\boldsymbol{L}}(\boldsymbol{e}_\infty \wedge \boldsymbol{e}_o)\,.$$

4.3.15.4 Point Pairs

Let $\boldsymbol{Q} \in \mathbb{G}^2_{4,1}$ be the outer-product representation of a point pair $\boldsymbol{a}, \boldsymbol{b} \in \mathbb{R}^3$, i.e. $\boldsymbol{Q} := \boldsymbol{A} \wedge \boldsymbol{B}$, where $\boldsymbol{A} := \mathcal{C}(\boldsymbol{a})$ and $\boldsymbol{B} := \mathcal{C}(\boldsymbol{b})$. Clearly,

$$\boldsymbol{L} := \boldsymbol{Q} \wedge \boldsymbol{e}_\infty = \boldsymbol{A} \wedge \boldsymbol{B} \wedge \boldsymbol{e}_\infty$$

is the outer-product representation of the line that passes through $\boldsymbol{a}$ and $\boldsymbol{b}$. The results of the remaining calculations that are necessary to extract the parameters of the point pair are stated without proof.

The outer-product representation of the plane that is perpendicular to $\boldsymbol{L}$ and lies halfway between the points is given by

$$\boldsymbol{P} := \boldsymbol{Q}^* \wedge \boldsymbol{e}_\infty \qquad \text{or} \qquad \boldsymbol{P}^* = \boldsymbol{Q} \cdot \boldsymbol{e}_\infty\,.$$

The point $\boldsymbol{X}$ on $\boldsymbol{L}$ that lies halfway between the two points $\boldsymbol{a}$ and $\boldsymbol{b}$ is thus given by

$$\boldsymbol{X} = \boldsymbol{P} \vee \boldsymbol{L} = \boldsymbol{P}^* \cdot \boldsymbol{L}.$$

Finally, the normalized inner-product representation of the sphere centered on $\boldsymbol{X}$ with a radius equal to half the distance between $\boldsymbol{a}$ and $\boldsymbol{b}$ is given by

$$\boldsymbol{S}^* := \boldsymbol{Q} \cdot \boldsymbol{L}^{-1}.$$

The distance $d \in \mathbb{R}$ between the points $\boldsymbol{a}$ and $\boldsymbol{b}$ is therefore

$$d = 2\sqrt{\boldsymbol{S}^* \cdot \boldsymbol{S}^*} = 2\sqrt{\frac{\boldsymbol{Q} \cdot \boldsymbol{Q}}{\boldsymbol{L} \cdot \boldsymbol{L}}}.$$

The center of the point pair is therefore $\boldsymbol{X}$, its direction can be extracted from $\boldsymbol{L}$, and the separation of the points is d.

As an aside, note that $\boldsymbol{Q}^*$ is an outer-product representation of an *imaginary* circle in a plane $\boldsymbol{P}$ centered on $\boldsymbol{X}$ with an *imaginary* radius $\mathrm{i}\, d/2$.

4.3.15.5 Circles

Let $\boldsymbol{C} \in \mathbb{G}^3_{4,1}$ be an outer-product representation of the circle that passes through $\boldsymbol{x}, \boldsymbol{y}, \boldsymbol{z} \in \mathbb{R}^3$, i.e. $\boldsymbol{C} := \boldsymbol{X} \wedge \boldsymbol{Y} \wedge \boldsymbol{Z}$, where $\boldsymbol{X} := \mathcal{C}(\boldsymbol{x})$, $\boldsymbol{Y} := \mathcal{C}(\boldsymbol{y})$, and $\boldsymbol{Z} := \mathcal{C}(\boldsymbol{z})$. Obviously, the plane in which the circle lies has an outer-product representation

$$\boldsymbol{P} := \boldsymbol{C} \wedge \boldsymbol{e}_\infty = \boldsymbol{A} \wedge \boldsymbol{B} \wedge \boldsymbol{C} \wedge \boldsymbol{e}_\infty.$$

If $\boldsymbol{S}_1^*, \boldsymbol{S}_2^* \in \mathbb{G}^1_{4,1}$ are the inner-product representations of two spheres that intersect in the circle $\boldsymbol{C}$, then the inner-product representation of $\boldsymbol{C}$ is

$$\boldsymbol{C}^* = \boldsymbol{S}_1^* \wedge \boldsymbol{S}_2^*.$$

The line $\boldsymbol{L}$ that passes through the centers of the spheres $\boldsymbol{S}_1$ and $\boldsymbol{S}_2$ therefore has an outer-product representation

$$\boldsymbol{L} := \boldsymbol{C}^* \wedge \boldsymbol{e}_\infty = \boldsymbol{S}_1^* \wedge \boldsymbol{S}_2^* \wedge \boldsymbol{e}_\infty.$$

Line $\boldsymbol{L}$ has to pass through the center of the circle, which implies that the center of the circle is the intersection of $\boldsymbol{L}$ and $\boldsymbol{P}$, which can be estimated via

$$\boldsymbol{U} := \boldsymbol{P} \vee \boldsymbol{L} = \boldsymbol{P}^* \cdot \boldsymbol{L}.$$

Finally, the normalized inner-product representation of a sphere centered on $\boldsymbol{U}$ with the same radius as the circle is

$$\boldsymbol{S}^* := \boldsymbol{C} \cdot \boldsymbol{P}^{-1},$$

so that the radius r of the circle is given by

$$r = \sqrt{\boldsymbol{S}^* \cdot \boldsymbol{S}^*} = \sqrt{\frac{\boldsymbol{C} \cdot \boldsymbol{C}}{\boldsymbol{P} \cdot \boldsymbol{P}}}.$$

The center of the circle is thus given by $\boldsymbol{U}$, its plane is $\boldsymbol{P}$, and its radius is r. Note that the radius may be imaginary.

As an aside, note that $\boldsymbol{C}^*$ is an outer-product representation of an *imaginary* point pair on the line $\boldsymbol{L}$, centered on $\boldsymbol{X}$ with an *imaginary* separation $2\,\mathrm{i}\,r$.

4.4 Conic Space

The geometric algebra of conformal space is a very potent representation of geometry and geometric transformations, but it is limited to the representation of spheres in all dimensions, with all radii, and conformal transformations. For example, a 1D sphere is a point pair, and a 2D sphere with an infinite radius is a line. The transformations most often employed are the Euclidean transformations. A scaling is also available in the form of a dilator, which scales isotropically. It is not, however, possible to represent conic sections, often referred to simply as *conics*, in terms of a geometric null space. Nor is a scaling with a preferred direction available, which implies that there exist no versors that represent affine transformation operators.

There are many applications where non-isotropic scaling plays an important role. One example is the calibration of cameras with non-square pixels. The projection of entities onto planes is also limited in $\mathbb{G}_{4,1}$ to those entities that are also representable in projective space, i.e. points and lines. The projection of a circle in 3D space onto a plane, which would result in an ellipse, cannot be represented in $\mathbb{G}_{4,1}$.

This provides a motivation to look for other vector spaces on which a geometric algebra can be constructed, such that the list of representable geometric entities and transformations is extended. A first step in this direction is presented in this section, where it shown how a geometric algebra can be constructed such that 2D conics are representable through geometric null spaces. A drawback of this space, however, is that no translation versors exist. It is, instead, comparable to projective space with the extended feature of a representation of conics.

It is shown later that when vectors represent conic sections in the geometric IPNS, the outer product of two vectors represents the intersection of the two corresponding conics. While investigating the properties of such bivectors when fitting them to data points, the author discovered an interesting property that led to a corner and junction detection algorithm [131, 132]. However, this will not be discussed further in this text.

4.4.1 Polynomial Embedding

Before the details of conic space are discussed, it is helpful to look at what needs to be achieved from a more general point of view. In terms of the geometric IPNS, a 1-blade $\boldsymbol{A} \in \mathbb{G}^1_{p,q}$ represents

$$\mathbb{NI}_G(\boldsymbol{A}) = \left\{\, \boldsymbol{x} \in \mathbb{R}^n \;:\; \mathcal{X}(\boldsymbol{x}) \cdot \boldsymbol{A} = 0 \,\right\},$$

where $\mathcal{X} : \mathbb{R}^n \to \mathbb{X} \subseteq \mathbb{R}^{p,q}$ is some bijective mapping. A particularly useful type of embedding is a polynomial embedding. That is, if $\boldsymbol{x} := \mathsf{x}^i\, \boldsymbol{e}_i$, then

$$\mathcal{X} : \boldsymbol{x} \mapsto \left[1, \mathsf{x}^1, \ldots, \mathsf{x}^n,\; (\mathsf{x}^1)^2, \ldots, (\mathsf{x}^n)^2,\; \ldots,\; (\mathsf{x}^1)^r, \ldots, (\mathsf{x}^n)^r\right] \in \mathbb{R}^{p,q}. \tag{4.77}$$

Defining $\boldsymbol{A} := [\mathsf{a}^1, \ldots, \mathsf{a}^{nr+1}] \in \mathbb{R}^{p,q}$ gives

$$\mathcal{X}(\boldsymbol{x}) \cdot \boldsymbol{A} = \mathsf{a}^1 + \sum_{j=1}^{r} \sum_{i=1}^{n} \left(\lambda^k\, \mathsf{a}^k\, (\mathsf{x}^i)^j\right), \qquad k := i + (j-1)n + 1, \tag{4.78}$$

where $\lambda^i := \boldsymbol{e}_i \cdot \boldsymbol{e}_i$, and $\boldsymbol{e}_i := \overline{\mathbb{R}}^{p,q}[i]$ is the signature of the ith basis vector of $\mathbb{R}^{p,q}$. Note that a^i denotes the ith component of a, whereas $(\mathsf{a}^i)^j$ denotes the jth power of the ith component. Therefore, the geometric IPNS of $\boldsymbol{A}$ is the set of roots of the polynomial given in (4.78). More generally, a polynomial with positive and negative powers could be considered.

With respect to $\mathbb{G}_{4,1}$, the embedding of a vector $\boldsymbol{x} \in \mathbb{R}^3$ with $\boldsymbol{x} := \mathsf{x}^i\, \boldsymbol{e}_i$ is defined as

$$\mathcal{C}(\boldsymbol{x}) := \boldsymbol{x} + \frac{1}{2}\, \boldsymbol{x}^2\, \boldsymbol{e}_\infty + \boldsymbol{e}_o.$$

A general vector $\boldsymbol{A} \in \mathbb{G}^1_{4,1}$ may be written as

$$\boldsymbol{A} := \mathsf{a}^1\, \boldsymbol{e}_1 + \mathsf{a}^2\, \boldsymbol{e}_2 + \mathsf{a}^3\, \boldsymbol{e}_3 + \mathsf{a}^4\, \boldsymbol{e}_\infty + \mathsf{a}^5\, \boldsymbol{e}_o,$$

such that

$$\mathcal{C}(\boldsymbol{x}) \cdot \boldsymbol{A} = \mathsf{a}^1\,\mathsf{x}^1 + \mathsf{a}^2\,\mathsf{x}^2 + \mathsf{a}^3\,\mathsf{x}^3 - \frac{1}{2}\,\mathsf{a}^5\,\boldsymbol{x}^2 - \mathsf{a}^4, \tag{4.79}$$

where $\boldsymbol{x}^2 = \sum_{i=1}^{3} (\mathsf{x}^i)^2$. Therefore, the geometric IPNS is the set of roots of the polynomial of (4.79). Here, the relation of geometric algebra to algebraic geometry becomes particularly clear, since this is exactly the definition of an affine variety in algebraic geometry (see [38]). The basic geometric entities that are representable in $\mathbb{G}_{4,1}$ are therefore those that can be expressed as the roots of this polynomial. Given these basic geometric entities, the other entities can be constructed by intersection, i.e. by taking the outer product of the basic entities.

The polynomial in (4.79) can also be written in terms of matrix products as

$$\mathcal{C}(\boldsymbol{x}) \cdot \boldsymbol{A} = \begin{bmatrix} \mathsf{x}^1\,\mathsf{x}^2\,\mathsf{x}^3\,1 \end{bmatrix} \frac{1}{2} \begin{bmatrix} -\mathsf{a}^5 & 0 & 0 & \mathsf{a}^1 \\ 0 & -\mathsf{a}^5 & 0 & \mathsf{a}^2 \\ 0 & 0 & -\mathsf{a}^5 & \mathsf{a}^3 \\ \mathsf{a}^1 & \mathsf{a}^2 & \mathsf{a}^3 & -2\,\mathsf{a}^4 \end{bmatrix} \begin{bmatrix} \mathsf{x}^1 \\ \mathsf{x}^2 \\ \mathsf{x}^3 \\ 1 \end{bmatrix}, \tag{4.80}$$

which is a special case of the representation of a projective conic. If we denote the column vector of the $\{\,\mathsf{x}^i\,\}$ by x and the symmetric matrix of the components $\{\,\mathsf{a}^i\,\}$ by A, then (4.80) can be written as $\mathsf{x}^{\mathsf{T}}\,\mathsf{A}\,\mathsf{x}$. In general, it is also possible to write this expression in terms of a matrix inner product,

$$\mathsf{x}^{\mathsf{T}}\,\mathsf{A}\,\mathsf{x} = (\mathsf{x}\,\mathsf{x}^{\mathsf{T}}) \cdot \mathsf{A}. \tag{4.81}$$

The matrix inner product is defined as follows. Let $\mathsf{a}^i{}_j$ denote the components of the matrix $\mathsf{A} \in \mathbb{R}^{n \times m}$ and $\mathsf{b}^i{}_j$ the components of the matrix $\mathsf{B} \in \mathbb{R}^{n \times m}$; then

$$\mathsf{A} \cdot \mathsf{B} := \sum_{i,j} \mathsf{a}^i{}_j\,\mathsf{b}^i{}_j .$$

Thus, everything required present to construct a geometric algebra is present. The set of matrices in $\mathbb{R}^{n \times m}$ forms a vector space, and an inner product is defined such that $\mathsf{A} \cdot \mathsf{A} \in \mathbb{R}$. The geometric algebra over the vector space of symmetric matrices in $\mathbb{R}^{4 \times 4}$ is thus a generalization of $\mathbb{G}_{4,1}$. As shown in (4.80), $\mathbb{G}_{4,1}$ is simply a special case of such an embedding.

4.4.2 Symmetric-Matrix Vector Space

Using the ideas of the previous subsection, the geometric algebra over the vector space of symmetric 3×3 matrices will be introduced here. Owing to the symmetry of the matrices, only the upper triangular part of a matrix is needed to define it uniquely.

Let $\mathsf{A} \in \mathbb{R}^{3\times 3}$ be a symmetric matrix, and denote by a_{ji} the component of A in row j and column i. The notation a_{ji} is here used instead of the correct notation $\mathsf{A}^j{}_i$ to increase the readability of the equations. We then define a mapping $\mathcal{M} : \mathbb{R}^{3\times 3} \to \mathbb{R}^6$ by

$$\mathcal{M} : \mathsf{A} \mapsto a_{13}\,\mathbf{e}_1 + a_{23}\,\mathbf{e}_2 + \frac{1}{\sqrt{2}}\,a_{33}\,\mathbf{e}_3 + \frac{1}{\sqrt{2}}\,a_{11}\,\mathbf{e}_4 + \frac{1}{\sqrt{2}}\,a_{22}\,\mathbf{e}_5 + a_{12}\,\mathbf{e}_6 . \tag{4.82}$$

Given a second symmetric matrix $\mathsf{B} \in \mathbb{R}^{3\times 3}$, it follows that

$$\mathsf{A} \cdot \mathsf{B} = 2\,\mathcal{M}(\mathsf{A}) \cdot \mathcal{M}(\mathsf{B}).$$

Thus the embedding $\mathcal{M}$ of symmetric 3×3 matrices in $\mathbb{R}^6$ is an isomorphism between $\mathbb{R}^{3\times 3}$ and $\mathbb{R}^6$.

It is well known that a symmetric 3×3 matrix $\mathsf{A} \in \mathbb{R}^{3\times 3}$ is a projective representation of a conic section, i.e. a quadratic polynomial. This can be seen quite easily as follows. We embed a 2D vector $[x\ y]^\mathsf{T} \in \mathbb{R}^{2\times 1}$ as $[x\ y\ 1]^\mathsf{T} \in \mathbb{R}^{3\times 1}$, which is just a homogenization of the 2D vector, as discussed in Sect. 4.2. Now

$$\begin{aligned} \mathsf{x}^\mathsf{T}\,\mathsf{A}\,\mathsf{x} &= (\mathsf{x}\,\mathsf{x}^\mathsf{T}) \cdot \mathsf{A} \\ &= \begin{bmatrix} x^2 & x\,y & x \\ y\,x & y^2 & y \\ x & y & 1 \end{bmatrix} \cdot \begin{bmatrix} a_{11} & a_{12} & a_{13} \\ a_{21} & a_{22} & a_{23} \\ a_{31} & a_{32} & a_{33} \end{bmatrix} \\ &= a_{11}\,x^2 + a_{22}\,y^2 + 2\,a_{12}\,x\,y + 2\,a_{13}\,x + 2\,a_{23}\,y + a_{33}. \end{aligned} \tag{4.83}$$

Therefore,

$$\begin{aligned} & \mathsf{x}^\mathsf{T}\,\mathsf{A}\,\mathsf{x} = 0 \\ \iff\ & a_{11}\,x^2 + a_{22}\,y^2 + 2\,a_{12}\,x\,y + 2\,a_{13}\,x + 2\,a_{23}\,y = -a_{33}. \end{aligned}$$

For example, if $a_{11} = a_{22} = 1$, $a_{33} = -1$, and all other matrix components are zero, then

$$\mathsf{x}^\mathsf{T}\,\mathsf{A}\,\mathsf{x} = 0 \qquad \iff \qquad x^2 + y^2 = 1,$$

which is satisfied for all points in $\mathbb{R}^2$ that lie on the unit circle. A complete analysis of all representable conic sections will be given later on.

A point in $[x\ y] \in \mathbb{R}^2$ is therefore represented in the vector space of symmetric 3×3 matrices as $\mathsf{x}\mathsf{x}^\mathsf{T}$, where $\mathsf{x} := [x\ y\ 1]^\mathsf{T}$. This is the embedding of points that is used for the construction of the geometric algebra $\mathbb{G}_6$. We define the mapping $\mathcal{D} : \mathbb{R}^2 \to \mathbb{D} \subset \mathbb{R}^6$ as

$$\mathcal{D} : \boldsymbol{x} \mapsto \boldsymbol{x} + \frac{1}{\sqrt{2}}\,\mathbf{e}_3 + \frac{1}{\sqrt{2}}\,x^2\,\mathbf{e}_4 + \frac{1}{\sqrt{2}}\,y^2\,\mathbf{e}_5 + x\,y\,\mathbf{e}_6, \tag{4.84}$$

where $\boldsymbol{x} := x\,\boldsymbol{e}_1 + y\,\boldsymbol{e}_2$. Then, if $\boldsymbol{A} := \mathcal{M}(\mathsf{A})$, where $\mathsf{A} \in \mathbb{R}^{3\times 3}$ is a symmetric matrix, $\mathcal{D}(\boldsymbol{x}) \cdot \boldsymbol{A} = 0$ if and only if $\boldsymbol{x}$ lies on the conic section represented by A. Hence,

$$\mathbb{NI}_G(\boldsymbol{A}) = \{\, \boldsymbol{x} \in \mathbb{R}^2 \;:\; \mathcal{D}(\boldsymbol{x}) \cdot \boldsymbol{A} = 0 \,\} \tag{4.85}$$

is a conic section.

4.4.3 The Geometric Algebra $\mathbb{G}_6$

The geometric algebra over the vector space of symmetric matrices $\mathbb{R}^6$ allows the representation of not just conic sections but also their intersections. Furthermore, conic sections can be constructed from a number of points.

4.4.3.1 Outer-Product Representation

Let $\boldsymbol{a}_1, \ldots, \boldsymbol{a}_5 \in \mathbb{R}^2$ be five different points, and let $\boldsymbol{A}_i := \mathcal{D}(\boldsymbol{a}_i)$ be their embedding in $\mathbb{G}_6$. Then the GOPNS of $\boldsymbol{A}_{\langle 5\rangle} := \bigwedge_{i=1}^{5} \boldsymbol{A}_i$ is the conic section that passes through the points $\{\, \boldsymbol{a}_i \,\}$. This can be seen as follows. Since

$$\mathbb{NO}_G(\boldsymbol{A}_{\langle 5\rangle}) = \{\, \boldsymbol{x} \in \mathbb{R}^2 \;:\; \mathcal{D}(\boldsymbol{x}) \wedge \boldsymbol{A}_{\langle 5\rangle} = 0 \,\}, \tag{4.86}$$

each of the points $\{\, \boldsymbol{a}_i \,\}$ lies on the entity represented by $\boldsymbol{A}_{\langle 5\rangle}$. Furthermore, $\mathbb{NO}_G(\boldsymbol{A}_{\langle 5\rangle}) = \mathbb{NI}_G(\boldsymbol{A}^*_{\langle 5\rangle})$, and $\boldsymbol{A}^*_{\langle 5\rangle} \in \mathbb{G}^1_6$ is a 1-blade. However, the GIPNS of a 1-blade is a conic section. Therefore, $\mathbb{NO}_G(\boldsymbol{A}_{\langle 5\rangle})$ has to be the conic section that passes through the $\{\, \boldsymbol{a}_i \,\}$.

The conic sections that can be constructed in this way are circles, ellipses, hyperbolas, parabolas, lines, parallel line pairs, and intersecting line pairs. All of these, apart from the parabola, are illustrated in Fig. 4.21. The main difference from $\mathbb{G}_{4,1}$ is that all of these entities are represented by the GOPNS of a 5-blade. Note, however, that the outer-product representation of a line passing through points $\boldsymbol{a}, \boldsymbol{b} \in \mathbb{R}^2$ is $\mathcal{D}(\boldsymbol{a}) \wedge \mathcal{D}(\boldsymbol{b}) \wedge \boldsymbol{e}_4 \wedge \boldsymbol{e}_5 \wedge \boldsymbol{e}_6$. This is how the line in Fig. 4.21(f) is represented.

The outer-product representations of blades of lower grade represent point sets. For example, the GOPNS of a 4-blade is a point quadruplet. This entity has to exist, since two 2D conics can intersect in at most four points. Accordingly, the GOPNS of a 3-blade is a point triplet, and that of a 2-blade is a point pair. How the point positions are extracted from a blade is not trivial. Later on it will be shown how the intersection points of two conic sections can be evaluated.

The construction of a conic section by use of the outer product of five points that lie on that conic section is also a very convenient way to evaluate the symmetric matrix that represents this conic. The symmetric matrix A that represents the conic that passes through $\{\, \boldsymbol{a}_i \,\}$ is given by

$$\mathsf{A} = \mathcal{M}^{-1}\left(\left(\bigwedge_{i=1}^{5} \mathcal{D}(\boldsymbol{a}_i)\right)^{*}\right). \tag{4.87}$$

This also demonstrates again that a conic section is uniquely defined by five points.

Note that the GOPNSs of 5-blades and 4-blades are automatically visualized by CLUCalc.

4.4.3.2 Inner-Product Representation

In the inner-product representation, a 1-blade represents a conic section, as shown earlier. Given two 1-blades $\boldsymbol{A}, \boldsymbol{B} \in \mathbb{G}_6^1$, the GIPNS of their outer product is the intersection of the corresponding blades. This follows in exactly the same way as for conformal, projective, and Euclidean space. Because

$$\mathbb{NI}_G(\boldsymbol{A} \wedge \boldsymbol{B}) = \left\{ \boldsymbol{x} \in \mathbb{R}^2 \ : \ \mathcal{D}(\boldsymbol{x}) \cdot (\boldsymbol{A} \wedge \boldsymbol{B}) = 0 \right\}$$

and

$$\boldsymbol{X} \cdot (\boldsymbol{A} \wedge \boldsymbol{B}) = (\boldsymbol{X} \cdot \boldsymbol{A})\,\boldsymbol{B} - (\boldsymbol{X} \cdot \boldsymbol{B})\,\boldsymbol{A},$$

where $\boldsymbol{X} := \mathcal{D}(\boldsymbol{x})$, the inner product of $\boldsymbol{X}$ and $\boldsymbol{A} \wedge \boldsymbol{B}$ is zero if and only if $\boldsymbol{X} \cdot \boldsymbol{A} = 0$ and $\boldsymbol{X} \cdot \boldsymbol{B} = 0$. This is only the case if $\boldsymbol{x}$ lies on both conics, and thus $\mathbb{NI}_G(\boldsymbol{A} \wedge \boldsymbol{B})$ is the intersection of the conics represented by $\boldsymbol{A}$ and $\boldsymbol{B}$. This follows analogously for the outer product of more than two vectors.

Even though the representation of the intersection of two conics is very simple, the evaluation of the actual intersection points is not. This is therefore the subject of a later section.

4.4.4 Rotation Operator

There does not seem to be a simple way to derive the rotation operator for conics represented in $\mathbb{G}_6$. The rotation operator is therefore simply stated here. Let $\boldsymbol{A} \in \mathbb{G}_6^1$ be an inner-product representation of a conic. The rotation operator for a rotation of the conic represented by $\boldsymbol{A}$ by an angle θ in an anticlockwise direction about the origin is given by

$$\boldsymbol{R} = \boldsymbol{R}_2\,\boldsymbol{R}_1\,, \tag{4.88}$$

where

$$\begin{aligned}
\boldsymbol{R}_1 &:= \cos\theta - \frac{1}{\sqrt{2}}\sin\theta\,(\boldsymbol{e}_4 - \boldsymbol{e}_5) \wedge \boldsymbol{e}_6\,,\\
\boldsymbol{R}_2 &:= \cos\theta/2 - \sin\theta/2\;\boldsymbol{e}_1 \wedge \boldsymbol{e}_2\,.
\end{aligned}$$

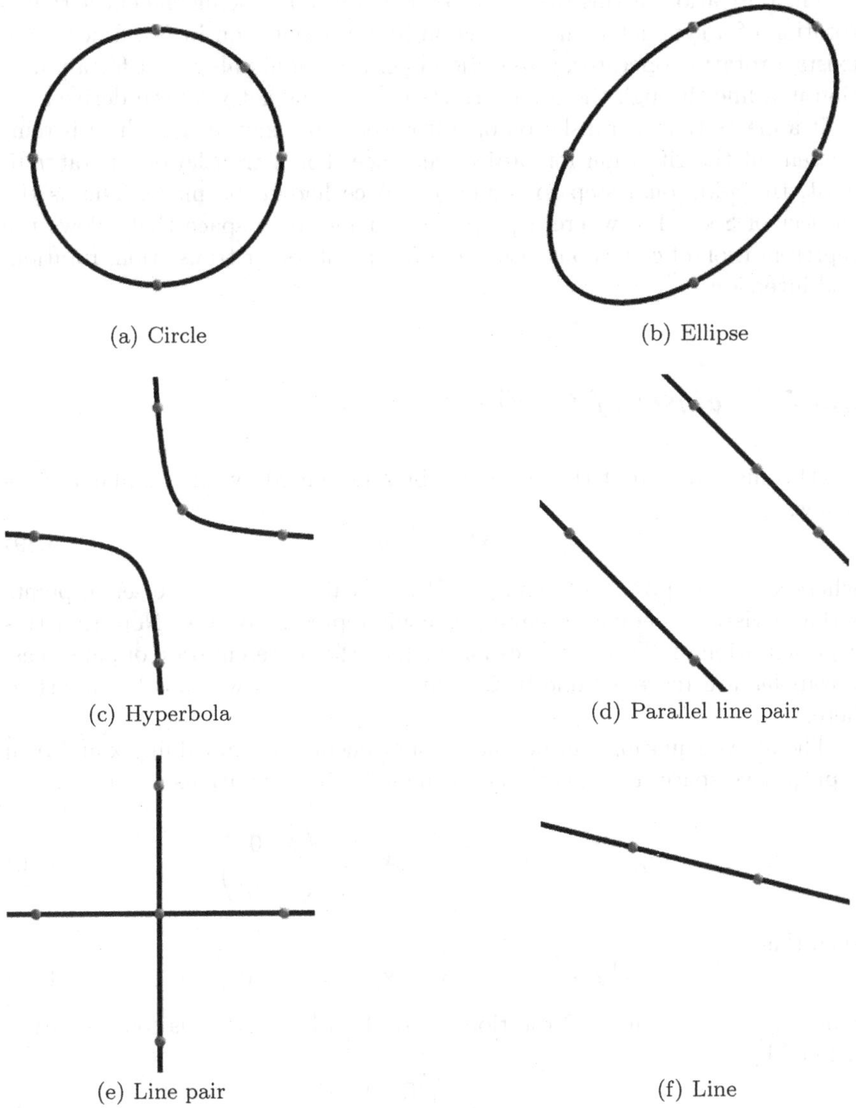

Fig. 4.21 Different types of projective conics

The rotor $\boldsymbol{R}$ therefore has components of grades zero, two, and four. Just like all versors, the rotor can be applied to any blade in $\mathbb{G}_6$ to rotate the representation of the blade. For the inner-product representation of a conic $\boldsymbol{A}$, the rotated conic $\boldsymbol{B}$ is given by

$$\boldsymbol{B} = \boldsymbol{R}\,\boldsymbol{A}\,\widetilde{\boldsymbol{R}} = \boldsymbol{R}_2\,\boldsymbol{R}_1\;\boldsymbol{A}\;\widetilde{\boldsymbol{R}}_1\,\widetilde{\boldsymbol{R}}_2.$$

The derivation of this rotor can be done by reproducing the effect that a rotation of a symmetric matrix has on its representation in $\mathbb{G}_6$. Since there exists a rotation operator, a reflection operator should also exist for a reflection in a line through the origin. However, this has not yet been derived.

It appears that a translation operator does not exist in $\mathbb{G}_6$, which is reminiscent of the situation for projective space. For a translation operator to exist, the additional step to conformal space had to be made. This is the subject of Sect. 4.5, where a proposal is made for a space that allows the representation of conics but also contains operators for translation, rotation, and inversion.

4.4.5 Analysis of Conics

A 2D conic centered at the origin can be represented by a 2×2 matrix A as follows:

$$\mathsf{x}^{\mathsf{T}} \mathsf{A} \mathsf{x} = \rho, \tag{4.89}$$

where $\mathsf{x} := [x_1, x_2]^{\mathsf{T}} \in \mathbb{R}^2$, and $\rho \in \mathbb{R}$ is a scale (radius). The set of points x that satisfy this equation lie on the conic represented by A. Note that this representation of 2D conics does *not* include the representation of parabolas. Parabolas are representable in $\mathbb{G}_6$ but their analysis will not be discussed here.

The above equation can be made homogeneous by embedding x and A in a projective space. For this purpose, the following definitions are made:

$$\mathsf{x}_H := [x_1, x_2, 1]^{\mathsf{T}}, \qquad \mathsf{A}_H := \begin{pmatrix} \mathsf{A} & 0 \\ 0 & -\rho \end{pmatrix}, \tag{4.90}$$

such that

$$\mathsf{x}^{\mathsf{T}} \mathsf{A} \mathsf{x} = \rho \quad \Longleftrightarrow \quad \mathsf{x}_H^{\mathsf{T}} \mathsf{A}_H \mathsf{x}_H = 0. \tag{4.91}$$

Similarly, for a given 2×2 rotation matrix R, a homogeneous counterpart is defined by

$$\mathsf{R}_H := \begin{pmatrix} \mathsf{R} & 0 \\ 0 & 1 \end{pmatrix}. \tag{4.92}$$

A conic rotated about the origin by R can be represented by rotating the vector that is multiplied from left and right by the conic matrix in the opposite direction. That is, a conic rotated by R is represented by those vectors x_H that satisfy

$$\mathsf{x}_H^{\mathsf{T}} \mathsf{R}_H \mathsf{A}_H \mathsf{R}_H^{\mathsf{T}} \mathsf{x}_H = \mathsf{x}_H^{\mathsf{T}} \begin{pmatrix} \mathsf{R} \mathsf{A} \mathsf{R}^{\mathsf{T}} & 0 \\ 0 & -\rho \end{pmatrix} = 0. \tag{4.93}$$

Diagonalizing A results in

$$\mathsf{A} = \mathsf{U}\,\Lambda\,\mathsf{U}^{\mathsf{T}}, \tag{4.94}$$

where U is a unitary matrix containing the eigenvectors of A in its columns, and Λ is a diagonal matrix with the eigenvalues of A on its diagonal. Since A is a real, symmetric matrix, the eigenvalues are real and the eigenvectors are orthogonal, and hence U is unitary. Therefore, U gives the rotation matrix by which the conic has been rotated, and Λ describes what type of conic A represents. The eigenvectors of A are denoted by u_1 and u_2 such that $\mathsf{U} = [\mathsf{u}_1, \mathsf{u}_2]$. Let

$$\mathsf{U}_H := \begin{pmatrix} \mathsf{U} & 0 \\ 0 & 1 \end{pmatrix}; \tag{4.95}$$

then the following relation holds:

$$\mathsf{U}_H^{\mathsf{T}}\,\mathsf{A}_H\,\mathsf{U}_H = \Lambda_H := \begin{bmatrix} \lambda_1 & 0 & 0 \\ 0 & \lambda_2 & 0 \\ 0 & 0 & -\rho \end{bmatrix}, \tag{4.96}$$

where λ_1 and λ_2 are the eigenvalues of A.

However, a conic need not be centered on the origin. Its origin may be translated to a point $\mathsf{t} \in \mathbb{R}^2$, by applying the inverse translation to the points that are multiplied by the conic. If we define a homogeneous translation matrix T_H by

$$\mathsf{T}_H := \begin{pmatrix} \mathsf{I} & -\mathsf{t} \\ -\mathsf{t}^{\mathsf{T}} & 1 \end{pmatrix}, \tag{4.97}$$

then those points x_H that satisfy

$$\mathsf{x}_H^{\mathsf{T}}\,\mathsf{T}_H^{\mathsf{T}}\,\mathsf{A}_H\,\mathsf{T}_H\,\mathsf{x}_H = 0 \tag{4.98}$$

lie on the conic represented by A_H, translated by the vector t.

Given an arbitrary, symmetric 3×3 matrix Q_H, how can the parameters of the conics be extracted? Using the above definitions, it is clear that Q_H may be written as

$$\mathsf{Q}_H = \mathsf{T}_H^{\mathsf{T}}\,\mathsf{A}_H\,\mathsf{T}_H. \tag{4.99}$$

Since the top left 2×2 submatrix of T_H is the identity matrix, the eigenvector matrix U of A can be evaluated from that part of Q_H. That is, Q_H has the form

$$\mathsf{Q}_H = \begin{pmatrix} \mathsf{A} & \mathsf{q} \\ \mathsf{q}^{\mathsf{T}} & p \end{pmatrix}, \tag{4.100}$$

where $\mathsf{q} \in \mathbb{R}^2$ is some vector and $p \in \mathbb{R}$ is a scalar. Now Λ_H and the translation vector t have to be extracted from Q_H. From

$$\mathsf{L}_H := \mathsf{U}_H^{\mathsf{T}}\,\mathsf{Q}_H\,\mathsf{U}_H = \mathsf{U}_H^{\mathsf{T}}\,\mathsf{T}_H^{\mathsf{T}}\,\mathsf{U}_H\,\Lambda\,\mathsf{U}_H^{\mathsf{T}}\,\mathsf{T}_H\,\mathsf{U}_H, \tag{4.101}$$

it follows that

$$\mathsf{S}_H := \mathsf{U}_H^\mathsf{T}\, \mathsf{T}_H\, \mathsf{U}_H = \begin{bmatrix} 1\; 0 & -\mathsf{t}^\mathsf{T}\,\mathsf{u}_1 \\ 0\; 1 & -\mathsf{t}^\mathsf{T}\,\mathsf{u}_2 \\ 0\; 0 & 1 \end{bmatrix} = \begin{bmatrix} 1\; 0\; s_1 \\ 0\; 1\; s_2 \\ 0\; 0\; 1 \end{bmatrix}, \tag{4.102}$$

where $s_1 := -\mathsf{t}^\mathsf{T}\,\mathsf{u}_1$ and $s_2 := -\mathsf{t}^\mathsf{T}\,\mathsf{u}_2$. Hence,

$$\mathsf{L}_H = \mathsf{S}_H^\mathsf{T}\, \Lambda_H\, \mathsf{S}_H = \begin{bmatrix} \lambda_1 & 0 & \lambda_1\, s_1 \\ 0 & \lambda_2 & \lambda_2\, s_2 \\ \lambda_1\, s_1 & \lambda_2\, s_2 & \lambda_1\, s_1^2 + \lambda_2\, s_2^2 - \rho \end{bmatrix}. \tag{4.103}$$

U and, thus, U_H can be evaluated from the upper left 2×2 submatrix of Q_H. The translation vector t can be evaluated from L_H via

$$\mathsf{t} = -s_1\,\mathsf{u}_1 - s_2\,\mathsf{u}_2. \tag{4.104}$$

Furthermore, ρ can be evaluated from L_H by first evaluating λ_1, λ_2, s_1, and s_2. Note also that if $\rho = 0$, the matrix Q_H is of rank two at most. This can be seen from the form of the matrix L_H, since the two matrices are related via a similarity transformation. If we multiply the first row of L_H by s_1 and the second row with s_2, the sum of these two rows is equal to the third row if $\rho = 0$.

The type of conic represented by Q_H can now be deduced from λ_1, λ_2, and ρ, which can all be evaluated from L_H. The following types of conics can be distinguished. Note that any scalar multiple of Q_H represents the same conic as Q_H. In particular, $-\mathsf{Q}_H$ represents the same conic as Q_H. Therefore, the following signatures of λ_1, λ_2, and ρ may also be inverted:

- *Point.* $\lambda_1, \lambda_2 > 0$ and $\rho = 0$.
- *Ellipse.* $\lambda_1, \lambda_2 > 0$ and $\rho > 0$.
- *Hyperbola.* $\lambda_1 > 0$ and $\lambda_2 < 0$ or vice versa, and $\rho > 0$.
- *Two intersecting lines.* $\lambda_1 > 0$ and $\lambda_2 < 0$ or vice versa, and $\rho = 0$.
- *Two parallel lines.* $\lambda_1 > 0$ and $\lambda_2 = 0$ or vice versa, and $\rho > 0$.
- *Line.* $\lambda_1 > 0$ and $\lambda_2 = 0$ or vice versa, and $\rho = 0$.

The axes or directions of the various entities are given by the eigenvectors u_1 and u_2, and the scales of the axes by the corresponding eigenvalues.

Note again that the above analysis does not take account of parabolas, which have to be treated separately. In fact, the case of a single line can be interpreted as a degenerate parabola.

4.4.5.1 Example

For example, in the case of two intersecting lines with $\lambda_1 > 0$, $\lambda_2 < 0$, and $\rho = 0$, the set of vectors x that satisfy the following equation lie on the conic:

$$\mathsf{x}^\mathsf{T}\,(\mathsf{u}_1,\,\mathsf{u}_2)\begin{pmatrix}|\lambda_1| & 0\\ 0 & -|\lambda_2|\end{pmatrix}(\mathsf{u}_1,\,\mathsf{u}_2)^\mathsf{T}\,\mathsf{x}=0$$
$$\iff |\lambda_1|\,\left(\mathsf{x}^\mathsf{T}\,\mathsf{u}_1\,\mathsf{u}_1^\mathsf{T}\,\mathsf{x}\right)-|\lambda_2|\,\left(\mathsf{x}^\mathsf{T}\,\mathsf{u}_2\,\mathsf{u}_2^\mathsf{T}\,\mathsf{x}\right)=0. \tag{4.105}$$

Since u_1 and u_2 are normalized and orthogonal, the solutions for x to the above equation are simply

$$\mathsf{x}=\pm\frac{1}{\sqrt{|\lambda_1|}}\,\mathsf{u}_1\pm\frac{1}{\sqrt{|\lambda_2|}}\,\mathsf{u}_2. \tag{4.106}$$

These solutions give the directions of the two lines. Their intersection point is given by t, which can be evaluated from the corresponding L_H.

4.4.6 Intersecting Lines and Conics

Intersecting a line with an arbitrary conic is quite simple. Suppose a line is given in parametric form as

$$\mathsf{x}(\alpha):=\mathsf{p}+\alpha\,\mathsf{r}, \tag{4.107}$$

where $\mathsf{p}\in\mathbb{R}^3$ gives the offset of the line from the origin and $\mathsf{r}\in\mathbb{R}^3$ is the direction of the line in homogeneous coordinates, i.e. the third component of p is unity and the third component of r is zero. Let $\mathsf{A}\in\mathbb{R}^{3\times 3}$ represent a projective conic as defined in the previous sections. Now, $\mathsf{x}(\alpha)$ lies on the conic if $\mathsf{x}^\mathsf{T}(\alpha)\,\mathsf{A}\,\mathsf{x}(\alpha)=0$. Expanding this equation gives

$$\begin{aligned}\mathsf{x}^\mathsf{T}(\alpha)\,\mathsf{A}\,\mathsf{x}(\alpha)&=(\mathsf{p}^\mathsf{T}+\alpha\,\mathsf{r}^\mathsf{T})\,\mathsf{A}\,(\mathsf{p}+\alpha\,\mathsf{r})\\&=\mathsf{r}^\mathsf{T}\,\mathsf{A}\,\mathsf{r}\,\alpha^2+\left(\mathsf{p}^\mathsf{T}\,\mathsf{A}\,\mathsf{r}+\mathsf{r}^\mathsf{T}\,\mathsf{A}\,\mathsf{p}\right)\,\alpha+\mathsf{p}^\mathsf{T}\,\mathsf{A}\,\mathsf{p}\\&=0.\end{aligned} \tag{4.108}$$

If we define

$$a:=\mathsf{r}^\mathsf{T}\,\mathsf{A}\,\mathsf{r},\quad b:=\mathsf{p}^\mathsf{T}\,\mathsf{A}\,\mathsf{r}+\mathsf{r}^\mathsf{T}\,\mathsf{A}\,\mathsf{p},\quad c:=\mathsf{p}^\mathsf{T}\,\mathsf{A}\,\mathsf{p}, \tag{4.109}$$

(4.108) becomes

$$a\,\alpha^2+b\,\alpha+c=0, \tag{4.110}$$

which has the well-known solutions

$$\alpha^{(1,2)} = \frac{-b \pm \sqrt{b^2 - 4ac}}{2a}. \tag{4.111}$$

If the term in the square root is negative, then the line does not intersect with the conic; if it is zero, the line intersects the conic in a single point; and if it is positive, the line intersects the conic in two points.

4.4.7 Intersection of Conics

In this subsection the following problem is considered: given two 1-blades $\boldsymbol{C}_1, \boldsymbol{C}_2 \in \mathbb{G}_6^1$ that are inner product representations of conics, what are the intersection points of the conics, of which there are at most four? We denote the space of intersection points by $\mathbb{W}$, which may be defined as

$$\mathbb{W} := \{\, \boldsymbol{X} = \mathcal{D}(\boldsymbol{x}) \;:\; \boldsymbol{x} \in \mathbb{R}^2,\; \boldsymbol{C}_1 \cdot \boldsymbol{X} = 0,\; \boldsymbol{C}_2 \cdot \boldsymbol{X} = 0 \,\}. \tag{4.112}$$

Note that $\mathbb{W}$ is in fact a vector space, since any linear combination of elements of $\mathbb{W}$ again lies in $\mathbb{W}$. That is, if $\boldsymbol{X}, \boldsymbol{Y} \in \mathbb{W}$, then $\boldsymbol{C}_1 \cdot (\alpha \boldsymbol{X} + \beta \boldsymbol{Y}) = 0$, $\forall\, \alpha, \beta \in \mathbb{R}$. It is also useful to define the set of Euclidean points in $\mathbb{R}^2$ that, when embedded with $\mathcal{D}$, lie in $\mathbb{W}$. This set is denoted by $\mathbb{W}_E$ and is defined as

$$\mathbb{W}_E := \{\, \boldsymbol{x} \in \mathbb{R}^2 \;:\; \boldsymbol{C}_1 \cdot \mathcal{D}(\boldsymbol{x}) = 0,\; \boldsymbol{C}_2 \cdot \mathcal{D}(\boldsymbol{x}) = 0 \,\}. \tag{4.113}$$

The expressions $\boldsymbol{C}_1 \cdot \boldsymbol{X}$ and $\boldsymbol{C}_2 \cdot \boldsymbol{X}$ are both polynomials of order two in the components of $\boldsymbol{x}$. By combining both expressions, it is possible to obtain a polynomial of order four whose roots are the intersection points of the two conics. The roots of a polynomial of order four can be found by evaluating the roots of a polynomial of order three and one of order two (see e.g. [26]). However, note that here there are two coupled polynomials of order four, and so the polynomial components will be rather complex. See [20] for a discussion of such an evaluation method.

The idea used in this book is to develop a method of evaluating the intersection of two conics that uses standard matrix algorithms, which can be applied directly to the matrices representing the conics.

4.4.7.1 Evaluating Degenerate Conics

A *degenerate conic* is a parallel line pair, an intersecting line pair, or a single line, for example as shown in Fig. 4.21. In this subsection, degenerate conics are analyzed, because they are the key to the evaluation of the intersection of two conics.

Lemma 4.3. *Let $\boldsymbol{C}_1, \boldsymbol{C}_2 \in \mathbb{G}^1_6$ be linearly independent inner-product representations of two conics that intersect in four points. That is, their Euclidean intersection set $\mathbb{W}_E$, as defined above, contains four elements and the intersection space $\mathbb{W}$ has dimension four. A conic $\boldsymbol{C} \in \mathbb{G}^1_6$ passes through these four intersection points if and only if it is a linear combination of $\boldsymbol{C}_1$ and $\boldsymbol{C}_2$. This is not necessarily the case if $|\mathbb{W}_E| < 4$.*

Proof. First of all, if $\boldsymbol{C}$ is a linear combination of $\boldsymbol{C}_1$ and $\boldsymbol{C}_2$, i.e. $\boldsymbol{C} = \alpha\,\boldsymbol{C}_1 + \beta\,\boldsymbol{C}_2$, $\alpha, \beta \in \mathbb{R}$, then for any $\boldsymbol{X} \in \mathbb{W}$ it follows that

$$\boldsymbol{C} \cdot \boldsymbol{X} = (\alpha\,\boldsymbol{C}_1 + \beta\,\boldsymbol{C}_2) \cdot \boldsymbol{X} = \alpha\,(\boldsymbol{C}_1 \cdot \boldsymbol{X}) + \beta\,(\boldsymbol{C}_2 \cdot \boldsymbol{X}) = 0. \tag{4.114}$$

Now for the other direction: that is, if $\boldsymbol{C} \cdot \boldsymbol{X} = 0$, it has to be shown that $\boldsymbol{C}$ is a linear combination of $\boldsymbol{C}_1$ and $\boldsymbol{C}_2$. This is done via a dimensional argument.

First of all, $\mathbb{G}^1_6$ is a 6-dimensional vector space. Since $\boldsymbol{C}_1$ and $\boldsymbol{C}_2$ are linearly independent, they span a 2D subspace of $\mathbb{G}^1_6$. If $\boldsymbol{C}_1$ and $\boldsymbol{C}_2$ intersect in four points, then $\dim(\mathbb{W}) = 4$, $\mathbb{W} \subset \mathbb{G}^1_6$, and $\text{span}\{\boldsymbol{C}_1,\, \boldsymbol{C}_2\} \perp \mathbb{W}$. Since $\boldsymbol{C} \perp \mathbb{W}$, $\boldsymbol{C}$ has to lie in $\text{span}\{\boldsymbol{C}_1,\, \boldsymbol{C}_2\}$. Hence, $\boldsymbol{C}$ has to be a linear combination of $\boldsymbol{C}_1$ and $\boldsymbol{C}_2$.

If $\dim(\mathbb{W}) < 4$, this argument does not hold anymore. That is, there do exist conics that pass through the same intersection points as $\boldsymbol{C}_1$ and $\boldsymbol{C}_2$, but cannot be written as a linear combination of $\boldsymbol{C}_1$ and $\boldsymbol{C}_2$.

□

This lemma gives the motivation for the following idea. If two conics $\boldsymbol{C}_1, \boldsymbol{C}_2 \in \mathbb{D}^2$ intersect in four points, then all conics that pass through these four points can be represented as linear combinations of $\boldsymbol{C}_1$ and $\boldsymbol{C}_2$. This also has to include degenerate conics, in particular those representing line pairs (2D cones). Given four points, there are three unique line pairs that contain these four points. If it is possible to evaluate the particular linear combinations of $\boldsymbol{C}_1$ and $\boldsymbol{C}_2$ that generate these degenerate conics, the evaluation of the intersection of two conics can be reduced to the intersection of line pairs. As will be seen later, if the two conics intersect in only two or three points, at least one degenerate conic can still be found, and the intersection points can be found by intersecting a conic with a degenerate one.

Lemma 4.4. *Let $\boldsymbol{C}_1, \boldsymbol{C}_2 \in \mathbb{G}^1_6$ denote two conics, and let $\mathsf{A} = \mathcal{M}^{-1}(\boldsymbol{C}_1)$ and $\mathsf{B} = \mathcal{M}^{-1}(\boldsymbol{C}_2)$ be their matrix representations. The 3×3 matrices A and B are of full rank if the conics are non-degenerate. Let B be of full rank; then $\mathsf{M} := \mathsf{B}^{-1}\,\mathsf{A}$ exists. If λ is a real eigenvalue of M, then $\mathsf{A} - \lambda\,\mathsf{B}$ represents a degenerate conic.*

Proof. Let $\mathsf{C} = \alpha\,\mathsf{A} + \beta\,\mathsf{B}$, $\alpha, \beta \in \mathbb{R}$. C represents a degenerate conic if and only if $\det(\mathsf{C}) = 0$. The goal is to find those α and β for which this is the case:

$$\begin{aligned} & \det(\alpha\,\mathsf{A} + \beta\,\mathsf{B}) = 0 \\ \iff & \alpha^3 \det\left(\mathsf{B}\,\mathsf{B}^{-1}\,(\mathsf{A} + \frac{\beta}{\alpha}\,\mathsf{B})\right) = 0 \\ \iff & \det(\mathsf{B})\,\det(\mathsf{B}^{-1}\mathsf{A} + \frac{\beta}{\alpha}\,\mathsf{I}) = 0 \\ \iff & \det(\mathsf{M} - \lambda\,\mathsf{I}) = 0, \end{aligned} \tag{4.115}$$

where $\mathsf{M} := \mathsf{B}^{-1}\mathsf{A}$ and $\lambda = -\beta/\alpha$. The values of λ that satisfy the last equation are just the eigenvalues of M. If λ is a real eigenvalue of M, then $\mathsf{C} = \mathsf{A} - \lambda\,\mathsf{B}$ represents a real, degenerate conic.

□

Clearly, the degenerate conics found in this way pass through the intersection points of $\boldsymbol{C}_1$ and $\boldsymbol{C}_2$, independently of how many intersection points these conics have.

Corollary 4.1. *Let $\boldsymbol{C}_1, \boldsymbol{C}_2 \in \mathbb{G}^1_6$ be the inner-product representations of two non-degenerate conics. There then exists at least one linear combination of $\boldsymbol{C}_1$ and $\boldsymbol{C}_2$ which represents a degenerate conic.*

Proof. Let $\mathsf{A} = \mathcal{M}^{-1}(\boldsymbol{C}_1)$ and $\mathsf{B} = \mathcal{M}^{-1}(\boldsymbol{C}_2)$ be the matrix representations of $\boldsymbol{C}_1$ and $\boldsymbol{C}_2$. Since the conics are non-degenerate, A and B are of full rank. Hence, $\mathsf{M} = \mathsf{B}^{-1}\,\mathsf{A}$ also has to be of full rank. Therefore, M has three non-zero eigenvalues. Furthermore, since complex eigenvalues of real matrices always appear in conjugate pairs, M must always have at least one real, non-zero eigenvalue. It thus follows from Lemma 4.4 that there always exists a linear combination of $\boldsymbol{C}_1$ and $\boldsymbol{C}_2$ that represents a degenerate conic.

□

It follows from Corollary 4.1 that if two conics intersect in at least two points, the degenerate conic will have to represent a line pair, or at least a line. Intersecting a line with a conic is quite simple, since this comes down to finding the roots of a quadratic equation. Therefore, in order to evaluate the intersection of the two conics, the line components of the lines represented by the degenerate conic have to be extracted, whence the intersection points can be easily evaluated.

4.4.7.2 Summary

The method developed here for evaluating the intersection of two non-degenerate conics represented by two symmetric 3×3 matrices A and B can be summarized as follows:

1. Find a degenerate conic as a linear combination of A and B that represents two lines by evaluating the eigenvalues of $\mathsf{M} = \mathsf{B}^{-1}\mathsf{A}$. If λ is a real eigenvalue of M, then $\mathsf{C} = \mathsf{A} - \lambda\,\mathsf{B}$ is a degenerate conic passing through the intersection points of A and B.

2. Analyze the degenerate conic C. If it represents two lines, extract the line parameters.
3. Intersect the lines found in the previous step with the conic represented by either A or B.

Good features of this method are that the only numerically sensitive calculation is the evaluation of eigenvectors and eigenvalues. However, many numerically stable algorithms already exist for such calculations. Furthermore, the matrices representing the conics are used directly, which makes the method fairly simple to apply.

4.5 Conformal Conic Space

In this section, a proposal for the extension of conic space is presented. Not all aspects of this new space are developed in detail. The discussion here is to be regarded, rather, as an initial idea for future research. The geometric algebra over this extended conic space contains versors for the *translation* of conics, which are not available in the standard conic space. The newly introduced space combines the ideas of conformal and conic space in a rather straight forward manner. Whether this is the only space that combines the representation of conics with translation versors is not clear.

When one is "designing" a new space, the goal is not only to represent a new type of geometric entity, but to construct a system that seamlessly integrates this representation with useful operators acting on the entities. In the case of conformal space, this is achieved in an ideal way:

1. Geometric entities are constructed from the outer products of points that lie on the relevant entities.
2. The algebraic representations of geometric entities are, at the same time, operators directly related to those entities. For example, a plane is also an operator for a reflection in just that plane.
3. Combinations of basic operators generate the Euclidean group.

To construct a space where the first point is satisfied is not too difficult, as was shown in the case of conic space. However, to ensure that the second point is satisfied, or at least that all operators necessary to generate the Euclidean group exist, is not trivial and not always possible. In the following, a space is constructed where the first point is satisfied, and the translation operation has an operator representation.

4.5.1 The Vector Space

Consider the vector space $\mathbb{R}^{5,3}$ with a canonical basis $\boldsymbol{e}_i := \overline{\mathbb{R}}^{5,3}[i]$, i.e.

$$e_i \cdot e_i = \begin{cases} +1, \; 1 \le i \le 5 \\ -1, \; 6 \le i \le 8 \, . \end{cases}$$

The following pairs of null basis vectors are defined:

$$\begin{aligned} \boldsymbol{n}_1 &:= \boldsymbol{e}_3 + \boldsymbol{e}_6 \, , \; \bar{\boldsymbol{n}}_1 := \frac{1}{2}\,(\boldsymbol{e}_6 - \boldsymbol{e}_3), \\ \boldsymbol{n}_2 &:= \boldsymbol{e}_4 + \boldsymbol{e}_7 \, , \; \bar{\boldsymbol{n}}_2 := \frac{1}{2}\,(\boldsymbol{e}_7 - \boldsymbol{e}_4), \\ \boldsymbol{n}_3 &:= \boldsymbol{e}_5 + \boldsymbol{e}_8 \, , \; \bar{\boldsymbol{n}}_3 := \frac{1}{2}\,(\boldsymbol{e}_8 - \boldsymbol{e}_5). \end{aligned} \tag{4.116}$$

These are basically three independent null bases of the same type as the $\boldsymbol{e}_\infty$, $\boldsymbol{e}_o$ of conformal space. Their properties are

$$\boldsymbol{n}_i \cdot \boldsymbol{n}_j = 0, \quad \bar{\boldsymbol{n}}_i \cdot \bar{\boldsymbol{n}}_j = 0, \quad \boldsymbol{n}_i \cdot \bar{\boldsymbol{n}}_j = -\delta_{ij}. \tag{4.117}$$

In fact, three linearly independent one-dimensional conformal spaces are constructed in this way. A point $\boldsymbol{x} \in \mathbb{R}^2$ defined by $\boldsymbol{x} := x\,\boldsymbol{e}_1 + y\,\boldsymbol{e}_2$ is then embedded using the operator $\mathcal{W} : \mathbb{R}^2 \to \mathbb{W} \subset \mathbb{R}^{5,3}$, which is defined by

$$\begin{aligned} \mathcal{W} \; : \; \boldsymbol{x} \; \mapsto \; & x\,\boldsymbol{e}_1 + y\,\boldsymbol{e}_2 + \frac{1}{2}\,x^2\,\boldsymbol{n}_1 + \bar{\boldsymbol{n}}_1 \\ & + \frac{1}{2}\,y^2\,\boldsymbol{n}_2 + \bar{\boldsymbol{n}}_2 \\ & + x\,y\,\boldsymbol{n}_3. \end{aligned} \tag{4.118}$$

This embedding generates null vectors, since $\mathcal{W}(\boldsymbol{x}) \cdot \mathcal{W}(\boldsymbol{x}) = 0$ just as in conformal space. Given a general vector

$$\boldsymbol{A} := a_1\,\boldsymbol{e}_1 + a_2\,\boldsymbol{e}_2 + a_3\,\boldsymbol{n}_1 + a_4\,\bar{\boldsymbol{n}}_1 + a_5\,\boldsymbol{n}_2 + a_6\,\bar{\boldsymbol{n}}_2 + a_7\,\boldsymbol{n}_3 + a_8\,\bar{\boldsymbol{n}}_3,$$

the inner product of an embedded point $\boldsymbol{x} \in \mathbb{R}^2$ and $\boldsymbol{A}$ gives

$$\mathcal{W}(\boldsymbol{x}) \cdot \boldsymbol{A} = -\frac{1}{2}\,a_4\,x^2 - \frac{1}{2}\,a_6\,y^2 - a_8\,x\,y + a_1\,x + a_2\,y - a_3 - a_5. \tag{4.119}$$

The geometric IPNS of $\boldsymbol{x}$ is therefore the same as the roots of this polynomial. This implies that the GIPNSs of vectors in $\mathbb{R}^{5,3}$ are conic sections.

4.5.2 The Geometric Algebra $\mathbb{G}_{5,3}$

The geometric algebra over $\mathbb{R}^{5,3}$ again allows the construction of conic sections by use of the outer product of points that lie on the conic section. It turns out that the conic section that passes through points $\boldsymbol{a}_1, \ldots, \boldsymbol{a}_5 \in \mathbb{R}^2$ is the GOPNS of

$$\boldsymbol{C} := \left(\bigwedge_{i=1}^{5} \mathcal{W}(\boldsymbol{a}_i) \right) \wedge (\bar{\boldsymbol{n}}_1 - \bar{\boldsymbol{n}}_2) \wedge \bar{\boldsymbol{n}}_3.$$

Furthermore, the line passing through $\boldsymbol{a}_1, \boldsymbol{a}_2 \in \mathbb{R}^2$ is the GOPNS of

$$\boldsymbol{L} := \mathcal{W}(\boldsymbol{a}_1) \wedge \mathcal{W}(\boldsymbol{a}_2) \wedge \boldsymbol{n}_1 \wedge \boldsymbol{n}_2 \wedge \boldsymbol{n}_3 \wedge (\bar{\boldsymbol{n}}_1 - \bar{\boldsymbol{n}}_2) \wedge \bar{\boldsymbol{n}}_3.$$

Representing conic sections in such a way was already possible in the conic space treated earlier. The reason for this rather complex embedding is to obtain a translation operator. The translation operator for a translation by a vector $\boldsymbol{t} := t_x\,\boldsymbol{e}_1 + t_y\,\boldsymbol{e}_2$ is split into four parts. Two of these simply represent 1D translations parallel to the basis vectors $\boldsymbol{e}_1$ and $\boldsymbol{e}_2$ using the definition of translators for conformal space. The other two parts ensure that the $x\,y$ component of an embedded point is transformed correctly under a translation. The individual parts of the translation operator are

$$\begin{aligned}
\boldsymbol{T}_{X_1} &:= 1 - \frac{1}{2}\,t_x\,\boldsymbol{e}_1\,\boldsymbol{n}_1, \\
\boldsymbol{T}_{X_2} &:= 1 - \frac{1}{2}\,t_x\,\boldsymbol{e}_2\,\boldsymbol{n}_3, \\
\boldsymbol{T}_{Y_1} &:= 1 - \frac{1}{2}\,t_y\,\boldsymbol{e}_2\,\boldsymbol{n}_2, \\
\boldsymbol{T}_{Y_2} &:= 1 - \frac{1}{2}\,t_y\,\boldsymbol{e}_1\,\boldsymbol{n}_3.
\end{aligned} \tag{4.120}$$

The complete translation operator is then given by

$$\boldsymbol{T} := \boldsymbol{T}_{Y_2}\,\boldsymbol{T}_{Y_1}\,\boldsymbol{T}_{X_2}\,\boldsymbol{T}_{X_1}. \tag{4.121}$$

It is not too difficult to verify that this translation operator does indeed translate points and conic sections in general. However, knowing the reflection operator for a reflection in an arbitrary line would be even more helpful, since then all translators and rotors could be generated right away. So far, the reflection operator of this space has not been found.

Note that the algebra $\mathbb{G}_{5,3}$ has an algebraic dimension $2^8 = 256$. In order to work with an algebra of this size efficiently on a computer, a highly optimized implementation is necessary. It is possible, though, that the entities and operators that are of actual interest in a particular application may lie in a much smaller subalgebra of $\mathbb{G}_{5,3}$. This would allow one to combine a general analytic treatment in a high-dimensional algebra with fast numerical calculations.

Chapter 5
Numerics

So far, it has been shown how geometric algebra can be used to describe geometry elegantly and succinctly. The goal of this chapter is to show how numerical evaluation methods can be applied to geometric-algebra expressions. It will be seen that such expressions have a straightforward representation as tensor contractions, to which standard linear-algebra evaluation methods may be applied. This chapter therefore starts with a discussion of the mapping from geometric-algebra expressions to tensor operations in Sect. 5.1. Next it is shown, in Sect. 5.2, how linear multivector equations can be solved using the tensor representation. The representation of uncertain multivectors is then introduced in Sect. 5.3, and its limits are investigated in Sect. 5.4, where an expression for the bias term in the error propagation of general bilinear functions is derived. Uncertain multivectors are particularly useful in the representation of uncertain geometric entities and uncertain transformations. Geometric algebra is ideally suited for the latter in particular. Sections 5.5, 5.6, and 5.7 give details of the representation of uncertain geometric entities in projective, conformal, and conic space, respectively.

Two linear least-squares estimation methods, the Gauss–Markov and the Gauss–Helmert method, are discussed in some detail in Sects. 5.8 and 5.9. The Gauss–Helmert method is particularly interesting, since it incorporates the uncertainty of the data into the optimization process. While the Gauss–Markov and Gauss–Helmert estimation methods are well known, it is a long way from the definition of a stochastic model to the corresponding estimation algorithm, which is why these estimation methods, which are used later on, are described here in some detail. In this way, the notation necessary for later chapters is introduced. The chapter concludes with a discussion, in Sect. 5.10, of how the Gauss–Helmert and Gauss–Markov methods may be applied to practical problems.

The foundations for this approach to solving general multivector equations were laid by the author and Sommer in [146], and were extended to uncertain algebraic entities in collaboration with W. Förstner [136].

C. Perwass, *Geometric Algebra with Applications in Engineering.*
Geometry and Computing.

5.1 Tensor Representation

This section introduces the idea that geometric-algebra operations are essentially bilinear functions and makes this notion explicit. In later chapters, this representation forms the foundation for the description of random multivector variables and statistical optimizations. Using this form, differentiation and integration in geometric algebra can be related right away to these operations in linear algebra.

Recall that the geometric algebra $\mathbb{G}_{p,q}$ has dimension 2^n. Let $\{\boldsymbol{E}_i\} := \overline{\mathbb{G}}_{p,q}$ denote the canonical algebraic basis, whereby $\boldsymbol{E}_1 \equiv 1$. For example, the algebraic basis of $\mathbb{G}_3$ is given in Table 5.1.

Table 5.1 Algebra basis of $\mathbb{G}_3$, where the geometric product of basis vectors is denoted by combining their indices, i.e. $\boldsymbol{e}_1\,\boldsymbol{e}_2 \equiv \boldsymbol{e}_{12}$

Type	No.	Basis Elements
Scalar	1	1
Vector	3	$\boldsymbol{e}_1, \boldsymbol{e}_2, \boldsymbol{e}_3$
2-Vector	3	$\boldsymbol{e}_{23}, \boldsymbol{e}_{31}, \boldsymbol{e}_{12}$
3-Vector	1	$\boldsymbol{e}_{123}$

A multivector $\boldsymbol{A} \in \mathbb{G}_{p,q}$ can then be written as $\boldsymbol{A} = a^i\,\boldsymbol{E}_i$, where a^i denotes the ith component of a vector $\mathsf{a} \in \mathbb{R}^{2^n}$ and a sum over the repeated index i is implied. Since the $\{\boldsymbol{E}_i\}$ form an algebraic basis, it follows that the geometric product of two basis blades has to result in a basis blade, i.e.

$$\boldsymbol{E}_i\,\boldsymbol{E}_j = {\Gamma^k}_{ij}\,\boldsymbol{E}_k, \qquad \forall\, i,j \in \{1,\ldots,2^2\}, \tag{5.1}$$

where ${\Gamma^k}_{ij} \in \mathbb{R}^{2^n \times 2^n \times 2^n}$ is a tensor encoding the geometric product. Here, the following notation is being used. The expression ${\Gamma^k}_{ij}$ denotes either the element of the tensor at the location (k,i,j) or the ordered set of all elements of the tensor. Which form is meant should be clear from the context in which the symbol is used. All indices in an expression that appear only once and are also not otherwise defined to take on a particular value indicate the ordered set over all values of this index. If an element has one undefined index, it denotes a column vector, and if it has two undefined indices it denotes a matrix, where the first index gives the row and the second index the column.

For example, the expression $b^j\,{\Gamma^k}_{ij}$ has two undefined indices k and i and thus represents a matrix with indices k and i. Since k is the first index, it denotes the row and i denotes the column of the resultant matrix. Similarly, $a^i\,b^j\,{\Gamma^k}_{ij}$ denotes a column vector with the index k.

Table 5.2 Tensor symbols for algebraic operations, and the corresponding Jacobi matrices. For tensors with two indices (i.e. matrices), the first index denotes the matrix row and the second index the matrix column

Operation	Tensor symbol	Jacobi matrices
Geometric Product	${\Gamma^k}_{ij}$	$\Gamma_R(\mathsf{a}) := a^i\,{\Gamma^k}_{ij}$ $\Gamma_L(\mathsf{b}) := b^j\,{\Gamma^k}_{ij}$
Outer Product	${\Lambda^k}_{ij}$	$\Lambda_R(\mathsf{a}) := a^i\,{\Lambda^k}_{ij}$ $\Lambda_L(\mathsf{b}) := b^j\,{\Lambda^k}_{ij}$
Inner Product	${\Theta^k}_{ij}$	$\Theta_R(\mathsf{a}) := a^i\,{\Theta^k}_{ij}$ $\Theta_L(\mathsf{b}) := b^j\,{\Theta^k}_{ij}$
Reverse	${R^j}_i$	$\mathsf{R} := {R^j}_i$
Dual	${D^j}_i$	$\mathsf{D} := {D^j}_i$

5.1.1 Component Vectors

If $\boldsymbol{A}, \boldsymbol{B}, \boldsymbol{C} \in \mathbb{G}_{p,q}$ are defined as $\boldsymbol{A} := a^i\,\boldsymbol{E}_i$, $\boldsymbol{B} := b^i\,\boldsymbol{E}_i$, and $\boldsymbol{C} := c^i\,\boldsymbol{E}_i$, then it follows from (5.1) that the components of $\boldsymbol{C}$ in the algebraic equation $\boldsymbol{C} = \boldsymbol{A}\,\boldsymbol{B}$ can be evaluated via

$$c^k = a^i\,b^j\,{\Gamma^k}_{ij}, \tag{5.2}$$

where a summation over i and j is again implied. Such a summation of tensor indices is called *contraction*. Equation (5.2) shows that the geometric product is simply a bilinear function. In fact, all products in geometric algebra that are of interest in this text can be expressed in this form, as will be discussed later on.

The geometric product can also be written purely in matrix notation, by defining the matrices

$$\Gamma_R(\mathsf{a}) := a^i\,{\Gamma^k}_{ij} \qquad \text{and} \qquad \Gamma_L(\mathsf{b}) := b^j\,{\Gamma^k}_{ij}. \tag{5.3}$$

The geometric product $\boldsymbol{A}\,\boldsymbol{B}$ can now be written as

$$a^i\,b^j\,{\Gamma^k}_{ij} = \Gamma_R(\mathsf{a})\,\mathsf{b} = \Gamma_L(\mathsf{b})\,\mathsf{a}. \tag{5.4}$$

Note that the matrices $\Gamma_R(\mathsf{a})$ and $\Gamma_L(\mathsf{b})$ are the two Jacobi matrices of the expression $c^k := a^i\,b^j\,{\Gamma^k}_{ij}$. That is,

$$\frac{\partial\,c^k}{\partial\,a^i} = b^j\,{\Gamma^k}_{ij} = \Gamma_L(\mathsf{b}) \qquad \text{and} \qquad \frac{\partial\,c^k}{\partial\,b^j} = a^i\,{\Gamma^k}_{ij} = \Gamma_R(\mathsf{a})\,. \tag{5.5}$$

At this point it is useful to introduce a notation that describes the mapping between multivectors and their corresponding component vectors. For this purpose, the operator $\mathcal{K}$ is introduced. For the algebra $\mathbb{G}_{p,q}$, $\mathcal{K}$ is the bijective mapping

$$\mathcal{K} : \mathbb{G}_{p,q} \longrightarrow \mathbb{R}^{2^{p+q}} \quad \text{and} \quad \mathcal{K}^{-1} : \mathbb{R}^{2^{p+q}} \longrightarrow \mathbb{G}_{p,q}. \tag{5.6}$$

For a multivector $\boldsymbol{A} \in \mathbb{G}_{p,q}$ with $\boldsymbol{A} := a^i \boldsymbol{E}_i$, the operator is defined as

$$\mathcal{K} : \boldsymbol{A} \mapsto \mathsf{a} \quad \text{and} \quad \mathcal{K}^{-1} : \mathsf{a} \mapsto \boldsymbol{A}. \tag{5.7}$$

It follows that

$$\mathcal{K}(\boldsymbol{A}\,\boldsymbol{B}) = \Gamma_R\,(\mathcal{K}(\boldsymbol{A}))\;\mathcal{K}(\boldsymbol{B}) = \Gamma_L\,(\mathcal{K}(\boldsymbol{B}))\;\mathcal{K}(\boldsymbol{A}) \tag{5.8}$$

and

$$\mathcal{K}^{-1}\big(\Gamma_R\,(\mathsf{a})\;\mathsf{b}\big) = \mathcal{K}^{-1}\big(\Gamma_L\,(\mathsf{b})\;\mathsf{a}\big) = \mathcal{K}^{-1}(\mathsf{a})\,\mathcal{K}^{-1}(\mathsf{b}). \tag{5.9}$$

The mapping $\mathcal{K}$ is therefore an isomorphism between $\mathbb{G}_{p,q}$ and $\mathbb{R}^{2^{(p+q)}}$.

In addition to the geometric product, the inner and outer products and the reverse and dual are important operations in geometric algebra. The corresponding operation tensors are therefore given specific symbols, which are listed in Table 5.2.

5.1.2 Example: Geometric Product in $\mathbb{G}_2$

A simple example of a product tensor is the geometric-product tensor of $\mathbb{G}_2$. The algebraic basis may be defined as

$$\boldsymbol{E}_1 := 1, \quad \boldsymbol{E}_2 := \boldsymbol{e}_1, \quad \boldsymbol{E}_3 := \boldsymbol{e}_2, \quad \boldsymbol{E}_4 := \boldsymbol{e}_{12}. \tag{5.10}$$

The geometric-product tensor $\Gamma^k{}_{ij} \in \mathbb{R}^{4\times4\times4}$ of $\mathbb{G}_2$ then takes on the form

$$\begin{aligned}
\Gamma^1{}_{ij} &= \begin{bmatrix} 1 & 0 & 0 & 0 \\ 0 & 1 & 0 & 0 \\ 0 & 0 & 1 & 0 \\ 0 & 0 & 0 & -1 \end{bmatrix}, \quad
\Gamma^2{}_{ij} = \begin{bmatrix} 0 & 1 & 0 & 0 \\ 1 & 0 & 0 & 0 \\ 0 & 0 & 0 & -1 \\ 0 & 0 & 1 & 0 \end{bmatrix}, \\
\Gamma^3{}_{ij} &= \begin{bmatrix} 0 & 0 & 1 & 0 \\ 0 & 0 & 0 & 1 \\ 1 & 0 & 0 & 0 \\ 0 & -1 & 0 & 0 \end{bmatrix}, \quad
\Gamma^4{}_{ij} = \begin{bmatrix} 0 & 0 & 0 & 1 \\ 0 & 0 & 1 & 0 \\ 0 & -1 & 0 & 0 \\ 1 & 0 & 0 & 0 \end{bmatrix},
\end{aligned} \tag{5.11}$$

where i is the row index and j is the column index. Let $\boldsymbol{A} = \boldsymbol{e}_1$ and $\boldsymbol{B} = \boldsymbol{e}_2$; then $\mathsf{a} = [0,\ 1,\ 0,\ 0]^\mathsf{T}$ and $\mathsf{b} = [0,\ 0,\ 1,\ 0]^\mathsf{T}$. Thus

$$\left.\begin{array}{l} b^j\, \Gamma^1{}_{ij} = [0,\ 0,\ 1,\ 0]^\mathsf{T} \\ b^j\, \Gamma^2{}_{ij} = [0,\ 0,\ 0,\ 1]^\mathsf{T} \\ b^j\, \Gamma^3{}_{ij} = [1,\ 0,\ 0,\ 0]^\mathsf{T} \\ b^j\, \Gamma^4{}_{ij} = [0,\ 1,\ 0,\ 0]^\mathsf{T} \end{array}\right\} \implies b^j\, \Gamma^k{}_{ij} = \begin{bmatrix} 0\,0\,1\,0 \\ 0\,0\,0\,1 \\ 1\,0\,0\,0 \\ 0\,1\,0\,0 \end{bmatrix}. \tag{5.12}$$

In the resultant matrix, k is the row index and i is the column index. It follows from (5.12) that

$$a^i\, b^j\, \Gamma^k{}_{ij} = [0,\ 0,\ 0,\ 1]^\mathsf{T} \quad \cong \quad \boldsymbol{e}_{12} = \boldsymbol{E}_4, \tag{5.13}$$

where $\cong$ denotes isomorphism.

5.1.3 Subspace Projection

Depending on the particular algebra, the corresponding product tensor can become very large. For example, in the algebra $\mathbb{G}_{4,1}$ of conformal space, the geometric-product tensor $\Gamma^k{}_{ij} \in \mathbb{R}^{32\times 32\times 32}$ has 32 768 components, most of which are zero. When one is performing actual computations with such a tensor on a computer, it can reduce the computational load considerably if the tensor is reduced to only those components that are actually needed. Furthermore, such a projection process can also be used to implement constraints, as will be seen later.

Let $\mathsf{m} \in \{1, \ldots, 2^{(p+q)}\}^r$ denote a vector of r indices that index those basis elements $\{\boldsymbol{E}_i\} \subset \mathbb{G}_{p,q}$ which are needed in a calculation. Also, let us define $\mathsf{e}_u := \mathcal{K}(\boldsymbol{E}_u)$ such that $e^i{}_u = \delta^i{}_u$, where $\delta^i{}_u$ denotes the Kronecker delta, defined as

$$\delta^i{}_u := \begin{cases} 1 \,:\, i = u\,, \\ 0 \,:\, i \neq u\,. \end{cases}$$

A corresponding projection matrix M is then defined as

$$M^j{}_i := \mathcal{K}\big(\boldsymbol{E}_{m^j}\big)^i = e^i{}_{m^j}, \qquad M^j{}_i \in \mathbb{R}^{r\times 2^{(p+q)}}, \tag{5.14}$$

where m^j denotes the jth element of m. A multivector $\boldsymbol{A} \in \mathbb{G}_{p,q}$ with $\mathsf{a} = \mathcal{K}(\boldsymbol{A})$ may be mapped to an r-dimensional component vector $\mathsf{a}_M \in \mathbb{R}^r$ related to those basis blades which are indexed by m, via

$$a^j_M = a^i\, M^j{}_i. \tag{5.15}$$

The reduced vector a^j_M can be mapped back to a component vector on the full basis through

$$a^j = a^i_M\, \tilde{M}^j{}_i, \tag{5.16}$$

where $\tilde{M}^j{}_i$ denotes the transpose of $M^j{}_i$.

Suppose $\boldsymbol{A}, \boldsymbol{B}, \boldsymbol{C} \in \mathbb{G}_{p,q}$ are related by $\boldsymbol{C} = \boldsymbol{A}\,\boldsymbol{B}$. In terms of component vectors $\mathsf{a} = \mathcal{K}(\boldsymbol{A})$, $\mathsf{b} = \mathcal{K}(\boldsymbol{B})$, and $\mathsf{c} = \mathcal{K}(\boldsymbol{C})$, this can be written as $c^k = a^i\, b^j\, \Gamma^k{}_{ij}$. If it can be assumed that only a subset of the elements of a^i and b^j are non-zero, the evaluation of this operation can be reduced as follows. Let $M^j_a{}_i$ and $M^j_b{}_i$ denote the projection matrices for a and b that map only the non-zero components to the reduced component vectors a_M and b_M, respectively. Clearly, only a subset of the components of the resultant vector c will be non-zero. The appropriate projection matrix and reduced component vector are denoted by $M^j_c{}_i$ and c_M. The geometric-product operation for the reduced component vectors then becomes

$$c^w_M = a^u_M\, b^v_M\, M^w_c{}_k\, \tilde{M}^i_a{}_u\, \tilde{M}^j_b{}_v\, \Gamma^k{}_{ij}. \tag{5.17}$$

This can be written equivalently as

$$c^w_M = a^u_M\, b^v_M\, \Gamma^w_M{}_{uv}, \qquad \Gamma^w_M{}_{uv} := M^w_c{}_k\, \tilde{M}^i_a{}_u\, \tilde{M}^j_b{}_v\, \Gamma^k{}_{ij}. \tag{5.18}$$

That is, $\Gamma^w_M{}_{uv}$ encodes the geometric product for the reduced component vectors. In applications, it is usually known which components of the constituent multivectors in an equation are non-zero. Therefore, reduced product tensors can be precalculated.

5.1.4 Example: Reduced Geometric Product

Consider again the algebra $\mathbb{G}_2$ with basis

$$\boldsymbol{E}_1 := 1, \quad \boldsymbol{E}_2 := \boldsymbol{e}_1, \quad \boldsymbol{E}_3 := \boldsymbol{e}_2, \quad \boldsymbol{E}_4 := \boldsymbol{e}_{12}.$$

The geometric-product tensor $\Gamma^k{}_{ij} \in \mathbb{R}^{4\times4\times4}$ of $\mathbb{G}_2$ is given by (5.11). Suppose $\boldsymbol{A}, \boldsymbol{B} \in \mathbb{G}^1_2$; that is, they are linear combinations of $\boldsymbol{E}_2$ and $\boldsymbol{E}_3$. It is clear that the result of the geometric product of $\boldsymbol{A}$ and $\boldsymbol{B}$ is a linear combination of $\boldsymbol{E}_1$ and $\boldsymbol{E}_4$, i.e. $\boldsymbol{C} = \boldsymbol{A}\,\boldsymbol{B} \in \mathbb{G}^+_2$. Therefore,

$$M^j_a{}_i = M^j_b{}_i = \begin{bmatrix} 0\,1\,0\,0 \\ 0\,0\,1\,0 \end{bmatrix}, \qquad M^j_c{}_i = \begin{bmatrix} 1\,0\,0\,0 \\ 0\,0\,0\,1 \end{bmatrix}. \tag{5.19}$$

The reduced geometric-product tensor is thus given by

$$\Gamma^w_M{}_{uv} := M^w_c{}_k\, \tilde{M}^i_a{}_u\, \tilde{M}^j_b{}_v\, \Gamma^k{}_{ij}, \tag{5.20}$$

where

$$\Gamma^1_M{}_{uv} = \begin{bmatrix} 1\,0 \\ 0\,1 \end{bmatrix}, \qquad \Gamma^2_M{}_{uv} = \begin{bmatrix} 0 & 1 \\ -1 & 0 \end{bmatrix}. \tag{5.21}$$

If $\boldsymbol{A} = \boldsymbol{e}_2 = \boldsymbol{E}_3$ and $\boldsymbol{B} = \boldsymbol{e}_1 = \boldsymbol{E}_2$, the reduced component vectors are $\mathsf{a}_M = [0, 1]^\mathsf{T}$ and $\mathsf{b}_M = [1, 0]^\mathsf{T}$. Thus

$$\left.\begin{array}{l} b^v_M \, \Gamma^1_{M\,uv} = [1, 0]^\mathsf{T} \\ b^v_M \, \Gamma^2_{M\,uv} = [0, -1]^\mathsf{T} \end{array}\right\} \Longrightarrow b^v_M \, \Gamma^w_{M\,uv} = \begin{bmatrix} 1 & 0 \\ 0 & -1 \end{bmatrix}. \tag{5.22}$$

It follows from this that

$$c^w_M = a^u_M \, b^v_M \, \Gamma^w_{M\,uv} = [0, -1]^\mathsf{T}. \tag{5.23}$$

Mapping c^w_M back to the non-reduced form gives

$$c^k = c^w_M \, \tilde{M}^k_{c\ w} = [0, 0, 0, -1]^\mathsf{T} \quad \cong \quad -\boldsymbol{e}_{12} = -\boldsymbol{E}_4. \tag{5.24}$$

5.1.5 Change of Basis

Let $\{\boldsymbol{F}_i\} \subset \mathbb{G}_{p,q}$ denote an algebraic basis of $\mathbb{G}_{p,q}$ which is different from the canonical algebraic basis $\{\boldsymbol{E}_i\}$ of $\mathbb{G}_{p,q}$. Given a multivector $\boldsymbol{A} \in \mathbb{G}_{p,q}$, its component vectors in the $\boldsymbol{E}$-basis and $\boldsymbol{F}$-basis are denoted by $\mathsf{a}_E = \mathcal{K}_E(\boldsymbol{A})$ and $\mathsf{a}_F = \mathcal{K}_F(\boldsymbol{A})$, respectively. The tensor that transforms a_E into a_F is given by

$$T^j{}_i := \mathcal{K}_E(\boldsymbol{F}_j)^i \qquad \Rightarrow \qquad a^j_F = a^i_E \, T^j{}_i. \tag{5.25}$$

The inverse of $T^j{}_i$ is denoted by $\bar{T}^j{}_i$, i.e. $a^j_E = a^i_F \, \bar{T}^j{}_i$. If the geometric-product tensor in the $\boldsymbol{E}$-basis is given by $\Gamma^k_{E\,ij}$, the corresponding product tensor in the $\boldsymbol{F}$-basis can be evaluated via

$$\Gamma^w_{F\,uv} = T^w{}_k \, \bar{T}^i{}_u \, \bar{T}^j{}_v \, \Gamma^k_{E\,ij}. \tag{5.26}$$

Such a change of basis finds an application, for example, if an implementation of $\mathbb{G}_{4,1}$ is given where the canonical basis blades are geometric products of the Minkowski basis $\{\boldsymbol{e}_1, \boldsymbol{e}_2, \boldsymbol{e}_3, \boldsymbol{e}_+, \boldsymbol{e}_-\}$. Here $\boldsymbol{e}_1$, $\boldsymbol{e}_2$, $\boldsymbol{e}_3$, and $\boldsymbol{e}_+$ square to $+1$ and $\boldsymbol{e}_-$ squares to -1. In many problem settings, however, it is essential to express constraints that involve $\boldsymbol{e}_\infty = \boldsymbol{e}_+ + \boldsymbol{e}_-$ and $\boldsymbol{e}_o = \frac{1}{2}(\boldsymbol{e}_- - \boldsymbol{e}_+)$. Therefore, component vectors and the product tensors have to be transformed into the algebraic basis constructed from $\{\boldsymbol{e}_1, \boldsymbol{e}_2, \boldsymbol{e}_3, \boldsymbol{e}_\infty, \boldsymbol{e}_o\}$.

5.2 Solving Linear Geometric Algebra Equations

In this section, it is shown how geometric-algebra equations of the form $\boldsymbol{A} \circ \boldsymbol{X} = \boldsymbol{B}$ can be solved for $\boldsymbol{X}$ numerically, where $\circ$ stands for any bilinear operation. Since the operation $\circ$ need not be invertible, there may exist a solution subspace. Furthermore, if $\circ$ is neither commutative nor anticommutative, there exists another important form of linear equation, namely $\boldsymbol{A} \circ \boldsymbol{X} + \boldsymbol{X} \circ \boldsymbol{B} = \boldsymbol{C}$. This form occurs when one is solving for versors, for example rotation and translation operators.

5.2.1 Inverse of a Multivector

Recall that not every multivector has an inverse. Consider, for example, the pair $\boldsymbol{A}, \boldsymbol{B} \in \mathbb{G}_n$, $n \geq 1$, defined as $\boldsymbol{A} := \frac{1}{2}(1 + \boldsymbol{e}_1)$ and $\boldsymbol{B} := \frac{1}{2}(1 - \boldsymbol{e}_1)$. Clearly,

$$\boldsymbol{A}\,\boldsymbol{A} = \boldsymbol{A}, \quad \boldsymbol{B}\,\boldsymbol{B} = \boldsymbol{B}, \quad \boldsymbol{A}\,\boldsymbol{B} = 0. \tag{5.27}$$

That is, $\boldsymbol{A}$ and $\boldsymbol{B}$ are idempotent. If $\boldsymbol{A}$ had an inverse $\boldsymbol{A}^{-1}$ such that $\boldsymbol{A}^{-1}\,\boldsymbol{A} = 1$, then

$$\boldsymbol{A}\,\boldsymbol{B} = 0 \iff \boldsymbol{A}^{-1}\,\boldsymbol{A}\,\boldsymbol{B} = 0 \iff \boldsymbol{B} = 0, \tag{5.28}$$

which contradicts the initial definition of $\boldsymbol{B}$. Hence, $\boldsymbol{A}$ cannot have an inverse, and the same is true for $\boldsymbol{B}$.

Given an arbitrary multivector, it is often necessary to test whether it has an inverse and, if it does, to evaluate the inverse. Evaluating the inverse of a multivector $\boldsymbol{A}$ is equivalent to solving the linear equation $\boldsymbol{A}\,\boldsymbol{X} = 1$ for $\boldsymbol{X}$. This is straightforward after this equation has been mapped to the tensor form. Let $\mathsf{a} := \mathcal{K}(\boldsymbol{A})$, $\mathsf{x} := \mathcal{K}(\boldsymbol{X})$, and $\mathsf{e}_1 := \mathcal{K}(1)$; then

$$\boldsymbol{A}\,\boldsymbol{X} = 1 \iff \mathsf{a}^i\,\mathsf{x}^j\,\Gamma^k{}_{ij} = \mathsf{e}_1^k \iff \Gamma_R(\mathsf{a})\,\mathsf{x} = \mathsf{e}_1. \tag{5.29}$$

Solving for x is now possible by inverting $\Gamma_R(\mathsf{a})$, if such an inverse exists. If $\Gamma_R(\mathsf{a})$ has no inverse, then $\boldsymbol{A}$ has no inverse. However, the pseudoinverse of $\Gamma_R(\mathsf{a})$ can still be used to find the pseudoinverse of $\boldsymbol{A}$. If $\Gamma_R(\mathsf{a})^{-1}$ exists, then

$$\boldsymbol{A}^{-1} = \mathcal{K}^{-1}\big(\Gamma_R(\mathsf{a})^{-1}\,\mathsf{e}_1\big). \tag{5.30}$$

The general equation $\boldsymbol{A}\,\boldsymbol{X} = \boldsymbol{B}$ can be solved for $\boldsymbol{X}$ in very much the same way. If $\mathsf{a} := \mathcal{K}(\boldsymbol{A})$ and $\mathsf{b} = \mathcal{K}(\boldsymbol{B})$, then

$$\boldsymbol{X} = \mathcal{K}^{-1}\big(\Gamma_R(\mathsf{a})^{-1}\,\mathsf{b}\big). \tag{5.31}$$

5.2.2 Versor Equation

Recall that a versor is an element of the Clifford group and, hence, every versor has an inverse. Every versor can be expressed as the geometric product of a set of (grade 1) vectors. As has been shown earlier, versors can be used to represent all kinds of transformations. In $\mathbb{G}_3$, versors can represent reflections and rotations, and in $\mathbb{G}_{4,1}$, reflections, inversions, rotations, translations, and dilations. In order to perform a transformation, a versor $\boldsymbol{V} \in \mathbb{G}_{p,q}$ is applied to a multivector $\boldsymbol{A} \in \mathbb{G}_{p,q}$ in the form

$$\boldsymbol{V}\,\boldsymbol{A}\,\boldsymbol{V}^{-1} = \boldsymbol{B}, \tag{5.32}$$

where $\boldsymbol{B} \in \mathbb{G}_{p,q}$. A versor is called *unitary* if $\boldsymbol{V}\,\widetilde{\boldsymbol{V}} = 1$, i.e. $\boldsymbol{V}^{-1} = \widetilde{\boldsymbol{V}}$.

A typical problem is to solve for the versor $\boldsymbol{V}$ given $\boldsymbol{A}$ and $\boldsymbol{B}$. This is in fact a linear problem, as can be seen by right-multiplying (5.32) by $\boldsymbol{V}$; that is,

$$\boldsymbol{V}\,\boldsymbol{A}\,\boldsymbol{V}^{-1} = \boldsymbol{B} \iff \boldsymbol{V}\,\boldsymbol{A} - \boldsymbol{B}\,\boldsymbol{V} = 0. \tag{5.33}$$

Now, let $\mathsf{a} = \mathcal{K}(\boldsymbol{A})$, $\mathsf{b} = \mathcal{K}(\boldsymbol{B})$, and $\mathsf{v} = \mathcal{K}(\boldsymbol{V})$; then (5.33) can be written as

$$\mathsf{v}^i\,\mathsf{a}^j\,\Gamma^k{}_{ij} - \mathsf{b}^r\,\mathsf{v}^s\,\Gamma^k{}_{rs} = \Gamma_L(\mathsf{a})\,\mathsf{v} - \Gamma_R(\mathsf{b})\,\mathsf{v} = \mathsf{Q}(\mathsf{a},\,\mathsf{b})\,\mathsf{v} = 0, \tag{5.34}$$

where $\mathsf{Q}(\mathsf{a},\,\mathsf{b}) := \Gamma_L(\mathsf{a}) - \Gamma_R(\mathsf{b})$. Therefore, the null space of $\mathsf{Q}(\mathsf{a},\,\mathsf{b})$ gives the solution for v. This null-space may, for example, be evaluated with the use of a singular-value decomposition (SVD) [167].

The SVD of $\mathsf{Q} \in \mathbb{R}^{m\times n}$ factorizes the matrix into two unitary matrices $\mathsf{U} \in \mathbb{R}^{m\times n}$ and $\mathsf{V} \in \mathbb{R}^{n\times n}$, and a diagonal matrix $\mathsf{W} \in \mathbb{R}^{n\times n}$, such that $\mathsf{Q} = \mathsf{U}\,\mathsf{W}\,\mathsf{V}^{\mathsf{T}}$. Typically, the diagonal elements in W are ordered from the largest to the smallest entry and the columns of U and V are ordered accordingly. The number of zero entries on the diagonal of W gives the dimension of the null-space of Q and the corresponding columns of V give an orthonormal basis of the right null-space of Q. If there is exactly one zero entry in W, then the corresponding column of V is the solution for v. If the dimension of the null-space is larger than one, then there is no unique solution for v.

If, on the other hand, the smallest diagonal entry of W is non-zero, the corresponding column vector of V is the *best* solution for v in a least-squares sense. In this case, the question of *which* measure has been minimized still has to be answered. For this purpose, (5.32) is written as

$$\boldsymbol{V}\,\boldsymbol{A}\,\boldsymbol{V}^{-1} - \boldsymbol{B} = \boldsymbol{C} \iff \boldsymbol{V}\,\boldsymbol{A} - \boldsymbol{B}\,\boldsymbol{V} = \boldsymbol{C}\,\boldsymbol{V}. \tag{5.35}$$

If we define $\boldsymbol{D} := \boldsymbol{C}\,\boldsymbol{V}$ and $\mathsf{d} := \mathcal{K}(\boldsymbol{D})$, the SVD method described above evaluates the $\boldsymbol{V}$ that minimizes $\mathsf{d}^{\mathsf{T}}\,\mathsf{d}$. Assuming that $\boldsymbol{A}$ and $\boldsymbol{B}$ are both versors or blades, then $\boldsymbol{C}$, which has to be of the same type, satisfies $\boldsymbol{C}\,\widetilde{\boldsymbol{C}} \in \mathbb{R}$. Recall that blades are a special type of versor. Furthermore, $\boldsymbol{D}$ has to be a versor, and thus $\boldsymbol{D}\,\boldsymbol{D}^{\dagger} \in \mathbb{R}$ and $\boldsymbol{D}\,\boldsymbol{D}^{\dagger} = \mathsf{d}^{\mathsf{T}}\,\mathsf{d}$. However, from (5.35) it follows that

$$\boldsymbol{D}\,\boldsymbol{D}^\dagger = \boldsymbol{C}\,\boldsymbol{V}\,\boldsymbol{V}^\dagger\,\boldsymbol{C}^\dagger = \rho\,\boldsymbol{C}\,\boldsymbol{C}^\dagger, \qquad \rho := \boldsymbol{V}\,\boldsymbol{V}^\dagger \in \mathbb{R}. \tag{5.36}$$

Therefore, minimizing $\mathsf{d}^\mathsf{T}\,\mathsf{d}$, minimizes $\boldsymbol{C}\,\boldsymbol{C}^\dagger$. Note that $\rho = 1$, since the column vectors of the matrix V are unit vectors and thus, also, the solution for $\boldsymbol{V}$ is a unit vector.

If $\boldsymbol{A}, \boldsymbol{B} \in \mathbb{G}_3^1$ are grade 1 vectors in Euclidean space, then $\boldsymbol{C} \in \mathbb{G}_3^1$, and $\boldsymbol{C}\,\boldsymbol{C}^\dagger$ is the Euclidean distance between $\boldsymbol{V}\,\boldsymbol{A}\,\boldsymbol{V}^{-1}$ and $\boldsymbol{B}$ squared. Hence, the SVD method estimates the versor that minimizes the squared Euclidean distance between $\boldsymbol{V}\,\boldsymbol{A}\,\boldsymbol{V}^{-1}$ and $\boldsymbol{B}$, which is typically the metric that is desired to be used.

When one is evaluating versors, there is typically not just a single pair of multivectors that are related by the versor as in (5.32). If the goal is, for example, to evaluate a rotor, then a single pair of points that are related by the rotor is not enough to evaluate the rotor uniquely. In general, two sets of multivectors are given, $\{\boldsymbol{A}_i\}$ and $\{\boldsymbol{B}_i\}$, say, and the goal is to find the versor $\boldsymbol{V}$ such that

$$\boldsymbol{V}\,\boldsymbol{A}_i\,\boldsymbol{V}^{-1} = \boldsymbol{B}_i \iff \mathsf{Q}(\mathsf{a}_i, \mathsf{b}_i)\,\mathsf{v} = 0, \quad \forall\, i, \tag{5.37}$$

where Q is defined as in (5.34). This may be achieved by evaluating the right null-space of the matrix constructed by stacking all matrices $\mathsf{Q}(\mathsf{a}_i, \mathsf{b}_i)$ on top of each other. Alternatively, the right null-space of the matrix $\sum_i \mathsf{Q}^\mathsf{T}(\mathsf{a}_i, \mathsf{b}_i)\,\mathsf{Q}(\mathsf{a}_i, \mathsf{b}_i)$ can be evaluated.

If $\{\boldsymbol{A}_i\}, \{\boldsymbol{B}_i\} \subset \mathbb{G}_3^1$ are Euclidean vectors, then evaluating $\boldsymbol{V}$ in this way results in the versor that minimizes the sum of the squared Euclidean distances between $\boldsymbol{V}\,\boldsymbol{A}_i\,\boldsymbol{V}^{-1}$ and $\boldsymbol{B}_i$ for all i.

It is very important to note at this point that even if we can evaluate a versor that relates two sets of multivectors, the solution versor will not necessarily be unique, independent of the number of multivector pairs used. This is the case if there exists a multivector $\boldsymbol{J}$, say, that commutes with the $\{\boldsymbol{A}_i\} \subset \mathbb{G}_{p,q}$, i.e. $\boldsymbol{J}\,\boldsymbol{A}_i = \boldsymbol{A}_i\,\boldsymbol{J}$ for all i, because then

$$\begin{aligned} &\boldsymbol{V}\,\boldsymbol{A}_i\,\boldsymbol{V}^{-1} = \boldsymbol{B}_i \\ \iff &\boldsymbol{V}\,\boldsymbol{A}_i\,\boldsymbol{J}\,\boldsymbol{J}^{-1}\,\boldsymbol{V}^{-1} = \boldsymbol{B}_i \\ \iff &\boldsymbol{V}\,\boldsymbol{J}\,\boldsymbol{A}_i\,\boldsymbol{J}^{-1}\,\boldsymbol{V}^{-1} = \boldsymbol{B}_i. \end{aligned} \tag{5.38}$$

Therefore, if $\boldsymbol{V}$ is a solution versor, then so is $\boldsymbol{V}\,\boldsymbol{J}$. This problem occurs, for example, when one is evaluating a rotor in $\mathbb{G}_3$ that relates two sets of multivectors $\{\boldsymbol{A}_i\}, \{\boldsymbol{B}_i\} \subset \mathbb{G}_3$. In $\mathbb{G}_3$, the pseudoscalar $\boldsymbol{I} \in \mathbb{G}_3$ commutes with all multivectors of $\mathbb{G}_3$. Hence, when one is solving for the versor that best relates the two sets of multivectors, two solutions will be found: the actual rotor $\boldsymbol{R}$, and $\boldsymbol{R}\,\boldsymbol{I}$. In order to make the solution unique, an appropriate subspace projection, as described in Sect. 5.1.3 has to be employed.

For this purpose, (5.34) is modified as follows. Let $\mathsf{M_+}^i{}_j \in \mathbb{R}^{4\times 8}$ denote the matrix that extracts the four components of the even subalgebra $\mathbb{G}_3^+$ from the full component vector $\mathsf{v} := \mathcal{K}(\boldsymbol{V})$, i.e. $\mathsf{v}_+{}^i := \mathsf{v}^j\, \mathsf{M}_+{}^i{}_j$ or $\mathsf{v}_+ = \mathsf{M}_+\, \mathsf{v}$. Then, the expression $\boldsymbol{V}\,\boldsymbol{A} - \boldsymbol{B}\,\boldsymbol{V} = 0$ with the additional constraint that $\boldsymbol{V} \in \mathbb{G}_3^+$ can be written in tensor form as

$$\begin{aligned} 0 &= \mathsf{v}_+{}^i\, \mathsf{a}^j\, \mathsf{M}_+{}_i{}^u\, \Gamma^k{}_{uj} - \mathsf{b}^r\, \mathsf{v}^s\, \mathsf{M}_+{}_s{}^v\, \Gamma^k{}_{rv} \\ &= \Gamma_L(\mathsf{a})\, \mathsf{M}_+^\mathsf{T}\, \mathsf{v}_+ - \Gamma_R(\mathsf{b})\, \mathsf{M}_+^\mathsf{T}\, \mathsf{v}_+ \\ &= \mathsf{Q}_+(\mathsf{a},\mathsf{b})\, \mathsf{v}_+, \end{aligned} \tag{5.39}$$

where

$$\mathsf{Q}_+(\mathsf{a},\mathsf{b}) := (\Gamma_L(\mathsf{a}) - \Gamma_R(\mathsf{b}))\, \mathsf{M}_+^\mathsf{T} = \mathsf{Q}(\mathsf{a},\mathsf{b})\, \mathsf{M}_+^\mathsf{T}. \tag{5.40}$$

The right multiplication of $\mathsf{Q}(\mathsf{a},\mathsf{b})$ by M_+^T basically picks out those columns of Q that are related to the even subalgebra $\mathbb{G}_3^+$. In this way, the additional constraint on the versor can easily be implemented.

5.2.3 Example: Inverse of a Multivector in $\mathbb{G}_2$

As an example of the evaluation of the inverse of a multivector, consider again $\mathbb{G}_2$, with the algebraic basis

$$\boldsymbol{E}_1 := 1, \quad \boldsymbol{E}_2 := \boldsymbol{e}_1, \quad \boldsymbol{E}_3 := \boldsymbol{e}_2, \quad \boldsymbol{E}_4 := \boldsymbol{e}_{12}.$$

The geometric-product tensor $\Gamma^k{}_{ij} \in \mathbb{R}^{4\times4\times4}$ of $\mathbb{G}_2$ is given by (5.11). Let $\boldsymbol{A} = 1 + \boldsymbol{e}_{12} = \boldsymbol{E}_1 + \boldsymbol{E}_4$, $\mathsf{a} = \mathcal{K}(\boldsymbol{A}) = (1\;0\;0\;1)^\mathsf{T}$, and $\mathsf{e}_1 = \mathcal{K}(1) = (1\;0\;0\;0)^\mathsf{T}$. Then

$$\Gamma_R(\mathsf{a}) = \begin{pmatrix} 1 & 0 & 0 & -1 \\ 0 & 1 & 1 & 0 \\ 0 & -1 & 1 & 0 \\ 1 & 0 & 0 & 1 \end{pmatrix}, \quad \Gamma_R(\mathsf{a})^{-1} = \begin{pmatrix} 0.5 & 0 & 0 & 0.5 \\ 0 & 0.5 & -0.5 & 0 \\ 0 & 0.5 & 0.5 & 0 \\ -0.5 & 0 & 0 & 0.5 \end{pmatrix}. \tag{5.41}$$

Thus

$$\Gamma_R(\mathsf{a})^{-1}\, \mathsf{e}_1 = (\tfrac{1}{2}\;0\;0\;-\tfrac{1}{2}) \iff \mathcal{K}^{-1}(\Gamma_R(\mathsf{a})^{-1}\, \mathsf{e}_1) = \frac{1}{2}\,(1 - \boldsymbol{e}_{12}). \tag{5.42}$$

On the other hand, if $\boldsymbol{A} = \frac{1}{2}\,(1 + \boldsymbol{e}_1)$, then $\mathsf{a} = \mathcal{K}(\boldsymbol{A}) = (\frac{1}{2}\;\frac{1}{2}\;0\;0)^\mathsf{T}$ and

$$\Gamma_R(\mathsf{a}) = \begin{pmatrix} 0.5 & 0.5 & 0 & 0 \\ 0.5 & 0.5 & 0 & 0 \\ 0 & 0 & 0.5 & 0.5 \\ 0 & 0 & 0.5 & 0.5 \end{pmatrix}, \tag{5.43}$$

which is clearly not of full rank, and thus $\boldsymbol{A}$ has no inverse.

5.3 Random Multivectors

The application of geometric algebra to "real-life" problems in computer vision or other areas of engineering invariably implies the necessity to deal with uncertain data. The goal of this section is to show how calculations with uncertain multivectors can be performed in geometric algebra. When we are talking about uncertain multivectors, it is always assumed in this text that the uncertainty is Gaussian; that is, a multivector's probability distribution is fully determined by its mean value and covariance matrix.

The first step is to express the operations between multivectors in tensor form, as described in Sect. 5.1. It is then straightforward to define appropriate covariance matrices and to apply error propagation to operations between uncertain multivectors. In the following, uncertain multivectors are first introduced and then error propagation for algebraic products is discussed.

5.3.1 Definition

Let $\underline{\mathsf{x}}$ and $\underline{\mathsf{y}}$ denote two vector-valued random variables with a joint probability distribution function (joint PDF) $f_{\mathsf{x},\mathsf{y}}$. The marginal PDFs are then given by

$$f_{\underline{\mathsf{x}}}(\underline{\mathsf{x}} = \mathsf{x}) = \int_{\mathbb{D}_{\mathsf{y}}} f_{\underline{\mathsf{x}},\underline{\mathsf{y}}}(\underline{\mathsf{x}} = \mathsf{x},\, \underline{\mathsf{y}} = \mathsf{y})\, d\mathsf{y}$$

and

$$f_{\underline{\mathsf{y}}}(\underline{\mathsf{y}} = \mathsf{y}) = \int_{\mathbb{D}_{\mathsf{x}}} f_{\underline{\mathsf{x}},\underline{\mathsf{y}}}(\underline{\mathsf{x}} = \mathsf{x},\, \underline{\mathsf{y}} = \mathsf{y})\, d\mathsf{x},$$

where $\mathbb{D}_{\mathsf{x}}$ and $\mathbb{D}_{\mathsf{y}}$ denote the domains of $\underline{\mathsf{x}}$ and $\underline{\mathsf{y}}$, respectively. The random vector variables $\underline{\mathsf{x}}$ and $\underline{\mathsf{y}}$ can be given in terms of scalar random variables $\{\, \underline{x}^i \,\}$ and $\{\, \underline{y}^i \,\}$ as

$$\underline{\mathsf{x}} = \left[\, \underline{x}^1,\, \underline{x}^2,\, \ldots,\, \underline{x}^n \,\right]^{\mathsf{T}}, \qquad \underline{\mathsf{y}} = \left[\, \underline{y}^1,\, \underline{y}^2,\, \ldots,\, \underline{y}^n \,\right]^{\mathsf{T}}.$$

The PDF $f_{\underline{\mathsf{x}}}(\underline{\mathsf{x}} = \mathsf{x})$ is therefore also a joint PDF of the set $\{\underline{x}^i\}$, which may be written as

$$f_{\underline{\mathsf{x}}}(\underline{\mathsf{x}} = \mathsf{x}) = f_{\underline{x}^1,\underline{x}^2,\ldots,\underline{x}^n}(\underline{x}^1 = x^1,\, \underline{x}^2 = x^2,\, \ldots,\, \underline{x}^n = x^n).$$

Therefore, also,

$$\begin{aligned} & f_{\underline{x}^1}(\underline{x}^1 = x^1) \\ & = \int_{\mathbb{D}_{\underline{x}^2}} \cdots \int_{\mathbb{D}_{\underline{x}^n}} f_{\underline{x}^1,\underline{x}^2,\ldots,\underline{x}^n}(\underline{x}^1 = x^1,\, \underline{x}^2 = x^2,\, \ldots,\, \underline{x}^n = x^n)\, dx^2 \ldots dx^n. \end{aligned}$$

The expectation value of $\underline{\mathsf{x}}$ is denoted by $\mathcal{E}(\underline{\mathsf{x}})$ and defined as

$$\mathcal{E}(\underline{\mathsf{x}}) := \int_{\mathbb{D}_{\mathsf{x}}} \mathsf{x}\, f_{\mathsf{x}}(\mathsf{x})\, d\mathsf{x}. \tag{5.44}$$

The expectation of $\underline{\mathsf{x}}$ is also called its *mean* and is written as $\bar{\mathsf{x}}$ for brevity. Similarly, each individual distribution $\underline{x}^i$ of $\underline{\mathsf{x}}$ has an expectation

$$\bar{x}^i := \mathcal{E}\left(\underline{x}^i\right) = \int_{\mathbb{D}_{\underline{x}^i}} x^i\, f_{\underline{x}^i}(x^i)\, dx^i.$$

The covariance matrix of $\underline{\mathsf{x}}$ is written as $\Sigma_{\mathsf{x},\mathsf{x}}$, with components $\Sigma^{ij}_{\mathsf{x},\mathsf{x}}$, where, as before, the first index gives the row and the second index the column of the matrix. In this text, the symbol $\mathcal{V}$ will be used to denote the covariance operator. The components of $\Sigma_{\mathsf{x},\mathsf{x}}$ are then given by

$$\Sigma^{ij}_{\mathsf{x},\mathsf{x}} = \mathcal{V}\left(\underline{x}^i,\, \underline{x}^j\right),$$

where

$$\mathcal{V}\left(\underline{x}^i,\, \underline{x}^j\right) := \int_{\mathbb{D}_{\underline{x}^i}} \int_{\mathbb{D}_{\underline{x}^j}} (x^i - \bar{x}^i)\,(x^j - \bar{x}^j)\, f_{\underline{x}^i \underline{x}^j}(x^i,\, x^j)\, dx^i\, dx^j. \tag{5.45}$$

The variance of some $\underline{x}^i$ is then given by $\mathcal{V}\left(\underline{x}^i,\, \underline{x}^i\right)$. Similarly, the cross-covariance between $\underline{x}^i$ and $\underline{y}^j$ is defined as

$$\Sigma^{ij}_{\mathsf{x},\mathsf{y}} := \mathcal{V}\left(\underline{x}^i,\, \underline{y}^j\right). \tag{5.46}$$

The cross-covariance matrix of $\underline{\mathsf{x}}$ and $\underline{\mathsf{y}}$ is therefore denoted by $\Sigma_{\mathsf{x},\mathsf{y}}$ and has components $\Sigma^{ij}_{\mathsf{x},\mathsf{y}}$. If $\Sigma^{ij}_{\mathsf{x},\mathsf{y}} = 0$, the random variables $\underline{x}^i$ and $\underline{y}^j$ are said to be *uncorrelated*. Note that this does not necessarily imply that they are independent.

Another important formula is that for the expectation value of a function of a number of random variables. Let $h : \mathbb{R}^n \to \mathbb{R}$; then

$$\begin{aligned} &\mathcal{E}\left(h(\underline{x}^1 = x^1,\, \ldots,\, \underline{x}^n = x^n)\right) \\ &= \int_{\mathbb{D}_{\underline{x}^1}} \cdots \int_{\mathbb{D}_{\underline{x}^n}} h(x^1,\, \ldots,\, x^n)\, f_{\underline{x}^1,\ldots,\underline{x}^n}(x^1,\, \ldots,\, x^n)\, dx^1 \ldots dx^n. \end{aligned} \tag{5.47}$$

Because the main application of this analysis is error propagation for algebraic products of Gaussian-distributed random multivector variables, which are bilinear functions, it suffices to consider an approximation for $\mathcal{E}(h(\underline{\mathsf{x}}))$. For this purpose, the Taylor expansion up to order two of $h(\mathsf{x})$ about the mean $(\bar{\mathsf{x}})$ is evaluated:

$$\begin{aligned} h(\mathsf{x}) \approx h(\bar{\mathsf{x}}) \\ &+ \sum_{i=1}^{n} \Delta x^i \, h_{x^i}(\bar{\mathsf{x}}) \\ &+ \frac{1}{2} \sum_{i,j=1}^{n} \Delta x^i \, \Delta x^j \, h_{x^i,x^j}(\bar{\mathsf{x}}), \end{aligned} \tag{5.48}$$

where $\Delta x^i := x^i - \bar{x}^i$,

$$h_{x^i}(\bar{\mathsf{x}}) := \big(\partial_{x^i} h\big)(\bar{\mathsf{x}}), \quad \text{and} \quad h_{x^i,x^j}(\bar{\mathsf{x}}) := \big(\partial_{x^i}\partial_{x^j} h\big)(\bar{\mathsf{x}}).$$

5.3.2 First-Order Error Propagation

Using the approximation for a function $h(\mathsf{x})$ given above, it is possible to give approximations for the expectation value and variance of $h(\underline{\mathsf{x}})$. The expectation of $h(\mathsf{x})$ can be approximated as follows:

$$\begin{aligned} \mathcal{E}\left(h(\underline{\mathsf{x}})\right) &= \int_{\mathbb{D}_{\mathsf{x}}} h(\mathsf{x})\, f_{\mathsf{x}}(\mathsf{x})\, d\mathsf{x} \\ &\approx h(\bar{\mathsf{x}}) \int_{\mathbb{D}_{\mathsf{x}}} f_{\mathsf{x}}(\mathsf{x})\, d\mathsf{x} \\ &+ \sum_{i=1}^{n} h_{x^i}(\bar{\mathsf{x}}) \int_{\mathbb{D}_{\underline{x}^i}} \Delta x^i \, f_{\underline{x}^i}(x^i)\, dx^i \\ &+ \frac{1}{2} \sum_{i,j=1}^{n} h_{x^i,x^j}(\bar{\mathsf{x}}) \int_{\mathbb{D}_{\underline{x}^i}} \int_{\mathbb{D}_{\underline{x}^j}} \Delta x^i \, \Delta x^j \, f_{\underline{x}^i,\underline{x}^j}(x^i,\, x^j)\, dx^i\, dx^j. \end{aligned} \tag{5.49}$$

By definition, $\int_{\mathbb{D}_{\mathsf{x}}} f_{\mathsf{x}}(\mathsf{x})\, d\mathsf{x} = 1$ and, because $\Delta x^i = x^i - \bar{x}^i$,

$$\int_{\mathbb{D}_{\underline{x}^i}} \Delta x^i \, f_{\underline{x}^i}(x^i)\, dx^i = 0.$$

Furthermore, by the definition of the covariance,

$$\int_{\mathbb{D}_{\underline{x}^i}} \int_{\mathbb{D}_{\underline{x}^j}} \Delta x^i \, \Delta x^j \, f_{\underline{x}^i,\underline{x}^j}(x^i,\, x^j)\, dx^i\, dx^j = \Sigma^{ij}_{\mathsf{x},\mathsf{x}}.$$

Substituting all the above equalities into (5.49) gives

$$\mathcal{E}\left(h(\underline{\mathsf{x}})\right) \approx h(\bar{\mathsf{x}}) + \frac{1}{2} \sum_{i,j=1}^{n} h_{x^i,x^j}(\bar{\mathsf{x}})\; \Sigma^{ij}_{\mathsf{x},\mathsf{x}}. \tag{5.50}$$

This shows that up to order two, the expectation value of a function of a random variable $\underline{\mathsf{x}}$ depends only on the expectation values and the covariance matrix of $\underline{\mathsf{x}}$. Up to first order, the expectation of $h(\underline{\mathsf{x}})$ is $h(\bar{\mathsf{x}})$, the value of the function for the expectation of $\underline{\mathsf{x}}$.

The variance of $h(\underline{\mathsf{x}})$ may be evaluated following the general relation between the variance and the expectation, i.e.

$$\Sigma_{\underline{A},\underline{B}} = \mathcal{V}(\underline{A},\, \underline{B}) = \mathcal{E}(\underline{A}\,\underline{B}) - \mathcal{E}(\underline{A})\ \mathcal{E}(\underline{B})\,. \tag{5.51}$$

An approximation for the expectation of $h(\underline{\mathsf{x}})$ is given in (5.50), and thus

$$\mathcal{E}^2\big(h(\mathsf{x})\big) \approx h^2(\bar{\mathsf{x}}) + h(\bar{\mathsf{x}}) \sum_{i,j=1}^{n} h_{x^i,x^j}(\bar{\mathsf{x}})\ \Sigma^{i,j}_{\mathsf{x},\mathsf{x}} + \frac{1}{4} \underbrace{\left(\sum_{i,j=1}^{n} h_{x^i,x^j}(\bar{\mathsf{x}})\ \Sigma^{i,j}_{\mathsf{x},\mathsf{x}} \right)^2}_{>\ \text{order } 2}\,. \tag{5.52}$$

The term of order greater than two will be neglected in the following. Next, the expectation of $h^2(\mathsf{x})$ has to be derived. Clearly,

$$\big(\partial_{x^i} h^2\big)(\mathsf{x}) = 2\ h(\mathsf{x})\ h_{x^i}(\mathsf{x}),$$

and

$$\big(\partial_{x^i}\, \partial_{x^j} h^2\big)(\mathsf{x}) = 2\ h_{x^i}(\mathsf{x})\ h_{x^j}(\mathsf{x}) + 2\ h(\mathsf{x})\ h_{x^i,\,x^j}(\mathsf{x}).$$

It therefore follows that

$$\mathcal{E}\big(h^2(\mathsf{x})\big) \approx h^2(\bar{\mathsf{x}}) + \sum_{i,j=1}^{n} \Big(h_{x^i}(\mathsf{x})\ h_{x^j}(\mathsf{x}) + h(\mathsf{x})\ h_{x^i,\,x^j}(\mathsf{x})\Big)\ \Sigma^{i,j}_{\mathsf{x},\mathsf{x}}. \tag{5.53}$$

The variance of $h(\underline{\mathsf{x}})$ is thus

$$\mathcal{V}\big(h(\underline{\mathsf{x}})\big) = \mathcal{E}\big(h^2(\underline{\mathsf{x}})\big) - \mathcal{E}^2\big(h(\underline{\mathsf{x}})\big) \approx \sum_{i,j=1}^{n} h_{x^i}(\bar{\mathsf{x}})\ h_{x^j}(\bar{\mathsf{x}})\ \Sigma^{i,j}_{\mathsf{x},\mathsf{x}}. \tag{5.54}$$

For a vector valued function $\mathsf{h} : \mathbb{R}^n \to \mathbb{R}^n$ it may be shown analogously that

$$\begin{aligned} \mathcal{V}\big(h^r(\underline{\mathsf{x}}),\, h^s(\underline{\mathsf{x}})\big) &= \Sigma^{r,s}_{\mathsf{h},\mathsf{h}} \\ &= \mathcal{E}\big(h^r(\underline{\mathsf{x}})\, h^s(\underline{\mathsf{x}})\big) - \mathcal{E}\big(h^r(\underline{\mathsf{x}})\big)\, \mathcal{E}\big(h^s(\underline{\mathsf{x}})\big) \\ &\approx \textstyle\sum_{i,j=1}^{n}\ h^r_{x^i}(\bar{\mathsf{x}})\, h^s_{x^j}(\bar{\mathsf{x}})\ \Sigma^{i,j}_{\mathsf{x},\mathsf{x}}. \end{aligned} \tag{5.55}$$

The terms $h^r_{x^i}(\bar{\mathsf{x}})$ are simply the components of the Jacobi matrix of $\mathsf{h}(\bar{\mathsf{x}})$. If H denotes the Jacobi matrix of $\mathsf{h}(\mathsf{x})$, i.e. $H^r{}_i = h^r_{x^i}(\bar{\mathsf{x}})$, where r denotes the row and i the column, then (5.55) can be written as

$$\Sigma_{\mathsf{h},\mathsf{h}} \approx \mathsf{H}\, \Sigma_{\mathsf{x},\mathsf{x}}\, \mathsf{H}^{\mathsf{T}}. \tag{5.56}$$

Let $\mathsf{h} : \mathbb{R}^{2n} \to \mathbb{R}^n$, which may be written as $\mathsf{h}(\mathsf{z})$ with $\mathsf{z} \in \mathbb{R}^{2n}$ or $\mathsf{h}(\mathsf{x}, \mathsf{y})$ with $\mathsf{x}, \mathsf{y} \in \mathbb{R}^n$. The relation between the respective covariance matrices is simply

$$\Sigma_{\mathsf{z,z}} = \begin{bmatrix} \Sigma_{\mathsf{x,x}} & \Sigma_{\mathsf{x,y}} \\ \Sigma_{\mathsf{y,x}} & \Sigma_{\mathsf{y,y}} \end{bmatrix}. \tag{5.57}$$

The Jacobi matrix H of $\mathsf{h}(\mathsf{z})$ is related to the Jacobi matrices $\mathsf{H_x} := \partial_\mathsf{x}\, \mathsf{h}(\mathsf{x}, \mathsf{y})$ and $\mathsf{H_y} := \partial_\mathsf{y}\, \mathsf{h}(\mathsf{x}, \mathsf{y})$ via

$$\mathsf{H} = [\,\mathsf{H_x}\,,\ \mathsf{H_y}\,]\,. \tag{5.58}$$

Substituting (5.57) and (5.58) into (5.56) therefore results in

$$\begin{aligned} \Sigma_{\mathsf{h,h}} &\approx [\,\mathsf{H_x}\,,\ \mathsf{H_y}\,] \begin{bmatrix} \Sigma_{\mathsf{x,x}} & \Sigma_{\mathsf{x,y}} \\ \Sigma_{\mathsf{y,x}} & \Sigma_{\mathsf{y,y}} \end{bmatrix} \begin{bmatrix} \mathsf{H_x^T} \\ \mathsf{H_y^T} \end{bmatrix} \\ &= \mathsf{H_x}\, \Sigma_{\mathsf{x,x}}\, \mathsf{H_x^T} + \mathsf{H_y}\, \Sigma_{\mathsf{y,y}}\, \mathsf{H_y^T} + \mathsf{H_x}\, \Sigma_{\mathsf{x,y}}\, \mathsf{H_y^T} + \mathsf{H_y}\, \Sigma_{\mathsf{y,x}}\, \mathsf{H_x^T}\,. \end{aligned} \tag{5.59}$$

5.3.3 Bilinear Functions

In the previous subsection, general expressions for the expectation value and covariance matrix of a function of a random vector variable were derived. The main interest of this text lies in bilinear functions, since that is exactly what algebraic products are. Therefore, let the $h(\mathsf{x}, \mathsf{y})$ used above be replaced by $g^k(\mathsf{x}, \mathsf{y}) = x^i y^j\, \Gamma^k{}_{ij}$. The Jacobi matrices of $\mathsf{g}(\mathsf{x}, \mathsf{y})$ are then given by

$$\partial_\mathsf{x}\, \mathsf{g}(\mathsf{x}, \mathsf{y}) = \Gamma_L\,(\mathsf{y}) \qquad \text{and} \qquad \partial_\mathsf{y}\, \mathsf{g}(\mathsf{x}, \mathsf{y}) = \Gamma_R\,(\mathsf{x})\,, \tag{5.60}$$

where

$$\Gamma_L\,(\mathsf{y}) = \begin{bmatrix} y^j\, \Gamma^1{}_{1j} & \cdots & y^j\, \Gamma^1{}_{nj} \\ \vdots & \ddots & \vdots \\ y^j\, \Gamma^n{}_{1j} & \cdots & y^j\, \Gamma^n{}_{nj} \end{bmatrix}, \quad \Gamma_R\,(\mathsf{x}) = \begin{bmatrix} x^i\, \Gamma^1{}_{i1} & \cdots & x^i\, \Gamma^1{}_{in} \\ \vdots & \ddots & \vdots \\ x^i\, \Gamma^n{}_{i1} & \cdots & x^i\, \Gamma^n{}_{in} \end{bmatrix}.$$

Furthermore, because

$$\begin{aligned} \partial_\mathsf{x}\, \partial_\mathsf{x}\, \mathsf{g}(\mathsf{x}, \mathsf{y}) &= \partial_\mathsf{y}\, \partial_\mathsf{y}\, \mathsf{g}(\mathsf{x}, \mathsf{y}) = 0 \\ \text{and} \quad \partial_\mathsf{x}\, \partial_\mathsf{y}\, \mathsf{g}(\mathsf{x}, \mathsf{y}) &= \partial_\mathsf{y}\, \partial_\mathsf{x}\, \mathsf{g}(\mathsf{x}, \mathsf{y}) = \Gamma^k{}_{ij}, \end{aligned} \tag{5.61}$$

a Taylor expansion of $\mathsf{g}(\mathsf{x}, \mathsf{y})$ up to order two is sufficient to represent the function exactly.

5.3.3.1 Expectation

It follows from (5.50) that

$$\mathcal{E}\left(g^k(\underline{\mathsf{x}},\underline{\mathsf{y}})\right) = g^k(\bar{\mathsf{x}},\bar{\mathsf{y}}) + \Sigma^{ij}_{\mathsf{x},\mathsf{y}}\, G^k{}_{ij}, \tag{5.62}$$

with an implicit sum over i and j. This equation is in fact exact, since any derivative of $g^k(\mathsf{x},\mathsf{y})$ of order higher than two is zero. If $\underline{\mathsf{x}}$ and $\underline{\mathsf{y}}$ are also assumed to be uncorrelated, then

$$\mathcal{E}\left(g^k(\underline{\mathsf{x}},\underline{\mathsf{y}})\right) = g^k(\bar{\mathsf{x}},\bar{\mathsf{y}}).$$

The function $g^k(\underline{\mathsf{x}},\underline{\mathsf{y}})$ is itself a random variable with a particular PDF. If we define $u^k := g^k(\underline{\mathsf{x}},\underline{\mathsf{y}})$ and $\underline{\mathsf{u}} := \mathsf{g}(\underline{\mathsf{x}},\underline{\mathsf{y}})$, ideal error propagation would mean that we evaluate f_{u}, the PDF of $\underline{\mathsf{u}}$. It is, however, in general, not possible to give f_{u} easily in terms of $f_{\mathsf{x},\mathsf{y}}$. However, an approximation is typically sufficient, where the expectation and covariance of $\underline{\mathsf{u}}$, i.e. the first two moments, are given in terms of the expectation and (cross-)covariances of $\underline{\mathsf{x}}$ and $\underline{\mathsf{y}}$.

5.3.3.2 Covariance Matrix $\Sigma_{\mathsf{u},\mathsf{u}}$

The covariance $\mathcal{V}(\underline{u}^r, \underline{u}^s)$ follows directly from (5.55) as

$$\begin{aligned}\mathcal{V}(\underline{u}^r, \underline{u}^s) &= \mathcal{E}(\underline{u}^r\, \underline{u}^s) - \mathcal{E}(\underline{u}^r)\, \mathcal{E}(\underline{u}^s)\\ &\approx \bar{y}^p\, \bar{y}^q\, \Sigma^{ij}_{\mathsf{x},\mathsf{x}}\, \Gamma^r{}_{ip}\, \Gamma^s{}_{jq} + \bar{x}^p\, \bar{x}^q\, \Sigma^{ij}_{\mathsf{y},\mathsf{y}}\, \Gamma^r{}_{pi}\, \Gamma^s{}_{qj}\\ &\quad + \bar{y}^p\, \bar{x}^q\, \Sigma^{ij}_{\mathsf{x},\mathsf{y}}\, \Gamma^r{}_{ip}\, \Gamma^s{}_{qj} + \bar{x}^p\, \bar{y}^q\, \Sigma^{ij}_{\mathsf{y},\mathsf{x}}\, \Gamma^r{}_{pi}\, \Gamma^s{}_{jq}\,.\end{aligned} \tag{5.63}$$

In matrix notation, $\Sigma_{\mathsf{u},\mathsf{u}} := \mathcal{V}(\underline{\mathsf{u}},\underline{\mathsf{u}})$ can be written analogously to (5.59) as

$$\begin{aligned}\Sigma_{\mathsf{u},\mathsf{u}} \approx\; &\Gamma_R(\bar{\mathsf{y}})\, \Sigma_{\mathsf{x},\mathsf{x}}\, \Gamma_R(\bar{\mathsf{y}})^{\mathsf{T}} + \Gamma_L(\bar{\mathsf{x}})\, \Sigma_{\mathsf{y},\mathsf{y}}\, \Gamma_L(\bar{\mathsf{x}})^{\mathsf{T}}\\ &+ \Gamma_L(\bar{\mathsf{x}})\, \Sigma_{\mathsf{y},\mathsf{x}}\, \Gamma_R(\bar{\mathsf{y}})^{\mathsf{T}} + \Gamma_R(\bar{\mathsf{y}})\, \Sigma_{\mathsf{x},\mathsf{y}}\, \Gamma_L(\bar{\mathsf{x}})^{\mathsf{T}}\,.\end{aligned} \tag{5.64}$$

5.3.3.3 Covariance Matrix $\Sigma_{\mathsf{u},\mathsf{x}}$

From (5.51), it follows that the components of $\Sigma_{\mathsf{u},\mathsf{x}} = \mathcal{V}(\underline{\mathsf{u}},\underline{\mathsf{x}})$ can be evaluated via

$$\Sigma^{rs}_{\mathsf{u},\mathsf{x}} = \mathcal{V}\left(g^r(\underline{\mathsf{x}},\underline{\mathsf{y}}),\, \underline{x}^s\right) = \mathcal{E}\left(g^r(\underline{\mathsf{x}},\underline{\mathsf{y}})\, \underline{x}^s\right) - \mathcal{E}\left(g^r(\underline{\mathsf{x}},\underline{\mathsf{y}})\right)\, \mathcal{E}(\underline{x}^s)\,. \tag{5.65}$$

The necessary derivatives are the following:

$$\partial_{x^j}\left(g^r(\mathsf{x},\mathsf{y})\, x^s\right) = y^p\, \Gamma^r{}_{jp}\, x^s + g^r(\mathsf{x},\mathsf{y})\, \delta^s{}_j, \tag{5.66}$$

where ${\delta^s}_j$ is the Kronecker delta, (i.e. unity if $j = s$ and zero otherwise);

$$\partial_{y^j}\left(g^r(\mathsf{x},\mathsf{y})\,x^s\right) = x^p\,{\Gamma^r}_{pj}\,x^s\,; \tag{5.67}$$

$$\partial_{x^i}\,\partial_{x^j}\left(g^r(\mathsf{x},\mathsf{y})\,x^s\right) = y^p\,\left({\Gamma^r}_{jp}\,{\delta^s}_i + {\Gamma^r}_{ip}\,{\delta^s}_j\right)\,; \tag{5.68}$$

$$\partial_{y^i}\,\partial_{y^j}\left(g^r(\mathsf{x},\mathsf{y})\,x^s\right) = 0\,; \tag{5.69}$$

$$\partial_{x^i}\,\partial_{y^j}\left(g^r(\mathsf{x},\mathsf{y})\,x^s\right) = {\Gamma^r}_{ij}\,x^s + x^p\,{\Gamma^r}_{pj}\,{\delta^s}_i\,. \tag{5.70}$$

Therefore,

$$\begin{aligned}
\mathcal{E}\left(g^r(\mathsf{x},\mathsf{y})\,x^s\right) &= g^r(\bar{\mathsf{x}},\bar{\mathsf{y}})\,\bar{x}^s \\
&\quad + \frac{1}{2}\,\Sigma^{ij}_{\mathsf{x},\mathsf{x}}\left(\bar{y}^p\,{\Gamma^r}_{jp}\,{\delta^s}_i + \bar{y}^p\,{\Gamma^r}_{ip}\,{\delta^s}_j\right) \\
&\quad + \Sigma^{ij}_{\mathsf{x},\mathsf{y}}\left({\Gamma^r}_{ij}\,\bar{x}^s + \bar{x}^p\,{\Gamma^r}_{jp}\,{\delta^s}_i\right) \\
&= g^r(\bar{\mathsf{x}},\bar{\mathsf{y}})\,\bar{x}^s \\
&\quad + \bar{y}^p\,\Sigma^{is}_{\mathsf{x},\mathsf{x}}\,{\Gamma^r}_{ip} + \bar{x}^p\,\Sigma^{is}_{\mathsf{y},\mathsf{x}}\,{\Gamma^r}_{pi} + \bar{x}^s\,\Sigma^{ij}_{\mathsf{x},\mathsf{y}}\,{\Gamma^r}_{ij}\,.
\end{aligned} \tag{5.71}$$

By substituting these results into (5.65), it may then be shown that

$$\Sigma^{rs}_{\mathsf{u},\mathsf{x}} = \bar{y}^p\,\Sigma^{is}_{\mathsf{x},\mathsf{x}}\,{\Gamma^p}_{ri} + \bar{x}^p\,\Sigma^{is}_{\mathsf{y},\mathsf{x}}\,{\Gamma^r}_{pi}\,. \tag{5.72}$$

In matrix notation, this becomes

$$\Sigma_{\mathsf{u},\mathsf{x}} = \Gamma_R(\bar{\mathsf{y}})\;\Sigma_{\mathsf{x},\mathsf{x}} + \Gamma_L(\bar{\mathsf{x}})\;\Sigma_{\mathsf{y},\mathsf{x}}\,. \tag{5.73}$$

Similarly, it can be shown that

$$\Sigma_{\mathsf{u},\mathsf{y}} = \Gamma_R(\bar{\mathsf{y}})\;\Sigma_{\mathsf{x},\mathsf{y}} + \Gamma_L(\bar{\mathsf{x}})\;\Sigma_{\mathsf{y},\mathsf{y}}\,, \tag{5.74}$$

and also

$$\Sigma_{\mathsf{u},\mathsf{z}} = \Gamma_R(\bar{\mathsf{y}})\;\Sigma_{\mathsf{x},\mathsf{z}} + \Gamma_L(\bar{\mathsf{x}})\;\Sigma_{\mathsf{y},\mathsf{z}}\,. \tag{5.75}$$

5.3.4 Summary

The goal of this section was to show how error propagation can be applied to geometric-algebra products, which have been shown to be bilinear functions. It was shown first how the expectation value and the covariance matrices of a bilinear function of two random variables are related to the original random variables. The resultant equations are summarized in the following.

Let $\underline{\mathsf{x}}$, $\underline{\mathsf{y}}$, and $\underline{\mathsf{z}}$ be three random variables with covariance matrices $\Sigma_{\mathsf{x,x}}$, $\Sigma_{\mathsf{y,y}}$, $\Sigma_{\mathsf{z,z}}$, $\Sigma_{\mathsf{x,y}}$, $\Sigma_{\mathsf{y,z}}$, and $\Sigma_{\mathsf{z,x}}$. The bilinear function $\mathsf{g}(\underline{\mathsf{x}}, \underline{\mathsf{y}})$ is again a random variable, which is denoted by $\underline{\mathsf{u}}$. In order to perform calculations with $\underline{\mathsf{u}}$, the covariance matrices $\Sigma_{\mathsf{u,u}}$, $\Sigma_{\mathsf{u,x}}$, $\Sigma_{\mathsf{u,y}}$, and $\Sigma_{\mathsf{u,z}}$ have to be evaluated. The last three of these covariance matrices are of importance if, for example, $\mathsf{g}(\underline{\mathsf{u}}, \underline{\mathsf{x}})$ or $\mathsf{g}(\underline{\mathsf{u}}, \underline{\mathsf{z}})$ has to be evaluated:

$$\begin{aligned}
\Sigma_{\mathsf{u,u}} \approx \quad & \Gamma_R(\bar{\mathsf{y}})\, \Sigma_{\mathsf{x,x}}\, \Gamma_R(\bar{\mathsf{y}})^{\mathsf{T}} + \Gamma_L(\bar{\mathsf{x}})\, \Sigma_{\mathsf{y,y}}\, \Gamma_L(\bar{\mathsf{x}})^{\mathsf{T}} \\
& + \Gamma_L(\bar{\mathsf{x}})\, \Sigma_{\mathsf{y,x}}\, \Gamma_R(\bar{\mathsf{y}})^{\mathsf{T}} + \Gamma_R(\bar{\mathsf{y}})\, \Sigma_{\mathsf{x,y}}\, \Gamma_L(\bar{\mathsf{x}})^{\mathsf{T}}\,. \\
\Sigma_{\mathsf{u,x}} &= \Gamma_R(\bar{\mathsf{y}})\, \Sigma_{\mathsf{x,x}} + \Gamma_L(\bar{\mathsf{x}})\, \Sigma_{\mathsf{y,x}}\,. \\
\Sigma_{\mathsf{u,y}} &= \Gamma_R(\bar{\mathsf{y}})\, \Sigma_{\mathsf{x,y}} + \Gamma_L(\bar{\mathsf{x}})\, \Sigma_{\mathsf{y,y}}\,. \\
\Sigma_{\mathsf{u,z}} &= \Gamma_R(\bar{\mathsf{y}})\, \Sigma_{\mathsf{x,z}} + \Gamma_L(\bar{\mathsf{x}})\, \Sigma_{\mathsf{y,z}}\,.
\end{aligned} \tag{5.76}$$

5.4 Validity of Error Propagation

There are two main problems in first-order error propagation of Gaussian distributed random variables:

1. A function of a Gaussian distributed random variable is not necessarily Gaussian, which implies that a characterization of the resultant random variable by its mean value and covariance may not be sufficient anymore.
2. First-order error propagation is an approximation even for Gaussian distributed random variables.

These two problems are discussed in some detail in the following.

5.4.1 Non-Gaussivity

In general, it is quite difficult to give an expression for the probability distribution function of a function of a number of Gaussian distributed random variables. An example of a simple algebraic operation is therefore analyzed numerically in this section to show the non-Gaussivity of the resultant random variable.

In [84], it was shown that bilinear functions of Gaussian distributed random variables are not Gaussian in general. In fact, the resultant distributions may even be bimodal. Because geometric-algebra products are bilinear functions, it is necessary to investigate this effect in algebraic products.

Consider for this purpose $\mathbb{G}_2$, with the algebraic basis $\overline{\mathbb{G}}_2 = \{1, e_1, e_2, e_{12}\}$, and a function $\boldsymbol{h} : \mathbb{G}_2^1 \times \mathbb{G}_2^1 \to \mathbb{G}_2^1$ defined as

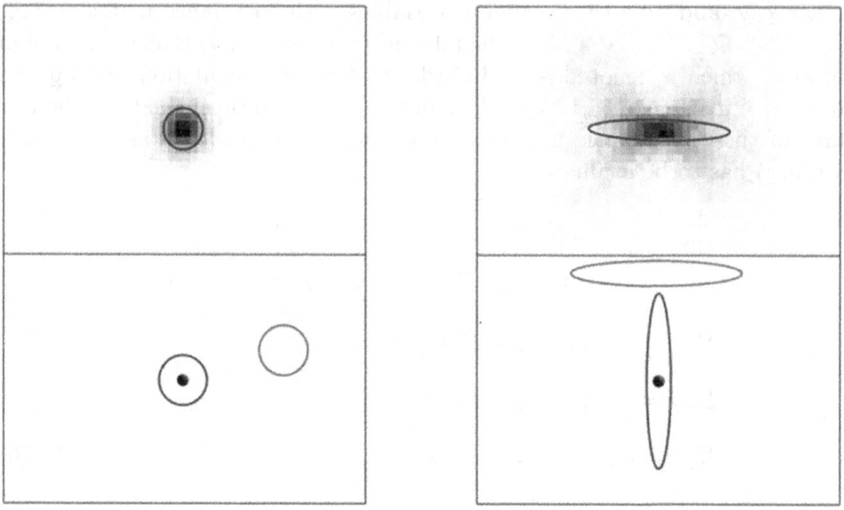

(a) Circularly Gaussian distributed 1-blades

(b) Elliptically Gaussian distributed 1-blades

Fig. 5.1 Initial Gaussian distributions of 1-blades (bottom), indicated by standard deviations, and the resultant distribution (top), shown as a histogram and with predicted standard deviation

$$\boldsymbol{h} : (\boldsymbol{x}, \boldsymbol{y}) \mapsto \boldsymbol{e}_1 \, \boldsymbol{x} \, \boldsymbol{y}. \tag{5.77}$$

For example, if $\boldsymbol{x} := x^i \, \boldsymbol{e}_i$ and $\boldsymbol{y} := y^i \, \boldsymbol{e}_i$, then

$$\begin{aligned} \boldsymbol{h}(\boldsymbol{x}, \boldsymbol{y}) &= \boldsymbol{e}_1 \left(x^1 \, y^1 + x^2 \, y^2 + (x^1 \, y^2 - x^2 \, y^1) \, \boldsymbol{e}_{12} \right) \\ &= (x^1 \, y^1 + x^2 \, y^2) \, \boldsymbol{e}_1 + (x^1 \, y^2 - x^2 \, y^1) \, \boldsymbol{e}_2 . \end{aligned}$$

The left multiplication by $\boldsymbol{e}_1$ has the effect that the result for $\boldsymbol{x} \, \boldsymbol{y}$, which contains grades zero and two, is mapped to a 1-blade. This multiplication does not change the distribution function.

Figure 5.1 shows the result of two numerical experiments to estimate the distribution of $\boldsymbol{h}(\boldsymbol{x}, \boldsymbol{y})$. The bottom image in each figure indicates the Gaussian distributions of the random variables $\underline{\boldsymbol{x}}$ (center) and $\underline{\boldsymbol{y}}$ (off-center) by their standard-deviation ellipses. The black dot indicates the origin of $\mathbb{R}^2$. The top images give the respective distributions of $\boldsymbol{h}(\underline{\boldsymbol{x}}, \underline{\boldsymbol{y}})$ as histograms evaluated numerically over 10 000 instances. The standard-deviation ellipses of the Gaussian distributions predicted by first-order error propagation are also shown. Note that the top images are drawn relative to the respective mean value of the distribution.

It can be seen right away that although in Fig. 5.1(a) the numerically estimated and predicted distributions agree, the numerically estimated distribution in Fig. 5.1(b) is not Gaussian anymore. This shows that, in general, the geometric product of two Gaussian distributed 1-blades is not Gaussian distributed. First-order error propagation of products in geometric algebra is therefore only a good approximation if the covariances of the respective multivectors are "small". What amount of error is tolerable depends on the actual application.

5.4.2 Error Propagation Bias

Consider again the bilinear function $\mathsf{g}(\mathsf{x},\mathsf{y})$ of Sect. 5.3.3, which is defined as $g^k(\mathsf{x},\mathsf{y}) = x^i\, y^j\, \Gamma^k{}_{ij}$. Recall that the error propagation equation for the covariance matrix of g as given in (5.64) is an approximation. The correct error propagation equation may be written as

$$\begin{aligned} \Sigma_{\mathsf{g},\mathsf{g}} = \;& \Gamma_R(\bar{\mathsf{y}})\, \Sigma_{\mathsf{x},\mathsf{x}}\, \Gamma_R(\bar{\mathsf{y}})^\mathsf{T} + \Gamma_L(\bar{\mathsf{x}})\, \Sigma_{\mathsf{y},\mathsf{y}}\, \Gamma_L(\bar{\mathsf{x}})^\mathsf{T} \\ & + \Gamma_L(\bar{\mathsf{x}})\, \Sigma_{\mathsf{y},\mathsf{x}}\, \Gamma_R(\bar{\mathsf{y}})^\mathsf{T} + \Gamma_R(\bar{\mathsf{y}})\, \Sigma_{\mathsf{x},\mathsf{y}}\, \Gamma_L(\bar{\mathsf{x}})^\mathsf{T} \\ & + \mathsf{B}(\mathsf{x},\mathsf{y}), \end{aligned} \tag{5.78}$$

where $\mathsf{B}(\mathsf{x},\mathsf{y})$ is an appropriate *bias term*. This bias term plays a particularly important role for distributions centered at the origin; this can be seen as follows.

The Jacobi matrices of $\mathsf{g}(\mathsf{x},\mathsf{y})$ are given by $\Gamma_R(\bar{\mathsf{y}}) = \bar{y}^j\, \Gamma^k{}_{ij}$ and $\Gamma_L(\bar{\mathsf{x}}) = \bar{x}^i\, \Gamma^k{}_{ij}$. Hence, if $\bar{\mathsf{x}} = \bar{\mathsf{y}} = 0$, then $\Gamma_R(\bar{\mathsf{y}}) = \Gamma_L(\bar{\mathsf{x}}) = 0$ and thus $\Sigma_{\mathsf{g},\mathsf{g}} = 0$, according to (5.64), independent of $\Sigma_{\mathsf{x},\mathsf{x}}$ and $\Sigma_{\mathsf{y},\mathsf{y}}$. However, geometrically speaking, only the relative positions of $\bar{\mathsf{x}}$ and $\bar{\mathsf{y}}$ should influence $\Sigma_{\mathsf{g},\mathsf{g}}$, and the origin should play no special role. This has to be recovered by the bias term. In Sect. 5.5, error propagation in projective spaces is discussed, where the origin is not part of the space, which remedies the situation to some extent, even without the bias term.

The bias term $\mathsf{B}(\mathsf{x},\mathsf{y})$ consists of two main parts. The first part is the term of order higher than two in (5.52) that previously neglected. The second part has not been discussed yet. It stems from an evaluation of $\mathcal{E}\big(g^r(\underline{\mathsf{x}},\underline{\mathsf{y}})\, g^s(\underline{\mathsf{x}},\underline{\mathsf{y}})\big)$ that is needed for the calculation of $\Sigma^{r,s}_{\mathsf{g},\mathsf{g}}$, as in

$$\Sigma^{r,s}_{\mathsf{g},\mathsf{g}} = \mathcal{E}\big(g^r(\underline{\mathsf{x}},\underline{\mathsf{y}})\, g^s(\underline{\mathsf{x}},\underline{\mathsf{y}})\big) - \mathcal{E}\big(g^r(\underline{\mathsf{x}},\underline{\mathsf{y}})\big)\, \mathcal{E}\big(g^s(\underline{\mathsf{x}},\underline{\mathsf{y}})\big). \tag{5.79}$$

The approximation to $\mathcal{E}\big(g^r(\underline{\mathsf{x}},\underline{\mathsf{y}})\, g^s(\underline{\mathsf{x}},\underline{\mathsf{y}})\big)$ used in (5.64) considered only the derivatives of $g^r(\underline{\mathsf{x}},\underline{\mathsf{y}})\, g^s(\underline{\mathsf{x}},\underline{\mathsf{y}})$ up to order two. However, to obtain an exact result, all derivatives up to order four have to be considered. Derivatives of order higher than four have no effect, since they are identically zero.

To keep the following analysis legible, we concatenate column vectors $\mathsf{x},\mathsf{y} \in \mathbb{R}^n$ into a single vector $\mathsf{z} \in \mathbb{R}^{2n}$ as $\mathsf{z} := \big[\mathsf{x}^\mathsf{T}, \mathsf{y}^\mathsf{T}\big]^\mathsf{T}$ and define the covariance matrix $\Sigma_{\mathsf{z},\mathsf{z}}$ accordingly. Furthermore, we define $h(\mathsf{z}) := g^r(\mathsf{x},\mathsf{y})\, g^s(\mathsf{x},\mathsf{y})$, which can be written exactly in terms of a Taylor series about the expectation value of $\underline{z}$ as

$$
\begin{aligned}
h(\mathsf{z}) = h(\bar{\mathsf{z}}) &+ \sum_{i=1}^{n} h_{z^i}(\bar{\mathsf{z}})\, \Delta z^i + \sum_{i,j=1}^{n} h_{z^i,z^j}(\bar{\mathsf{z}})\, \Delta z^i\, \Delta z^j \\
&+ \sum_{i,j,k=1}^{n} h_{z^i,z^j,z^k}(\bar{\mathsf{z}})\, \Delta z^i\, \Delta z^j\, \Delta z^k \\
&+ \sum_{i,j,k,l=1}^{n} h_{z^i,z^j,z^k,z^l}(\bar{\mathsf{z}})\, \Delta z^i\, \Delta z^j\, \Delta z^k\, \Delta z^l,
\end{aligned} \tag{5.80}
$$

where $\Delta z^i := z^i - \bar{z}^i$. Analogously to (5.49), the expectation $\mathcal{E}\big(h(\underline{z})\big)$ therefore becomes

$$
\begin{aligned}
&\mathcal{E}\,(h(\underline{z})) = \int_{\mathbb{D}_{\mathsf{z}}} h(\mathsf{z})\, f_{\mathsf{z}}(\mathsf{z})\, d\mathsf{z} \\
&\approx h(\bar{\mathsf{z}}) \int f_{\mathsf{z}}(\mathsf{z})\, d\mathsf{x} + \sum_{i=1}^{2n} h_{z^i}(\bar{\mathsf{z}}) \int \Delta z^i\, f_{z^i}(z^i)\, dz^i \\
&+ \frac{1}{2!} \sum_{i,j=1}^{2n} h_{z^i,z^j}(\bar{\mathsf{z}}) \int\!\!\int \Delta z^i\, \Delta z^j\, f_{z^i,z^j}(z^i,\, z^j)\, dz^i\, dz^j. \\
&+ \frac{1}{3!} \sum_{i,j,k=1}^{2n} h_{z^i,z^j,z^k}(\bar{\mathsf{z}}) \int\!\!\int\!\!\int \Delta z^i\, \Delta z^j\, \Delta z^k\, f_{z^i,z^j,z^k}(z^i,\, z^j,\, z^k)\, dz^i\, dz^j\, dz^k \\
&+ \frac{1}{4!} \sum_{i,j,k,l=1}^{2n} h_{z^i,z^j,z^k,z^l}(\bar{\mathsf{z}}) \\
&\qquad \int\!\!\int\!\!\int\!\!\int \Delta z^i\, \Delta z^j\, \Delta z^k\, \Delta z^l\, f_{z^i,z^j,z^k,z^l}(z^i,\, z^j,\, z^k,\, z^l)\, dz^i\, dz^j\, dz^k\, dz^l.
\end{aligned} \tag{5.81}
$$

It was shown earlier that $\int f_{\mathsf{z}}(\mathsf{z})\, d\mathsf{z} = 1$, $\int \Delta z^i\, f_{z^i}(z^i)\, dz^i = 0$, and

$$
\Sigma_{\mathsf{z},\mathsf{z}}^{i,j} = \int\!\!\int \Delta z^i\, \Delta z^j\, f_{z^i,z^j}(z^i,\, z^j)\, dz^i\, dz^j.
$$

In [104], p. 120, Koch noted that

$$
\int\!\!\int\!\!\int \Delta z^i\, \Delta z^j\, \Delta z^k\, f_{z^i,z^j,z^k}(z^i,\, z^j,\, z^k)\, dz^i\, dz^j\, dz^k = 0
$$

and

$$
\begin{aligned}
&\int\!\!\int\!\!\int\!\!\int \Delta z^i\, \Delta z^j\, \Delta z^k\, \Delta z^l\, f_{z^i,z^j,z^k,z^l}(z^i,\, z^j,\, z^k,\, z^l)\, dz^i\, dz^j\, dz^k\, dz^l \\
&= \Sigma_{\mathsf{z},\mathsf{z}}^{i,j}\, \Sigma_{\mathsf{z},\mathsf{z}}^{k,l} + \Sigma_{\mathsf{z},\mathsf{z}}^{i,k}\, \Sigma_{\mathsf{z},\mathsf{z}}^{j,l} + \Sigma_{\mathsf{z},\mathsf{z}}^{j,k}\, \Sigma_{\mathsf{z},\mathsf{z}}^{i,l}.
\end{aligned}
$$

What is left is the evaluation of $h_{z^i,z^j,z^k,z^l}(\bar{\mathsf{z}})$. In order to do this, the z^i have to be expressed using x^i and y^i again. By definition, $x^i = z^i$ and $y^i = z^{n+i}$, for $i \in \{1,\dots,n\}$. Because $h_{x^i,x^j,x^k}(\bar{\mathsf{z}}) = 0$ and $h_{y^i,y^j,y^k}(\bar{\mathsf{z}}) = 0$, the evaluation of $h_{z^i,z^j,z^k,z^l}(\bar{\mathsf{z}})$ with $i,j,k,l \in \{1,\dots,2n\}$ reduces to the evaluation of $h_{x^i,x^j,y^k,y^l}(\bar{\mathsf{x}},\bar{\mathsf{y}})$ with $i,j,k,l \in \{1,\dots,n\}$. It may be shown by a straightforward calculation that

$$\begin{aligned} h^{r\,s}_{x^i,x^j,y^k,y^l}(\bar{\mathsf{x}},\bar{\mathsf{y}}) &= \Gamma^r{}_{jl}\,\Gamma^s{}_{ik} + \Gamma^r{}_{ik}\,\Gamma^s{}_{jl} \\ &\quad + \Gamma^r{}_{il}\,\Gamma^s{}_{jk} + \Gamma^r{}_{jk}\,\Gamma^s{}_{il}. \end{aligned} \tag{5.82}$$

In the last term of (5.81), the sum goes over $h_{z^i,z^j,z^k,z^l}(\bar{\mathsf{z}})$ for $i,j,k,l \in \{1,\dots,2n\}$. However, for a given i, j, k, and l, all permutations of the four derivatives of $h(\mathsf{z})$ result in the same derivative. In the sum

$$\sum_{i,j,k,l=1}^{n} h^{r\,s}_{x^i,x^j,y^k,y^l}(\bar{\mathsf{x}},\bar{\mathsf{y}}),$$

the interchanges of x^i with x^j and y^k with y^l have already been accounted for. The remaining interchanges of derivatives that have to be considered are all *combinations* (not permutations) of two derivatives from the ordered set $\{\,\partial_{x^i},\,\partial_{x^j},\,\partial_{y^k},\,\partial_{y^l}\,\}$, of which there are six. That is,

$$\begin{aligned} &h^{r\,s}_{x^i,x^j,y^k,y^l}(\bar{\mathsf{x}},\bar{\mathsf{y}}) = h^{r\,s}_{x^i,y^k,x^j,y^l}(\bar{\mathsf{x}},\bar{\mathsf{y}}) = h^{r\,s}_{x^i,y^k,y^l,x^j}(\bar{\mathsf{x}},\bar{\mathsf{y}}) \\ &= h^{r\,s}_{y^k,x^i,x^j,y^l}(\bar{\mathsf{x}},\bar{\mathsf{y}}) = h^{r\,s}_{y^k,x^i,y^l,x^j}(\bar{\mathsf{x}},\bar{\mathsf{y}}) = h^{r\,s}_{y^k,y^l,x^i,x^j}(\bar{\mathsf{x}},\bar{\mathsf{y}}). \end{aligned}$$

The factor of the last term of (5.81) therefore becomes $6\,(1/4!) = 1/4$. The whole term becomes

$$\begin{aligned} &\frac{1}{4!}\sum_{i,j,k,l=1}^{2n} h_{z^i,z^j,z^k,z^l}(\bar{\mathsf{z}}) \\ &\int\!\!\int\!\!\int\!\!\int \Delta z^i\,\Delta z^j\,\Delta z^k\,\Delta z^l\, f_{z^i,z^j,z^k,z^l}(z^i,\,z^j,\,z^k,\,z^l)\,dz^i\,dz^j\,dz^k\,dz^l \\ &\quad = \frac{1}{4}\sum_{i,j,k,l=1}^{n} \Big(\big(\Gamma^r{}_{jl}\,\Gamma^s{}_{ik} + \Gamma^r{}_{ik}\,\Gamma^s{}_{jl} + \Gamma^r{}_{il}\,\Gamma^s{}_{jk} + \Gamma^r{}_{jk}\,\Gamma^s{}_{il}\big) \\ &\qquad\qquad \times \big(\Sigma^{ij}_{\mathsf{x,x}}\,\Sigma^{kl}_{\mathsf{y,y}} + \Sigma^{ik}_{\mathsf{x,y}}\,\Sigma^{jl}_{\mathsf{x,y}} + \Sigma^{jk}_{\mathsf{x,y}}\,\Sigma^{il}_{\mathsf{x,y}}\big)\Big). \end{aligned} \tag{5.83}$$

The complete bias term $\mathsf{B}(\mathsf{x},\mathsf{y})$ for a bilinear function is therefore given by

$$B^{r\,s}(\mathsf{x},\mathsf{y}) = \frac{1}{4} \sum_{i,j,k,l=1}^{n} \Big(\big(\Gamma^{r}{}_{jl}\,\Gamma^{s}{}_{ik} + \Gamma^{r}{}_{ik}\,\Gamma^{s}{}_{jl} + \Gamma^{r}{}_{il}\,\Gamma^{s}{}_{jk} + \Gamma^{r}{}_{jk}\,\Gamma^{s}{}_{il} \big) \times \big(\Sigma^{i\,j}_{\mathsf{x},\mathsf{x}}\,\Sigma^{k\,l}_{\mathsf{y},\mathsf{y}} + \Sigma^{i\,k}_{\mathsf{x},\mathsf{y}}\,\Sigma^{j\,l}_{\mathsf{x},\mathsf{y}} + \Sigma^{j\,k}_{\mathsf{x},\mathsf{y}}\,\Sigma^{i\,l}_{\mathsf{x},\mathsf{y}} \big) \Big) - \sum_{i,j=1}^{n} \Big(\big(G^{r}{}_{ij}\,\Sigma^{i\,j}_{\mathsf{x},\mathsf{y}} \big) \big(G^{s}{}_{ij}\,\Sigma^{i\,j}_{\mathsf{x},\mathsf{y}} \big) \Big) . \tag{5.84}$$

Clearly, $\mathsf{B}(\mathsf{x},\mathsf{y})$ does not actually depend on $\bar{\mathsf{x}}$ and $\bar{\mathsf{y}}$, but only on their covariance matrices. Hence, the bias term does not vanish if $\bar{\mathsf{x}} = \bar{\mathsf{y}} = 0$.

5.4.2.1 Example

As a very simple example, consider the scalar multiplication of two stochastically independent scalar random variables $\underline{x}$ and $\underline{y}$, i.e. $\underline{z} = \underline{x}\,\underline{y}$. The corresponding bilinear product tensor $G^{k}{}_{ij} \in \mathbb{R}^{1\times1\times1}$ is given simply by $G^{k}{}_{ij} = 1$. The covariance matrices $\Sigma_{\mathsf{x},\mathsf{x}}$ and $\Sigma_{\mathsf{y},\mathsf{y}}$ each consist of a single scalar value and may be denoted by σ_x and σ_y, respectively. The cross-covariance $\Sigma_{\mathsf{x},\mathsf{y}}$ is equal to zero, by definition. If $\bar{x} = \bar{y} = 0$, then

$$\sigma_z = B(x,y) = \sigma_x\,\sigma_y \ ,$$

which is what is expected.

5.4.3 Conclusions

It has been shown in this section that there are two main problems when error propagation of bilinear functions of Gaussian distributed random variables is considered: the distribution of the resultant random variable may not be Gaussian, and first-order error propagation neglects a bias term that cannot always be neglected even for small variances. The first problem cannot be remedied. For each bilinear function considered, it should be checked whether the resultant distribution can be approximated well by a Gaussian distribution for the expected magnitudes of the variances. The second problem, on the other hand, can be remedied easily using the bias term given in (5.84), if the problem at hand can be split into separate bilinear functions. The resultant error propagation is then exact. However, in the estimation methods presented later on, the inclusion of the bias term would make an analytic solution intractable.

5.5 Uncertainty in Projective Space

In the preceding parts of this book, we have introduced the representation of multivectors by component vectors, which leads to a representation of Gaussian distributed random multivector variables by a mean component vector and a corresponding covariance matrix. In this section, this is made more specific by a discussion of uncertain multivectors in the various geometries introduced in Chap. 4.

In Sect. 4.2 projective space was introduced. The homogenization of a vector $\mathsf{x} \in \mathbb{R}^n$ was defined there as

$$\mathcal{H} : \mathsf{x} \in \mathbb{R}^n \mapsto \mathsf{u} = \begin{bmatrix} \mathsf{x} \\ 1 \end{bmatrix} \in \mathbb{R}^{n+1}. \tag{5.85}$$

A vector $\mathcal{H}(\mathsf{x})$ is therefore an element of the affine plane denoted by $\mathbb{A}^n \subset \mathbb{R}^{n+1}$. The inverse mapping $\mathcal{H}^{-1}$ is defined only for elements on the affine plane, as

$$\mathcal{H}^{-1} : \mathsf{u} = \begin{bmatrix} \mathsf{x} \\ 1 \end{bmatrix} \in \mathbb{A}^n \mapsto \mathsf{x} \in \mathbb{R}^n. \tag{5.86}$$

An affine projection operator $\mathcal{A}$ that projects elements in $\mathbb{R}^{n+1}$ into $\mathbb{A}^n$ was also defined in Sect. 4.2:

$$\mathcal{A} : \mathsf{u} = \begin{bmatrix} \mathsf{x} \\ u^{n+1} \end{bmatrix} \in \mathbb{R}^{n+1} \mapsto \begin{bmatrix} \dfrac{\mathsf{x}}{u^{n+1}} \\ 1 \end{bmatrix} \in \mathbb{A}^n. \tag{5.87}$$

5.5.1 Mapping

The projective space that is of particular interest in this text is the projective space of $\mathbb{R}^3$, which is represented in $\mathbb{R}^4$. First, the mapping of uncertain vectors between Euclidean and projective space is discussed.

5.5.1.1 The Embedding $\mathcal{H}$

Given a Gaussian distributed random vector variable $\underline{\mathsf{x}}$ with a mean $\bar{\mathsf{x}} \in \mathbb{R}^3$ and a covariance matrix $\Sigma_{\mathsf{x},\mathsf{x}} \in \mathbb{R}^{3\times 3}$, the question is what the mean and the covariance matrix of $\underline{\mathsf{u}} := \mathcal{H}(\underline{\mathsf{x}})$ are. Since $\partial^2_{\mathsf{x}} \mathcal{H}(\mathsf{x}) = 0$, it follows from (5.50) that

$$\bar{\mathsf{u}} = \mathcal{E}\big(\mathcal{H}(\underline{\mathsf{x}})\big) = \mathcal{H}(\bar{\mathsf{x}}).$$

The covariance matrix $\Sigma_{u,u}$ can be evaluated with the help of (5.56). In order to apply this equation, the Jacobi matrix $J_{\mathcal{H}}(x) := \partial_x \mathcal{H}(x)$ has to be evaluated. This is given by

$$J_{\mathcal{H}}(x) = \partial_x u = \begin{bmatrix} \frac{\partial u^1}{\partial x^1} & \cdots & \frac{\partial u^1}{\partial x^3} \\ \vdots & \ddots & \vdots \\ \frac{\partial u^4}{\partial x^1} & \cdots & \frac{\partial u^4}{\partial x^3} \end{bmatrix} = \begin{bmatrix} 1\,0\,0 \\ 0\,1\,0 \\ 0\,0\,1 \\ 0\,0\,0 \end{bmatrix} . \tag{5.88}$$

If $I \in \mathbb{R}^{3\times3}$ denotes the identity matrix and $0 \in \mathbb{R}^3$ the zero column vector, then this may also be written as $J_{\mathcal{H}}(x) = [\,I,\ 0\,]^{\mathsf{T}}$. Because $\partial_x^2 \mathcal{H}(x) = 0$, the covariance matrix $\Sigma_{u,u} \in \mathbb{R}^{4\times4}$ is given exactly by

$$\Sigma_{u,u} = J_{\mathcal{H}}(x)\, \Sigma_{x,x}\, J_{\mathcal{H}}(x)^{\mathsf{T}} = \begin{bmatrix} \Sigma_{x,x} & 0 \\ 0^{\mathsf{T}} & 0 \end{bmatrix} . \tag{5.89}$$

The Jacobi matrix of the inverse mapping $x = \mathcal{H}^{-1}(u)$ is given simply by

$$J_{\mathcal{H}^-}(u) = \partial_u x = \begin{bmatrix} 1\,0\,0\,0 \\ 0\,1\,0\,0 \\ 0\,0\,1\,0 \end{bmatrix} . \tag{5.90}$$

5.5.1.2 The Projection $\mathcal{A}$

The projection of a vector $u \in \mathbb{R}^4$ into $\mathbb{A}^3$ is more problematic, because a division by an element of u is present. This implies that the Taylor series of the projection operator $\mathcal{A}$ has an infinite number of terms, so that first-order error propagation cannot be exact. Let $u \in \mathbb{R}^4$, with a covariance matrix $\Sigma_{u,u} \in \mathbb{R}^{4\times4}$, and define

$$v := \mathcal{A}(u) = \left[\frac{u^1}{u^4},\ \frac{u^2}{u^4},\ \frac{u^3}{u^4},\ 1 \right]^{\mathsf{T}} .$$

The expectation $\bar{v} = \mathcal{E}(\mathcal{A}(\underline{u}))$ follows from (5.50) as $\bar{v} = [\,\bar{v}^1,\ \bar{v}^2,\ \bar{v}^3,\ \bar{v}^4\,]$, with $\bar{v}^4 = 1$ and, for $i \in \{1, 2, 3\}$,

$$\bar{v}^i = \left(\frac{\bar{u}^i}{\bar{u}^4} - \left(\frac{1}{\bar{u}^4}\right)^2 \Sigma_{u,u}^{i\,4} \right) . \tag{5.91}$$

If the covariances $\Sigma_{u,u}^{i\,4}$ are sufficiently small, these terms may be neglected. The Jacobi matrix of $\mathcal{A}$, $J_{\mathcal{A}}(u)$, is

$$\mathsf{J}_{\mathcal{A}}(\mathsf{u}) = \partial_{\mathsf{u}}\,\mathsf{v} = \begin{bmatrix} \frac{\partial v^1}{\partial u^1} & \cdots & \frac{\partial v^1}{\partial u^4} \\ \vdots & \ddots & \vdots \\ \frac{\partial v^4}{\partial u^1} & \cdots & \frac{\partial v^4}{\partial u^4} \end{bmatrix} = \frac{1}{u^4} \begin{bmatrix} 1\,0\,0 & -\frac{u^1}{u^4} \\ 0\,1\,0 & -\frac{u^2}{u^4} \\ 0\,0\,1 & -\frac{u^3}{u^4} \\ 0\,0\,0 & 0 \end{bmatrix}. \tag{5.92}$$

If $\mathsf{I} \in \mathbb{R}^{4\times 4}$ denotes the identity matrix, then this can also be written as

$$\mathsf{J}_{\mathcal{A}}(\mathsf{u}) = \frac{1}{u^4}\,\mathsf{I} - \left(\frac{1}{u^4}\right)^2 \mathsf{u}\,(\partial_{u^4}\,\mathsf{u})^{\mathsf{T}}.$$

5.5.2 Random Homogeneous Vectors

The embedding of the covariance matrix $\Sigma_{\mathsf{x},\mathsf{x}} \in \mathbb{R}^{3\times 3}$ of a Euclidean vector $\mathsf{x} \in \mathbb{R}^3$ in projective space as given in (5.89) shows that the variance along the homogeneous dimension is zero. When the error-propagated covariance matrices of bilinear functions of such homogeneous vectors are evaluated, however, the variance of the homogeneous dimension need not be zero anymore. The question, therefore, is what a variance for the homogeneous dimension means. Because homogeneous vectors are *representations* of Euclidean vectors, it has to be shown what the covariance matrix of a homogeneous vector represents in Euclidean space.

It can be seen quite easily that only that part of the covariance of a homogeneous vector u that is perpendicular to u is of importance:

$$\mathsf{J}_{\mathcal{A}}(\mathsf{u})\,\mathsf{u} = \left(\frac{1}{u^4}\,\mathsf{I} - \left(\frac{1}{u^4}\right)^2 \mathsf{u}\,(\partial_{u^4}\,\mathsf{u})^{\mathsf{T}}\right)\mathsf{u} = \frac{1}{u^4}\,\mathsf{u} - \left(\frac{1}{u^4}\right)^2 u^4\,\mathsf{u} = 0. \tag{5.93}$$

When u is projected onto the affine plane via $\mathsf{v} := \mathcal{A}(\mathsf{u})$, the covariance matrix $\Sigma_{\mathsf{u},\mathsf{u}}$ is mapped via

$$\Sigma_{\mathsf{v},\mathsf{v}} = \mathsf{J}_{\mathcal{A}}(\mathsf{u})\;\Sigma_{\mathsf{u},\mathsf{u}}\;\mathsf{J}_{\mathcal{A}}(\mathsf{u})^{\mathsf{T}}.$$

Hence, any component of $\Sigma_{\mathsf{u},\mathsf{u}}$ along u is mapped to zero and does not influence $\Sigma_{\mathsf{v},\mathsf{v}}$. Mapping $\Sigma_{\mathsf{v},\mathsf{v}}$ to the corresponding Euclidean covariance matrix is then simply an extraction of the top left 3×3 submatrix.

5.5.2.1 Example

As a simple example, consider a homogeneous vector u with a covariance matrix $\Sigma_{\mathsf{u},\mathsf{u}}$ defined by

$$\Sigma_{\mathsf{u},\mathsf{u}} := \begin{bmatrix} \sigma_1 & & & \\ & \sigma_2 & & \\ & & \sigma_3 & \\ & & & \sigma_4 \end{bmatrix}.$$

The covariance matrix of $\mathsf{v} = \mathcal{A}(\mathsf{u})$ may then be shown to be

$$\begin{aligned} \Sigma_{\mathsf{v},\mathsf{v}} &= \mathsf{J}_{\mathcal{A}}(\mathsf{u})\, \Sigma_{\mathsf{u},\mathsf{u}}\, \mathsf{J}_{\mathcal{A}}(\mathsf{u})^{\mathsf{T}} \\ &= \left(\frac{1}{u^4}\right)^2 \begin{bmatrix} \sigma_1 & & & \\ & \sigma_2 & & \\ & & \sigma_3 & \\ & & & 0 \end{bmatrix} + \sigma_4 \left(\frac{1}{u^4}\right)^4 \begin{bmatrix} u^1 \\ u^2 \\ u^3 \\ 0 \end{bmatrix} \left[u^1 , u^2 , u^3 , 0 \right] . \end{aligned}$$

That is, only the top left 3×3 submatrix of $\Sigma_{\mathsf{v},\mathsf{v}}$ is non-zero. This submatrix is also equivalent to the covariance matrix of the corresponding projection of v into Euclidean space.

5.5.3 Conditioning

In [93], Heuel argued that covariance matrices of homogeneous vectors should always be projected onto the subspace perpendicular to their corresponding homogeneous vector, since any part parallel to the homogeneous vector has no relevance in the corresponding Euclidean space. In this text, it is argued not only that this is not necessary but also that an additional bias may be introduced when the procedure suggested by Heuel is followed.

To show this, the matrix forms of the projection and rejection operators will be given. In matrix notation, the projection operator $\mathcal{P}_{\mathsf{u}}(\mathsf{x})$ is expressed as

$$\mathcal{P}_{\mathsf{u}}(\mathsf{x}) = \left(\mathsf{u}\,(\mathsf{u}^{\mathsf{T}}\,\mathsf{u})^{-1}\,\mathsf{u}^{\mathsf{T}}\right)\mathsf{x} = \mathsf{P}_{\mathsf{u}}\,\mathsf{x}, \tag{5.94}$$

where

$$\mathsf{P}_{\mathsf{u}} := \mathsf{u}\,(\mathsf{u}^{\mathsf{T}}\,\mathsf{u})^{-1}\,\mathsf{u}^{\mathsf{T}} .$$

This is equivalent to the projection operator in Definition 3.22. Similarly, the rejection operator $\mathcal{P}_{\mathsf{u}}^{\perp}(\mathsf{x})$ takes the following form in matrix notation:

$$\mathcal{P}_{\mathsf{u}}^{\perp}(\mathsf{x}) = \left(\mathsf{I} - \mathsf{u}\,(\mathsf{u}^{\mathsf{T}}\,\mathsf{u})^{-1}\,\mathsf{u}^{\mathsf{T}}\right)\mathsf{x} = \mathsf{P}_{\mathsf{u}}^{\perp}\,\mathsf{x}, \tag{5.95}$$

where I is the identity matrix and

$$\mathsf{P}_{\mathsf{u}}^{\perp} := \mathsf{I} - \mathsf{u}\,(\mathsf{u}^{\mathsf{T}}\,\mathsf{u})^{-1}\,\mathsf{u}^{\mathsf{T}} .$$

Because $\mathsf{J}_{\mathcal{A}}(\mathsf{u})\,\mathsf{u} = 0$, it follows that

$$\mathsf{J}_{\mathcal{A}}(\mathsf{u})\, \mathsf{P}_{\mathsf{u}}^{\perp} = \mathsf{J}_{\mathcal{A}}(\mathsf{u}) - \mathsf{J}_{\mathcal{A}}(\mathsf{u})\, \mathsf{u}\, (\mathsf{u}^{\mathsf{T}}\, \mathsf{u})^{-1}\, \mathsf{u}^{\mathsf{T}} = \mathsf{J}_{\mathcal{A}}(\mathsf{u}). \tag{5.96}$$

This implies, for the covariance matrix $\Sigma_{\mathsf{u},\mathsf{u}}$ of a homogeneous vector u, that

$$\mathsf{J}_{\mathcal{A}}(\mathsf{u})\, \Sigma_{\mathsf{u},\mathsf{u}}\, \mathsf{J}_{\mathcal{A}}(\mathsf{u})^{\mathsf{T}} = \mathsf{J}_{\mathcal{A}}(\mathsf{u})\, \mathsf{P}_{\mathsf{u}}^{\perp}\, \Sigma_{\mathsf{u},\mathsf{u}}\, (\mathsf{P}_{\mathsf{u}}^{\perp})^{\mathsf{T}}\, \mathsf{J}_{\mathcal{A}}(\mathsf{u})^{\mathsf{T}}. \tag{5.97}$$

That is, it does not matter whether a covariance matrix is projected onto the subspace perpendicular to its corresponding homogeneous vector or not – in both cases, it represents the same covariance matrix in Euclidean space.

5.5.3.1 Bilinear Functions

The same is true for error propagation in the case of bilinear functions. Let $\mathsf{g}(\mathsf{u},\mathsf{v})$ denote a bilinear function of two homogeneous vectors $\mathsf{u},\mathsf{v} \in \mathbb{R}^4$, with corresponding covariance matrices $\Sigma_{\mathsf{u},\mathsf{u}}$ and $\Sigma_{\mathsf{v},\mathsf{v}}$, respectively. The Jacobi matrices of $\mathsf{g}(\mathsf{u},\mathsf{v})$ are $\Gamma_L(\mathsf{v}) = \partial_{\mathsf{u}}\, \mathsf{g}(\mathsf{u},\mathsf{v})$ and $\Gamma_R(\mathsf{u}) = \partial_{\mathsf{v}}\, \mathsf{g}(\mathsf{u},\mathsf{v})$, so that the covariance matrix of $\mathsf{w} := \mathsf{g}(\mathsf{u},\mathsf{v})$ is given by

$$\Sigma_{\mathsf{w},\mathsf{w}} = \Gamma_L(\mathsf{v})\, \Sigma_{\mathsf{u},\mathsf{u}}\, \Gamma_L(\mathsf{v})^{\mathsf{T}} + \Gamma_R(\mathsf{u})\, \Sigma_{\mathsf{v},\mathsf{v}}\, \Gamma_R(\mathsf{u})^{\mathsf{T}}\,.$$

Since $\mathsf{g}(\mathsf{u},\mathsf{v})$ is a bilinear function, it follows that

$$\Gamma_L(\mathsf{v})\, \mathsf{u} = \Gamma_R(\mathsf{u})\, \mathsf{v} = \mathsf{w}\,.$$

Therefore,

$$\Gamma_L(\mathsf{v})\, \mathsf{P}_{\mathsf{u}}^{\perp} = \Gamma_L(\mathsf{v}) - \Gamma_L(\mathsf{v})\, \frac{\mathsf{u}\,\mathsf{u}^{\mathsf{T}}}{\mathsf{u}^{\mathsf{T}}\,\mathsf{u}} = \Gamma_L(\mathsf{v}) - \frac{\mathsf{w}\,\mathsf{u}^{\mathsf{T}}}{\mathsf{u}^{\mathsf{T}}\,\mathsf{u}}$$

and

$$\Gamma_R(\mathsf{u})\, \mathsf{P}_{\mathsf{v}}^{\perp} = \Gamma_R(\mathsf{u}) - \Gamma_R(\mathsf{u})\, \frac{\mathsf{v}\,\mathsf{v}^{\mathsf{T}}}{\mathsf{v}^{\mathsf{T}}\,\mathsf{v}} = \Gamma_R(\mathsf{u}) - \frac{\mathsf{w}\,\mathsf{v}^{\mathsf{T}}}{\mathsf{v}^{\mathsf{T}}\,\mathsf{v}}\,.$$

Because $\mathsf{J}_{\mathcal{A}}(\mathsf{w})\, \mathsf{w} = 0$,

$$\mathsf{J}_{\mathcal{A}}(\mathsf{w})\, \Gamma_L(\mathsf{v})\, \mathsf{P}_{\mathsf{u}}^{\perp} = \mathsf{J}_{\mathcal{A}}(\mathsf{w})\, \Gamma_L(\mathsf{v}) \quad \text{and} \quad \mathsf{J}_{\mathcal{A}}(\mathsf{w})\, \Gamma_R(\mathsf{u})\, \mathsf{P}_{\mathsf{v}}^{\perp} = \mathsf{J}_{\mathcal{A}}(\mathsf{w})\, \Gamma_R(\mathsf{u})\,.$$

It therefore follows that

$$\begin{aligned} \mathsf{J}_{\mathcal{A}}(\mathsf{w})\, \Sigma_{\mathsf{w},\mathsf{w}}\, \mathsf{J}_{\mathcal{A}}(\mathsf{w})^{\mathsf{T}} = &\ \mathsf{J}_{\mathcal{A}}(\mathsf{w})\, \Gamma_L(\mathsf{v})\, \mathsf{P}_{\mathsf{u}}^{\perp}\, \Sigma_{\mathsf{u},\mathsf{u}}\, (\mathsf{P}_{\mathsf{u}}^{\perp})^{\mathsf{T}}\, \Gamma_L(\mathsf{v})^{\mathsf{T}}\, \mathsf{J}_{\mathcal{A}}(\mathsf{w})^{\mathsf{T}} \\ &+ \mathsf{J}_{\mathcal{A}}(\mathsf{w})\, \Gamma_R(\mathsf{u})\, \mathsf{P}_{\mathsf{v}}^{\perp}\, \Sigma_{\mathsf{v},\mathsf{v}}\, (\mathsf{P}_{\mathsf{v}}^{\perp})^{\mathsf{T}}\, \Gamma_R(\mathsf{u})^{\mathsf{T}}\, \mathsf{J}_{\mathcal{A}}(\mathsf{w})^{\mathsf{T}} \\ = &\ \mathsf{J}_{\mathcal{A}}(\mathsf{w})\, \Gamma_L(\mathsf{v})\, \Sigma_{\mathsf{u},\mathsf{u}}\, \Gamma_L(\mathsf{v})^{\mathsf{T}}\, \mathsf{J}_{\mathcal{A}}(\mathsf{w})^{\mathsf{T}} \\ &+ \mathsf{J}_{\mathcal{A}}(\mathsf{w})\, \Gamma_R(\mathsf{u})\, \Sigma_{\mathsf{v},\mathsf{v}}\, \Gamma_R(\mathsf{u})^{\mathsf{T}}\, \mathsf{J}_{\mathcal{A}}(\mathsf{w})^{\mathsf{T}}\,. \end{aligned} \tag{5.98}$$

Hence, the covariance matrix $\Sigma_{\mathsf{w,w}}$ represents the same covariance matrix in Euclidean space, independent of whether the covariance matrices $\Sigma_{\mathsf{u,u}}$ and $\Sigma_{\mathsf{v,v}}$ are projected onto the subspaces perpendicular to u and v, respectively, or not. This shows that it is not necessary to project covariance matrices of homogeneous vectors onto the subspaces perpendicular to their respective homogeneous vectors prior to any calculations.

5.5.3.2 Scaling

Apart from the projection of the covariance matrices discussed above, Heuel also suggested in [93] that the Euclidean parts of homogeneous vectors should be scaled such that their magnitudes were much smaller than the homogeneous components and the covariance matrices should be scaled accordingly. After this scaling, the projection of the covariance matrices should be performed as discussed above. Heuel argued that this conditioning reduced the magnitude of the error propagation bias. In the following, it is shown that this procedure can lead to incorrect error propagation results.

Generally speaking, it is not desirable that a scaling of the Euclidean parts of homogeneous vectors can change the relative magnitude of the error propagation bias, because this would imply that the homogenization of Euclidean vectors at different scales leads to different results. Such a property would make error propagation unusable for practical applications, where an overall scale should have no effect.

A simple example that demonstrates the scale invariance of error propagation if the covariance matrices are *not* projected is given in the following. For this purpose, consider two vectors $\boldsymbol{x}, \boldsymbol{y} \in \mathbb{R}^1$ with associated covariance matrices $\Sigma_{\boldsymbol{x},\boldsymbol{x}}$ and $\Sigma_{\boldsymbol{y},\boldsymbol{y}}$, which are homogenized as $\boldsymbol{u} := \mathcal{H}(\boldsymbol{x}) \in \mathbb{G}_2^1$ and $\boldsymbol{v} := \mathcal{H}(\boldsymbol{y}) \in \mathbb{G}_2^1$, with $\boldsymbol{u} = u_1\,\boldsymbol{e}_1 + u_2\,\boldsymbol{e}_2$ and $\boldsymbol{v} = v_1\,\boldsymbol{e}_1 + v_2\,\boldsymbol{e}_2$. Note that subscript indices have been used here for the components of the vectors in order to increase the readability of the following equations. The outer product of $\boldsymbol{u}$ and $\boldsymbol{v}$ therefore results in

$$\boldsymbol{w} := \boldsymbol{u} \wedge \boldsymbol{v} = (u_1\,v_2 - u_2\,v_1)\,\boldsymbol{e}_{12} = w\,\boldsymbol{e}_{12}\,,$$

where $w := u_1\,v_2 - u_2\,v_1$. In terms of component vectors, this outer-product operation can be regarded as a bilinear function that maps vectors $\mathsf{u} := [\,u_1\,,\;u_2\,]$ and $\mathsf{v} := [\,v_1\,,\;v_2\,]$ to the scalar value w. The Jacobi vectors of this function are therefore

$$\partial_{\mathsf{u}}\,w = [\,v_2\,,\;-v_1\,] \qquad \text{and} \qquad \partial_{\mathsf{v}}\,w = [\,-u_2\,,\;u_1\,]\;. \tag{5.99}$$

Let the covariance matrices of u and v be given by

$$\Sigma_{\mathsf{u},\mathsf{u}} := \begin{bmatrix} \alpha_1 & \\ & \alpha_2 \end{bmatrix} \quad \text{and} \quad \Sigma_{\mathsf{v},\mathsf{v}} := \begin{bmatrix} \beta_1 & \\ & \beta_2 \end{bmatrix} .$$

Assuming that u and v are independent, the variance of w, denoted by σ_w^2, is given by first-order error propagation as

$$\begin{aligned} \sigma_w^2 &= (\partial_{\mathsf{u}}\, w)\ \Sigma_{\mathsf{u},\mathsf{u}}\ (\partial_{\mathsf{u}}\, w)^{\mathsf{T}} + (\partial_{\mathsf{v}}\, w)\ \Sigma_{\mathsf{v},\mathsf{v}}\ (\partial_{\mathsf{v}}\, w)^{\mathsf{T}} \\ &= \alpha_1\, v_2^2 + \alpha_2\, v_1^2 + \beta_1\, u_2^2 + \beta_2\, u_1^2 . \end{aligned} \tag{5.100}$$

Now, suppose the Euclidean parts of u and v are uniformly scaled by a scalar factor $\tau \in \mathbb{R}$, i.e. $\mathsf{u}' := [\tau\, u_1\,,\ u_2]$ and $\mathsf{v}' := [\tau\, v_1\,,\ v_2]$. The corresponding covariance matrices are

$$\Sigma_{\mathsf{u}',\mathsf{u}'} := \begin{bmatrix} \tau^2\,\alpha_1 & \\ & \alpha_2 \end{bmatrix} \quad \text{and} \quad \Sigma_{\mathsf{v}',\mathsf{v}'} := \begin{bmatrix} \tau^2\,\beta_1 & \\ & \beta_2 \end{bmatrix} .$$

Then

$$w' = u_1'\, v_2' - u_2'\, v_1' = \tau\, u_1\, v_2 - u_2\, \tau\, v_1 = \tau\ w\ ,$$

and the corresponding variance $\sigma_{w'}^2$ is

$$\sigma_{w'}^2 = \tau^2\, \alpha_1\, v_2^2 + \alpha_2\, \tau^2\, v_1^2 + \tau^2\, \beta_1\, u_2^2 + \beta_2\, \tau^2\, u_1^2 = \tau^2\, \sigma_w^2\ , \tag{5.101}$$

as would be expected. That is, the ratio σ_w / w is scale-invariant:

$$\frac{\sigma_{w'}}{w'} = \frac{\sigma_w}{w} .$$

Hence, when the Euclidean vectors are scaled, the relative standard deviation of w stays constant.

If the covariance matrices are projected, this scale invariance is not preserved. The rejection matrices are

$$\mathsf{P}_{\mathsf{u}}^{\perp} = \mathsf{I} - \frac{\mathsf{u}\,\mathsf{u}^{\mathsf{T}}}{\mathsf{u}^{\mathsf{T}}\,\mathsf{u}} \quad \text{and} \quad \mathsf{P}_{\mathsf{v}}^{\perp} = \mathsf{I} - \frac{\mathsf{v}\,\mathsf{v}^{\mathsf{T}}}{\mathsf{v}^{\mathsf{T}}\,\mathsf{v}}.$$

If we apply these rejections, σ_w^2 is evaluated as

$$\begin{aligned} \sigma_w^2 = &(\partial_{\mathsf{u}}\, w)\, \mathsf{P}_{\mathsf{u}}^{\perp}\ \Sigma_{\mathsf{u},\mathsf{u}}\ (\mathsf{P}_{\mathsf{u}}^{\perp})^{\mathsf{T}}\, (\partial_{\mathsf{u}}\, w)^{\mathsf{T}} \\ &+ (\partial_{\mathsf{v}}\, w)\, \mathsf{P}_{\mathsf{v}}^{\perp}\ \Sigma_{\mathsf{v},\mathsf{v}}\ (\mathsf{P}_{\mathsf{v}}^{\perp})^{\mathsf{T}}\, (\partial_{\mathsf{v}}\, w)^{\mathsf{T}} . \end{aligned} \tag{5.102}$$

One component of this expression is

$$\begin{aligned}(\partial_{\mathsf{u}}\, w)\, \frac{\mathsf{u}\, \mathsf{u}^{\mathsf{T}}}{\mathsf{u}^{\mathsf{T}}\, \mathsf{u}}\, \Sigma_{\mathsf{u},\mathsf{u}}\, \frac{\mathsf{u}\, \mathsf{u}^{\mathsf{T}}}{\mathsf{u}^{\mathsf{T}}\, \mathsf{u}}\, (\partial_{\mathsf{u}}\, w)^{\mathsf{T}} \\ = \left(\frac{1}{\mathsf{u}^{\mathsf{T}}\, \mathsf{u}}\right)^2 \left(v_2^2\, a_1 + v_1^2\, a_3 - 2\, v_1\, v_2\, a_2\right),\end{aligned}$$

where

$$a_1 := \alpha_1\, u_1^4 + \alpha_2\, u_1^2\, u_2^2\,, \quad a_2 := \alpha_1\, u_1^3\, u_2 + \alpha_2\, u_1\, u_2^3\,, \quad a_3 := \alpha_1\, u_1^2\, u_2^2 + \alpha_2\, u_2^4\,.$$

If the Euclidean components of u and v are again scaled by τ, then

$$a_1' = \tau^2\, \alpha_1\, \tau^4\, u_1^4 + \alpha_2\, \tau^2\, u_1^2\, u_2^2.$$

That is a_1' cannot be written as a scaled version of a_1, which implies also that $\sigma_{w'}^2$ is not a scaled version of σ_w^2. Hence, the ratio σ_w / w is not scale-invariant. Therefore, if the covariance matrices are projected, the relative standard deviation depends on the overall scale of the Euclidean vectors.

5.6 Uncertainty in Conformal Space

A Euclidean vector $\boldsymbol{x} \in \mathbb{R}^3$ defined as $\boldsymbol{x} := x^i\, \boldsymbol{e}_i$ is embedded in the geometric algebra of conformal space $\mathbb{G}_{4,1}$ (see Sect. 4.3) as

$$\boldsymbol{u} := \mathcal{C}(\boldsymbol{x}) = \boldsymbol{x} + \frac{1}{2}\, \boldsymbol{x}^2\, \boldsymbol{e}_\infty + \boldsymbol{e}_o.$$

Given a Gaussian distributed random vector variable $\underline{\boldsymbol{x}}$ with a mean $\bar{\boldsymbol{x}} \in \mathbb{R}^3$ and a covariance matrix $\Sigma_{\boldsymbol{x},\boldsymbol{x}} \in \mathbb{R}^{3\times 3}$, the mean and the covariance matrix of $\mathcal{C}(\underline{\boldsymbol{x}})$ have to be evaluated. From (5.50), it follows that

$$\bar{\boldsymbol{u}} = \mathcal{E}\big(\mathcal{C}(\underline{\boldsymbol{x}})\big) = \mathcal{C}(\bar{\boldsymbol{x}}) + \sum_{i=1}^{3} \Sigma_{\boldsymbol{x},\boldsymbol{x}}^{i\,i}\, \boldsymbol{e}_\infty = \mathcal{C}(\bar{\boldsymbol{x}}) + \mathrm{tr}(\Sigma_{\boldsymbol{x},\boldsymbol{x}})\, \boldsymbol{e}_\infty\,, \tag{5.103}$$

where $\mathrm{tr}(\Sigma_{\boldsymbol{x},\boldsymbol{x}})$ denotes the trace of $\Sigma_{\boldsymbol{x},\boldsymbol{x}}$. The IPNS of the vector $\bar{\boldsymbol{u}}$ is therefore a sphere with an imaginary radius equal to the trace of $\Sigma_{\boldsymbol{x},\boldsymbol{x}}$ (see (4.46)). Depending on the actual application, the trace of $\Sigma_{\boldsymbol{x},\boldsymbol{x}}$ may be negligible compared with $\bar{\boldsymbol{x}}^2$, in which case the trace term may be neglected.

To evaluate the covariance matrix of $\mathcal{C}(\underline{\boldsymbol{x}})$, the Jacobi matrix of $\boldsymbol{u} = \mathcal{C}(\boldsymbol{x})$ has to be known. For this purpose, the component vector representation of $\boldsymbol{u}$ is used, i.e.

$$\mathsf{u} := \mathcal{K}(\boldsymbol{u}) = \left[x^1\,,\, x^2\,,\, x^3\,,\, \frac{1}{2}\, \boldsymbol{x}^2\,,\, 1\right]. \tag{5.104}$$

Note that the basis of u is not Euclidean. The operator $\mathcal{K}$ used here implicitly does not map $\boldsymbol{u}$ to a vector of dimension 2^n, but only maps those components that may be non-zero. This is equivalent to a mapping of the kind described in Sect. 5.1.3.

The Jacobi matrix of $\mathsf{u} := \mathcal{C}(\mathsf{x})$ is then given by

$$\mathsf{J}_{\mathcal{C}}(\mathsf{x}) := \partial_{\mathsf{x}} \mathcal{C}(\mathsf{x}) = \begin{bmatrix} 1 & 0 & 0 \\ 0 & 1 & 0 \\ 0 & 0 & 1 \\ x^1 & x^2 & x^3 \\ 0 & 0 & 0 \end{bmatrix}, \tag{5.105}$$

and thus

$$\Sigma_{\mathsf{u},\mathsf{u}} = \mathsf{J}_{\mathcal{C}}(\mathsf{x})\, \Sigma_{\mathsf{x},\mathsf{x}}\, \mathsf{J}_{\mathcal{C}}(\mathsf{x})^{\mathsf{T}} . \tag{5.106}$$

The inverse mapping is defined only for null vectors, i.e. 1-vectors in $\mathbb{G}_{4,1}$ that actually do represent a point in Euclidean space. For these types of vectors, the inverse mapping is basically the same as that for projective space. That is, given a vector $\boldsymbol{u} \in \mathbb{G}^1_{4,1}$ with $\boldsymbol{u}^2 = 0$ and $\boldsymbol{u} = u^1\, \boldsymbol{e}_1 + u^2\, \boldsymbol{e}_2 + u^3\, \boldsymbol{e}_3 + u^4\, \boldsymbol{e}_\infty + u^5\, \boldsymbol{e}_o$, the corresponding Euclidean vector $\boldsymbol{x}$ is given by

$$\boldsymbol{x} = \mathcal{C}^{-1}(\boldsymbol{u}) = \frac{u^1}{u^5}\, \boldsymbol{e}_1 + \frac{u^2}{u^5}\, \boldsymbol{e}_2 + \frac{u^3}{u^5}\, \boldsymbol{e}_3 .$$

The expectation of $\mathcal{C}^{-1}(\underline{\boldsymbol{u}})$ follows from (5.50) as

$$\bar{\boldsymbol{x}} = \sum_{i=1}^{3} \left(\frac{\bar{u}^i}{\bar{u}^5} - \left(\frac{1}{\bar{u}^5}\right)^2 \Sigma^{i\,5}_{\mathsf{u},\mathsf{u}} \right) \boldsymbol{e}_i \,. \tag{5.107}$$

If the covariances $\Sigma^{i\,5}_{\mathsf{u},\mathsf{u}}$ are sufficiently small, these terms may be neglected. The Jacobi matrix $\mathsf{J}_{\mathcal{C}^-}(\mathsf{u}) := \partial_{\mathsf{u}}\, \mathcal{C}(\mathsf{u})$ is given by

$$\mathsf{J}_{\mathcal{C}^-}(\mathsf{u}) = \partial_{\mathsf{u}}\, \mathcal{C}^{-1}(\mathsf{u}) = \frac{1}{u^5} \begin{bmatrix} 1\,0\,0\,0 & -\dfrac{1}{u^5} \\ 0\,1\,0\,0 & -\dfrac{1}{u^5} \\ 0\,0\,1\,0 & -\dfrac{1}{u^5} \end{bmatrix}, \tag{5.108}$$

so that

$$\Sigma_{\mathsf{x},\mathsf{x}} = \mathsf{J}_{\mathcal{C}^-}(\mathsf{u})\, \Sigma_{\mathsf{u},\mathsf{u}}\, \mathsf{J}_{\mathcal{C}^-}(\mathsf{u})^{\mathsf{T}} . \tag{5.109}$$

5.6.1 Blades and Operators

The goal of this subsection is to show that covariance matrices are an appropriate representation of the uncertainty of blades and operators in geometric algebra. The fundamental problem is, that although covariance matrices describe the uncertainty of an entity through a linear subspace, the subspace spanned by entities of the same type may not be linear.

For example, Heuel [93] described the evaluation of general homographies by writing the homography matrix H as a vector h and solving for it, given appropriate constraints. It is then also possible to evaluate a covariance matrix $\Sigma_{\mathsf{h},\mathsf{h}}$ for h. While this is fine for general homographies, Heuel also noted that it is problematic for constrained transformations such as rotations, since the necessary constraints on h are non-linear. The basic problem here is that the subspace of vectors h that represent rotation matrices is not linear. Hence, a covariance matrix for h is not well suited to describing the uncertainty of the corresponding rotation matrix. The question, therefore, is whether the representation of geometric entities and operators in geometric algebra allows a description of uncertainty via covariance matrices.

5.6.1.1 Points

The description of the uncertainty of a point via a covariance matrix in conformal space is not exact. This is due to the squared term in the embedding of a Euclidean point. That is, if $\boldsymbol{x} \in \mathbb{R}^3$, then the embedding in conformal space is

$$\boldsymbol{X} = \mathcal{C}(\boldsymbol{x}) = \boldsymbol{x} + \frac{1}{2}\boldsymbol{x}^2\,\boldsymbol{e}_\infty + \boldsymbol{e}_o\,.$$

As shown in Sect. 4.3.1, the embedded points lie on a cone, which implies that the embedded covariance matrices can represent only the *tangential* uncertainty at the embedded point. However, a homogeneous point of the form $\boldsymbol{X} \wedge \boldsymbol{e}_\infty$ does not contain a squared term anymore, since

$$\boldsymbol{X} \wedge \boldsymbol{e}_\infty = \boldsymbol{x} \wedge \boldsymbol{e}_\infty + \boldsymbol{e}_o \wedge \boldsymbol{e}_\infty\,,$$

and the corresponding covariance matrix describes the uncertainty exactly.

5.6.1.2 Blades

Because a covariance matrix only approximates the uncertainty of a point in conformal space, this is also the case for point pairs, circles, and spheres. In all these cases, the covariance matrix is tangential to the actual subspace that the entity lies in. The situation is different for those blades that represent homogeneous points, lines, and planes, since in these instances the squared terms are canceled.

As shown earlier, the uncertainty of homogeneous points can be appropriately represented by a covariance matrix. The same is true for planes, as can be seen by considering the inner-product representation of a plane $\boldsymbol{P}$, defined by

$$\boldsymbol{P} := \boldsymbol{a} + \alpha\, \boldsymbol{e}_\infty ,$$

where $\boldsymbol{a}$ is the normal of the plane, and α is the plane's orthogonal separation from the origin. A random plane variable may therefore be written as

$$\underline{\boldsymbol{P}} := (\bar{\boldsymbol{a}} + \bar{\alpha}\, \boldsymbol{e}_\infty) + (\underline{\boldsymbol{n}} + \underline{\nu}\, \boldsymbol{e}_\infty),$$

where $\underline{\boldsymbol{n}}$ and $\underline{\nu}$ are zero-mean random variables. The uncertainty of $\underline{\boldsymbol{P}}$ is therefore linear in the zero-mean random variables, and, thus, a covariance matrix is an appropriate representation of the uncertainty of the plane.

Next, consider a line $\boldsymbol{L}$, which may be represented in conformal space as $\boldsymbol{L} = \boldsymbol{X} \wedge \boldsymbol{Y} \wedge \boldsymbol{e}_\infty$ (see Table 4.3). The six components of $\boldsymbol{L}$ are the well-known Plücker coordinates, which have to satisfy the Plücker condition in order to describe a line. In geometric algebra, the Plücker condition is equivalent to demanding that $\boldsymbol{L}$ is a blade, i.e. that it can be factorized into the outer product of three vectors. This is equivalent to demanding that $\boldsymbol{L}\,\widetilde{\boldsymbol{L}} \in \mathbb{R}$.

To describe the uncertainty of a line $\boldsymbol{L}$ with a covariance matrix, the sum of the component vector of $\boldsymbol{L}$ with any component vector in the linear subspace spanned by the covariance matrix should satisfy the Plücker condition. We will show in the following that this is not the case, in general. A covariance matrix can therefore only approximate the uncertainty of a line.

Consider two random vector variables $\underline{\boldsymbol{x}} := \bar{\boldsymbol{x}} + \underline{\boldsymbol{n}}$ and $\underline{\boldsymbol{y}} := \bar{\boldsymbol{y}} + \underline{\boldsymbol{m}}$. Here, $\bar{\boldsymbol{x}}, \bar{\boldsymbol{y}} \in \mathbb{R}^3$ are the expectation values of $\underline{\boldsymbol{x}}$ and $\underline{\boldsymbol{y}}$, respectively, and $\underline{\boldsymbol{n}}$ and $\underline{\boldsymbol{m}}$ are Gaussian distributed, zero-mean, random vector variables. If we define $\underline{\boldsymbol{X}} := \mathcal{C}(\underline{\boldsymbol{x}})$ and $\underline{\boldsymbol{Y}} := \mathcal{C}(\underline{\boldsymbol{y}})$, it follows that

$$\underline{\boldsymbol{X}} \wedge \boldsymbol{e}_\infty = (\bar{\boldsymbol{x}} + \underline{\boldsymbol{n}}) \wedge \boldsymbol{e}_\infty + \boldsymbol{e}_o \wedge \boldsymbol{e}_\infty , \quad \underline{\boldsymbol{Y}} \wedge \boldsymbol{e}_\infty = (\bar{\boldsymbol{y}} + \underline{\boldsymbol{m}}) \wedge \boldsymbol{e}_\infty + \boldsymbol{e}_o \wedge \boldsymbol{e}_\infty .$$

Hence, $\underline{\boldsymbol{L}} := \underline{\boldsymbol{X}} \wedge \underline{\boldsymbol{Y}} \wedge \boldsymbol{e}_\infty$ can be expressed as

$$\begin{aligned}
\underline{\boldsymbol{L}} &= (\bar{\boldsymbol{x}} + \underline{\boldsymbol{n}}) \wedge (\bar{\boldsymbol{y}} + \underline{\boldsymbol{m}}) \wedge \boldsymbol{e}_\infty + \big((\bar{\boldsymbol{x}} + \underline{\boldsymbol{n}}) - (\bar{\boldsymbol{y}} + \underline{\boldsymbol{m}})\big) \wedge \boldsymbol{e}_o \wedge \boldsymbol{e}_\infty \\
&= \quad \big(\bar{\boldsymbol{x}} \wedge \bar{\boldsymbol{y}} \wedge \boldsymbol{e}_\infty + (\bar{\boldsymbol{x}} - \bar{\boldsymbol{y}}) \wedge \boldsymbol{e}_o \wedge \boldsymbol{e}_\infty\big) \\
&\quad + (\bar{\boldsymbol{y}} \wedge \boldsymbol{e}_\infty + \boldsymbol{e}_o \wedge \boldsymbol{e}_\infty) \wedge \underline{\boldsymbol{n}} \\
&\quad - (\bar{\boldsymbol{x}} \wedge \boldsymbol{e}_\infty + \boldsymbol{e}_o \wedge \boldsymbol{e}_\infty) \wedge \underline{\boldsymbol{m}} \\
&\quad + \underline{\boldsymbol{n}} \wedge \underline{\boldsymbol{m}} \wedge \boldsymbol{e}_\infty .
\end{aligned}$$

This shows that the random line variable, $\underline{\boldsymbol{L}}$, can be written as the sum of an expectation part and a zero-mean random-variable part. However, the random variable part is not linear in the random variables $\underline{\boldsymbol{n}}$ and $\underline{\boldsymbol{m}}$, because of the term $\underline{n} \wedge \underline{m} \wedge e_\infty$. A covariance matrix can therefore only approximate the uncertainty of a line, in general. In the special case, where $\underline{L}$ is constructed from one random variable and one constant vector, i.e. $\underline{L} := \underline{X} \wedge Y \wedge e_\infty$, a covariance matrix is an exact representation of the uncertainty of the line.

5.6.1.3 Versors

The operators that are of main interest in geometric algebra are versors. By definition, versors are geometric products of a number of 1-vectors (see Sect. 3.3). For example, a general rotation may be constructed from the geometric product of two vectors in conformal space that are inner-product representations of planes (see Sect. 4.3.4). The rotation axis, in this case, is the intersection line of the two planes.

Let $\underline{P} = \bar{P} + \underline{N}$ and $\underline{Q} = \bar{Q} + \underline{M}$ denote two random plane variables, where $\bar{P}, \bar{Q} \in \mathbb{G}^1_{4,1}$ are the expectation values of $\underline{P}$ and $\underline{Q}$, respectively, and $\underline{N}, \underline{M}$ are Gaussian distributed, zero-mean, random plane variables. A random general rotor variable $\underline{R}$ is given by

$$\underline{R} := \underline{P}\,\underline{Q} = (\bar{P} + \underline{N})\,(\bar{Q} + \underline{M}) = \bar{P}\,\bar{Q} + \bar{P}\,\underline{M} + \underline{N}\,\bar{Q} + \underline{N}\,\underline{M}\,.$$

Just as for lines, a general random rotor variable can be written as the sum of a constant general rotor and a random variable. However, the random variable part is not linear in the random variables $\underline{N}$ and $\underline{M}$, due to the term $\underline{N}\,\underline{M}$. Hence, a covariance matrix is only an approximation of the uncertainty of a general rotor.

Nevertheless, it is quite instructive to visualize the uncertainty of a general rotor, as approximated by its covariance matrix. We constructed an uncertain general rotor by evaluating the geometric product of two uncertain planes, and obtained the rotor's covariance matrix by error propagation. It turns out that the covariance matrix of the rotor can be of rank six at most. The effect on the rotation operation when such an uncertain rotor is transformed separately along the six eigenvectors of its covariance matrix is shown in Fig. 5.2.

Despite the fact, that covariance matrices are only approximations of the uncertainty of random rotor variables, this method of representing uncertain rotations appears to be more robust than using rotation matrices. This was found in synthetic experiments in Sect. 6.2.6, as well as in the pose estimation procedure, that will be presented in Chap. 8.

Furthermore, note that the subalgebra of rotors for rotations about the origin is isomorphic to quaternion algebra (see Sect. 3.8.3) and the subalgebra of motors is isomorphic to the algebra of dual quaternions [34, 40]. Compared with quaternions and dual quaternions, not only does the geometric algebra of conformal space allow the description of the operators themselves, but the operators can also be applied to any geometric entity that can be expressed

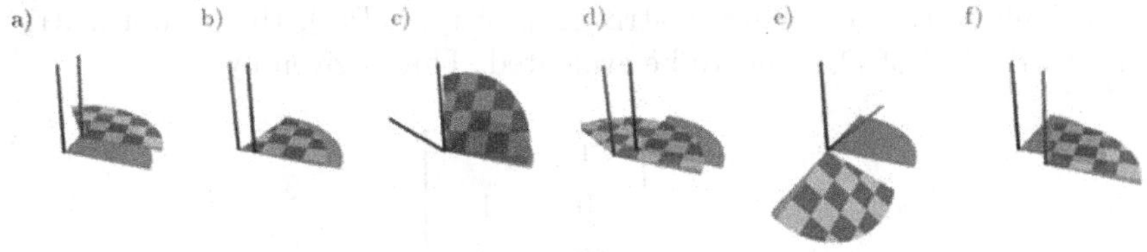

Fig. 5.2 Effect of adding each of the six eigenvectors of the covariance matrix of a general rotor to the rotor's component vector. In each of the images, the darker rotor is the initial one

in the algebra. In contrast, when quaternions are used, only points can be represented by pure quaternions (i.e. quaternions with no scalar part), and in the case of dual quaternions, only lines can be represented.

5.7 Uncertainty in Conic Space

Conic space was introduced in Sect. 4.4. It is basically a vector space over symmetric 3×3 matrices. The embedding of a vector $\mathsf{x} \in \mathbb{R}^2$ in $\mathbb{R}^6$ was defined in (4.84) as

$$\mathcal{D}(\mathsf{x}) = \left[x^1 ,\; x^2 ,\; \frac{1}{\sqrt{2}} ,\; \frac{1}{\sqrt{2}} (x^1)^2 ,\; \frac{1}{\sqrt{2}} (x^2)^2 ,\; (x^1\, x^2) \right]^{\mathsf{T}} ,$$

where $\mathsf{x} = \left[x^1 ,\; x^2 \right]$. With the help of (5.50), the mean value of $\bar{\mathsf{u}} := \mathcal{D}(\underline{\mathsf{x}})$ can be evaluated as

$$\bar{\mathsf{u}} = \begin{bmatrix} \bar{x}^1 \\ \bar{x}^2 \\ \frac{1}{\sqrt{2}} \\ \frac{1}{\sqrt{2}} (\bar{x}^1)^2 \\ \frac{1}{\sqrt{2}} (\bar{x}^2)^2 \\ \bar{x}^1\, \bar{x}^2 \end{bmatrix} + \begin{bmatrix} 0 \\ 0 \\ 0 \\ \frac{1}{\sqrt{2}}\, \Sigma^{1\,1}_{\mathsf{x},\mathsf{x}} \\ \frac{1}{\sqrt{2}}\, \Sigma^{2\,2}_{\mathsf{x},\mathsf{x}} \\ \Sigma^{1\,2}_{\mathsf{x},\mathsf{x}} \end{bmatrix} . \tag{5.110}$$

Again, if the covariances in $\Sigma_{\mathsf{x},\mathsf{x}}$ are much smaller than the components of x, then they can be neglected in the above equation. Just as for conformal space, if the covariance terms are not neglected, then $\bar{\mathsf{u}}$ does not represent a point in its IPNS.

To evaluate the covariance matrix $\Sigma_{\mathsf{u},\mathsf{u}}$ of $\underline{\mathsf{u}} := \mathcal{D}(\underline{\mathsf{x}})$, the Jacobi matrix $\mathsf{J}_{\mathcal{D}}(\mathsf{x}) := \partial_{\mathsf{x}} \mathcal{D}(\mathsf{x})$ of $\mathcal{D}(\mathsf{x})$ has to be evaluated. This is given by

$$\mathsf{J}_{\mathcal{D}}(\mathsf{x}) = \begin{bmatrix} 1 & 0 \\ 0 & 1 \\ 0 & 0 \\ \sqrt{2}\, x^1 & 0 \\ 0 & \sqrt{2}\, x^2 \\ x^2 & x^1 \end{bmatrix}, \tag{5.111}$$

so that

$$\Sigma_{\mathsf{u},\mathsf{u}} = \mathsf{J}_{\mathcal{D}}(\mathsf{x})\; \Sigma_{\mathsf{x},\mathsf{x}}\; \mathsf{J}_{\mathcal{D}}(\mathsf{x})\,. \tag{5.112}$$

The inverse mapping $\mathcal{D}^{-1}(\mathsf{u})$ is defined only for vectors in $\mathbb{R}^6$ that represent points in their IPNS in $\mathbb{R}^2$. Given a vector $\mathsf{u} = \left[\, u^1,\; u^2,\; u^3,\; u^4,\; u^5,\; u^6 \,\right]^{\mathsf{T}}$ that satisfies this condition, the inverse mapping $\mathcal{D}^{-1}(\mathsf{u})$ is given by

$$\mathsf{x} = \mathcal{D}^{-1}(\mathsf{u}) = \left[\, \frac{u^1}{u^3}\,,\; \frac{u^2}{u^3} \,\right]^{\mathsf{T}}. \tag{5.113}$$

The expectation $\bar{\mathsf{x}} = \mathcal{E}\big(\mathcal{D}^{-1}(\underline{\mathsf{u}})\big)$ is then given by

$$\bar{\mathsf{x}} = \mathcal{E}\big(\mathcal{D}^{-1}(\underline{\mathsf{u}})\big) = \begin{bmatrix} \dfrac{\bar{u}^1}{\bar{u}^3} \\ \dfrac{\bar{u}^2}{\bar{u}^3} \end{bmatrix} - \left(\frac{1}{\bar{u}^3}\right)^2 \begin{bmatrix} \Sigma^{1\,3}_{\mathsf{u},\mathsf{u}} \\ \Sigma^{2\,3}_{\mathsf{u},\mathsf{u}} \end{bmatrix}. \tag{5.114}$$

The covariance matrix $\Sigma_{\mathsf{x},\mathsf{x}}$ of $\underline{\mathsf{x}} := \mathcal{D}^{-1}(\underline{\mathsf{u}})$ is evaluated with the help of the Jacobi matrix of $\mathcal{D}^{-1}(\mathsf{u})$, which is given by

$$\mathsf{J}_{\mathcal{D}^-}(\mathsf{u}) := \partial_{\mathsf{u}} \mathcal{D}^{-1}(\mathsf{u}) = \frac{1}{u^3} \begin{bmatrix} 1\;0\; -\dfrac{1}{u^3}\;0\;0 \\ 0\;1\; -\dfrac{1}{u^3}\;0\;0 \end{bmatrix}, \tag{5.115}$$

so that

$$\Sigma_{\mathsf{x},\mathsf{x}} = \mathsf{J}_{\mathcal{D}^-}(\mathsf{u})\; \Sigma_{\mathsf{u},\mathsf{u}}\; \mathsf{J}_{\mathcal{D}^-}(\mathsf{u})^{\mathsf{T}}\,. \tag{5.116}$$

5.8 The Gauss–Markov Model

The Gauss–Markov model is a standard linear stochastic model for least-squares estimation (see e.g. [104]). In this section, a least-squares estimation method based on this model is presented.

The fundamental setup of the problem is as follows. Let $\{\mathsf{b}_i\} \subset \mathbb{R}^{N_{b_i}}$ denote a set of M observations or data vectors, and let $\mathsf{p} \in \mathbb{R}^{N_p}$ denote a parameter vector. In addition, let $\mathsf{q}_i : \mathbb{R}^{N_{b_i}} \times \mathbb{R}^{N_p} \to \mathbb{R}^{N_{g_i}}$ be a constraint or query function that satisfies $\mathsf{q}_i(\mathsf{b}_i, \mathsf{p}) = 0$ if and only if the parameter vector p is chosen such that it satisfies the given constraint. For example, q_i could be a function that results in zero if a line represented by p passes through a point represented by b_i. Note that different data vectors may be of different dimensions and have different corresponding constraint functions q_i. This could, for example, be the case if we are seeking a line represented by p that passes through some given points and intersects some given lines. The two constraints need different constraint functions and data vectors.

Given a set of constraint equations q_i and corresponding data points $\{\mathsf{b}_i\}$, the goal is to find the parameter vector p such that all constraints are satisfied, or at least minimized.

5.8.1 Linearization

Suppose an initial estimate $\widehat{\mathsf{p}}$ of the true parameter vector p is given, so that $\mathsf{p} = \widehat{\mathsf{p}} + \Delta\mathsf{p}$. The goal then becomes to estimate the correct $\Delta\mathsf{p}$.

The Gauss-Markov model is a linear stochastic model and thus the first step is to linearize the constraint functions $\mathsf{q}_i(\mathsf{b}_i, \mathsf{p})$, which can be achieved via a Taylor expansion about $\widehat{\mathsf{p}}$ up to order one:

$$\mathsf{q}_i(\mathsf{b}_i, \mathsf{p}) = \mathsf{q}_i(\mathsf{b}_i, \widehat{\mathsf{p}} + \Delta\mathsf{p}) \approx \mathsf{q}_i(\mathsf{b}_i, \widehat{\mathsf{p}}) + (\partial_{\mathsf{p}}\, \mathsf{q}_i)(\mathsf{b}_i, \widehat{\mathsf{p}})\, \Delta\mathsf{p}. \tag{5.117}$$

For brevity, the following definitions are made:

$$\mathsf{c}_{q_i} := -\mathsf{q}_i(\mathsf{b}_i, \widehat{\mathsf{p}}) \qquad \text{and} \qquad \mathsf{U}_i := (\partial_{\mathsf{p}}\, \mathsf{q}_i)(\mathsf{b}_i, \widehat{\mathsf{p}}) \;. \tag{5.118}$$

Note that the Jacobi matrix U_i depends on b_i and $\widehat{\mathsf{p}}$ in general. All of the c_{q_i} and U_i can be combined into

$$\mathsf{A} := \begin{bmatrix} \mathsf{U}_1 \\ \vdots \\ \mathsf{U}_M \end{bmatrix} \qquad \text{and} \qquad \mathsf{c}_q := \begin{bmatrix} \mathsf{c}_{q1} \\ \vdots \\ \mathsf{c}_{qM} \end{bmatrix}, \tag{5.119}$$

so that all linearized constraint equations can be combined into the equation

$$\mathsf{A}\, \Delta\mathsf{p} = \mathsf{c}_q \;. \tag{5.120}$$

The immediate least-squares solution to this equation is

$$\Delta\mathsf{p} = \left(\mathsf{A}^\mathsf{T}\, \mathsf{A}\right)^{-1} \mathsf{A}^\mathsf{T}\, \mathsf{c}_q. \tag{5.121}$$

However, in general, the problem is somewhat more complicated.

5.8.2 Constraints on Parameters Alone

In many problems there is, in addition to the constraint functions q_i between the data vectors and the parameter vector, also a constraint function $\mathsf{h} : \mathbb{R}^{N_p} \to \mathbb{R}^{N_h}$ on the parameter vector p alone. For example, this occurs if p contains the six Plücker coordinates of a line in 3D space. A 6D vector represents a line in this way only if it also satisfies the Plücker condition. This constraint may then be modeled by a function $\mathsf{h}(\mathsf{p})$, such that $\mathsf{h}(\mathsf{p}) = 0$ if and only if p satisfies the Plücker condition.

To include this additional constraint on the parameters alone, h has to be linearized as well:

$$\mathsf{h}(\mathsf{p}) = \mathsf{h}(\widehat{\mathsf{p}} + \Delta\mathsf{p}) \approx \mathsf{h}(\widehat{\mathsf{p}}) + (\partial_{\mathsf{p}}\, \mathsf{h})(\widehat{\mathsf{p}})\; \Delta\mathsf{p}\,. \tag{5.122}$$

If we define

$$\mathsf{c}_h := -\mathsf{h}(\widehat{\mathsf{p}}) \qquad \text{and} \qquad \mathsf{H} := (\partial_{\mathsf{p}}\, \mathsf{h})(\widehat{\mathsf{p}})^{\mathsf{T}}\,, \tag{5.123}$$

the constraint $\mathsf{h}(\mathsf{p}) = 0$ can be approximated by

$$\mathsf{H}^{\mathsf{T}}\, \Delta\mathsf{p} = \mathsf{c}_h\,. \tag{5.124}$$

5.8.3 Least-Squares Estimation

The constraint equations $\mathsf{q}_i(\mathsf{b}_i,\, \mathsf{p}) = 0$ and $\mathsf{h}(\mathsf{p}) = 0$ could now be combined into a single equation system,

$$\begin{bmatrix} \mathsf{A} \\ \mathsf{H} \end{bmatrix} \Delta\mathsf{p} = \begin{bmatrix} \mathsf{c}_q \\ \mathsf{c}_h \end{bmatrix}\,, \tag{5.125}$$

which could be solved as before. The problem with this approach is that while it suffices that the expression $\mathsf{A}\,\Delta\mathsf{p} - \mathsf{c}_q$ is minimized, the constraint $\mathsf{H}^{\mathsf{T}}\,\Delta\mathsf{p} = \mathsf{c}_h$ must be satisfied exactly. For example, if p is a 6D vector of Plücker coordinates, it has to be ensured that the Plücker condition is satisfied exactly, whereas the line need not pass exactly through all data points. Therefore, the method of Lagrange multipliers is employed.

Instead of minimizing $\|\mathsf{A}\,\Delta\mathsf{p} - \mathsf{c}_q\|^2$, a new function Φ is defined, which takes its minimum if $\|\mathsf{A}\,\Delta\mathsf{p} - \mathsf{c}_q\|^2$ is minimal *and* $\mathsf{H}^{\mathsf{T}}\,\Delta\mathsf{p} - \mathsf{c}_h = 0$:

$$\Phi(\Delta\mathsf{p},\, \mathsf{m}) := \frac{1}{2}\,(\mathsf{A}\,\Delta\mathsf{p} - \mathsf{c}_q)^{\mathsf{T}}\,(\mathsf{A}\,\Delta\mathsf{p} - \mathsf{c}_q) + (\mathsf{H}^{\mathsf{T}}\,\Delta\mathsf{p} - \mathsf{c}_h)^{\mathsf{T}}\,\mathsf{m}\,, \tag{5.126}$$

where $\mathsf{m} \in \mathbb{R}^{N_h}$ is a vector of Lagrange multipliers, which are regarded as additional parameters. The function $\Phi(\Delta\mathsf{p}, \mathsf{m})$ attains a minimum if its derivatives with respect to $\Delta\mathsf{p}$ and m are zero. Its derivative with respect to m is zero if the constraint on the parameters is satisfied, i.e. $\mathsf{H}^\mathsf{T}\,\Delta\mathsf{p} - \mathsf{c}_h = 0$. The derivatives are, explicitly,

$$\begin{aligned}(\partial_{\Delta\mathsf{p}}\,\Phi)(\Delta\mathsf{p},\, \mathsf{m}) &= \mathsf{A}^\mathsf{T}\,\mathsf{A}\,\Delta\mathsf{p} + \mathsf{H}\,\mathsf{m} - \mathsf{A}^\mathsf{T}\,\mathsf{c}_q\,, \\ (\partial_{\mathsf{m}}\,\Phi)(\Delta\mathsf{p},\, \mathsf{m}) &= \mathsf{H}^\mathsf{T}\,\Delta\mathsf{p} - \mathsf{c}_h\,.\end{aligned} \tag{5.127}$$

Both of these equations have to be zero when $\Phi(\Delta\mathsf{p}, \mathsf{m})$ is minimal (or maximal). This can be written in a single equation system,

$$\begin{bmatrix} \mathsf{A}^\mathsf{T}\,\mathsf{A} & \mathsf{H} \\ \mathsf{H}^\mathsf{T} & 0 \end{bmatrix} \begin{bmatrix} \Delta\mathsf{p} \\ \mathsf{m} \end{bmatrix} = \begin{bmatrix} \mathsf{A}^\mathsf{T}\,\mathsf{c}_q \\ \mathsf{c}_h \end{bmatrix}\,. \tag{5.128}$$

If we define $\mathsf{N} := \mathsf{A}^\mathsf{T}\,\mathsf{A}$ and $\mathsf{c}_n := \mathsf{A}^\mathsf{T}\,\mathsf{c}_q$, this can be written in the standard form

$$\begin{bmatrix} \mathsf{N} & \mathsf{H} \\ \mathsf{H}^\mathsf{T} & 0 \end{bmatrix} \begin{bmatrix} \Delta\mathsf{p} \\ \mathsf{m} \end{bmatrix} = \begin{bmatrix} \mathsf{c}_n \\ \mathsf{c}_h \end{bmatrix}\,. \tag{5.129}$$

This equation system is often called the *normal equations*. If the matrix on the left is invertible, then this equation can be solved directly for $\Delta\mathsf{p}$. Otherwise, the pseudoinverse may be used.

Similarly to the derivation of the covariance matrix $\Sigma_{\mathsf{p},\mathsf{p}}$ in Sect. 5.9.3, it may be shown that for the Gauss–Markov model also,

$$\begin{bmatrix} \Sigma_{\mathsf{p},\mathsf{p}} & \cdot \\ \cdot & \cdot \end{bmatrix} = \begin{bmatrix} \mathsf{N} & \mathsf{H} \\ \mathsf{H}^\mathsf{T} & 0 \end{bmatrix}^{-1}\,, \tag{5.130}$$

where the dots indicate non-zero parts of the matrix.

Often, a weighting of the least-squares constraint is known. That is, the term to be minimized is given by

$$\|(\mathsf{A}\,\Delta\mathsf{p} - \mathsf{c}_q)^\mathsf{T}\,\mathsf{W}\,(\mathsf{A}\,\Delta\mathsf{p} - \mathsf{c}_q)\|^2\,, \tag{5.131}$$

where W is a weight matrix, which may be interpreted as the inverse covariance matrix of the residual of the Taylor expansion in (5.117). In this case, the matrix N and the vector c_n become

$$\mathsf{N} = \mathsf{A}^\mathsf{T}\,\mathsf{W}\,\mathsf{A} \qquad \text{and} \qquad \mathsf{c}_n = \mathsf{A}^\mathsf{T}\,\mathsf{W}\,\mathsf{c}_q\,. \tag{5.132}$$

5.8.4 Numerical Calculation

To solve the normal equation system given in (5.129), the matrix $\mathsf{N} = \mathsf{A}^\mathsf{T}\,\mathsf{A}$ and the vector $\mathsf{c}_n = \mathsf{A}^\mathsf{T}\,\mathsf{c}_q$ have to be evaluated. From (5.119), it follows that

$$\mathsf{N} = \left[\mathsf{U}_1^\mathsf{T},\, \cdots,\, \mathsf{U}_M^\mathsf{T}\right] \begin{bmatrix} \mathsf{U}_1 \\ \cdots \\ \mathsf{U}_M \end{bmatrix} = \sum_{i=1}^{M} \mathsf{U}_i^\mathsf{T}\,\mathsf{U}_i \tag{5.133}$$

and

$$\mathsf{c}_n = \left[\mathsf{U}_1^\mathsf{T},\, \cdots,\, \mathsf{U}_M^\mathsf{T}\right] \begin{bmatrix} \mathsf{c}_{q_1} \\ \vdots \\ \mathsf{c}_{q_M} \end{bmatrix} = \sum_{i=1}^{M} \mathsf{U}_i^\mathsf{T}\,\mathsf{c}_{q_i}\,. \tag{5.134}$$

If the initial constraint functions q_i are linear, then the solution for $\Delta\mathsf{p}$ in (5.129) is provably its best linear unbiased estimator (BLUE). However, in most problems the constraint functions q_i are multilinear or, more generally, non-linear. In this case a solution may be obtained by updating the initial estimate $\widehat{\mathsf{p}}$ with the estimated $\Delta\mathsf{p}$ and using the result as a new starting value for a new iteration step.

5.8.5 Generalization

Instead of regarding the parameter vector p as completely free, it may also be regarded as a random vector variable with an associated covariance matrix $\Sigma_{\mathsf{p},\mathsf{p}}$. In this case, the problem is to find the parameter vector p that minimizes $\|\mathsf{A}\,\Delta\mathsf{p} - \mathsf{c}_q\|^2$, satisfies $\mathsf{H}^\mathsf{T}\,\Delta\mathsf{p} - \mathsf{c}_h = 0$ but only differs minimally from an initial estimate $\widehat{\mathsf{p}}$ according to its covariance matrix. The function $\Phi(\Delta\mathsf{p},\, \mathsf{m})$ therefore becomes

$$\begin{aligned} \Phi(\Delta\mathsf{p},\, \mathsf{m}) = & \frac{1}{2}\,\Delta\mathsf{p}^\mathsf{T}\,\Sigma_{\mathsf{p},\mathsf{p}}^{-1}\,\Delta\mathsf{p} \\ & +\frac{1}{2}\,(\mathsf{A}\,\Delta\mathsf{p} - \mathsf{c}_q)^\mathsf{T}\,(\mathsf{A}\,\Delta\mathsf{p} - \mathsf{c}_q) \\ & +(\mathsf{H}^\mathsf{T}\,\Delta\mathsf{p} - \mathsf{c}_h)^\mathsf{T}\,\mathsf{m}\,. \end{aligned} \tag{5.135}$$

Components of $\Delta\mathsf{p}$ that have a large variance may therefore vary by a larger amount than components with a small variance. The term $\Delta\mathsf{p}^\mathsf{T}\,\Sigma_{\mathsf{p},\mathsf{p}}^{-1}\,\Delta\mathsf{p}$ may therefore be regarded as a *regularization* term. If $\Sigma_{\mathsf{p},\mathsf{p}}$ is a diagonal matrix, then zero entries on the diagonal of $\Sigma_{\mathsf{p},\mathsf{p}}^{-1}$ relate to free parameters in $\Delta\mathsf{p}$, and entries with very large values relate to fixed parameters. This offers a very convenient method to dynamically select which parameters should be estimated without changing the actual minimization equations.

For example, p may contain parameters that describe various properties of a camera, such as its focal length and orientation. Then, by simply changing the appropriate values in $\Sigma_{\mathsf{p},\mathsf{p}}$, one can use the minimization method to estimate the focal length, or regard it as fixed.

By calculating the derivatives of $\Phi(\Delta\mathsf{p}, \mathsf{m})$ from (5.135), we find the matrix N and vector c_n of the normal equation system of (5.129) to be

$$\mathsf{N} = \Sigma_{\mathsf{p},\mathsf{p}}^{-1} + \mathsf{A}^{\mathsf{T}}\,\mathsf{A} \qquad \text{and} \qquad \mathsf{c}_n = \mathsf{A}^{\mathsf{T}}\,\mathsf{c}_q\,. \tag{5.136}$$

5.9 The Gauss–Helmert Model

The *Gauss–Helmert* model was introduced by Helmert in 1872 ([86], p.215), as the general case of least-squares adjustment. It is also called the *mixed model* [104]. The Gauss–Helmert model is a linear, stochastic model, which will not itself be discussed here. Rather, a least-squares estimation technique which is based on this model will be described.

The Gauss–Helmert model extends the Gauss–Markov model described in the previous section by adjusting not only the parameters but also the data vectors. The constraints between parameters and data vectors are again given by functions $\mathsf{q}_i(\mathsf{b}_i, \mathsf{p})$ that are zero if the constraint is satisfied. The idea is now not to minimize $\|\mathsf{q}_i(\mathsf{b}_i, \mathsf{p})\|^2$ but to minimally vary the $\{\,\mathsf{b}_i\,\}$ such that all $\mathsf{q}_i(\mathsf{b}_i, \mathsf{p})$ are exactly zero. The advantage of this approach is that the covariance matrices of the data vectors can be taken into account. That is, the data vectors can be varied more in directions of high variance than in directions of low variance. This method also allows the estimation of a covariance matrix for the parameter vector, which reflects not only the uncertainties of the separate data points but also the distribution of the data.

5.9.1 The Constraints

5.9.1.1 Constraints on the Observations

We denote the vector of parameters by $\mathsf{p} \in \mathbb{R}^{N_p}$. It is assumed that there are M observations. The ith observation is represented by the vector $\mathsf{b}_i \in \mathbb{R}^{N_{b_i}}$. Note that the observation vectors for different observations can be of different dimensions. For each observation, there exists a constraint of the form

$$\mathsf{q}_i(\mathsf{b}_i, \mathsf{p}) = 0\ \in \mathbb{R}^{N_{q_i}}; \tag{5.137}$$

that is, the dimension of q_i can also differ for different observations. Let

$$\mathsf{b}_i = \widehat{\mathsf{b}}_i + \Delta\mathsf{b}_i\,, \qquad \mathsf{p} = \widehat{\mathsf{p}} + \Delta\mathsf{p}, \tag{5.138}$$

where $\widehat{\mathsf{b}}_i$ is the measured data vector, b_i is the true data vector, and $\Delta\mathsf{b}_i$ gives the error in the estimate. Similarly, $\widehat{\mathsf{p}}$ is an estimate of the true p, and $\Delta\mathsf{p}$ is the sought adjustment of $\widehat{\mathsf{p}}$. The constraint functions q_i are now linearized by means of a Taylor expansion up to order one about $(\widehat{\mathsf{b}}_i, \widehat{\mathsf{p}})$,

$$\begin{aligned} \mathsf{q}_i(\mathsf{b}_i, \mathsf{p}) &= \mathsf{q}_i(\widehat{\mathsf{b}}_i + \Delta\mathsf{b}_i, \widehat{\mathsf{p}} + \Delta\mathsf{p}) \\ &\approx \mathsf{q}_i(\widehat{\mathsf{b}}, \widehat{\mathsf{p}}) \\ &\quad + (\partial_{\mathsf{b}_i} \mathsf{q}_i)(\widehat{\mathsf{b}}_i, \widehat{\mathsf{p}})\, \Delta\mathsf{b}_i \\ &\quad + (\partial_{\mathsf{p}} \mathsf{q}_i)(\widehat{\mathsf{b}}_i, \widehat{\mathsf{p}})\, \Delta\mathsf{p}\,. \end{aligned} \tag{5.139}$$

The Jacobi matrices of q_i are written as

$$\begin{aligned} \underset{N_{q_i}\times N_{\mathsf{b}_i}}{\mathsf{V}_i(\widehat{\mathsf{p}})} &:= (\partial_{\mathsf{b}_i} \mathsf{q}_i)(\widehat{\mathsf{b}}_i, \widehat{\mathsf{p}}), \\ \underset{N_{q_i}\times N_p}{\mathsf{U}_i(\widehat{\mathsf{b}}_i)} &:= (\partial_{\mathsf{p}} \mathsf{q}_i)(\widehat{\mathsf{b}}_i, \widehat{\mathsf{p}}), \\ \mathsf{c}_{q_i} &:= -\mathsf{q}_i(\widehat{\mathsf{b}}_i, \widehat{\mathsf{p}}). \end{aligned} \tag{5.140}$$

The linearized constraint equation then becomes

$$\underset{N_{q_i}\times N_p}{\mathsf{U}_i(\widehat{\mathsf{b}}_i)} \;\; \underset{N_p\times 1}{\Delta\mathsf{p}} \; + \; \underset{N_{q_i}\times N_{\mathsf{b}_i}}{\mathsf{V}_i(\widehat{\mathsf{p}})} \;\; \underset{N_{\mathsf{b}_i}\times 1}{\Delta\mathsf{b}_i} \;=\; \underset{N_{q_i}\times 1}{\mathsf{c}_{q_i}}\,. \tag{5.141}$$

All these linearized constraint equations may be combined into a single equation system as follows:

$$\underset{(\sum_i N_{q_i})\times N_p}{\begin{bmatrix} \mathsf{U}_1(\widehat{\mathsf{b}}_1) \\ \vdots \\ \mathsf{U}_M(\widehat{\mathsf{b}}_M) \end{bmatrix}} \; \underset{N_p\times 1}{\Delta\mathsf{p}} + \underset{(\sum_i N_{q_i})\times(\sum_i N_{\mathsf{b}_i})}{\begin{bmatrix} \mathsf{V}_1(\widehat{\mathsf{p}}) & & 0 \\ & \ddots & \\ 0 & & \mathsf{V}_M(\widehat{\mathsf{p}}) \end{bmatrix}} \; \underset{(\sum_i N_{\mathsf{b}_i})\times 1}{\begin{bmatrix} \Delta\mathsf{b}_1 \\ \vdots \\ \Delta\mathsf{b}_M \end{bmatrix}} = \underset{(\sum_i N_{q_i})\times 1}{\begin{bmatrix} \mathsf{c}_{g1} \\ \vdots \\ \mathsf{c}_{gM} \end{bmatrix}}\,. \tag{5.142}$$

For brevity, the following definitions are made:

$$N_q := \sum_{i=1}^{M} N_{q_i}\,, \qquad N_{\mathsf{b}} := \sum_{i=1}^{M} N_{\mathsf{b}_i}\,, \tag{5.143}$$

$$\underset{N_q \times N_p}{\mathsf{A}} := \begin{bmatrix} \mathsf{U}_1(\widehat{\mathsf{b}}_1) \\ \vdots \\ \mathsf{U}_M(\widehat{\mathsf{b}}_M) \end{bmatrix}, \qquad \underset{N_q \times N_{\mathsf{b}}}{\mathsf{B}^\mathsf{T}} := \begin{bmatrix} \mathsf{V}_1(\widehat{\mathsf{p}}) & & 0 \\ & \ddots & \\ 0 & & \mathsf{V}_M(\widehat{\mathsf{p}}) \end{bmatrix}, \tag{5.144}$$

$$\underset{N_{\mathsf{b}} \times 1}{\Delta\mathsf{b}} := \begin{bmatrix} \Delta\mathsf{b}_1 \\ \vdots \\ \Delta\mathsf{b}_M \end{bmatrix}, \qquad \underset{N_q \times 1}{\mathsf{c}_q} := \begin{bmatrix} \mathsf{c}_{g_1} \\ \vdots \\ \mathsf{c}_{g_M} \end{bmatrix}. \tag{5.145}$$

By substituting these definitions in (5.142), the combination of all constraint equations can be written as

$$\underset{N_q \times N_p}{\mathsf{A}} \; \underset{N_p \times 1}{\Delta\mathsf{p}} + \underset{N_q \times N_{\mathsf{b}}}{\mathsf{B}^\mathsf{T}} \; \underset{N_{\mathsf{b}} \times 1}{\Delta\mathsf{b}} = \underset{N_q \times 1}{\mathsf{c}_q} \; . \tag{5.146}$$

5.9.1.2 Constraints on the Parameters Alone

In many applications, there also exists a constraint on the parameters alone. As mentioned in the discussion of the Gauss-Markov model, an example is provided by a parameter vector that represents a line through 6D Plücker coordinates. In order to represent a line, a 6D vector has to satisfy the Plücker condition, which can be modeled as a constraint function h on the parameters alone. A correct parameter vector p then satisfies

$$\mathsf{h}(\mathsf{p}) = 0 \; \in \mathbb{R}^{N_h} \; . \tag{5.147}$$

Just as before, this constraint function may be linearized by means of a Taylor expansion up to order one:

$$\begin{aligned} \mathsf{h}(\mathsf{p}) &= \mathsf{h}(\widehat{\mathsf{p}} + \Delta\mathsf{p}) \\ &= \mathsf{h}(\widehat{\mathsf{p}}) + (\partial_{\mathsf{p}}\, \mathsf{h})(\widehat{\mathsf{p}})\, \Delta\mathsf{p} \; . \end{aligned} \tag{5.148}$$

If we define

$$\mathsf{H}^\mathsf{T}(\widehat{\mathsf{p}}) := (\partial_{\mathsf{p}}\, \mathsf{h})(\widehat{\mathsf{p}}), \qquad \mathsf{c}_h = -\mathsf{h}(\widehat{\mathsf{p}}), \tag{5.149}$$

(5.148) becomes

$$\underset{N_h \times N_p}{\mathsf{H}^\mathsf{T}(\widehat{\mathsf{p}})} \; \underset{N_p \times 1}{\Delta\mathsf{p}} = \underset{N_h \times 1}{\mathsf{c}_h} \; . \tag{5.150}$$

5.9.1.3 Summary of Constraints

The constraint equations are

$$\begin{array}{ccccccc} \underset{N_q \times N_p}{\mathsf{A}} & \underset{N_p \times 1}{\Delta \mathsf{p}} & + & \underset{N_q \times N_\mathsf{b}}{\mathsf{B}^\mathsf{T}} & \underset{N_\mathsf{b} \times 1}{\Delta \mathsf{b}} & = & \underset{N_q \times 1}{\mathsf{c}_q} \ , \\ \underset{N_h \times N_p}{\mathsf{H}^\mathsf{T}(\widehat{\mathsf{p}})} & \underset{N_p \times 1}{\Delta \mathsf{p}} & & & & = & \underset{N_h \times 1}{\mathsf{c}_h} \ . \end{array} \tag{5.151}$$

These can be written in a single matrix equation,

$$\begin{bmatrix} \mathsf{A} & \mathsf{B}^\mathsf{T} \\ \mathsf{H}^\mathsf{T} & 0 \end{bmatrix} \begin{bmatrix} \Delta \mathsf{p} \\ \Delta \mathsf{b} \end{bmatrix} = \begin{bmatrix} \mathsf{c}_q \\ \mathsf{c}_h \end{bmatrix} . \tag{5.152}$$

This equation system could be solved directly, but this would not ensure that the constraint equation on the parameters alone was satisfied exactly. Instead, the method of Lagrange multipliers is applied, as for the Gauss–Markov model.

5.9.2 Least-Squares Minimization

For each observation vector b_i, it is assumed that a covariance matrix $\Sigma_{\mathsf{b}_i,\mathsf{b}_i}$ is given. The covariance matrix for the total observation vector b is therefore

$$\Sigma_{\mathsf{b},\mathsf{b}} := \begin{bmatrix} \Sigma_{\mathsf{b}_1,\mathsf{b}_1} & & 0 \\ & \ddots & \\ 0 & & \Sigma_{\mathsf{b}_M,\mathsf{b}_M} \end{bmatrix} . \tag{5.153}$$

For brevity, we define

$$\mathsf{W} := \Sigma_{\mathsf{b},\mathsf{b}} . \tag{5.154}$$

The goal is to find the optimal values for $\Delta \mathsf{p}$ and $\Delta \mathsf{b}$ in a least-squares sense, which is done by minimizing

$$\Delta \mathsf{b}^\mathsf{T} \, \mathsf{W} \, \Delta \mathsf{b}$$

subject to the constraints described in the previous subsection. This can be formulated mathematically by using the method of Lagrange multipliers. For this purpose, we define a new function Φ, which takes its minimum if $\Delta \mathsf{b}^\mathsf{T} \, \mathsf{W} \, \Delta \mathsf{b}$ is minimal and the constraint functions q_i and h are satisfied:

$$\begin{aligned}\Phi(\Delta \mathsf{p}, \Delta \mathsf{b}, \mathsf{m}, \mathsf{n}) := \;& \frac{1}{2} \Delta \mathsf{b}^\mathsf{T} \mathsf{W}^{-1} \Delta \mathsf{b} \\ & - \left(\mathsf{A} \Delta \mathsf{p} + \mathsf{B}^\mathsf{T} \Delta \mathsf{b} - \mathsf{c}_q\right)^\mathsf{T} \mathsf{m} \\ & + \left(\mathsf{H}^\mathsf{T} \Delta \mathsf{p} - \mathsf{c}_h\right)^\mathsf{T} \mathsf{n},\end{aligned} \tag{5.155}$$

where $\mathsf{m} \in \mathbb{R}^{N_q}$ and $\mathsf{n} \in \mathbb{R}^{N_h}$. The function Φ attains a minimum or maximum value if all of its partial derivatives are zero. That is,

$$\frac{\partial \Phi}{\partial \Delta \mathsf{b}} = 0 \iff \mathsf{W}^{-1} \Delta \mathsf{b} - \mathsf{B}\,\mathsf{m} = 0, \tag{5.156}$$

$$\frac{\partial \Phi}{\partial \Delta \mathsf{p}} = 0 \iff -\mathsf{A}^\mathsf{T} \mathsf{m} + \mathsf{H}\,\mathsf{n} = 0, \tag{5.157}$$

$$\frac{\partial \Phi}{\partial \mathsf{m}} = 0 \iff \mathsf{A} \Delta \mathsf{p} + \mathsf{B}^\mathsf{T} \Delta \mathsf{b} - \mathsf{c}_q = 0, \tag{5.158}$$

$$\frac{\partial \Phi}{\partial \mathsf{n}} = 0 \iff \mathsf{H}^\mathsf{T} \Delta \mathsf{p} - \mathsf{c}_h = 0, \tag{5.159}$$

From (5.156), it follows that

$$\Delta \mathsf{b} = \mathsf{W}\,\mathsf{B}\,\mathsf{m}. \tag{5.160}$$

Substituting this result into (5.158) gives

$$\mathsf{A} \Delta \mathsf{p} + \mathsf{B}^\mathsf{T} \mathsf{W}\,\mathsf{B}\;\mathsf{m} - \mathsf{c}_q = 0. \tag{5.161}$$

If we define

$$\mathsf{D}_{bb} := \mathsf{B}^\mathsf{T} \mathsf{W}\,\mathsf{B},$$

(5.161) can be rewritten as

$$\mathsf{m} = -\mathsf{D}_{bb}^{-1} \mathsf{A} \Delta \mathsf{p} + \mathsf{D}_{bb}^{-1} \mathsf{c}_q. \tag{5.162}$$

Substituting this equation into (5.157) results in

$$\mathsf{A}^\mathsf{T} \mathsf{D}_{bb}^{-1} \mathsf{A} \Delta \mathsf{p} - \mathsf{A}^\mathsf{T} \mathsf{D}_{bb}^{-1} \mathsf{c}_q + \mathsf{H}\,\mathsf{n} = 0, \tag{5.163}$$

which can also be written as

$$\mathsf{A}^\mathsf{T} \mathsf{D}_{bb}^{-1} \mathsf{A} \Delta \mathsf{p} + \mathsf{H}\,\mathsf{n} = \mathsf{A}^\mathsf{T} \mathsf{D}_{bb}^{-1} \mathsf{c}_q. \tag{5.164}$$

Equations (5.164) and (5.159) can now be combined into a single equation system as

$$\begin{bmatrix} \mathsf{A}^\mathsf{T} \mathsf{D}_{bb}^{-1} \mathsf{A} & \mathsf{H} \\ \mathsf{H}^\mathsf{T} & 0 \end{bmatrix} \begin{bmatrix} \Delta \mathsf{p} \\ \mathsf{n} \end{bmatrix} = \begin{bmatrix} \mathsf{A}^\mathsf{T} \mathsf{D}_{bb}^{-1} \mathsf{c}_q \\ \mathsf{c}_h \end{bmatrix}. \tag{5.165}$$

If we define

$$\underset{N_p \times N_p}{\mathsf{N}} := \underset{N_p \times N_q}{\mathsf{A}^\mathsf{T}} \underset{N_q \times N_q}{\mathsf{D}_{bb}^{-1}} \underset{N_q \times N_p}{\mathsf{A}}$$

and

$$\underset{N_p \times 1}{\mathsf{c}_n} := \underset{N_p \times N_q}{\mathsf{A}^\mathsf{T}} \underset{N_q \times N_q}{\mathsf{D}_{bb}^{-1}} \underset{N_q \times 1}{\mathsf{c}_q} ,$$

(5.165) becomes

$$\underset{(N_p+N_h)\times(N_p+N_h)}{\begin{bmatrix} \mathsf{N} & \mathsf{H} \\ \mathsf{H}^\mathsf{T} & 0 \end{bmatrix}} \underset{(N_p+N_h)\times 1}{\begin{bmatrix} \Delta\mathsf{p} \\ \mathsf{n} \end{bmatrix}} = \underset{(N_p+N_h)\times 1}{\begin{bmatrix} \mathsf{c}_n \\ \mathsf{c}_h \end{bmatrix}} . \tag{5.166}$$

5.9.3 Derivation of the Covariance Matrix $\Sigma_{\Delta\mathsf{p},\Delta\mathsf{p}}$

One useful property of the Gauss-Helmert model is that the covariance matrix of the parameter vector can be estimated. In this subsection, it is shown that the covariance matrix of p, which is equivalent to that of $\Delta\mathsf{p}$, is given by

$$\Sigma_{\Delta\mathsf{p},\Delta\mathsf{p}} = \mathsf{N}^{-1}\left(\mathsf{I} - \mathsf{H}\left(\mathsf{H}^\mathsf{T}\mathsf{N}^{-1}\mathsf{H}\right)^{-1}\mathsf{H}^\mathsf{T}\mathsf{N}^{-1}\right). \tag{5.167}$$

$\Sigma_{\Delta\mathsf{p},\Delta\mathsf{p}}$ may be calculated more easily via

$$\begin{bmatrix} \mathsf{N} & \mathsf{H} \\ \mathsf{H}^\mathsf{T} & 0 \end{bmatrix}^{-1} = \begin{bmatrix} \Sigma_{\Delta\mathsf{p},\Delta\mathsf{p}} & \cdot \\ \cdot & \cdot \end{bmatrix} , \tag{5.168}$$

where the dots are placeholders for other parts of the matrix. First of all, we give a formula for the inverse of a square matrix

$$\begin{bmatrix} \mathsf{N} & \mathsf{H} \\ \mathsf{H}^\mathsf{T} & 0 \end{bmatrix}^{-1}$$
$$= \begin{bmatrix} \mathsf{N}^{-1} - \mathsf{N}^{-1}\mathsf{H}\,(\mathsf{H}^\mathsf{T}\mathsf{N}^{-1}\mathsf{H})^{-1}\,\mathsf{H}^\mathsf{T}\mathsf{N}^{-1} & \mathsf{N}^{-1}\mathsf{H}\,(\mathsf{H}^\mathsf{T}\mathsf{N}^{-1}\mathsf{H})^{-1} \\ (\mathsf{H}^\mathsf{T}\mathsf{N}^{-1}\mathsf{H})^{-1}\,\mathsf{H}^\mathsf{T}\mathsf{N}^{-1} & (\mathsf{H}^\mathsf{T}\mathsf{N}^{-1}\mathsf{H})^{-1} \end{bmatrix} . \tag{5.169}$$

By writing (5.166) as

$$\begin{bmatrix} \Delta\mathsf{p} \\ \mathsf{n} \end{bmatrix} = \begin{bmatrix} \mathsf{N} & \mathsf{H} \\ \mathsf{H}^\mathsf{T} & 0 \end{bmatrix}^{-1} \begin{bmatrix} \mathsf{c}_n \\ \mathsf{c}_h \end{bmatrix} , \tag{5.170}$$

it follows with the help of (5.169) that

$$\begin{aligned}\Delta \mathsf{p} = \quad & \left(\mathsf{N}^{-1} - \mathsf{N}^{-1}\,\mathsf{H}\,(\mathsf{H}^\mathsf{T}\,\mathsf{N}^{-1}\,\mathsf{H})^{-1}\,\mathsf{H}^\mathsf{T}\,\mathsf{N}^{-1}\right)\mathsf{c}_n \\ & + \left(\mathsf{N}^{-1}\,\mathsf{H}\,(\mathsf{H}^\mathsf{T}\,\mathsf{N}^{-1}\,\mathsf{H})^{-1}\right)\mathsf{c}_h\,.\end{aligned} \tag{5.171}$$

Equation (5.149) has defined $\mathsf{c}_h := -\mathsf{h}(\widehat{\mathsf{p}})$. Since $\widehat{\mathsf{p}}$ is a constant vector, it does not have an associated covariance matrix. However, c_n depends on $\Delta\mathsf{b} = \mathsf{b} - \widehat{\mathsf{b}}$ and thus on b, which has an associated covariance matrix $\Sigma_{\mathsf{b},\mathsf{b}}$. It therefore follows from the rules of error propagation that

$$\underset{N_p \times N_p}{\Sigma_{\Delta\mathsf{p},\Delta\mathsf{p}}} = \underset{N_p \times N_\mathsf{b}}{\left(\frac{\partial \Delta\mathsf{p}}{\partial \mathsf{b}}\right)} \; \underset{N_\mathsf{b} \times N_\mathsf{b}}{\Sigma_{\mathsf{b},\mathsf{b}}} \; \underset{N_\mathsf{b} \times N_p}{\left(\frac{\partial \Delta\mathsf{p}}{\partial \mathsf{b}}\right)^\mathsf{T}}\,. \tag{5.172}$$

Using the chain rule for differentiation, this can be written as

$$\begin{aligned}\underset{N_p \times N_p}{\Sigma_{\Delta\mathsf{p},\Delta\mathsf{p}}} &= \underset{N_p \times N_{\mathsf{c}_n}}{\left(\frac{\partial \Delta\mathsf{p}}{\partial \mathsf{c}_n}\right)} \; \underset{N_{\mathsf{c}_n} \times N_\mathsf{b}}{\left(\frac{\partial \mathsf{c}_n}{\partial \mathsf{b}}\right)} \; \underset{N_\mathsf{b} \times N_\mathsf{b}}{\Sigma_{\mathsf{b},\mathsf{b}}} \; \underset{N_\mathsf{b} \times N_{\mathsf{c}_n}}{\left(\frac{\partial \mathsf{c}_n}{\partial \mathsf{b}}\right)^\mathsf{T}} \; \underset{N_{\mathsf{c}_n} \times N_p}{\left(\frac{\partial \Delta\mathsf{p}}{\partial \mathsf{c}_n}\right)^\mathsf{T}} \\ &= \underset{N_p \times N_{\mathsf{c}_n}}{\left(\frac{\partial \Delta\mathsf{p}}{\partial \mathsf{c}_n}\right)} \; \underset{N_{\mathsf{c}_n} \times N_{\mathsf{c}_n}}{\Sigma_{\mathsf{c}_n,\mathsf{c}_n}} \; \underset{N_{\mathsf{c}_n} \times N_p}{\left(\frac{\partial \Delta\mathsf{p}}{\partial \mathsf{c}_n}\right)^\mathsf{T}}\,.\end{aligned} \tag{5.173}$$

From (5.171), it follows that

$$\left(\frac{\partial \Delta\mathsf{p}}{\partial \mathsf{c}_n}\right) = \mathsf{N}^{-1} - \mathsf{N}^{-1}\,\mathsf{H}\,(\mathsf{H}^\mathsf{T}\,\mathsf{N}^{-1}\,\mathsf{H})^{-1}\,\mathsf{H}^\mathsf{T}\,\mathsf{N}^{-1}\,. \tag{5.174}$$

Now an expression for $\Sigma_{\mathsf{c}_n,\mathsf{c}_n}$ has to be found. From (5.173), it is clear that

$$\underset{N_{\mathsf{c}_n} \times N_{\mathsf{c}_n}}{\Sigma_{\mathsf{c}_n,\mathsf{c}_n}} = \underset{N_{\mathsf{c}_n} \times N_\mathsf{b}}{\left(\frac{\partial \mathsf{c}_n}{\partial \mathsf{b}}\right)} \; \underset{N_\mathsf{b} \times N_\mathsf{b}}{\Sigma_{\mathsf{b},\mathsf{b}}} \; \underset{N_\mathsf{b} \times N_{\mathsf{c}_n}}{\left(\frac{\partial \mathsf{c}_n}{\partial \mathsf{b}}\right)^\mathsf{T}}\,. \tag{5.175}$$

Furthermore,

$$\mathsf{c}_n = \mathsf{A}^\mathsf{T}\,\mathsf{D}_{bb}^{-1}\,\mathsf{c}_q, \tag{5.176}$$

where

$$\begin{aligned}\mathsf{c}_q &= \mathsf{A}\,\Delta\mathsf{p} + \mathsf{B}^\mathsf{T}\,\Delta\mathsf{b} \\ &= \mathsf{A}\,\Delta\mathsf{p} + \mathsf{B}^\mathsf{T}\,(\mathsf{b} - \widehat{\mathsf{b}}).\end{aligned} \tag{5.177}$$

It therefore follows that

$$\left(\frac{\partial \mathsf{c}_n}{\partial \mathsf{b}}\right) = \mathsf{A}^\mathsf{T}\,\mathsf{D}_{bb}^{-1}\,\mathsf{B}^\mathsf{T}. \tag{5.178}$$

Substituting this expression into (5.175), it is found after some calculation that

$$\Sigma_{\mathsf{c}_n,\mathsf{c}_n} = \mathsf{N}. \tag{5.179}$$

Substituting this result into (5.173) gives

$$\begin{aligned}
\Sigma_{\Delta\mathsf{p},\Delta\mathsf{p}} &= \left(\mathsf{N}^{-1} - \mathsf{N}^{-1}\,\mathsf{H}\,(\mathsf{H}^\mathsf{T}\,\mathsf{N}^{-1}\,\mathsf{H})^{-1}\,\mathsf{H}^\mathsf{T}\,\mathsf{N}^{-1}\right)\,\mathsf{N} \\
&\qquad \times \left(\mathsf{N}^{-1} - \mathsf{N}^{-1}\,\mathsf{H}\,(\mathsf{H}^\mathsf{T}\,\mathsf{N}^{-1}\,\mathsf{H})^{-1}\,\mathsf{H}^\mathsf{T}\,\mathsf{N}^{-1}\right) \\
&= \mathsf{N}^{-1} + \mathsf{N}^{-1}\,\mathsf{H}\,(\mathsf{H}^\mathsf{T}\,\mathsf{N}^{-1}\,\mathsf{H})^{-1}\,\mathsf{H}^\mathsf{T}\,\mathsf{N}^{-1}\,\mathsf{H}\,(\mathsf{H}^\mathsf{T}\,\mathsf{N}^{-1}\,\mathsf{H})^{-1}\,\mathsf{H}^\mathsf{T}\,\mathsf{N}^{-1} \\
&\quad -2\,\mathsf{N}^{-1}\,\mathsf{H}\,(\mathsf{H}^\mathsf{T}\,\mathsf{N}^{-1}\,\mathsf{H})^{-1}\,\mathsf{H}^\mathsf{T}\mathsf{N}^{-1} \\
&= \mathsf{N}^{-1} - \mathsf{N}^{-1}\,\mathsf{H}\,(\mathsf{H}^\mathsf{T}\,\mathsf{N}^{-1}\,\mathsf{H})^{-1}\,\mathsf{H}^\mathsf{T}\,\mathsf{N}^{-1}.
\end{aligned} \tag{5.180}$$

If we compare this result with (5.169), it is clear that

$$\begin{bmatrix} \mathsf{N} & \mathsf{H} \\ \mathsf{H}^\mathsf{T} & 0 \end{bmatrix}^{-1} = \begin{bmatrix} \Sigma_{\Delta\mathsf{p},\Delta\mathsf{p}} & \cdot \\ \cdot & \cdot \end{bmatrix}, \tag{5.181}$$

where the dots represent some other non-zero parts of the matrix.

5.9.4 Numerical Evaluation

In the first step, (5.166) has to be solved for $\Delta\mathsf{p}$ and n. Then n can be used to evaluate the residue $\Delta\mathsf{b}$ and thus b. The assumption is that a fairly good estimate for p and b, denoted by $\widehat{\mathsf{p}}$ and $\widehat{\mathsf{b}}$, already exists. In order to solve for $\Delta\mathsf{p}$ and n, the following matrix has to be inverted:

$$\underbrace{\begin{bmatrix} \mathsf{N} & \mathsf{H} \\ \mathsf{H}^\mathsf{T} & 0 \end{bmatrix}}_{(N_p+N_h)\times(N_p+N_h)} . \tag{5.182}$$

Therefore, the first step is to construct this matrix from the various observations. However, there are a number of block-diagonal matrices here which can become very large for a large number of observations. It will therefore be shown how N may be evaluated without generating huge matrices at intermediate steps. After all, N is only an $N_p \times N_p$ matrix. That is, its size depends only on the number of parameters that have to be evaluated.

5.9.4.1 Construction of Equations

Clearly, D_{bb} is a block-diagonal matrix, since the matrices that it is calculated from, B and W, are block-diagonal matrices. Each block on its diagonal gives the constraints for one observation. If we define

$$\mathsf{W}_i := \Sigma_{\mathsf{b}_i,\mathsf{b}_i}, \tag{5.183}$$

it follows that

$$\mathsf{N} = \sum_{i=1}^{M} \mathsf{U}_i^\mathsf{T} \left(\mathsf{V}_i\, \mathsf{W}_i\, \mathsf{V}_i^\mathsf{T}\right)^{-1} \mathsf{U}_i \tag{5.184}$$

and

$$\mathsf{c}_n = \sum_{i=1}^{M} \mathsf{U}_i^\mathsf{T} \left(\mathsf{V}_i\, \mathsf{W}_i\, \mathsf{V}_i^\mathsf{T}\right)^{-1} \mathsf{c}_{q_i}\,. \tag{5.185}$$

The linear system of equations that has to be solved may therefore be written as

$$\begin{aligned} &\begin{bmatrix} \sum_{i=1}^{M} \mathsf{U}_i^\mathsf{T} \left(\mathsf{V}_i\, \mathsf{W}_i\, \mathsf{V}_i^\mathsf{T}\right)^{-1} \mathsf{U}_i & \mathsf{H} \\ \mathsf{H}^\mathsf{T} & 0 \end{bmatrix} \begin{bmatrix} \Delta\mathsf{p} \\ \mathsf{n} \end{bmatrix} \\ &\qquad\qquad = \begin{bmatrix} \sum_{i=1}^{M} \mathsf{U}_i^\mathsf{T} \left(\mathsf{V}_i\, \mathsf{W}_i\, \mathsf{V}_i^\mathsf{T}\right)^{-1} \mathsf{c}_{q_i} \\ \mathsf{c}_h \end{bmatrix}. \end{aligned} \tag{5.186}$$

Hence, N and c_n can be calculated from

$$\mathsf{N} := \sum_{i=1}^{M} \mathsf{U}_i^\mathsf{T} \left(\mathsf{V}_i\, \mathsf{W}_i\, \mathsf{V}_i^\mathsf{T}\right)^{-1} \mathsf{U}_i\,, \qquad \mathsf{c}_n := \sum_{i=1}^{M} \mathsf{U}_i^\mathsf{T} \left(\mathsf{V}_i\, \mathsf{W}_i\, \mathsf{V}_i^\mathsf{T}\right)^{-1} \mathsf{c}_{q_i}\,. \tag{5.187}$$

5.9.4.2 Back-Substitution

Suppose now that $\Delta\mathsf{p}$ and n have been calculated from (5.186). By substituting (5.162) into (5.156), an expression for $\Delta\mathsf{b}$ in terms of $\Delta\mathsf{p}$ is obtained:

$$\begin{aligned} &\Delta\mathsf{b} = \mathsf{W}\,\mathsf{B}\,\mathsf{D}_{bb}^{-1}\left(\mathsf{c}_q - \mathsf{A}\,\Delta\mathsf{p}\right) \\ \iff\ &\Delta\mathsf{b}_i = \mathsf{W}_i\,\mathsf{V}_i^\mathsf{T}\left(\mathsf{V}_i\,\mathsf{W}_i\,\mathsf{V}_i^\mathsf{T}\right)^{-1}\left(\mathsf{c}_{q_i} - \mathsf{U}_i\,\Delta\mathsf{p}\right). \end{aligned} \tag{5.188}$$

The estimates for p ($\widehat{\mathsf{p}}$) and b ($\widehat{\mathsf{b}}$) can now be updated via

$$\widehat{\mathsf{p}}' = \widehat{\mathsf{p}} + \Delta\mathsf{p}\,, \qquad \widehat{\mathsf{b}}_i' = \widehat{\mathsf{b}}_i + \Delta\mathsf{b}_i\,. \tag{5.189}$$

If the constraint functions q_i and h are linear, then this gives the BLUE for p and b_i. However, in general the constraint functions are non-linear, in which case these estimates have to be regarded as an iteration step.

5.9.5 Generalization

In addition to regarding the data vectors as random variables with associated covariance matrices, the parameter vector can also be regarded as a random variable with a covariance matrix. Just as for the Gauss–Markov model, this extends the function Φ by a term $\Delta\mathsf{p}^{\mathsf{T}}\,\Sigma_{\mathsf{p},\mathsf{p}}^{-1}\,\Delta\mathsf{p}$, i.e.

$$\begin{aligned}\Phi(\Delta\mathsf{p},\,\Delta\mathsf{b},\,\mathsf{m},\,\mathsf{n}) := \;& \frac{1}{2}\,\Delta\mathsf{b}^{\mathsf{T}}\,\mathsf{W}^{-1}\,\Delta\mathsf{b} + \frac{1}{2}\,\Delta\mathsf{p}^{\mathsf{T}}\,\Sigma_{\mathsf{p},\mathsf{p}}^{-1}\,\Delta\mathsf{p} \\ & - \left(\mathsf{A}\,\Delta\mathsf{p} + \mathsf{B}^{\mathsf{T}}\,\Delta\mathsf{b} - \mathsf{c}_q\right)^{\mathsf{T}}\mathsf{m} \\ & + \left(\mathsf{H}^{\mathsf{T}}\,\Delta\mathsf{p} - \mathsf{c}_h\right)^{\mathsf{T}}\mathsf{n}\,,\end{aligned} \tag{5.190}$$

The added term acts as a regularization, which can be used to selectively solve for only a subset of parameters. This can be achieved if $\Sigma_{\mathsf{p},\mathsf{p}}^{-1}$ is a diagonal matrix with zero entries on the diagonal for those parameters that can vary freely, and large entries for fixed parameters.

The only effect of the additional term in Φ on the normal equations is that there is a different matrix N, which is now

$$\mathsf{N} = \Sigma_{\mathsf{p},\mathsf{p}}^{-1} + \mathsf{A}^{\mathsf{T}}\,\mathsf{D}_{bb}^{-1}\,\mathsf{A}\,. \tag{5.191}$$

5.10 Applying the Gauss–Markov and Gauss–Helmert Models

The initial estimate $\widehat{\mathsf{p}}$ has to be close to the true parameter vector p for the Gauss–Helmert model to converge, if the constraint equations q_i are non-linear. The Gauss–Markov model, on the other hand, shows a more robust behavior. However, it does not take account of the uncertainties in the data vectors. The best results can therefore be achieved by a combined application of both methods. First, the Gauss–Markov method is used to robustly obtain a good estimate of the starting parameters for use with the Gauss–Helmert method, which then takes account of the uncertainties in the data vectors and returns a covariance matrix for the parameter vector.

In the remainder of this section, problems that can occur with an iterative application of the Gauss–Helmert model are discussed, which should clarify the reason why a good initial parameter estimate is necessary.

5.10.1 Iterative Application of Gauss–Helmert Method

An iterative application of the Gauss–Helmert model is necessary only if the constraint functions q_i and h are non-linear. In the applications discussed later in this section, the constraint functions are typically multilinear. Because the Gauss–Helmert method adapts not only the parameter vector but also the data vectors in each iteration step, it is easy to imagine that this method might not converge for non-linear constraint functions. This will be demonstrated with a simple example in the following.

Consider the problem of fitting a line to a set of points in $\mathbb{R}^2$. The data vectors $\mathsf{b}_i \in \mathbb{R}^3$ are 2D vectors embedded in a projective space $\mathbb{R}^3$, representing the positions of the points, with associated covariance matrices $\Sigma_{\mathsf{b}_i,\mathsf{b}_i} \in \mathbb{R}^{3\times3}$, which are assumed to be diagonal for simplicity. That is, the b_i and $\Sigma_{\mathsf{b}_i,\mathsf{b}_i}$ are of the form

$$\mathsf{b}_i = \begin{bmatrix} b_i^1 \,,\; b_i^2 \,,\; 1 \end{bmatrix} \qquad \text{and} \qquad \Sigma_{\mathsf{b}_i,\mathsf{b}_i} = \begin{bmatrix} \sigma_1^2 & & \\ & \sigma_2^2 & \\ & & 0 \end{bmatrix} .$$

The parameter vector $\mathsf{p} \in \mathbb{R}^3$ represents a line in projective space by its normal and its orthogonal separation from the origin, i.e. $\mathsf{p} = \begin{bmatrix} \mathsf{u}^\mathsf{T} \,,\; -d \end{bmatrix}^\mathsf{T}$, where $\mathsf{u} \in \mathbb{R}^2$ is the normal of the line and $d \in \mathbb{R}$ is the orthogonal separation from the origin.

The constraint function q between the data and parameter vectors is therefore

$$\mathsf{q}(\mathsf{b}_i,\, \mathsf{p}) := \mathsf{b}_i^\mathsf{T}\,\mathsf{p} \,,$$

which results in zero if b_i represents a point that lies on the line represented by p. Note that this is equivalent to the IPNS representation of a line given in Sect. 4.2.3. The Jacobi matrices of q are

$$\mathsf{V}_i(\widehat{\mathsf{p}}) := \widehat{\mathsf{p}}^\mathsf{T} \qquad \text{and} \qquad \mathsf{U}_i(\widehat{\mathsf{b}}_i) := \widehat{\mathsf{b}}_i^\mathsf{T} \,,$$

and $\mathsf{c}_{q_i} := -\widehat{\mathsf{b}}_i^\mathsf{T}\,\widehat{\mathsf{p}}$.

The constraint equation h on the parameters ensures that $\|\mathsf{p}\|^2 = 1$, since otherwise the scale of p would be another degree of freedom. That is,

$$\mathsf{h}(\mathsf{p}) := \frac{1}{2}\,\mathsf{p}^\mathsf{T}\,\mathsf{p} - 1 \,,$$

with a Jacobi matrix

$$\mathsf{H}^\mathsf{T} := \mathsf{p}^\mathsf{T} \,,$$

and $\mathsf{c}_h := \frac{1}{2}\,\widehat{\mathsf{p}}^\mathsf{T}\,\widehat{\mathsf{p}} - 1$. The matrix N and vector c_n of the normal equations therefore become

$$\mathsf{N} = \sum_{i=1}^{M} \widehat{\mathsf{b}}_i \, (\widehat{\mathsf{p}}^\mathsf{T} \, \Sigma_{\mathsf{b}_i,\mathsf{b}_i} \, \widehat{\mathsf{p}})^{-1} \, \widehat{\mathsf{b}}_i^\mathsf{T} \qquad \text{and} \qquad \mathsf{c}_n = -\sum_{i=1}^{M} \widehat{\mathsf{b}}_i \, (\widehat{\mathsf{p}}^\mathsf{T} \, \Sigma_{\mathsf{b}_i,\mathsf{b}_i} \, \widehat{\mathsf{p}})^{-1} \, \widehat{\mathsf{b}}_i^\mathsf{T} \, \widehat{\mathsf{p}} \, .$$

If a solution for $\Delta\mathsf{p}$ has been evaluated, $\Delta\mathsf{b}_i$ can be found by back-substitution using (5.188), which gives

$$\begin{aligned} \Delta\mathsf{b}_i &= \Sigma_{\mathsf{b}_i,\mathsf{b}_i} \, \widehat{\mathsf{p}} \, (\widehat{\mathsf{p}}^\mathsf{T} \, \Sigma_{\mathsf{b}_i,\mathsf{b}_i} \, \widehat{\mathsf{p}})^{-1} \, (-\widehat{\mathsf{b}}_i^\mathsf{T} \, \widehat{\mathsf{p}} - \widehat{\mathsf{b}}_i^\mathsf{T} \, \Delta\mathsf{p}) \\ &= -\, \Sigma_{\mathsf{b}_i,\mathsf{b}_i} \, \widehat{\mathsf{p}} \, (\widehat{\mathsf{p}}^\mathsf{T} \, \Sigma_{\mathsf{b}_i,\mathsf{b}_i} \, \widehat{\mathsf{p}})^{-1} \, \widehat{\mathsf{b}}_i^\mathsf{T} \, (\widehat{\mathsf{p}} + \Delta\mathsf{p}) \, . \end{aligned}$$

From (5.141), it follows that

$$\mathsf{c}_{q_i} - \mathsf{U}_i(\widehat{\mathsf{b}}_i) \, \Delta\mathsf{p} = \mathsf{V}_i(\widehat{\mathsf{p}}) \, \Delta\mathsf{b}_i \qquad \Longleftrightarrow \qquad -\widehat{\mathsf{b}}_i^\mathsf{T} \, (\widehat{\mathsf{p}} + \Delta\mathsf{p}) = \widehat{\mathsf{p}}^\mathsf{T} \, \Delta\mathsf{b}_i \, .$$

Hence,

$$\Delta\mathsf{b}_i = -\, \Sigma_{\mathsf{b}_i,\mathsf{b}_i} \, \widehat{\mathsf{p}} \, (\widehat{\mathsf{p}}^\mathsf{T} \, \Sigma_{\mathsf{b}_i,\mathsf{b}_i} \, \widehat{\mathsf{p}})^{-1} \, \widehat{\mathsf{p}}^\mathsf{T} \, \Delta\mathsf{b}_i \, .$$

If $\Sigma_{\mathsf{b}_i,\mathsf{b}_i}$ is a scaled identity matrix, then the term $\widehat{\mathsf{p}} \, (\widehat{\mathsf{p}}^\mathsf{T} \, \Sigma_{\mathsf{b}_i,\mathsf{b}_i} \, \widehat{\mathsf{p}})^{-1} \, \widehat{\mathsf{p}}^\mathsf{T}$ is a projection matrix that projects onto the subspace spanned by $\widehat{\mathsf{p}}$. Thus, $\widehat{\mathsf{b}}_i$ is adjusted along the direction of $\widehat{\mathsf{p}}$. In this particular example, only the Euclidean part, i.e. the first two dimensions of $\widehat{\mathsf{b}}_i$, is adjusted along the normal of the initial estimate of the line. That is, the data points are adjusted perpendicular to the initial line. The data point adjustments can differ considerably depending on the initial line, which, in fact, can also result in different line solutions that the iterations converge to.

Figure 5.3 shows several different results for the estimation of a line using the Gauss–Helmert method for several different initial estimates $\widehat{\mathsf{p}}$ of the line. The true line is indicated by p, and the initial data points are the green points distributed about the true line. The adjustments of the data points are drawn as red lines away from the initial data points. After a single application of the Gauss–Helmert method, the adjusted data points (red) come to lie on a line, which implies that further iteration steps will not change the result.

In Fig. 5.3(a), the initial estimate of the line is perpendicular to the true line, which is the worst possible initial estimate. The data points have been adjusted along the normal of the initial estimate such that they all lie along a line. This is exactly what the Gauss–Helmert method should do; however, in this setup it results in a completely wrong solution for the line. If the initial line is only slightly tilted, as in Figs. 5.3(b) and 5.3(d), the final estimate is much better. If the initial estimate is close to the true line, as in Fig. 5.3(c), the final solution is quite close to the true one.

The first iteration step of the Gauss–Markov method does in fact result in the same estimate for the line as in the case of the Gauss–Helmert method. To obtain a good estimate of the line with either method, the estimation has to be iterated. In the case of the Gauss–Helmert method, it is essential that the data vectors are *not* updated after each iteration, since, otherwise, the method

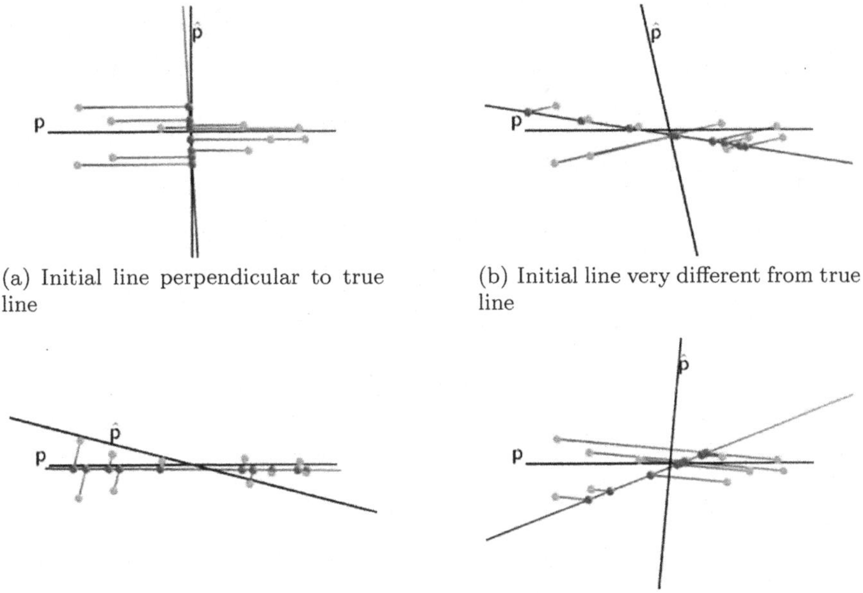

(a) Initial line perpendicular to true line

(b) Initial line very different from true line

(c) Initial line close to true line

(d) Initial line very different to true line.

Fig. 5.3 Several different results for estimation of a line by the Gauss–Helmert method after a single step starting from different initial estimates $\widehat{\mathsf{p}}$. The true line is denoted by p, and the initial data points are green

may diverge. Nevertheless, it was found that, in more complex settings, the Gauss–Markov method was more robust than the Gauss–Helmert method.

The reason for this behavior may be that the Gauss–Helmert method implicitly assumes that an adjustment of the data vectors, in a direction perpendicular to the parameter vector, can satisfy the constraint equation exactly. In a non-linear setting, this may, however, not be possible, which could result in a diverging estimation. For the estimation problems considered in this text, the best results were obtained, by calculating an initial estimate with the Gauss–Markov method, and then improving on this estimate with the Gauss–Helmert method.

Part II
Applications

Chapter 6
Uncertain Geometric Entities and Operators

In computer vision, one has to deal almost invariably with uncertain data and thus also with uncertain geometric entities. Appropriate methods to deal with this uncertainty therefore play an important role. Building mainly on the results presented in Chap. 5, the construction and estimation of geometric entities and transformation operators are discussed in the following.

A particular advantage of the approach presented here stems from the linear representation of geometric entities and transformations and from the fact that algebraic operations are simply bilinear functions. This allows the simple construction of geometric constraints using the symbolic power of the algebra, which can be expressed equivalently as multi-linear functions, such that the whole body of linear algebra can be applied. The solutions to many problems, such as the estimation of the best fit of a line, plane, circle, or sphere through a set of points, or the best rotation between two point sets (in a least-squares sense), reduce to the evaluation of the null space of a matrix. By applying the Gauss–Helmert model (see Sect. 5.9), it is also possible to evaluate the uncertainty of the estimated entity.

This chapter builds on previous work by Förstner et al. [74] and Heuel [93], where uncertain points, lines, and planes were treated in a unified manner. The linear estimation of rotation operators in geometric algebra was first discussed in [146], albeit without taking account of uncertainty. The description of uncertain circles and 2D conics in geometric algebra was introduced in [136], and the estimation of uncertain general operators was introduced in [138]. Applications and extensions of these initial developments were presented in [139].

6.1 Construction

The first application of the combination of uncertainty with geometric algebra is the construction of uncertain geometric entities and operators. The basic

C. Perwass, *Geometric Algebra with Applications in Engineering.*
Geometry and Computing.

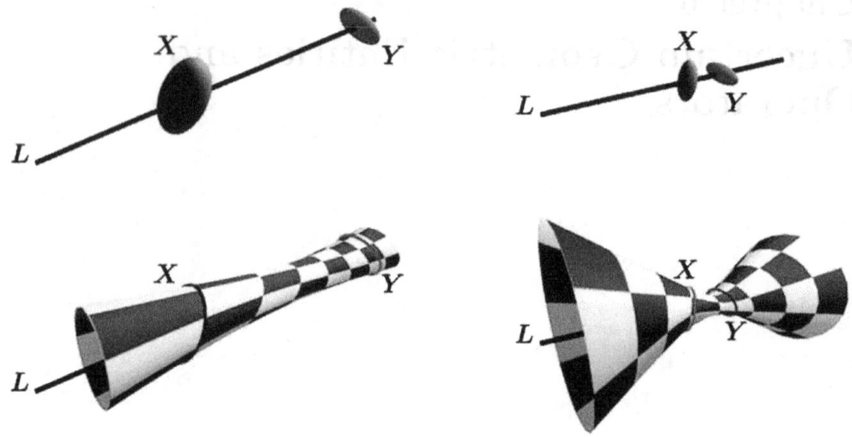

Fig. 6.1 Examples of lines constructed from uncertain points, and their standard-deviation envelopes

idea is that given uncertain primitives, typically points with an associated covariance matrix, we can construct higher-level geometric entities such as lines, planes, and circles, and also operators, using algebraic operations. The associated covariance matrices of the resultant entities can be calculated by propagating the covariances of the primitives through the algebraic products.

Since all algebraic products are representable as bilinear functions, there is no fundamental difference between generating uncertain geometric entities and generating uncertain operators in this way. In the following, a number of examples are given for both cases.

6.1.1 Geometric Entities in Conformal Space

Figure 6.1 shows two examples of the construction of an uncertain line from two uncertain points $\boldsymbol{X}$ and $\boldsymbol{Y}$. In the top images, the uncertain points and the mean line $\boldsymbol{L} = \boldsymbol{X} \wedge \boldsymbol{Y} \wedge \boldsymbol{e}_\infty$ are shown. The uncertainty of the points is visualized as a covariance ellipsoid. The resultant standard-deviation envelope of the line $\boldsymbol{L}$ is visualized below. Clearly, as the uncertain points move closer together, the standard-deviation envelope increases in size away from the points faster.

While the construction of an uncertain line is also possible in projective space, the implicit representation of a circle is available only in CGA. An uncertain circle can be constructed just as easily as an uncertain line. Two examples of this are visualized in Fig. 6.2. The top images show again the initial uncertain points $\boldsymbol{X}$, $\boldsymbol{Y}$, and $\boldsymbol{Z}$ with their covariance ellipsoids, and

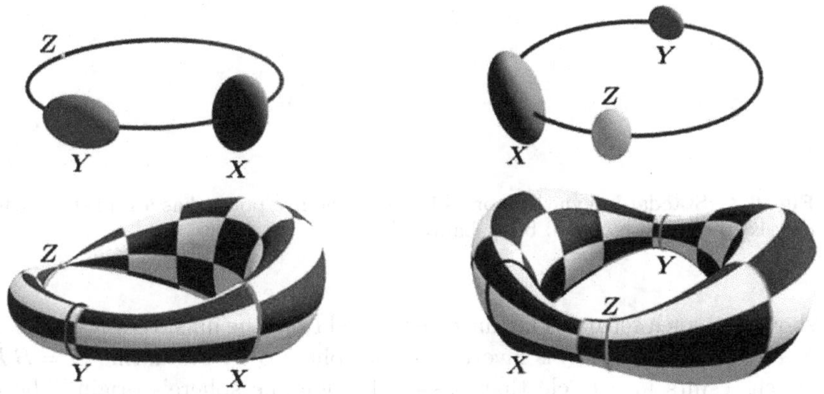

Fig. 6.2 Examples of circles constructed from uncertain points, and their standard-deviation envelopes

the bottom images show the standard-deviation envelopes of the circles $\boldsymbol{C} = \boldsymbol{X} \wedge \boldsymbol{Y} \wedge \boldsymbol{Z}$. In the left image, the point $\boldsymbol{Z}$ has zero covariance, which is why the standard deviation of the circle is also zero at that point.

The visualizations of the standard-deviation envelopes of lines and circles show that they are typically much more complex structures than simply a cylindrical tube around the mean line or circle. Nevertheless, these complex structures are represented by a simple covariance matrix. It is therefore much better to use covariance matrices to represent the uncertainty of lines and circles than an ad hoc uncertainty model, such as, for example a cylindrical tube.

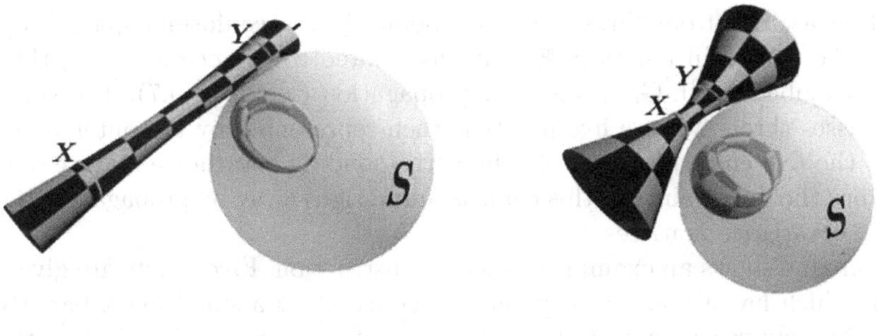

Fig. 6.3 Examples of the inversion of an uncertain line, which generates an uncertain circle, with zero variance at the center of the inversion sphere

Clearly, all operators available in CGA can be applied to uncertain geometric entities when one propagates the covariances of the entities. A particularly interesting case is the inversion of an uncertain line in a sphere. Figure 6.3

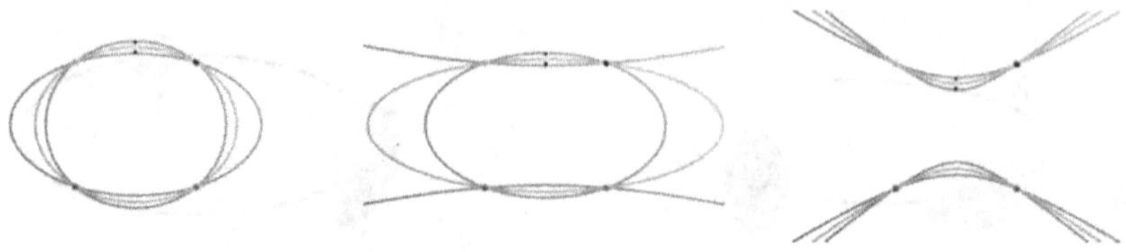

Fig. 6.4 Standard-deviation conics if one of the five points has a rank-1 covariance matrix (indicated by small black bars)

shows two such setups. The line constructed from the uncertain points $\boldsymbol{X}$ and $\boldsymbol{Y}$ as $\boldsymbol{L} = \boldsymbol{X} \wedge \boldsymbol{Y} \wedge \boldsymbol{e}_\infty$ is inverted in the sphere $\boldsymbol{S}$ by evaluating $\boldsymbol{C} = \boldsymbol{S}\,\boldsymbol{L}\,\widetilde{\boldsymbol{S}}$, which results in a circle that passes through the sphere's origin. The corresponding standard-deviation envelopes are shown in the figure. Note that the covariance of the circle at the origin of the inversion sphere is zero. This should be expected, since the point on the line that is mapped to the inversion sphere's origin is the point at infinity. Because all lines have to pass through the point at infinity, this is a constant point with no uncertainty.

Conversely, given an uncertain circle whose mean passes through the origin of the inversion sphere, but whose covariance at the origin of the inversion sphere is not zero, the inversion of the circle represents a line in terms of its mean only. The geometric entities that lie in the standard-deviation envelope of the inversion are circles.

6.1.2 Geometric Entities in Conic Space

Constructing a conic from five uncertain points in $\mathbb{G}_6$ is very similar to constructing a circle from three uncertain points in the conformal space $\mathbb{G}_{4,1}$. Given five uncertain points in $\mathbb{R}^2$ with associated covariance matrices, they are first embedded in $\mathbb{G}_6$ using error propagation (see Sect. 5.7). The conic that passes through these five points is then represented by the outer product of the five embedded points. The corresponding covariance matrix that describes the uncertainty of this conic is found right away by propagating the points' covariance matrices.

Figure 6.4 shows an example of such a construction. Five points are given, one of which has a non-zero covariance, indicated by a small black bar. By taking the outer product of the embedding of these five points in $\mathbb{G}_6^1$, the mean conic, represented by the black conic, and also the covariance matrix of the conic can be estimated. In this case the covariance matrix is of rank one, which generates two conics that have a probability $\exp(-\frac{1}{2})$ of occurring, represented by the gray conics. Figure 6.4 shows another example, where two of the five points have non-zero covariances, indicated by small black bars. The covariance matrix of the corresponding conic is of rank two, which generates a whole set of standard-deviation conics. As can be seen in this figure, the area swept by this set of "standard-deviation conics" has a highly non-linear shape, which is in no way adequately represented by a tube around the mean conic.

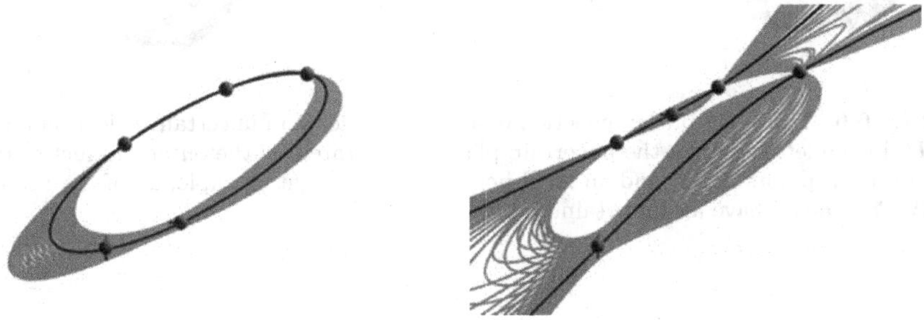

Fig. 6.5 Standard-deviation conics if two of the five points have rank-1 covariance matrices (indicated by small black bars)

6.1.3 Operators in Conformal Space

6.1.3.1 Reflection

Just as uncertain geometric entities can be constructed as described in the previous subsection, it is also possible to construct uncertain operators. Two simple examples of this are shown in Fig. 6.6, where, in the first step, an uncertain plane $\boldsymbol{P}$ is constructed from three uncertain points $\boldsymbol{X}$, $\boldsymbol{Y}$, and $\boldsymbol{Z}$ as $\boldsymbol{P} = \boldsymbol{X} \wedge \boldsymbol{Y} \wedge \boldsymbol{Z} \wedge \boldsymbol{e}_\infty$. This uncertain plane also represents the reflection operator for a reflection in $\boldsymbol{P}$. A circle $\boldsymbol{C}$ with zero covariance is then reflected in the uncertain plane by evaluating $\boldsymbol{C}' = \boldsymbol{P}\,\boldsymbol{C}\,\widetilde{\boldsymbol{P}}$. Using error propagation, the uncertainty of the reflection plane $\boldsymbol{P}$ induces an uncertainty in $\boldsymbol{C}'$. The corresponding standard-deviation envelopes of the reflected circles $\boldsymbol{C}'$ are shown in the figure.

An interesting application of such an uncertain reflection is shown in Fig. 6.7. Again an uncertain plane $\boldsymbol{P}$ is constructed from three uncertain points $\boldsymbol{X}$, $\boldsymbol{Y}$, and $\boldsymbol{Z}$ as $\boldsymbol{P} = \boldsymbol{X} \wedge \boldsymbol{Y} \wedge \boldsymbol{Z} \wedge \boldsymbol{e}_\infty$. An ideal line $\boldsymbol{L}$ is taken to represent a laser beam that is reflected off a mirror represented by $\boldsymbol{P}$. The uncertainty of $\boldsymbol{P}$ may represent the expected vibrations of the mirror. When $\boldsymbol{L}$ is reflected via $\boldsymbol{L}' = \boldsymbol{P}\,\boldsymbol{L}\,\widetilde{\boldsymbol{P}}$, the uncertainty of the reflection plane induces an uncertainty in the reflected line, which represents the reflected light ray.

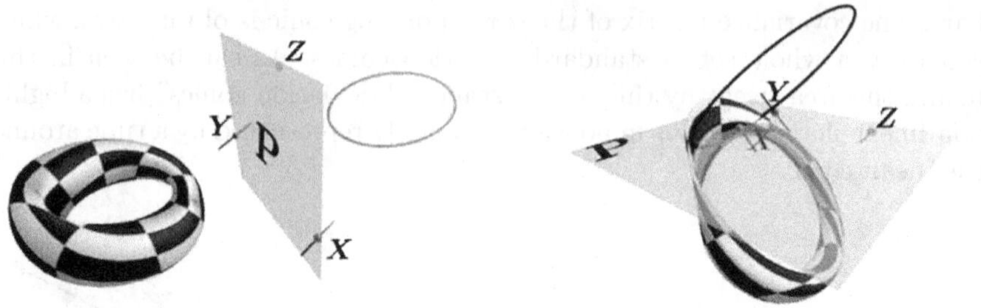

Fig. 6.6 Examples of the reflection of an ideal circle in an uncertain reflection plane $\boldsymbol{P}$. In the left example, the uncertain plane is generated by the outer product of two uncertain points $\boldsymbol{X}$, $\boldsymbol{Y}$ and an ideal point $\boldsymbol{Z}$. In the right example, all of the vectors $\boldsymbol{X}$, $\boldsymbol{Y}$, and $\boldsymbol{Z}$ have an uncertainty

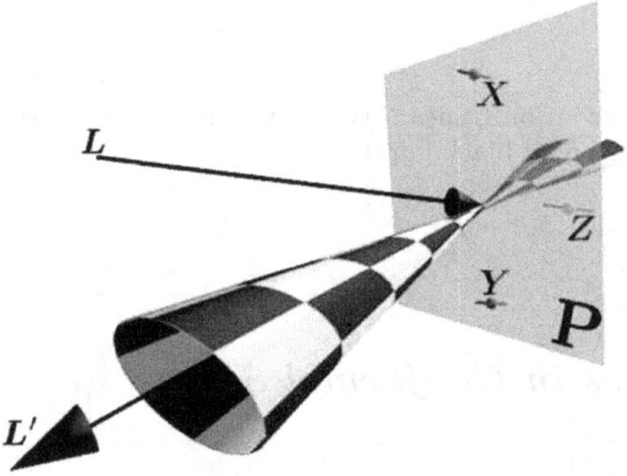

Fig. 6.7 Reflection of an ideal line in an uncertain plane, which results in an uncertain line

The standard-deviation envelope of $\boldsymbol{L}'$ then represents the expected deviation of the reflected laser beam from its mean.

6.1.3.2 Rotation

Another interesting example is the construction and application of an uncertain motor, i.e. a general Euclidean transformation. Consider four statistically independent points $\boldsymbol{X}, \boldsymbol{Y}, \boldsymbol{Z}, \boldsymbol{W} \in \mathbb{G}^1_{4,1}$ with associated covariances, as shown in Fig. 6.8(a). From these four points, two planes are constructed as $\boldsymbol{P} := \boldsymbol{X} \wedge \boldsymbol{Y} \wedge \boldsymbol{Z} \wedge \boldsymbol{e}_\infty$ and $\boldsymbol{Q} := \boldsymbol{X} \wedge \boldsymbol{Y} \wedge \boldsymbol{W} \wedge \boldsymbol{e}_\infty$. The motor $\boldsymbol{M} := \boldsymbol{P}\boldsymbol{Q}$ then represents a rotation about the intersection axis of $\boldsymbol{P}$ and $\boldsymbol{Q}$ by an angle that is twice the angle between the planes (see Sect. 4.3.11). This motor is visualized in Fig. 6.8(b) by its axis with a standard-deviation envelope and a wedge indicating the mean rotation plane and angle. The uncertainty in the rotation angle is not visualized. Rotating the ideal line $\boldsymbol{L}$ with this uncertain motor $\boldsymbol{M}$ results in an uncertain line $\boldsymbol{H}$, as shown in the figure. This construction is a good example for showing how error propagation is applied.

We denote the component vectors of $\boldsymbol{X}$, $\boldsymbol{Y}$, $\boldsymbol{Z}$, and $\boldsymbol{W}$ by x, y, z, and w, respectively. The corresponding covariance matrices are denoted by $\Sigma_{\mathsf{x,x}}$, $\Sigma_{\mathsf{y,y}}$, $\Sigma_{\mathsf{z,z}}$, and $\Sigma_{\mathsf{w,w}}$. First, the line that both planes have in common is calculated, i.e. $\boldsymbol{K} := \boldsymbol{X} \wedge \boldsymbol{Y} \wedge \boldsymbol{e}_\infty$. This contains the outer product $\boldsymbol{A} := \boldsymbol{X} \wedge \boldsymbol{Y}$, which is written in terms of the component vectors as

$$a^k = x^i\, y^j\, \Lambda^k{}_{ij}\,, \qquad \Lambda_L(\mathsf{y}) = y^j\, \Lambda^k{}_{ij}\,, \qquad \Lambda_R(\mathsf{x}) = x^i\, \Lambda^k{}_{ij}\,,$$

where $\Lambda_L(\mathsf{y})$ and $\Lambda_R(\mathsf{x})$ are the left and right Jacobi matrices of the outer-product operation. Since vectors $\boldsymbol{X}$ and $\boldsymbol{Y}$ statistically independent, the covariance matrix $\Sigma_{\mathsf{a,a}}$ of a is given by

$$\Sigma_{\mathsf{a,a}} = \Lambda_L(\mathsf{y})\, \Sigma_{\mathsf{x,x}}\, \Lambda_L(\mathsf{y})^{\mathsf{T}} + \Lambda_R(\mathsf{x})\, \Sigma_{\mathsf{y,y}}\, \Lambda_R(\mathsf{x})^{\mathsf{T}}\,.$$

The line $\boldsymbol{K}$, with component vector k, is evaluated via

$$k^r = a^i\, e^j_\infty\, \Lambda^r{}_{ij}\,, \qquad \Lambda_L(\mathsf{e}_\infty) = e^j_\infty\, \Lambda^r{}_{ij}\,,$$

where $\mathsf{e}_\infty := \mathcal{K}(\boldsymbol{e}_\infty)$ is the component vector of $\boldsymbol{e}_\infty$ and $\Lambda_L(\mathsf{e}_\infty)$ is the left Jacobi matrix of the outer-product operation. The covariance matrix of k is then given by

$$\Sigma_{\mathsf{k,k}} = \Lambda_L(\mathsf{e}_\infty)\, \Sigma_{\mathsf{a,a}}\, \Lambda_L(\mathsf{e}_\infty)^{\mathsf{T}}\,,$$

because e_∞ has zero covariance. The planes $\boldsymbol{P}$ and $\boldsymbol{Q}$ are then given by $\boldsymbol{P} = \boldsymbol{Z} \wedge \boldsymbol{K}$ and $\boldsymbol{Q} = \boldsymbol{W} \wedge \boldsymbol{K}$. Hence,

$$p^r = z^i\, k^j\, \Lambda^r{}_{ij} \qquad \text{and} \qquad q^r = w^i\, k^j\, \Lambda^r{}_{ij}\,,$$

with covariance matrices

$$\Sigma_{\mathsf{p,p}} = \Lambda_L(\mathsf{k})\, \Sigma_{\mathsf{z,z}}\, \Lambda_L(\mathsf{k})^{\mathsf{T}} + \Lambda_R(\mathsf{z})\, \Sigma_{\mathsf{k,k}}\, \Lambda_R(\mathsf{z})^{\mathsf{T}}\,,$$
$$\Sigma_{\mathsf{q,q}} = \Lambda_L(\mathsf{k})\, \Sigma_{\mathsf{w,w}}\, \Lambda_L(\mathsf{k})^{\mathsf{T}} + \Lambda_R(\mathsf{w})\, \Sigma_{\mathsf{k,k}}\, \Lambda_R(\mathsf{w})^{\mathsf{T}}\,.$$

Because $\boldsymbol{P}$ and $\boldsymbol{Q}$ both depend on $\boldsymbol{K}$, they are no longer statistically independent. The covariance matrix $\Sigma_{\mathsf{p,q}}$ capturing this relationship is needed later on and can be estimated as follows. The covariance between $\boldsymbol{P}$ and $\boldsymbol{K}$ is given by

$$\Sigma_{\mathsf{p,k}} = \Lambda_R(\mathsf{z})\, \Sigma_{\mathsf{k,k}}$$

and thus

$$\Sigma_{\mathsf{p,q}} = \Sigma_{\mathsf{p,k}}\, \Lambda_R(\mathsf{w})^{\mathsf{T}}\,,$$

which can be written in a single equation as

$$\Sigma_{\mathsf{p,q}} = \Lambda_R(\mathsf{z})\, \Sigma_{\mathsf{k,k}}\, \Lambda_R(\mathsf{w})^{\mathsf{T}}\,.$$

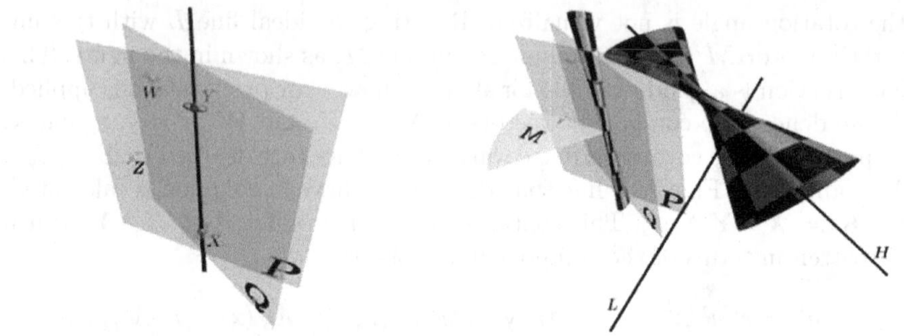

(a) Construction of uncertain planes $\boldsymbol{P}$ and $\boldsymbol{Q}$ from uncertain points $\boldsymbol{X}$, $\boldsymbol{Y}$, $\boldsymbol{Z}$, and $\boldsymbol{W}$

(b) Application of an uncertain motor $\boldsymbol{M} = \boldsymbol{P}\boldsymbol{Q}$ to a line $\boldsymbol{L}$, which results in $\boldsymbol{L}'$

Fig. 6.8 Example of construction and application of an uncertain motor

The motor $\boldsymbol{M} = \boldsymbol{P}\boldsymbol{Q}$, with component vector m and associated covariance matrix $\Sigma_{\mathsf{m},\mathsf{m}}$, is given by

$$m^r = p^i\, q^j\, \Gamma^r{}_{ij}\,, \quad \Gamma_L(\mathsf{q}) = q^j\, \Gamma^r{}_{ij}\,, \quad \Gamma_R(\mathsf{p}) = p^i\, \Gamma^r{}_{ij}\,,$$

where $\Gamma_L(\mathsf{q})$ and $\Gamma_R(\mathsf{p})$ are the left and right Jacobi matrices of the geometric product. The covariance matrix $\Sigma_{\mathsf{m},\mathsf{m}}$ is

$$\begin{aligned}\Sigma_{\mathsf{m},\mathsf{m}} = \;& \Gamma_L(\mathsf{q})\, \Sigma_{\mathsf{p},\mathsf{p}}\, \Gamma_L(\mathsf{q})^{\mathsf{T}} + \Gamma_R(\mathsf{p})\, \Sigma_{\mathsf{q},\mathsf{q}}\, \Gamma_R(\mathsf{p})^{\mathsf{T}} \\ &+ \Gamma_L(\mathsf{q})\, \Sigma_{\mathsf{p},\mathsf{q}}\, \Gamma_R(\mathsf{p})^{\mathsf{T}} + \Gamma_R(\mathsf{p})\, \Sigma_{\mathsf{p},\mathsf{q}}^{\mathsf{T}}\, \Gamma_L(\mathsf{q})^{\mathsf{T}}\,.\end{aligned}$$

The rotation of an ideal line $\boldsymbol{L}$ with the motor $\boldsymbol{M}$ is given by $\boldsymbol{H} = \boldsymbol{M}\,\boldsymbol{L}\,\widetilde{\boldsymbol{M}}$, which can be calculated in two steps. First, $\boldsymbol{B} := \boldsymbol{M}\,\boldsymbol{L}$ is evaluated as

$$b^r = m^i\, \ell^j\, \Gamma^r{}_{ij}\,, \quad \Gamma_L(\boldsymbol{\ell}) = \ell^j\, \Gamma^r{}_{ij}\,,$$

so that

$$\Sigma_{\mathsf{b},\mathsf{b}} = \Gamma_L(\boldsymbol{\ell})\, \Sigma_{\mathsf{m},\mathsf{m}}\, \Gamma_L(\boldsymbol{\ell})^{\mathsf{T}} \qquad \text{and} \qquad \Sigma_{\mathsf{b},\mathsf{m}} = \Gamma_L(\boldsymbol{\ell})\, \Sigma_{\mathsf{m},\mathsf{m}}\,.$$

The component vector h of the rotated line $\boldsymbol{H}$ is then given by

$$h^r = b^i\, m^j\, G^r{}_{ij}\,, \qquad G^r{}_{ij} := \Gamma^r{}_{ik}\, R^k{}_j\,,$$

where $R^k{}_j$ represents the reverse operation. The Jacobi matrices are

$$\mathsf{G}_L(\mathsf{m}) := m^j\, G^r{}_{ij} \qquad \text{and} \qquad \mathsf{G}_R(\mathsf{b}) := b^i\, G^r{}_{ij}\,.$$

Hence,

$$\begin{aligned}\Sigma_{\mathsf{h,h}} = \;& \mathsf{G}_L(\mathsf{m})\, \Sigma_{\mathsf{b,b}}\, \mathsf{G}_L(\mathsf{m})^\mathsf{T} + \mathsf{G}_R(\mathsf{b})\, \Sigma_{\mathsf{m,m}}\, \mathsf{G}_R(\mathsf{b})^\mathsf{T} \\ &+ \mathsf{G}_L(\mathsf{m})\, \Sigma_{\mathsf{b,m}}\, \mathsf{G}_R(\mathsf{b})^\mathsf{T} + \mathsf{G}_R(\mathsf{b})\, \Sigma_{\mathsf{b,m}}^\mathsf{T}\, \mathsf{G}_L(\mathsf{m})^\mathsf{T} \,.\end{aligned}$$

6.2 Estimation

A fundamental problem that often occurs is the evaluation of a geometric entity based on the measurement of a number of geometric entities of a different type. One of the simplest examples of this problem is probably to find the line $\boldsymbol{L}$ that best fits a given set of points $\{\boldsymbol{X}_n\}$. Additionally, the covariance matrix of the estimated line is to be evaluated. This can be achieved using the Gauss–Helmert model as described in Sect. 5.9. The Gauss–Helmert model has been used in the context of projective geometry for the estimation of geometric entities by Heuel [93]. Its application to the estimation of geometric entities in geometric algebra was introduced by Perwass and Förstner in [136] and was extended further to the estimation of operators in geometric algebra in [138, 139].

The Gauss–Helmert model allows the estimation of a parameter vector with an associated covariance matrix, given a set of data vectors with covariance matrices, a constraint function between the data and parameter vectors, and possibly a constraint function on the parameters alone. The resultant parameter vector is the solution to a system of linear equations that depends on the Jacobi matrices of the constraint functions and on the data and the covariance matrices, as discussed in Sect. 5.9.

6.2.1 Estimation of Geometric Entities

Consider again the example of fitting a line $\boldsymbol{L}$ through a set of points $\{\, \boldsymbol{X}_n \,\}$. In terms of the Gauss–Helmert model, the parameter vector is the component vector $\boldsymbol{\ell}$ of $\boldsymbol{L}$ that is to be estimated, and the data vectors $\{\mathsf{x}_n\}$ are the component vectors of the points $\{\boldsymbol{X}_n\}$. The constraint function $\boldsymbol{Q}(\boldsymbol{X}_n,\, \boldsymbol{L})$ between the data and the parameters has to be zero if a point lies on the line. The constraint function on the parameters alone $\boldsymbol{H}(\boldsymbol{L})$ has to be zero if $\boldsymbol{L}$ represents a normalized line, i.e. $\boldsymbol{\ell}$ satisfies the Plücker condition and $\boldsymbol{\ell}^\mathsf{T}\, \boldsymbol{\ell} = 1$.

The appropriate constraint between the data and parameter vectors can be read off from Table 4.6, i.e.

$$\boldsymbol{Q}(\boldsymbol{X}_n,\, \boldsymbol{L}) = \boldsymbol{X}_n \wedge \boldsymbol{L}\,, \quad \text{or} \quad q^k(\mathsf{x}_n,\, \boldsymbol{\ell}) = x_n^i\, \ell^j\, \Lambda^k{}_{ij}\,.$$

The Plücker and normalization conditions on $\boldsymbol{L}$ can be combined into the function

$$\boldsymbol{H}(\boldsymbol{L}) = \boldsymbol{L}\,\widetilde{\boldsymbol{L}} - 1\,, \quad \text{or} \quad h^k(\boldsymbol{\ell}) = \ell^{i_1}\,\ell^j\,R^{i_2}{}_j\,\Gamma^k{}_{i_1 i_2} - \delta^k{}_1\,,$$

where $\delta^k{}_j$ is the Kronecker delta, and the index 1 is assumed to be the index of the scalar component of the corresponding multivector. In fact, this type of constraint is applicable to all blades that represent geometric entities and to all operators that represent Euclidean transformations. That is, a multivector $\boldsymbol{A} \in \mathbb{G}_{4,1}$ represents a geometric entity or Euclidean transformation operator if and only if $\boldsymbol{A}\,\widetilde{\boldsymbol{A}} \in \mathbb{G}^0_{4,1}$, i.e. it is a scalar. The free scale can be fixed by enforcing $\boldsymbol{A}\,\widetilde{\boldsymbol{A}} = 1$.

The Jacobi matrices of q are

$$\mathsf{Q}^k_{nj} = x^i_n\,\Lambda^k{}_{ij} \quad \text{and} \quad \bar{\mathsf{Q}}^k{}_i = \ell^j\,\Lambda^k{}_{ij}$$

and the Jacobi matrix of h is

$$\mathsf{H}^k{}_j = \ell^{i_1}\,\left(R^{i_2}{}_{i_1}\,\Gamma^k{}_{j i_2} + R^{i_2}{}_j\,\Gamma^k{}_{i_1 i_2}\right)\,.$$

These results can be substituted directly into the normal-equation formulas of the Gauss–Markov model (see Sect. 5.8.4) or the Gauss–Helmert model (see Sect. 5.9.4) to estimate the best line through the set of points in a least-squares sense.

Table 6.1, which is based on Table 4.6, lists the constraint functions $\boldsymbol{Q}$ between geometric entities that result in a zero vector if one geometric entity is completely contained within the other. For example, the constraint between two lines is zero only if the multivectors describe the same line up to a scale. The constraint function $\boldsymbol{H}$ remains the same for all parameter types.

Note that the constraints $\boldsymbol{Q}$ between two entity types can be used in two directions: instead of fitting a line to a set of points, a point can be fitted to a set of lines using the same constraint. This particular example can be used for triangulation, where the best intersection of a set of projection rays has to be evaluated. Similarly, the best intersection line of a set of projective planes can found. In Table 6.1, the symbols $\underline{\times}$ and $\overline{\times}$ denote the commutator and the anticommutator product, respectively, which are defined as $\boldsymbol{A} \underline{\times} \boldsymbol{B} = \frac{1}{2}(\boldsymbol{A}\,\boldsymbol{B} - \boldsymbol{B}\,\boldsymbol{A})$ and $\boldsymbol{A} \overline{\times} \boldsymbol{B} = \frac{1}{2}(\boldsymbol{A}\,\boldsymbol{B} + \boldsymbol{B}\,\boldsymbol{A})$ for $\boldsymbol{A}, \boldsymbol{B} \in \mathbb{G}_{p,q}$. In terms of the component vectors, this becomes

$$\boldsymbol{A} \underline{\times} \boldsymbol{B} \iff \frac{1}{2}\,\mathsf{a}^i\,\mathsf{b}^j\,(\Gamma^k{}_{ij} - \Gamma^k{}_{ji}), \qquad \boldsymbol{A} \overline{\times} \boldsymbol{B} \iff \frac{1}{2}\,\mathsf{a}^i\,\mathsf{b}^j\,(\Gamma^k{}_{ij} + \Gamma^k{}_{ji}). \tag{6.1}$$

Table 6.1 Constraints between data and parameters that are zero if the corresponding geometric entities are contained in one another

Data	Parameter				
	Point $\boldsymbol{X}$	Line $\boldsymbol{L}$	Plane $\boldsymbol{P}$	Circle $\boldsymbol{C}$	Sphere $\boldsymbol{S}$
Points $\{\boldsymbol{Y}_n\}$	$\boldsymbol{X}\wedge\boldsymbol{Y}_n$	$\boldsymbol{L}\wedge\boldsymbol{Y}_n$	$\boldsymbol{P}\wedge\boldsymbol{Y}_n$	$\boldsymbol{C}\wedge\boldsymbol{Y}_n$	$\boldsymbol{S}\wedge\boldsymbol{Y}_n$
Lines $\{\boldsymbol{K}_n\}$	$\boldsymbol{X}\wedge\boldsymbol{K}_n$	$\boldsymbol{L}\underline{\times}\boldsymbol{K}_n$	$\boldsymbol{P}\overline{\times}\boldsymbol{K}_n$		
Planes $\{\boldsymbol{O}_n\}$	$\boldsymbol{X}\wedge\boldsymbol{O}_n$	$\boldsymbol{L}\overline{\times}\boldsymbol{O}_n$	$\boldsymbol{P}\underline{\times}\boldsymbol{O}_n$		
Circles $\{\boldsymbol{B}_n\}$	$\boldsymbol{X}\wedge\boldsymbol{B}_n$			$\boldsymbol{C}\underline{\times}\boldsymbol{B}_n$	$\boldsymbol{S}\overline{\times}\boldsymbol{B}_n$
Spheres $\{\boldsymbol{R}_n\}$	$\boldsymbol{X}\wedge\boldsymbol{R}_n$			$\boldsymbol{C}\overline{\times}\boldsymbol{R}_n$	$\boldsymbol{S}\underline{\times}\boldsymbol{R}_n$

6.2.2 Versor Equation

There are many possibilities to define constraint equations that depend on transformation operators. The most common one is the *versor equation* that was introduced in Sect. 3.3 and whose numerical solution was treated in Sect. 5.2.2. Another important example arises in the case of monocular pose estimation, which is treated in Chap. 8. In this subsection, the estimation of a general Euclidean transformation that relates two point sets is discussed as an example of the versor equation.

Let $\boldsymbol{M} \in \mathbb{G}_{4,1}$ denote a motor (a general Euclidean transformation) and let $\{\,\boldsymbol{X}_n\,\} \subset \mathbb{G}^1_{4,1}$ and $\{\,\boldsymbol{Y}_n\,\} \subset \mathbb{G}^1_{4,1}$ denote two sets of points, such that

$$\boldsymbol{Y}_n = \boldsymbol{M}\,\boldsymbol{X}_n\,\widetilde{\boldsymbol{M}} \quad \forall\, n\,. \tag{6.2}$$

Hence, the constraint function between the data vectors and the parameter vector $\boldsymbol{M}$ can be written as

$$\boldsymbol{Q}(\boldsymbol{X}_n,\,\boldsymbol{Y}_n,\,\boldsymbol{M}) = \boldsymbol{M}\,\boldsymbol{X}_n - \boldsymbol{Y}_n\,\boldsymbol{M}\,. \tag{6.3}$$

Note that this implies that $\boldsymbol{X}_n$ and $\boldsymbol{Y}_n$ are not scaled, or at least they are scaled by the same factor. Typically, this can be safely assumed for points, since they are embedded from Euclidean space, in which the data has been measured. For other geometric entities such as lines this is generally not the case; this situation will be treated later.

Let $\mathsf{m} := \mathcal{K}(\boldsymbol{M})$, $\mathsf{x}_n := \mathcal{K}(\boldsymbol{X}_n)$, and $\mathsf{y}_n := \mathcal{K}(\boldsymbol{Y}_n)$ denote the respective component vectors; then (6.3) can be written as

$$\begin{aligned} & q^k(\mathsf{x}_n,\,\mathsf{y}_n,\,\mathsf{m}) = m^i\,x^j_n\,\Gamma^k{}_{ij} - y^i_n\,m^j\,\Gamma^k{}_{ij} \\ \iff\; & q^k(\mathsf{x}_n,\,\mathsf{y}_n,\,\mathsf{m}) = m^i\,\left(x^j_n\,\Gamma^k{}_{ij} - y^j_n\,\Gamma^k{}_{ji}\right)\,. \end{aligned} \tag{6.4}$$

In matrix notation, this becomes

$$\mathsf{q}(\mathsf{x}_n, \mathsf{y}_n, \mathsf{m}) = \mathsf{Q}_n\, \mathsf{m}\ , \qquad \mathsf{Q}_n := x_n^j\, {\Gamma^k}_{ij} - y_n^j\, {\Gamma^k}_{ji}\ . \tag{6.5}$$

Therefore, the solution for m lies in the right null space of Q_n. If the algebraic product tensors are reduced to the minimal number of components necessary for the specific calculation, then $\mathsf{Q}_n \in \mathbb{R}^{16\times 8}$ has a 4-dimensional right null space.

Given at least three point pairs, the motor components m can be evaluated by finding the combined right null space of the corresponding $\{\,\mathsf{Q}_n\,\}$. If $\boldsymbol{M}$ is known to represent only a rotation about some axis and not a general Euclidean transformation (which also contains a translation along the rotation axis), then m has only seven components (see Table 4.5), and two pairs of points suffice to calculate m. The solution for m is given by the right null space of

$$\mathsf{Q} = \begin{bmatrix} \mathsf{Q}_1 \\ \vdots \\ \mathsf{Q}_N \end{bmatrix}, \qquad \text{or} \qquad \mathsf{Q} = \sum_{n=1}^{N} \mathsf{Q}_n^\mathsf{T}\, \mathsf{Q}_n\ .$$

When a singular-value decomposition is used to find the right null space, no additional constraints on m are necessary, since the SVD algorithm already ensures that the resultant null space vector is normalized. The remaining constraint that $\boldsymbol{M}\,\widetilde{\boldsymbol{M}} \in \mathbb{G}^0_{4,1}$ is a scalar is already implicit in the constraint equation, since (6.2) and (6.3) are equivalent only if $\boldsymbol{M}\,\widetilde{\boldsymbol{M}} = 1$. Note again that this approach works for all combinations of operators and geometric entities, if the geometric entities have the same scale. It can even be used to estimate an operator that relates two sets of *operators*.

6.2.2.1 Gauss–Helmert Estimation

Given covariance matrices $\Sigma_{\mathsf{x}_n,\mathsf{x}_n}$ and $\Sigma_{\mathsf{y}_n,\mathsf{y}_n}$ for the $\{\,\mathsf{x}_n\,\}$ and $\{\,\mathsf{y}_n\,\}$, the initial solution for m given by the right null space of Q is a good starting value for the Gauss–Helmert estimation method (see Sect. 5.9). To apply Gauss–Helmert estimation, the Jacobi matrices of the constraint functions $\boldsymbol{Q}$ and $\boldsymbol{H}$ have to be known. The constraint function $\boldsymbol{H}(\boldsymbol{M})$ on the parameters alone is again given by

$$\boldsymbol{H}(\boldsymbol{M}) = \boldsymbol{M}\,\widetilde{\boldsymbol{M}} - 1\ . \tag{6.6}$$

To bring the constraint function $\boldsymbol{Q}$ into the form in which it is needed for the Gauss–Helmert estimation, the data vector pairs x_n, y_n are combined into single vectors

$$\mathsf{z}_n := \begin{bmatrix} \mathsf{x}_n \\ \mathsf{y}_n \end{bmatrix}, \qquad \Sigma_{\mathsf{z}_n,\mathsf{z}_n} := \begin{bmatrix} \Sigma_{\mathsf{x}_n,\mathsf{x}_n} & \\ & \Sigma_{\mathsf{y}_n,\mathsf{y}_n} \end{bmatrix}.$$

The Jacobi matrices of $\mathsf{q}(\mathsf{z}_n,\,\mathsf{m})$ are then

$$\begin{aligned} \mathsf{U}_n &:= (\partial_{\mathsf{m}}\,\mathsf{q})(\mathsf{z}_n,\,\mathsf{m}) = \mathsf{Q}_n\,, \\ \mathsf{V}_n &:= (\partial_{\mathsf{z}_n}\,\mathsf{q})(\mathsf{z}_n,\,\mathsf{m}) = \Big[\, m^i\, \Gamma^k{}_{ij}\,,\; -m^i\, \Gamma^k{}_{ji} \,\Big]\,. \end{aligned} \tag{6.7}$$

The constraint function $\boldsymbol{H}(\boldsymbol{M})$ can be expressed in terms of component vectors as

$$h^r(\mathsf{m}) = m^i\, m^j\, R^k{}_j\, \Gamma^r{}_{ik} - \delta^r{}_1\,,$$

where $\delta^k{}_j$ is the Kronecker delta, and the index 1 is assumed to be the index of the scalar component of the corresponding multivector. This is equivalent to the Plücker condition mentioned in Sect. 6.2.1. The corresponding Jacobi matrix is

$$\mathsf{H}^r{}_k = m^i\, (R^j{}_i\, \Gamma^r{}_{kj} + R^j{}_k\, \Gamma^r{}_{ij})\,.$$

These Jacobi matrices can be substituted directly into the normal equations as given in Sect. 5.9.4.

6.2.3 Projective Versor Equation

The projective versor equation relates two multivectors up to a scale. For example, given $\boldsymbol{A}, \boldsymbol{B} \in \mathbb{G}_{p,q}$ and a versor $\boldsymbol{V} \in \mathbb{G}_{p,q}$, then $\boldsymbol{A}$, $\boldsymbol{B}$ satisfy the projective-versor-equation constraint if

$$\boldsymbol{B} \simeq \boldsymbol{V}\,\boldsymbol{A}\,\widetilde{\boldsymbol{V}}\,,$$

where "$\simeq$" denotes equality up to a scalar factor. This type of equation often occurs in projective spaces, such as, for example, conformal space. This is because a multivector in conformal space represents the same geometric entity independent of its scale. This versor equation can be made into a constraint equation by writing it as

$$\boldsymbol{Q}(\boldsymbol{A},\,\boldsymbol{B},\,\boldsymbol{V}) = \boldsymbol{B} \underline{\times} (\boldsymbol{V}\,\boldsymbol{A}\,\widetilde{\boldsymbol{V}})\,. \tag{6.8}$$

This is directly related to the diagonal entries of the containment relations listed in Table 6.1. If $\boldsymbol{A}, \boldsymbol{B} \in \mathbb{G}^1_{p,q}$ are 1-vectors, then the commutator product is equivalent to the outer product.

More generally, all of the containment relations in Table 6.1 can be used in this way to generate constraints on versors. For example, one relation that is of particular importance for monocular pose estimation,is the point–line constraint. Let $\boldsymbol{X} \in \mathbb{G}^1_{4,1}$ represent a point, $\boldsymbol{L} \in \mathbb{G}^3_{4,1}$ a line, and $\boldsymbol{M} \in \mathbb{G}_{4,1}$ a motor (a general Euclidean transformation). If the goal is to find the versor $\boldsymbol{V}$ that maps the point $\boldsymbol{X}$ onto the line $\boldsymbol{L}$, then the corresponding constraint equation is

$$\boldsymbol{Q}(\boldsymbol{X}, \boldsymbol{L}, \boldsymbol{M}) = \boldsymbol{L} \wedge (\boldsymbol{M}\, \boldsymbol{X}\, \widetilde{\boldsymbol{M}}) . \tag{6.9}$$

The main difference from the standard versor equation is that this constraint equation cannot be made linear in $\boldsymbol{M}$. In terms of the component vectors $\boldsymbol{\ell} = \mathcal{K}(\boldsymbol{L})$, $\mathsf{x} = \mathcal{K}(\boldsymbol{X})$, and $\mathsf{m} = \mathcal{K}(\boldsymbol{M})$, the constraint equation becomes

$$q^r(\boldsymbol{\ell}, \mathsf{x}, \mathsf{m}) := \ell^i\, x^j\, \mathsf{m}^r\, \mathsf{m}^s\, V^p{}_{ijrs} , \tag{6.10}$$

where

$$V^p{}_{ijrs} := \Lambda^p{}_{iu}\, \Gamma^k{}_{rj}\, \Gamma^u{}_{kt}\, R^t{}_s .$$

Let $\Sigma_{\boldsymbol{\ell},\boldsymbol{\ell}}$ and $\Sigma_{\mathsf{x},\mathsf{x}}$ denote the covariance matrices of the data vectors $\boldsymbol{\ell}$ and x. These quantities can be combined into a single vector z with a covariance matrix $\Sigma_{\mathsf{z},\mathsf{z}}$ as

$$\mathsf{z} := \begin{bmatrix} \boldsymbol{\ell} \\ \mathsf{x} \end{bmatrix} , \qquad \Sigma_{\mathsf{z},\mathsf{z}} := \begin{bmatrix} \Sigma_{\boldsymbol{\ell},\boldsymbol{\ell}} & \\ & \Sigma_{\mathsf{x},\mathsf{x}} \end{bmatrix} .$$

The Jacobi matrices of $\mathsf{q}(\mathsf{z}, \mathsf{m})$ are then given by

$$\begin{aligned} \mathsf{U} &:= (\partial_{\mathsf{m}}\, \mathsf{q})(\mathsf{z}, \mathsf{m}) = \ell^i\, x^j\, \mathsf{m}^r\, \left(V^p{}_{ijrs} + V^p{}_{ijsr}\right) , \\ \mathsf{V} &:= (\partial_{\mathsf{z}}\, \mathsf{q})(\mathsf{z}, \mathsf{m}) = \left[\, x^j\, \mathsf{m}^r\, \mathsf{m}^s\, V^p{}_{ijrs} ,\; \ell^i\, \mathsf{m}^r\, \mathsf{m}^s\, V^p{}_{ijrs} \right] . \end{aligned} \tag{6.11}$$

Because the constraint equation $\mathsf{q}(\mathsf{z}, \mathsf{m})$ is quadratic in m, an initial estimate for m cannot be obtained by evaluating the null space of some matrix. Instead, the Gauss–Markov estimation (see Sect. 5.8) may be used for this purpose. The result of the Gauss–Markov estimation may then be used as an initial estimate for the Gauss–Helmert estimation. The Gauss–Helmert estimation then takes account of the covariances of the data vectors.

6.2.4 Constraint Metrics

The only requirement on the constraint function $\boldsymbol{Q}$ between the data and parameter vectors noted so far is that it results in zero if the corresponding constraint is satisfied. However, this requirement alone is not sufficient to ensure a successful minimization. It is also important that $\boldsymbol{Q}$ is at least convex and has a linear relation to the true ideal constraint. For example, when one is fitting a geometric entity through a set of points, the constraint function should ideally represent the Euclidean distance between the points and the geometric entity. Whether a constraint function satisfies these requirements has to be checked in each case. In the following, some of the most common cases are discussed in some detail.

6.2.4.1 Point–Line Metric

The point–line metric is the metric of the point–line constraint, which is given by

$$\boldsymbol{X} \wedge \boldsymbol{L} = 0 \,,$$

where $\boldsymbol{X} \in \mathbb{G}^1_{4,1}$ represents a point and $\boldsymbol{L} \in \mathbb{G}^3_{4,1}$ a line. In the following, it is shown that $\Delta := \|\boldsymbol{X} \wedge \boldsymbol{L}\|$ is proportional to the Euclidean distance of the point represented by $\boldsymbol{X}$ from the line represented by $\boldsymbol{L}$.

For this purpose, consider three points $\boldsymbol{a}, \boldsymbol{b}, \boldsymbol{x} \in \mathbb{R}^3$ in 3D Euclidean space. The line passes through the points $\boldsymbol{a}$ and $\boldsymbol{b}$, while $\boldsymbol{x}$ is the test point, whose distance from the line is to be calculated. Let $\boldsymbol{b} = \boldsymbol{a} + \alpha\, \boldsymbol{r}$ and $\boldsymbol{c} = \boldsymbol{a} + \beta\, \boldsymbol{r} + \boldsymbol{d}$, where $\alpha, \beta \in \mathbb{R}$, $\boldsymbol{r} := (\boldsymbol{b} - \boldsymbol{a})/\|\boldsymbol{b} - \boldsymbol{a}\|$ is the unit direction vector of the line, and $\boldsymbol{d}$ is the orthogonal distance vector of $\boldsymbol{x}$ from the line, and hence $\boldsymbol{d} \cdot \boldsymbol{r} = 0$.

We define $\boldsymbol{A} := \mathcal{C}(\boldsymbol{a}) = \boldsymbol{a} + \dfrac{1}{2}\boldsymbol{a}^2\, \boldsymbol{e}_\infty + \boldsymbol{e}_o$, $\boldsymbol{B} := \mathcal{C}(\boldsymbol{b})$, and $\boldsymbol{X} := \mathcal{C}(\boldsymbol{x})$. The line $\boldsymbol{L}$ through $\boldsymbol{A}$ and $\boldsymbol{B}$ is the given by $\boldsymbol{L} := \boldsymbol{A} \wedge \boldsymbol{B} \wedge \boldsymbol{e}_\infty$. To calculate the distance measure $\Delta = \|\boldsymbol{X} \wedge \boldsymbol{L}\|$, the outer product $\boldsymbol{X} \wedge \boldsymbol{L}$ has to be evaluated:

$$\begin{aligned} \boldsymbol{X} \wedge \boldsymbol{L} &= \boldsymbol{A} \wedge \boldsymbol{B} \wedge \boldsymbol{X} \wedge \boldsymbol{e}_\infty \\ &= \boldsymbol{a} \wedge \boldsymbol{b} \wedge \boldsymbol{c} \wedge \boldsymbol{e}_\infty + (\boldsymbol{a} \wedge \boldsymbol{b} + \boldsymbol{b} \wedge \boldsymbol{c} + \boldsymbol{c} \wedge \boldsymbol{a}) \wedge \boldsymbol{e}_o \wedge \boldsymbol{e}_\infty \\ &= \alpha\, \boldsymbol{r} \wedge \boldsymbol{d} \wedge (\boldsymbol{a} + \boldsymbol{e}_o) \wedge \boldsymbol{e}_\infty \,, \end{aligned} \tag{6.12}$$

which is independent of β, the translation of the test point $\boldsymbol{x}$ parallel to the direction of the line. Thus,

$$\Delta = \|\boldsymbol{X} \wedge \boldsymbol{L}\| = \|\boldsymbol{d}\|\ \|\alpha\|\ \|\boldsymbol{r} \wedge \hat{\boldsymbol{d}} \wedge (\boldsymbol{a} + \boldsymbol{e}_o) \wedge \boldsymbol{e}_\infty\| \propto \|\boldsymbol{d}\| \,, \tag{6.13}$$

where $\hat{\boldsymbol{d}} := \boldsymbol{d}/\|\boldsymbol{d}\|$. That is, for fixed points $\boldsymbol{a}$ and $\boldsymbol{b}$, the value of $\Delta = \|\boldsymbol{X} \wedge \boldsymbol{L}\|$ depends linearly on the Euclidean distance $\|\boldsymbol{d}\|$ of the point $\boldsymbol{X}$ from the line $\boldsymbol{L}$. Hence, when the best line $\boldsymbol{L}$ that passes through a set of points $\{\,\boldsymbol{X}_n\,\}$ is estimated by minimizing $\|\boldsymbol{X}_n \wedge \boldsymbol{L}\|$, the Euclidean distance between the points and the line is minimized.

6.2.4.2 Point–Circle Metric

The metric of the point–circle constraint is somewhat more complex. The point–circle constraint is given by $\boldsymbol{X} \wedge \boldsymbol{C}$, where $\boldsymbol{X} \in \mathbb{G}^1_{4,1}$ represents a point and $\boldsymbol{C} \in \mathbb{G}^3_{4,1}$ represents a circle. If the point $\boldsymbol{X}$ lies on the circle $\boldsymbol{C}$, then $\boldsymbol{X} \wedge \boldsymbol{C} = 0$. The question to be discussed here is the form of the distance measure

$$\Delta := \|\boldsymbol{X} \wedge \boldsymbol{C}\| \,,$$

if $\boldsymbol{X}$ does not lie on the circle. Figure 6.9 shows the contours of equal Δ in a plane perpendicular to the plane of the circle that passes through the circle's

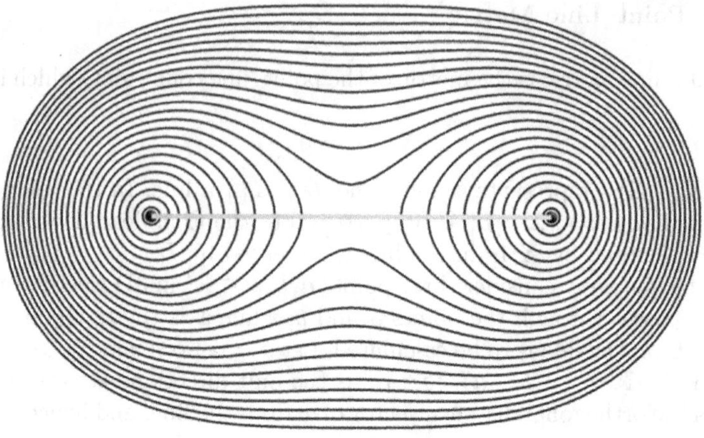

Fig. 6.9 Isometries of the point–circle constraint. Close to the circle, the metric approximates the Euclidean metric

center. The circle is indicated by the gray line, and its intersections with the plane are drawn as black points. This metric is rotationally symmetric about an axis perpendicular to the plane of the circle that passes through the circle's center.

It can be seen that close to the circle, the isometries approximate the Euclidean metric, while further away this is no longer the case. In general, it may be said that points on the inside of the circle are given less weight than points with the same Euclidean distance on the outside.

In the following, the evaluation of the isometries is outlined. First of all, it simplifies the equations considerably if the IPNS representation of a circle is used. This does not change Δ, since

$$\Delta = \|\boldsymbol{X} \wedge \boldsymbol{C}\| = \|(\boldsymbol{X} \cdot \boldsymbol{C}^*)\boldsymbol{I}\| = \|\boldsymbol{X} \cdot \boldsymbol{C}^*\| \,;$$

multiplication by the unit pseudoscalar does not change the magnitude of a multivector. The inner-product representation of a circle can be parameterized as the intersection of two inner-product spheres. That is, we define

$$\boldsymbol{S}_1 := y\, \boldsymbol{e}_2 + \frac{1}{2}\,(y^2 - r^2)\, \boldsymbol{e}_\infty + \boldsymbol{e}_o \quad \text{and} \quad \boldsymbol{S}_2 := -y\, \boldsymbol{e}_2 + \frac{1}{2}\,(y^2 - r^2)\, \boldsymbol{e}_\infty + \boldsymbol{e}_o \,,$$

such that $\boldsymbol{C}^* = \boldsymbol{S}_1 \wedge \boldsymbol{S}_2$ is the intersection of the two spheres. The circle lies in the $\boldsymbol{e}_1$–$\boldsymbol{e}_2$ plane and is centered on the origin. Since any circle can be moved into this position using a Euclidean transformation (an isometric transformation), this setup describes a circle without loss of generality. Equating $\boldsymbol{C}^*$ gives

$$\boldsymbol{C}^* = y\,(y^2 - r^2)\, \boldsymbol{e}_2\, \boldsymbol{e}_\infty + 2\, y\, \boldsymbol{e}_2\, \boldsymbol{e}_o \,.$$

If we define $\boldsymbol{X} := a\,\boldsymbol{e}_1 + b\,\boldsymbol{e}_2 + \frac{1}{2}\,(a^2+b^2)\,\boldsymbol{e}_\infty + \boldsymbol{e}_o$, it follows that

$$\boldsymbol{X}\cdot\boldsymbol{C}^* = \left((a^2+b^2)\,y + y\,(y^2-r^2)\right)\boldsymbol{e}_2 + b\,y\,(y^2-r^2)\,\boldsymbol{e}_\infty + 2\,b\,y\,\boldsymbol{e}_o\,.$$

Straightforward calculation then gives

$$\begin{aligned}\Delta^2 &= \|\boldsymbol{X}\cdot\boldsymbol{C}^*\|^2\\ &= p^2\,A + q^2\,A + 2\,p\,q\,A + p\,B + q\,C + D\,,\end{aligned}$$

where $p := a^2$, $q := b^2$, and

$$\begin{aligned}A &:= y^2\,,\\ B &:= 2\,y^2\,(y^2-r^2)\,,\\ C &:= 2\,y^2\left((y^2-r^2)^2 + (y^2-r^2) + 1\right),\\ D &:= y^2\,(y^2-r^2)^2\,.\end{aligned}$$

The isometric contour for a distance d is therefore a root of the polynomial $\Delta^2 - d^2$, which is quadratic in p and q. A parameterized expression for p in terms of q or vice versa can thus be easily found.

6.2.4.3 Versor Equation Metric

Here, the metric of the versor constraint equation given in Sect. 6.2.2 is discussed for points. That is, let $\boldsymbol{X}, \boldsymbol{Y} \in \mathbb{G}^1_{4,1}$ represent two points such that $\boldsymbol{Y} = \boldsymbol{V}\,\boldsymbol{X}\,\widetilde{\boldsymbol{V}}$, where $\boldsymbol{V} \in \mathbb{G}_{4,1}$ is a unitary versor, i.e. $\boldsymbol{V}\,\widetilde{\boldsymbol{V}} = 1$. The constraint equation for $\boldsymbol{V}$ is then

$$\boldsymbol{Q}(\boldsymbol{X},\boldsymbol{Y},\boldsymbol{V}) = \boldsymbol{V}\,\boldsymbol{X} - \boldsymbol{Y}\,\boldsymbol{V}\,.$$

If $\boldsymbol{V}$ does not satisfy the equation $\boldsymbol{Q}(\boldsymbol{X},\boldsymbol{Y},\boldsymbol{V}) = 0$, then the distance measure is

$$\Delta := \|\boldsymbol{V}\,\boldsymbol{X} - \boldsymbol{Y}\,\boldsymbol{V}\| = \|(\boldsymbol{V}\,\boldsymbol{X}\,\widetilde{\boldsymbol{V}} - \boldsymbol{Y})\,\boldsymbol{V}\|\,.$$

What is minimized is Δ^2, which can be written as

$$\begin{aligned}\Delta^2 &= (\boldsymbol{V}\,\boldsymbol{X}\,\widetilde{\boldsymbol{V}} - \boldsymbol{Y})\,\boldsymbol{V}\,\boldsymbol{V}^\dagger\,(\boldsymbol{V}\,\boldsymbol{X}\,\widetilde{\boldsymbol{V}} - \boldsymbol{Y})^\dagger\\ &= (\boldsymbol{V}\,\boldsymbol{X}\,\widetilde{\boldsymbol{V}} - \boldsymbol{Y})\,(\boldsymbol{V}\,\boldsymbol{X}\,\widetilde{\boldsymbol{V}} - \boldsymbol{Y})^\dagger\\ &= \|\boldsymbol{V}\,\boldsymbol{X}\,\widetilde{\boldsymbol{V}} - \boldsymbol{Y}\|^2\,.\end{aligned}$$

We define $\boldsymbol{Z} := \boldsymbol{V}\,\boldsymbol{X}\,\widetilde{\boldsymbol{V}}$ to be the transformed vector $\boldsymbol{X}$, which can be written as $\boldsymbol{Z} = \boldsymbol{z} + \frac{1}{2}\,\boldsymbol{z}^2\,\boldsymbol{e}_\infty + \boldsymbol{e}_o$. Since $\boldsymbol{X}$ and $\boldsymbol{Y}$ are assumed to be of the same scale, $\boldsymbol{Y}$ may be written as $\boldsymbol{Y} = \boldsymbol{y} + \frac{1}{2}\,\boldsymbol{y}^2\,\boldsymbol{e}_\infty + \boldsymbol{e}_o$. Thus,

$$\boldsymbol{V}\,\boldsymbol{X}\,\widetilde{\boldsymbol{V}} - \boldsymbol{Y} = (\boldsymbol{z} - \boldsymbol{y}) + \frac{1}{2}\,(\boldsymbol{z}^2 - \boldsymbol{y}^2)\,\boldsymbol{e}_\infty\,,$$

from which it follows that

$$\begin{aligned}\Delta^2 &= \left((\boldsymbol{z} - \boldsymbol{y}) + \frac{1}{2}\,(\boldsymbol{z}^2 - \boldsymbol{y}^2)\,\boldsymbol{e}_\infty\right)\left((\boldsymbol{z} - \boldsymbol{y}) - (\boldsymbol{z}^2 - \boldsymbol{y}^2)\,\boldsymbol{e}_o\right)\\ &= (\boldsymbol{z} - \boldsymbol{y})^2 + (\boldsymbol{z}^2 - \boldsymbol{y}^2)^2\,,\end{aligned}$$

because $\boldsymbol{e}_\infty^\dagger = -2\,\boldsymbol{e}_o$. The Euclidean metric is $(\boldsymbol{z} - \boldsymbol{y})^2$, which shows that Δ^2 approximates the Euclidean metric if $\boldsymbol{z} \approx \boldsymbol{y}$. If $\boldsymbol{X}$ and $\boldsymbol{Y}$ are related by a rotation about the origin, then $\|\boldsymbol{x}\| = \|\boldsymbol{y}\|$. Hence, if $\boldsymbol{V}$ is constrained to be some rotation about the origin, then $\|\boldsymbol{z}\| = \|\boldsymbol{x}\| = \|\boldsymbol{y}\|$ and thus $\Delta^2 = (\boldsymbol{z} - \boldsymbol{y})^2$, the Euclidean metric.

In Sect. 4.3.4 it was shown that $\boldsymbol{X} \cdot \boldsymbol{Y} = (\boldsymbol{x} - \boldsymbol{y})^2$, the Euclidean distance between $\boldsymbol{x}$ and $\boldsymbol{y}$ if $\boldsymbol{X}$ and $\boldsymbol{Y}$ are scaled appropriately. Therefore, if the versor that minimizes the Euclidean distance between two points $\boldsymbol{X}$ and $\boldsymbol{Y}$ is to be found, then the constraint function should be

$$\boldsymbol{Q}(\boldsymbol{X}, \boldsymbol{Y}, \boldsymbol{V}) = (\boldsymbol{V}\,\boldsymbol{X}\,\widetilde{\boldsymbol{V}}) \cdot \boldsymbol{Y}\,.$$

However, this constraint equation cannot be made linear in $\boldsymbol{V}$.

6.2.5 Estimation of a 3D Circle

To demonstrate the applicability of the approach discussed above for the estimation of geometric entities, a synthetic experiment on fitting a 3D circle to a set of uncertain data points is presented here.

The uncertain data to which a circle was to be fitted were generated as follows. Let C denote a "true" circle of radius one, oriented arbitrarily in 3D space. A set of $N = 10$ points $\{\mathsf{a}_n \in \mathbb{R}^3\}$ on the true circle within a given angle range was selected randomly. For each of these points, a covariance matrix $\Sigma_{\mathsf{a}_n,\mathsf{a}_n}$ was generated randomly, within a certain range. For each of the a_n, $\Sigma_{\mathsf{a}_n,\mathsf{a}_n}$ was used to generate a Gaussian-distributed random error vector r_n. The data points $\{\mathsf{b}_n\} \subset \mathbb{R}^3$, with corresponding covariance matrices $\Sigma_{\mathsf{b}_n,\mathsf{b}_n}$, were then defined as $\mathsf{b}_n = \mathsf{a}_n + \mathsf{r}_n$ and $\Sigma_{\mathsf{b}_n,\mathsf{b}_n} = \Sigma_{\mathsf{a}_n,\mathsf{a}_n}$. The standard deviation of the set $\{\|\mathsf{r}_n\|\}$ is denoted by σ_r. For each angle range, 30 sets of true points $\{\mathsf{a}_n\}$ and, for each of these sets, 40 sets of data points $\{\mathsf{b}_n\}$ were generated.

A circle was then fitted to each of the data point sets. A circle estimate is denoted by $\hat{C}$, and the shortest vector between a true point a_n and $\hat{C}$ by d_n. For each $\hat{C}$, two quality measures were evaluated: the Euclidean root mean

Table 6.2 Results of circle estimation for SVD method (SVD) and Gauss–Helmert method (GH)

σ_r	Angle range	$\bar{\Delta}_\Sigma$ ($\bar{\sigma}_\Sigma$)		$\bar{\Delta}_E$ ($\bar{\sigma}_E$)	
		SVD	GH	SVD	GH
0.07	10°	2.13 (0.90)	1.26 (0.52)	0.047 (0.015)	0.030 (0.009)
	60°	1.20 (0.44)	0.92 (0.31)	0.033 (0.010)	0.028 (0.009)
	180°	1.38 (0.56)	0.97 (0.36)	0.030 (0.009)	0.025 (0.008)
0.15	10°	2.17 (0.90)	1.15 (0.51)	0.100 (0.032)	0.057 (0.019)
	60°	1.91 (0.99)	1.35 (0.68)	0.083 (0.033)	0.069 (0.028)
	180°	1.21 (0.44)	0.90 (0.30)	0.070 (0.022)	0.058 (0.018)

squared (RMS) distance δ_E and the Mahalanobis RMS distance δ_Σ:

$$\delta_E := \sqrt{\frac{1}{N} \sum_n \mathsf{d}_n^\mathsf{T} \mathsf{d}_n}\,, \qquad \delta_\Sigma := \sqrt{\frac{1}{N} \sum_n d_n^\mathsf{T} \Sigma_{a_n,a_n}^{-1} d_n}\,.$$

The latter measure uses the covariance matrices as local metrics for the distance measure; δ_Σ is a unitless value that is > 1, $= 1$, or < 1 if d_n lies outside, on, or inside the standard-deviation error ellipsoid represented by Σ_{a_n,a_n}, respectively.

For each true point set, the mean and standard deviation of δ_E and δ_Σ over all data point sets is denoted by Δ_E, σ_E and Δ_Σ, σ_Σ, respectively. Finally, the means of Δ_E, σ_E and Δ_Σ, σ_Σ over all true point sets are denoted by $\bar{\Delta}_E$, $\bar{\sigma}_E$ and $\bar{\Delta}_\Sigma$, $\bar{\sigma}_\Sigma$. These quality measures were evaluated for the circle estimates by the SVD and the Gauss–Helmert method. The results for two different values of σ_r and three different angle ranges are given in Table 6.2. In all cases, 10 data points were used.

It can be seen that for both levels of noise (σ_r), the Gauss–Helmert method always performs better in terms of the mean quality and the mean standard deviation than the SVD method. It is also interesting to note that the Euclidean measure $\bar{\Delta}_E$ is approximately doubled when σ_r is doubled, while the "stochastic" measure $\bar{\Delta}_\Sigma$ increases only slightly. This is to be expected, since an increase in σ_r implies larger values in the $\Sigma_{\mathsf{a}_n,\mathsf{a}_n}$. Note that $\bar{\Delta}_\Sigma < 1$ implies that the estimated circle lies mostly inside the standard-deviation ellipsoids of the true points.

6.2.6 Estimation of a General Rotor

As an example of the estimation of an operator, a synthetic experiment on the estimation of a general rotor is presented here. This implements the versor equation as discussed in Sect. 6.2.2.

For the evaluation of a general rotor, the "true" points $\{\mathsf{a}_n\} \subset \mathbb{R}^3$ were a cloud of Gaussian-distributed points about the origin with standard deviation 0.8. These points were then transformed by a "true" general rotation $\boldsymbol{R} \in \mathbb{G}_{4,1}$, i.e. a rotation about an arbitrary axis. Given the set $\{\mathsf{a}'_n\}$ of rotated true points, noise was added to generate the data points $\{\mathsf{b}_n\}$ in just the same way as for the circle experiment of Sect. 6.2.5. For each of 40 sets of true points, 40 data point sets were generated and a general rotor $\hat{\boldsymbol{R}}$ was estimated. Using $\hat{\boldsymbol{R}}$, the true points were rotated to give $\{\hat{\mathsf{a}}'_n\}$. The distance vectors $\{\mathsf{d}_n\}$ were then defined as $\mathsf{d}_n := \mathsf{a}'_n - \hat{\mathsf{a}}'_n$. From the $\{\mathsf{d}_n\}$, the same quality measures as for the circle experiment were evaluated.

Table 6.3 Results of estimation of a general rotor for a standard method (Std), the SVD method (SVD), and the Gauss–Helmert method (GH)

σ_r	$\bar{\Delta}_\Sigma$ ($\bar{\sigma}_\Sigma$)			$\bar{\Delta}_E$ ($\bar{\sigma}_E$)		
	Std	SVD	GH	Std	SVD	GH
0.09	1.44 (0.59)	1.47 (0.63)	0.68 (0.22)	0.037 (0.011)	0.037 (0.012)	0.024 (0.009)
0.18	1.47 (0.62)	1.53 (0.67)	0.72 (0.25)	0.078 (0.024)	0.079 (0.026)	0.052 (0.019)

The results of the Gauss–Helmert method, the initial SVD estimate, and the results of a standard approach described in [12] are compared in Table 6.3. Since the quality measures did not give significantly different results for rotation angles between 3 and 160 degrees, the means of the respective values over all rotation angles are shown in the table. The rotation axis always pointed parallel to the z-axis and was moved one unit away from the origin parallel to the x-axis. In all experiments, 10 points were used.

It can be seen that for both levels of noise (σ_r), the Gauss–Helmert method always performs significantly better in terms of the mean quality and the mean standard deviation than the other two methods. Just as for the circle, the Euclidean measure $\bar{\Delta}_E$ is approximately doubled when σ_r is doubled, while the "stochastic" measure $\bar{\Delta}_\Sigma$ increases only slightly. Note that $\bar{\Delta}_\Sigma < 1$ implies that the points $\{\hat{\mathsf{a}}'_n\}$ lie mostly inside the standard-deviation ellipsoids of the $\{\mathsf{a}'_n\}$.

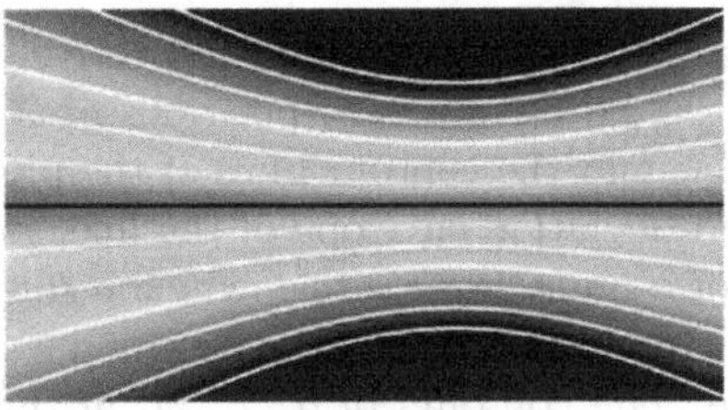

(a) Point-on-line test

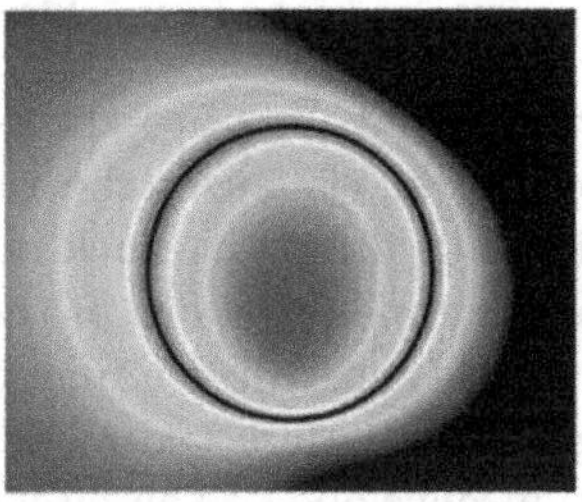

(b) Point-on-circle test

Fig. 6.10 Plot of the value of $\mathsf{q}^\mathsf{T}\ \Sigma_{\mathsf{q},\mathsf{q}}^{-1}\ \mathsf{q}$ for positions of a test point in a plane

6.3 Hypothesis Testing

Given uncertain geometric entities, a question such as "does point $\boldsymbol{X}$ lie on line $\boldsymbol{L}$" is not applicable, since the probability that this occurs for ideal points and lines is infinitesimal. Instead, statistical hypothesis testing, as described, for example, in [104], has to be used. To apply hypothesis testing to geometric-algebra relations, the method described by Heuel [93] and Förstner et al. [74] for projective geometry can be employed.

The basic idea is that the hypothesis H_0 "$\boldsymbol{X}$ lies on $\boldsymbol{L}$" is tested against the hypothesis H_1 "$\boldsymbol{X}$ does *not* lie on $\boldsymbol{L}$". In order to perform the hypothesis test, a probability ρ that H_0 will be rejected even though it is true has to be chosen. Furthermore, it is assumed that a vector-valued distance measure $\mathsf{q} \in \mathbb{R}^n$ with an associated covariance matrix $\Sigma_{\mathsf{q},\mathsf{q}} \in \mathbb{R}^{n\times n}$ is given, which is zero if $\boldsymbol{X}$ is incident with the line $\boldsymbol{L}$. Then the hypothesis H_0 can be rejected if

$$\mathsf{q}^\mathsf{T}\ \Sigma_{\mathsf{q},\mathsf{q}}^{-1}\ \mathsf{q} > \chi^2_{1-\rho;n}\ , \tag{6.14}$$

where $\chi^2_{1-\rho;n}$ is the $(1-\rho)$-quantile of the χ^2_n distribution for n degrees of freedom. Note that if $\Sigma_{\mathsf{q},\mathsf{q}}$ is not of full rank, its pseudoinverse can be used in the above equation.

In the present example, the distance measure q is simply the component vector of the containment constraint $\boldsymbol{Q}(\boldsymbol{X}, \boldsymbol{L})$ between a point and a line given in Table 4.6. The corresponding covariance matrix $\Sigma_{\mathsf{q},\mathsf{q}}$ can be evaluated using error propagation. That is, if $\mathsf{x} := \mathcal{K}(\boldsymbol{X})$ and $\boldsymbol{\ell} := \mathcal{K}(\boldsymbol{L})$, then

$$q^r = x^i\ \ell^j\ \Lambda^r{}_{ij}\ ,$$

and the Jacobi matrices are

$$Q^r_{Lp} = \ell^j\ \Lambda^r{}_{pj} \qquad \text{and} \qquad Q^r_{Rp} = x^i\ \Lambda^r{}_{ip}\ .$$

Therefore, the covariance matrix of q is

$$\Sigma_{\mathsf{q},\mathsf{q}} = \mathsf{Q}_L\, \Sigma_{\mathsf{x},\mathsf{x}}\, \mathsf{Q}_L^\mathsf{T} + \mathsf{Q}_R\, \Sigma_{\boldsymbol{\ell},\boldsymbol{\ell}}\, \mathsf{Q}_R^\mathsf{T}\,,$$

where $\Sigma_{\mathsf{x},\mathsf{x}}$ and $\Sigma_{\boldsymbol{\ell},\boldsymbol{\ell}}$ are the covariance matrices of x and $\boldsymbol{\ell}$, respectively. Figure 6.10(a) visualizes the value of $\mathsf{q}^\mathsf{T}\, \Sigma_{\mathsf{q},\mathsf{q}}^{-1}\, \mathsf{q}$ for different positions of $\boldsymbol{X}$. The line $\boldsymbol{L}$ is drawn in black in the center, and the contour lines of the plot are simply scaled versions of the standard-deviation envelope of the line.

The same construction may be done for a point and a circle, which results in Fig. 6.10(b). Here the mean circle is drawn in black, and the standard-deviation envelope of the circle in white. The contours of $\mathsf{q}^\mathsf{T}\, \Sigma_{\mathsf{q},\mathsf{q}}^{-1}\, \mathsf{q}$ in this plot are not scaled versions of the standard-deviation envelope.

Chapter 7
The Inversion Camera Model

In this chapter, a camera model is discussed which encompasses the standard pinhole camera model, a lens distortion model, and catadioptric cameras with parabolic mirrors. This generalized camera model is particularly well suited to being represented in conformal geometric algebra (CGA), since it is based on an inversion operation.

The pinhole camera model is the oldest perspective model which realistically represents a 3D scene on a 2D plane, where "realistically" means appearing to human perception as the original scene would. According to [25], "focused perspective was discovered around 1425 by the sculptor and architect Brunelleschi (1377–1446), developed by the painter and architect Leone Battista (1404–1472), and finally perfected by Leonardo da Vinci (1452–1519)." This concept of *focused perspective* also lies at the heart of projective space, which becomes apparent in Fig. 4.8, for example. In Sect. 7.1, the pinhole camera model and its representation in geometric algebra are briefly discussed. Figure 7.2 shows the mathematical construction of a pinhole camera. The point $\boldsymbol{A}_4$ in that figure is called the optical center, and $\boldsymbol{P}$ represents the image plane. If a point $\boldsymbol{X}$ is to be projected onto the image plane, the line through $\boldsymbol{X}$ and $\boldsymbol{A}_4$, the projective ray, is intersected with the image plane $\boldsymbol{P}$. The intersection point is the resultant projection. Owing to its mathematical simplicity and its good approximation to real cameras in many cases, the pinhole camera model is widely used in computer vision.

However, the pinhole camera model is only an idealized model of a real camera. Depending on the application, it may therefore be necessary to take account of distortion in the imaging process due to the particular form of the imaging system used. This is typically the case if high accuracy is needed or if wide-angle lens systems are employed, or both. The problem is that all real lens systems differ and their distortion cannot be modeled easily mathematically. The best approximations are the *thin-lens* and *thick-lens* models. However, even those are only approximations to a real lens system. Therefore, for an exact rectification of an image taken with a distorting lens system, the individual lens system has to be measured. This process is time-consuming

C. Perwass, *Geometric Algebra with Applications in Engineering.* Geometry and Computing.

Fig. 7.1 Typical image taken with a catadioptric camera with a parabolic mirror

and tedious, which is why lens distortion is typically approximated by a simple mathematical model. Various lens distortion models have been suggested for this purpose, such as the widely used polynomial model [85], the bicubic model [103], the rational model [33], and the division model [71].

Another type of imaging system that is particularly useful for navigation applications is the catadioptric camera, since such cameras allow a 360 degree view in a single image. The type of catadioptric camera considered here is the central catadioptric camera, where a standard camera looks at a parabolic mirror. A typical image generated by such an imaging system is shown in Fig. 7.1. Geyer and Daniilidis showed in [78] how such systems can be modeled quite easily mathematically. The calibration of central catadioptric cameras is discussed, for example, in [77, 176].

In this chapter, a novel camera model, the *inversion camera model*, is introduced, which combines the above-mentioned models: the pinhole camera model, a lens distortion model, and a model for central catadioptric cameras with parabolic mirror. As is shown later on, the lens distortion model is just the division model introduced by Fitzgibbon in [71], and the catadioptric-camera model is the same as that presented first by Geyer and Daniilidis in [78]. However, the present author has found that both of these models can be represented in much the same way using inversion in a sphere. This also extends the division model to lenses with an angular field of view of 180 degrees or more and leads to a projective model for central catadioptric cameras [170].

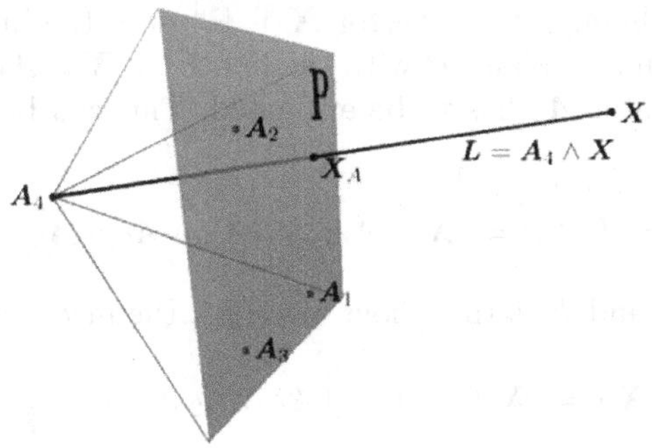

Fig. 7.2 Model of a pinhole camera in projective space

Inversion in a sphere can be represented as a (tri-)linear function in the geometric algebra of conformal space, which makes this algebra an ideal mathematical framework for working with the inversion camera model. The induced image distortion can be expressed as a multivector, and a covariance matrix can be associated with it, which makes it directly applicable to statistical *linear* estimation methods as presented in Chap. 5. This is demonstrated in Chap. 8, where results for the simultaneous estimation of object pose, camera focal length, and lens distortion are presented.

The plan of this chapter is as follows. First, a brief overview of the representation of the pinhole camera model in projective space is given in Sect. 7.1. A detailed discussion of this representation together with treatments of the fundamental matrix and of the trifocal and the quadfocal tensor can be found in [130]. In Sect. 7.2, a *geometric* introduction to the fundamental idea of the inversion model is given. This is detailed mathematically in Sect. 7.3, where the exact form of the implicit lens distortion model is derived. In Sect. 7.4, it is shown that the inversion camera model cannot fully model the imaging process of a fisheye lens. The application of the model to parabolic-mirror imaging systems is discussed, finally, in Sect. 7.5.

To keep the inversion camera model as simple as possible, not all representable degrees of freedom were considered. This is briefly discussed in Sect. 7.6.

7.1 The Pinhole Camera Model

The geometric algebra of projective space (cf. section 4.2) is very useful for representing projections in the pinhole camera model. Figure 7.2 shows such a setup. Consider four homogeneous vectors $\boldsymbol{A}_1, \boldsymbol{A}_2, \boldsymbol{A}_3, \boldsymbol{A}_4 \in \mathbb{R}^4$ that form a basis of $\mathbb{R}^4$. The homogeneous vector $\boldsymbol{A}_4$ represents the optical center of the pinhole camera, and $\boldsymbol{P} := \boldsymbol{A}_1 \wedge \boldsymbol{A}_2 \wedge \boldsymbol{A}_3$ represents the image plane. In order to project a homogeneous vector $\boldsymbol{X} \in \mathbb{R}^4$ onto the image plane, the intersection of the image plane $\boldsymbol{P}$ with the line $\boldsymbol{L} := \boldsymbol{X} \wedge \boldsymbol{A}_4$ connecting $\boldsymbol{X}$ with the optical center $\boldsymbol{A}_4$ has to be evaluated. This can be done with the meet operation,

$$\boldsymbol{X}_A = \boldsymbol{L} \vee \boldsymbol{P} = (\boldsymbol{X} \wedge \boldsymbol{A}_4) \vee (\boldsymbol{A}_1 \wedge \boldsymbol{A}_2 \wedge \boldsymbol{A}_3) .$$

Since the join of $\boldsymbol{L}$ and $\boldsymbol{P}$ is the whole space $\mathbb{R}^4$, the meet operation can be expressed as

$$\boldsymbol{X}_A = (\boldsymbol{X} \wedge \boldsymbol{A}_4)^* \cdot (\boldsymbol{A}_1 \wedge \boldsymbol{A}_2 \wedge \boldsymbol{A}_3) . \tag{7.1}$$

In this way, a closed algebraic expression for the projection of a point onto the image plane is constructed. The inner and outer camera calibration is contained in the vectors $\{ \boldsymbol{A}_i \}$. The relation of this expression to the camera matrix can be derived quite easily.

From the rules for the inner product (cf. section 3.2.7), it follows that

$$\begin{aligned} \boldsymbol{X}_A &= (\boldsymbol{X} \wedge \boldsymbol{A}_4)^* \cdot (\boldsymbol{A}_1 \wedge \boldsymbol{A}_2 \wedge \boldsymbol{A}_3) \\ &= \quad (\boldsymbol{X} \wedge \boldsymbol{A}_2 \wedge \boldsymbol{A}_3 \wedge \boldsymbol{A}_4)^* \, \boldsymbol{A}_1 \\ &\quad + (\boldsymbol{X} \wedge \boldsymbol{A}_3 \wedge \boldsymbol{A}_1 \wedge \boldsymbol{A}_4)^* \, \boldsymbol{A}_2 \\ &\quad + (\boldsymbol{X} \wedge \boldsymbol{A}_1 \wedge \boldsymbol{A}_2 \wedge \boldsymbol{A}_4)^* \, \boldsymbol{A}_3 . \end{aligned} \tag{7.2}$$

Each of the dualized 4-vector expressions can also be written as

$$(\boldsymbol{X} \wedge \boldsymbol{A}_2 \wedge \boldsymbol{A}_3 \wedge \boldsymbol{A}_4)^* = \boldsymbol{X} \cdot (\boldsymbol{A}_2 \wedge \boldsymbol{A}_3 \wedge \boldsymbol{A}_4)^* .$$

The expression $(\boldsymbol{A}_2 \wedge \boldsymbol{A}_3 \wedge \boldsymbol{A}_4)^*$ is simply the reciprocal vector $\boldsymbol{A}^1$ of $\boldsymbol{A}_1$, as follows from the definition of reciprocal vectors in Sect. 3.5. Therefore, (7.2) can also be written as

$$\boldsymbol{X}_A = \sum_{i=1}^{3} (\boldsymbol{X} \cdot \boldsymbol{A}^i) \, \boldsymbol{A}_i . \tag{7.3}$$

Suppose $\boldsymbol{X}$ is given in some basis $\{ \boldsymbol{Z}_i \} \subset \mathbb{R}^4$ as $\boldsymbol{X} = x^i \, \boldsymbol{Z}_i$; then

$$\boldsymbol{X}_A = \sum_{i=1}^{3} \sum_{j=1}^{4} x^j \, (\boldsymbol{Z}_j \cdot \boldsymbol{A}^i) \, \boldsymbol{A}_i . \tag{7.4}$$

This can be expressed as the matrix equation

$$\mathsf{x}_A = \mathsf{K} \, \mathsf{x} , \qquad \mathsf{K}^i{}_j := \boldsymbol{Z}_j \cdot \boldsymbol{A}^i , \tag{7.5}$$

where $\mathsf{K} \in \mathbb{R}^{3\times4}$ is the camera matrix and x_A is the component vector of $\boldsymbol{X}_A$ in the bases $\{ \boldsymbol{A}_1, \boldsymbol{A}_2, \boldsymbol{A}_3 \}$.

Similarly, the geometric construction of the fundamental matrix and the trifocal tensor can be expressed as geometric-algebra expressions, and their relations to the camera matrix can be derived. A construction with geometric algebra also simplifies the analysis of the constraints on the fundamental matrix and the trifocal tensor, as shown in [109, 145, 144]. A detailed discussion of these topics can also be found in [130].

7.2 Definition of the Inversion Camera Model

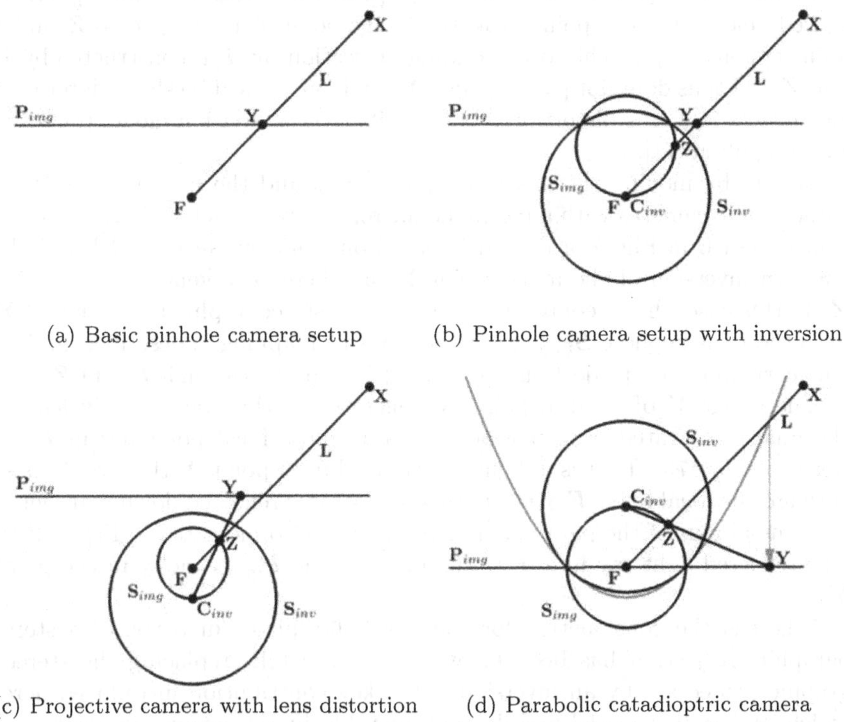

(a) Basic pinhole camera setup (b) Pinhole camera setup with inversion

(c) Projective camera with lens distortion (d) Parabolic catadioptric camera

Fig. 7.3 Various cameras representable by the inversion camera model

The basic setup of the camera model is shown schematically in Fig. 7.3. Figure 7.3(a) shows the setup of the pinhole camera model. Point $\boldsymbol{F}$ is the focal point or optical center, point $\boldsymbol{X}$ is a world point, and $\boldsymbol{Y}$ is the image of $\boldsymbol{X}$ on the image plane $\boldsymbol{P}_{img}$. In a typical problem setup, the image point $\boldsymbol{Y}$ is given and the projection ray $\boldsymbol{L}$ has to be evaluated. This is, for example, the case for monocular pose estimation, where correspondences between image points and points on an object are assumed to be given and the corresponding projection rays have to be evaluated in order to estimate the object's pose, i.e. position and orientation.

If the pinhole camera's internal calibration is given, the projection ray $\boldsymbol{L}$ can immediately be evaluated in the camera's coordinate frame. In the inversion camera model, this pinhole camera setup is represented as shown in Fig. 7.3(b). The sphere $\boldsymbol{S}_{inv}$, with center $\boldsymbol{C}_{inv}$, is used to perform an inversion of the image plane $\boldsymbol{P}_{img}$, which results in the sphere $\boldsymbol{S}_{img}$. In particular, the image point $\boldsymbol{Y}$ is mapped to point $\boldsymbol{Z}$. In Fig. 7.3(b), the center $\boldsymbol{C}_{inv}$ of the inversion sphere $\boldsymbol{S}_{inv}$ coincides with the focal point $\boldsymbol{F}$. In this case the inversion of $\boldsymbol{Y}$ in $\boldsymbol{S}_{inv}$ results again in a point on the projection ray $\boldsymbol{L}$. Therefore, this setup is equivalent to the standard pinhole camera setup.

Figure 7.3(c) demonstrates what happens when the inversion sphere is moved below the focal point. Now the image point $\boldsymbol{Y}$ is mapped to $\boldsymbol{Z}$ under an inversion in $\boldsymbol{S}_{inv}$. The corresponding projection ray $\boldsymbol{L}$ is constructed by $\boldsymbol{F}$ and $\boldsymbol{Z}$ and thus does not pass through $\boldsymbol{Y}$ anymore. It will be shown later that this results in a lens distortion model similar to the division model proposed by Fitzgibbon [71].

Simply by moving the inversion sphere $\boldsymbol{S}_{inv}$ and the image plane $\boldsymbol{P}_{img}$, catadioptric cameras with a parabolic mirror can be modeled. This construction is shown in Fig. 7.3(d), and is based on work by Geyer and Daniilidis [78]. An inversion of the image point $\boldsymbol{Y}$ in sphere $\boldsymbol{S}_{inv}$ generates the point $\boldsymbol{Z}$. In this case, this is equivalent to an inverse stereographic projection of $\boldsymbol{Y}$ onto the image sphere $\boldsymbol{S}_{img}$, which is how this mapping is described in [78]. The corresponding projection ray $\boldsymbol{L}$ is again the line through $\boldsymbol{F}$ and $\boldsymbol{Z}$.

The image $\boldsymbol{Y}$ of a world point $\boldsymbol{X}$ generated in this way is equivalent to the image generated by a parabolic mirror whose focal point lies in $\boldsymbol{F}$, as was shown in [78]. That is, a light ray emitted from point $\boldsymbol{X}$ that would pass through the focal point $\boldsymbol{F}$ of the parabolic mirror is reflected down, parallel to the central axis of the parabolic mirror. This is also indicated in Fig. 7.3(d). The reflected light ray intersects the image plane $\boldsymbol{P}_{img}$ exactly in the point $\boldsymbol{Y}$.

Whereas the construction for the parabolic mirror in terms of a stereographic projection has been known for some while, replacing the stereographic projection by an inversion is the key contribution introduced here, which makes this model readily representable in the geometric algebra of conformal space. In the following, the mathematical details of the inversion camera model will be discussed.

7.3 From Pinhole to Lens

In this section, the mathematical details of the inversion camera model with respect to lens distortion are discussed. In all calculations, a right-handed coordinate system (Fig. 7.4) is assumed, where $\boldsymbol{e}_1$ points towards the right, parallel to the horizontal image plane direction, $\boldsymbol{e}_1$ points upwards, parallel to the vertical image plane direction, and $\boldsymbol{e}_3$ points from the image plane center towards the focal point or optical center. This implies that objects that are in front of the camera will have a negative $\boldsymbol{e}_3$ coordinate.

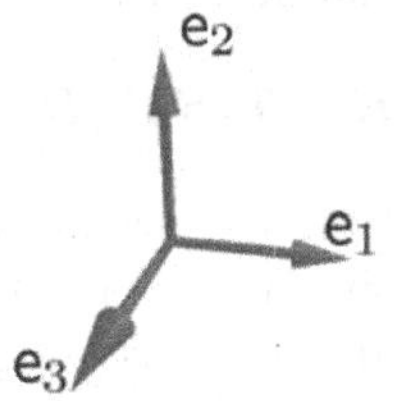

Fig. 7.4 Right-handed camera coordinate system

In the following, the main interest lies in the rectification of images with lens distortion. For example, given an image as in Fig. 7.5(a), a transformation of the image is to be found such that it is equivalent to an image of the same scene taken by a pinhole camera. An example of such a *rectified* image is shown in Fig. 7.5(b).

(a) Image with lens distortion (b) Rectified image

Fig. 7.5 Example of an image with lens distortion, and rectification

7.3.1 Mathematical Formulation

The geometric setup of the inversion camera model, as presented in the previous section, can be modeled algebraically in CGA as follows. Like all transformations in geometric algebra, the image point transformation in the inversion camera model will be represented by a versor $\boldsymbol{K} \in \mathbb{G}_{4,1}$. That is, if $\boldsymbol{Y} \in \mathbb{G}^1_{4,1}$ represents an image point, then $\boldsymbol{Z} := \boldsymbol{K}\,\boldsymbol{Y}\,\widetilde{\boldsymbol{K}}$ is the transformed image point. As can be seen in Fig. 7.3, the point $\boldsymbol{Z}$ will in general not lie on the image plane; however, the goal is to find a $\boldsymbol{K}$ such that $\boldsymbol{Z}$ lies on the "correct" projection ray. The transformed image point in the image plane can then be found by intersecting the projection ray with the image plane.

One of the simplest forms that $\boldsymbol{K}$ can take is

$$\boldsymbol{K} = \boldsymbol{T}_s\,\boldsymbol{S}\,\widetilde{\boldsymbol{T}}_s\;\boldsymbol{D}, \tag{7.6}$$

where $\boldsymbol{S}$ is a sphere centered on the origin, $\boldsymbol{T}_s$ is a translator, and $\boldsymbol{D}$ a dilator (an isotropic scaling operator). This form has also been found to behave well numerically. The dilator scales the image, which has the same effect as varying the focal length if the inversion sphere $\boldsymbol{S}_{inv} := \boldsymbol{T}_s\,\boldsymbol{S}\,\widetilde{\boldsymbol{T}}_s$ is centered on the focal point (see Fig. 7.3(b)). If the inversion sphere is not centered on the focal point, the dilator also influences the distortion. In the following, the transformation $\boldsymbol{K}\,\boldsymbol{Y}\,\widetilde{\boldsymbol{K}}$ is analyzed in some detail.

To simplify matters, it is assumed that $\boldsymbol{T}_s$ translates the inversion sphere along only the $\boldsymbol{e}_3$ axis, and that $\boldsymbol{S}$ is a sphere of radius r centered on the origin. Hence,

$$\boldsymbol{S} := \boldsymbol{e}_o - \frac{1}{2}\,r^2\,\boldsymbol{e}_\infty, \qquad \boldsymbol{T}_s := 1 - \frac{1}{2}\,\tau_s\,\boldsymbol{e}_3\,\boldsymbol{e}_\infty. \tag{7.7}$$

It is then straight forward to show that

$$\begin{aligned}\boldsymbol{S}_{inv} &= \boldsymbol{T}_s\,\boldsymbol{S}\,\widetilde{\boldsymbol{T}}_s\\ &= \tau_s\,\boldsymbol{e}_3 + \frac{1}{2}\,(\tau_s^2 - r^2)\,\boldsymbol{e}_\infty + \boldsymbol{e}_o\\ &= s_1\,\boldsymbol{e}_3 + \frac{1}{2}\,s_2\,\boldsymbol{e}_\infty + \boldsymbol{e}_o,\end{aligned} \tag{7.8}$$

where $s_1 := \tau_s$ and $s_2 := \tau_s^2 - r^2$. The inversion sphere $\boldsymbol{S}_{inv}$ can thus be regarded as a vector with two free parameters, that influence the sphere's position along $\boldsymbol{e}_3$ and its radius.

Recall that the dilation operator $\boldsymbol{D}$ for a scaling by a factor $d \in \mathbb{R}$ is given by

$$\boldsymbol{D} = 1 + \frac{1-d}{1+d}\,\boldsymbol{E}, \tag{7.9}$$

where $\boldsymbol{E} := \boldsymbol{e}_\infty \wedge \boldsymbol{e}_o$. For brevity, we define $\tau_d := -(1-d)/(1+d)$, so that $\boldsymbol{D} = 1 - \tau_d\,\boldsymbol{E}$. The image point transformation operator $\boldsymbol{K}$ is then given by

$$\begin{aligned}\boldsymbol{K} &= \boldsymbol{S}_{inv}\,\boldsymbol{D}\\ &= s_1\,\boldsymbol{e}_3 + \frac{1}{2}\,s_2\,(1-\tau_d)\,\boldsymbol{e}_\infty(1+\tau_d)\,\boldsymbol{e}_o - \tau_d\,s_1\,\boldsymbol{e}_3\,\boldsymbol{E}\\ &= k_1\,\boldsymbol{e}_3 + k_2\,\boldsymbol{e}_\infty + k_3\,\boldsymbol{e}_o + k_4\,\boldsymbol{e}_3\,\boldsymbol{E},\end{aligned} \tag{7.10}$$

where

$$k_1 := s_1, \quad k_2 := \frac{1}{2}\, s_2\,(1 - \tau_d), \quad k_3 := 1 + \tau_d, \quad k_4 := -\tau_d\, s_1. \tag{7.11}$$

In the model setup, the image plane $\boldsymbol{P}_{img}$ passes through the origin and is perpendicular to $\boldsymbol{e}_3$. That is, the image points lie in the $\boldsymbol{e}_1$–$\boldsymbol{e}_2$ plane. An image point is denoted in Euclidean space by $\boldsymbol{y} \in \mathbb{R}^3$ and its embedding in conformal space by $\boldsymbol{Y} := \mathcal{C}(\boldsymbol{y}) \in \mathbb{G}^1_{4,1}$; that is,

$$\boldsymbol{Y} = \boldsymbol{y} + \frac{1}{2}\,\boldsymbol{y}^2\,\boldsymbol{e}_\infty + \boldsymbol{e}_o. \tag{7.12}$$

After a straightforward, if tedious, calculation, it may be shown that

$$\begin{aligned} \boldsymbol{K}\,\boldsymbol{Y}\,\widetilde{\boldsymbol{K}} = \quad & (k_4^2 - k_1^2 + 2k_2k_3)\,\boldsymbol{y} \\ & - 2\left[(k_1 + k_4)\,k_2 + (k_1 - k_4)\,k_3\,y_\infty\right]\,\boldsymbol{e}_3 \\ & - \left[2\,(k_2 + k_3\,y_\infty)\,k_2 - (k_1 - k_4)^2\,y_\infty\right]\,\boldsymbol{e}_\infty \\ & - \left[(k_1 + k_4)^2 + 2k_3^2 y_\infty\right]\,\boldsymbol{e}_o, \end{aligned} \tag{7.13}$$

where $y_\infty := \frac{1}{2}\,\boldsymbol{y}^2$. Substituting the definitions of the $\{k_i\}$ given in (7.11) into (7.13) results in

$$\begin{aligned} \boldsymbol{K}\,\boldsymbol{Y}\,\widetilde{\boldsymbol{K}} = & - (1 - \tau_d^2)\,(s_1^2 - s_2)\,\boldsymbol{y} \\ & - (1 - \tau_d)^2\,\left[s_1\,s_2 + 2\,s_1\,d^2\,y_\infty\right]\,\boldsymbol{e}_3 \\ & - (1 - \tau_d)^2\,\left[\tfrac{1}{2}\,s_2^2 + (s_2 - s_1\,d)\,s_2\,d\,y_\infty\right]\,\boldsymbol{e}_\infty \\ & - (1 - \tau_d)^2\,\left[s_1^2 + 2\,d^2\,y_\infty\right]\,\boldsymbol{e}_o, \end{aligned} \tag{7.14}$$

where the identity $d = (1 + \tau_d)/(1 - \tau_d)$ has been used. Since the dilation operator $\boldsymbol{D}$ and the inversion operator $\boldsymbol{S}_{inv}$ map points into points, the vector $\boldsymbol{Z} = \boldsymbol{K}\,\boldsymbol{Y}\,\widetilde{\boldsymbol{K}}$ has to represent a point. $\boldsymbol{Z}$ can be mapped into its normal form by dividing it by its $\boldsymbol{e}_o$ component. Recall that this does not change the Euclidean point that $\boldsymbol{Z}$ represents. However, in the normal form, the Euclidean point that $\boldsymbol{Z}$ represents can be read off directly:

$$\begin{aligned} \boldsymbol{Z} = \quad & \boldsymbol{K}\,\boldsymbol{Y}\,\widetilde{\boldsymbol{K}} \\ \simeq \quad & \frac{d\,(s_1^2 - s_2)}{s_1^2 + 2\,d^2\,y_\infty}\,\boldsymbol{y} + \frac{s_1\,s_2 + 2\,s_1\,d^2\,y_\infty}{s_1^2 + 2\,d^2\,y_\infty}\,\boldsymbol{e}_3 \\ + \quad & \frac{\frac{1}{2}\,s_2^2 + (s_2 - s_1\,d)\,s_2\,d\,y_\infty}{s_1^2 + 2\,d^2\,y_\infty}\,\boldsymbol{e}_\infty + \boldsymbol{e}_o, \end{aligned} \tag{7.15}$$

where "$\simeq$" denotes equality up to a scalar factor. Hence, the Euclidean point $\boldsymbol{z} \in \mathbb{R}^3$ represented by $\boldsymbol{Z}$ is

$$\boldsymbol{z} = \frac{d\,(s_1^2 - s_2)}{s_1^2 + d^2\,\boldsymbol{y}^2}\,\boldsymbol{y} + \frac{s_1\,s_2 + s_1\,d^2\,\boldsymbol{y}^2}{s_1^2 + d^2\,\boldsymbol{y}^2}\,\boldsymbol{e}_3, \tag{7.16}$$

where $y_\infty = \frac{1}{2}\,\boldsymbol{y}^2$ has been used. The point $\boldsymbol{z}$ will, in general, not lie in the image plane, but on the sphere which results from the inversion of the image plane in the inversion sphere. Examples of this mapping are shown in Fig. 7.6 for various types of distortion. In these examples, the dilator was set to unity ($d = 1$). The circles drawn on the images give the location of the unit transformation, that is, the location of those image points whose location is not changed under the transformation.

To calculate the corresponding transformation of the image points in the image plane, the intersection point $\boldsymbol{v}$ of the projection ray through $\boldsymbol{z}$ with the image plane has to be evaluated. The result of this calculation is shown in Fig. 7.6 for various locations of the inversion sphere $\boldsymbol{S}_{inv}$.

Assuming the focal point $\boldsymbol{f} \in \mathbb{R}^3$ to be defined as $\boldsymbol{f} := \boldsymbol{e}_3$, the intersection point $\boldsymbol{y}_d$ is given by

$$\boldsymbol{y}_d = \boldsymbol{f} - \frac{\boldsymbol{f} - \boldsymbol{z}}{(\boldsymbol{f} - \boldsymbol{z}) \cdot \boldsymbol{e}_3}. \tag{7.17}$$

This results in

$$\boldsymbol{y}_d = \frac{-(s_1^2 - s_2)\,d}{s_1\,(s_2 - s_1) + (s_1 - 1)\,d^2\,\boldsymbol{y}^2}\,\boldsymbol{y} = \frac{\beta}{1 + \alpha\,\boldsymbol{y}^2}\,\boldsymbol{y}, \tag{7.18}$$

where

$$\alpha := \frac{(s_1 - 1)\,d^2}{s_1\,(s_2 - s_1)}, \qquad \beta := \frac{-(s_1^2 - s_2)\,d}{s_1\,(s_2 - s_1)}. \tag{7.19}$$

Note that $\boldsymbol{v}/\beta$ corresponds to the division model proposed by Fitzgibbon in [71]. Typically, lens distortion models are used to remove the distortion in an image independently of the focal length or field of view (FOV) of the imaging system that generated the image. This is usually done either by requiring that lines which appear curved in the image have to be straight, or by enforcing multiview constraints given a number of images of the same scene. The rectified image can then be used for any other type of application. For this purpose, and for lenses with an FOV of at most 180°, the inversion model is equivalent to the division model.

However, here the applicability of the inversion model as a *camera model* is being investigated. That is to say, the lens distortion of a camera system is modeled directly in the context of a constraint equation. For example, in the monocular pose estimation problem, the transformation operator $\boldsymbol{M}$ has to be evaluated such that a model point $\boldsymbol{X}$ comes to lie on the projection ray of a corresponding image point $\boldsymbol{Y}$. If $\boldsymbol{F}$ denotes the focal point, this can be formulated in CGA as

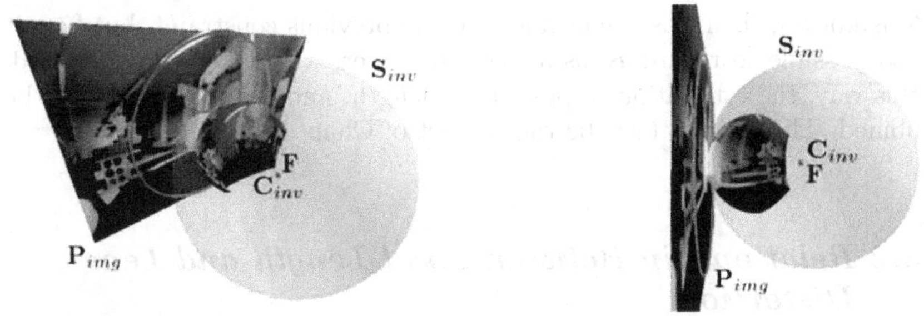

(a) Pinhole setup

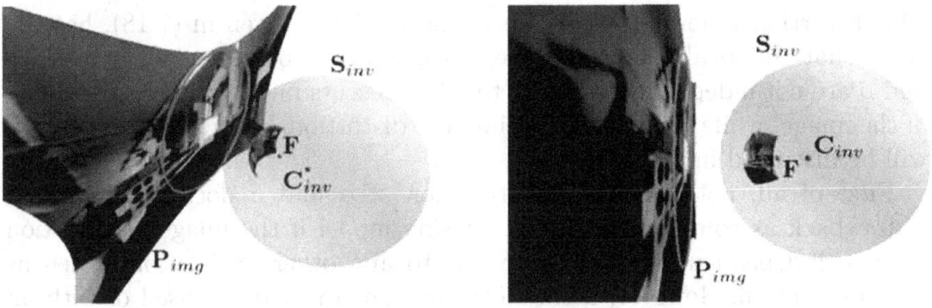

(b) Trapezoidal distortion

(c) Cushion distortion

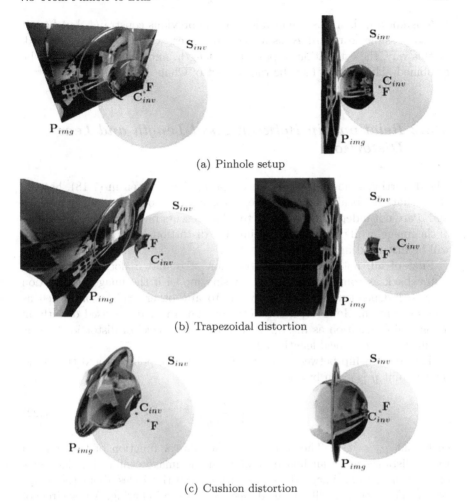

Fig. 7.6 Visualization of the inversion camera model for various types of lens distortion. See the text for more detail

$$(\boldsymbol{Y} \wedge \boldsymbol{F} \wedge \boldsymbol{e}_\infty) \wedge (\boldsymbol{M}\,\boldsymbol{X}\,\widetilde{\boldsymbol{M}}) = 0, \tag{7.20}$$

where it is assumed that the internal camera calibration is known. Also, any lens distortion has to be rectified before image points can be used in this constraint equation.

At this point, the inversion camera model can be used immediately in the constraint equation as

$$\big((\boldsymbol{K}\,\boldsymbol{Y}\,\widetilde{\boldsymbol{K}}) \wedge \boldsymbol{F} \wedge \boldsymbol{e}_\infty\big) \wedge (\boldsymbol{M}\,\boldsymbol{X}\,\widetilde{\boldsymbol{M}}) = 0. \tag{7.21}$$

If K is known, then this is equivalent to the previous constraint, but here it is also possible to regard K as a free parameter, which is to be estimated. In this way, the estimation of pose, focal length, and lens distortion can be combined. This will, in fact, be the subject of Chap. 8.

7.3.2 Relationship Between Focal Length and Lens Distortion

The distortion generated by the inversion model as given in (7.18), has the effect that the focal length and the distortion are not independent, since α and β are not independent. The factor β represents mainly an overall scaling of the image, while α influences mainly the distortion. The exact relationship will be discussed in the following.

First of all, note that the interrelation of α and β does not represent a drawback as compared with the division model if the image rectification is done independently of and previous to any other calculations, such as pose estimation. However, if the inversion camera model is used directly in a constraint equation as in (7.21), then not every level of distortion can be rectified for every focal length or FOV.

The relationship between the transformed image point $\boldsymbol{y}_d$ and the initial image point $\boldsymbol{y}$ is given by the factor

$$\omega := \frac{\beta}{1 + \alpha\, \boldsymbol{y}^2}, \tag{7.22}$$

such that $\boldsymbol{y}_d = \omega\, \boldsymbol{y}$. The factor ω is therefore a function of the squared radial distance $\boldsymbol{y}^2$ of an image point from the image center. The locations of constant ω in an image therefore form concentric circles about the image center. These circles will be called *isocircles* in the following. An isocircle of particular interest in the analysis is the one that touches the upper and lower vertical borders of the image, i.e. its radius is equal to half the vertical extent of the image. This particular isocircle will be called the *vertical* isocircle (Fig. 7.7), and its radius will be denoted by ρ_v.

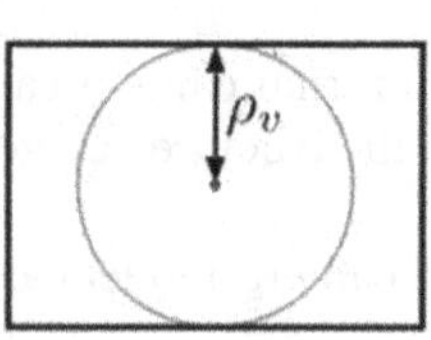

Fig. 7.7 Schematic representation of a vertical isocircle of radius ρ_v

The value of ω for image points on the vertical isocircle is directly related to the vertical field of view (vFOV). Note that the relation to the focal length is more complex if lens distortion is present, since the focal length is now a function of $\boldsymbol{y}^2$. That is, the focal length depends on the position of an image point in the image. It is therefore more useful to define an *effective* focal length (EFL) as the focal length for the image points on the vertical isocircle.

The value of ω for image points on the vertical isocircle is denoted by ω_v and is given by

$$\omega_v = \frac{\alpha}{1 + \beta\,\rho_v^2}\,. \tag{7.23}$$

The Euclidean position vector $\boldsymbol{f}$ of the focal point is parameterized in the following as $\boldsymbol{f} = \tau_f\, \boldsymbol{e}_3$. That is, if the image is neither scaled nor distorted, τ_f is the focal length. The EFL, however, is related to ω_v, as is shown in the following. The vFOV θ_v is related to ω_v via

$$\tan\left(\frac{1}{2}\theta_v\right) = \frac{\omega_v\,\rho_v}{\tau_f} \quad \Longleftrightarrow \quad \theta_v = 2\,\tan^{-1}\left(\frac{\omega_v\,\rho_v}{\tau_f}\right) \tag{7.24}$$

(see Fig. 7.8).

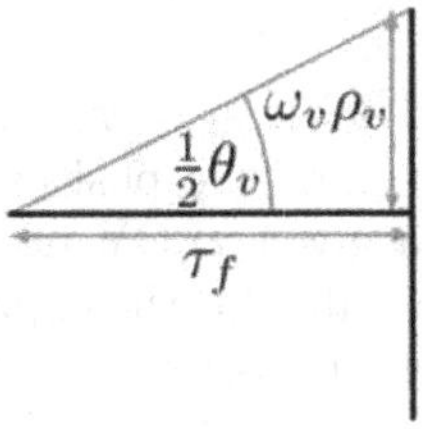

Fig. 7.8 Schematic representation of vertical field of view (vFOV) θ_v

Since the physical size of a CCD chip in the image plane is constant, with a vertical extension $2\,\rho_v$, the EFL f_e is given in terms of the vFOV by

$$f_e = \frac{\rho_v}{\tan(\frac{1}{2}\,\theta_v)} = \frac{\tau_f}{\omega_v}\,. \tag{7.25}$$

The goal now is to find the relation between the vFOV and the possible range of distortion. The range of distortion is given by the range of the possible *diagonal* field of view (dFOV) for a given vFOV. Now, ω_v depends on d, s_1, and s_2, which in turn depend on τ_s, τ_d, and r. It is therefore necessary to find the possible values of τ_s, τ_d, and r for a given ω_v. Solving (7.23) for d, given ω_v, results in

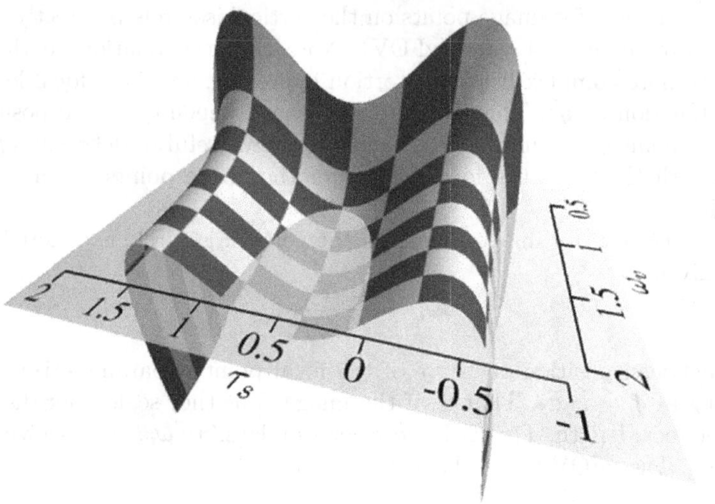

Fig. 7.9 Plot of the polynomial in (7.29)

$$d^{(1,2)} = \frac{1}{2\,(\tau_f - s_1)\,\omega_v\,\rho_v^2}\left[r^2 \pm \sqrt{r^4 - 4\,(\tau_f - s_1)\,(\tau_f\,s_1^2 - s_1\,s_2)\,\omega_v^2\,\rho_v^2}\right]. \tag{7.26}$$

Recall that $\tau_d = -(1-d)/(1+d)$, $s_1 = \tau_s$, and $s_2 = \tau_s^2 - r^2$. Fixing the inversion sphere radius r leaves τ_s, the position of the inversion sphere, as the only free parameter for a given vFOV (ω_v). Clearly, τ_s can take only values that result in a positive value for the expression under the square root in (7.26). Substituting the definitions for s_1 and s_2 into this expression and equating it to zero gives

$$\begin{aligned} 0 &= r^4 - 4\,(\tau_f - s_1)\,(\tau_f\,s_1^2 - s_1\,s_2)\,\omega_v^2\,\rho_v^2 \\ \iff 0 &= -\tau_s^4 + 2\,\tau_f\,\tau_s^3 - (\tau_f - r^2)\,\tau_s^2 - \tau_f\,r^2\,\tau_s + \frac{r^4}{4\,\omega_v^2\,\rho_v^2}. \end{aligned} \tag{7.27}$$

Solving for τ_s gives the following solutions:

$$\tau_s^{(1,2,3,4)} = \frac{1}{2}\left[\tau_f \pm \sqrt{\tau_f^2 + 2\,r^2\left(1 \pm \sqrt{1 + \frac{1}{\omega_v^2\,\rho_v^2}}\right)}\right]. \tag{7.28}$$

Figure 7.9 shows a plot of the polynomial given in (7.27), in τ_s and ω_v that is,

$$p(\tau_s, \omega_v) := -\tau_s^4 + 2\,\tau_f\,\tau_s^3 - (\tau_f - r^2)\,\tau_s^2 - \tau_f\,r^2\,\tau_s + \frac{r^4}{4\,\omega_v^2\,\rho_v^2}, \tag{7.29}$$

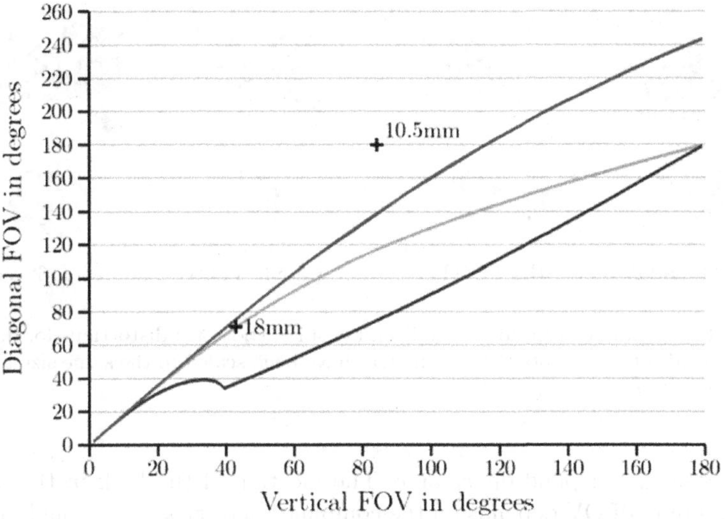

Fig. 7.10 Vertical vs. diagonal field of view for the pinhole model (middle, green graph), maximum trapezoidal distortion (top, red graph), and maximum cushion distortion (bottom, blue graph). The radius of the inversion sphere is 0.5

with $\tau_f = 1$ and $r = 1$. For better visualization of the roots of this polynomial, the zero plane has been plotted transparently. It can be seen immediately from Fig. 7.9 that all four roots of (7.28) are not always present. If all roots exist, they satisfy

$$\tau_s^{(-,+)} \leq \tau_s^{(-,-)} \leq \tau_s^{(+,-)} \leq \tau_s^{(+,+)}, \tag{7.30}$$

where, for example, $\tau_s^{(-,+)}$ denotes the root of (7.28) where the minus sign is chosen for the first "$\pm$" and the plus sign for the second "$\pm$". Valid choices for τ_s then lie in either $[\tau_s^{(-,+)}, \tau_s^{(-,-)}]$ or $[\tau_s^{(+,-)}, \tau_s^{(+,+)}]$. If only two roots of (7.28) exist, then τ_s has to lie in $[\tau_s^{(-,+)}, \tau_s^{(+,+)}]$.

For each τ_s in these ranges, a value for d can be calculated from (7.26) such that image points on the vertical iso-circle are scaled by the given ω_v. This has the effect that the image distortion can be varied while the vFOV is kept constant. Finding an analytic expression for the τ_s that maximizes or minimizes the dFOV for a given vFOV appears to be intractable. A numerical approach, however, yields the plots shown in Fig. 7.10.

Here $\tau_f = 1$, $r = 0.5$, and the image plane size was assumed to be $23.7 \times 15.6\,mm$. The middle (green) line shows the relation between the vFOV and dFOV for a standard pinhole setup. The top (red) line gives the relation for the maximum trapezoidal distortion and the bottom (blue) line gives the relation for the maximum cushion distortion. It appears that the maximum

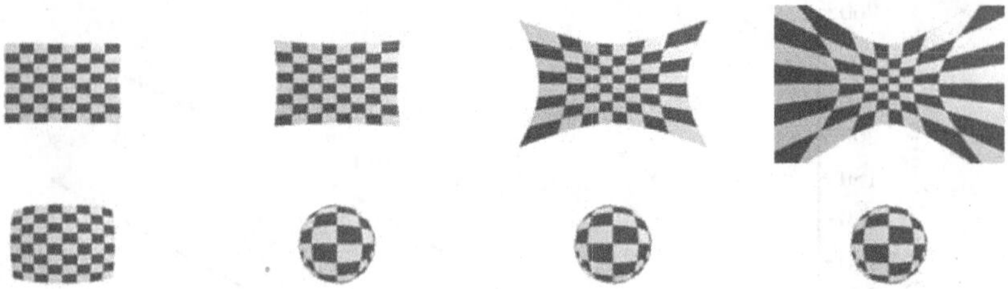

(a) 20° vFOV (b) 43° vFOV (c) 84° vFOV (d) 116° vFOV

Fig. 7.11 Maximal trapezoidal (top) and cushion (bottom) distortions for different vertical fields of view. Note that the images have been scaled to the same size to allow better comparison

dFOV does not depend on τ_f or r. The location of the kink in the plot of the minimum dFOV depends on the combination of τ_f and r, though. Figure 7.11 visualizes the maximum amounts of the two different types of distortion for different vFOVs.

To check whether the inversion camera model could model actual lenses, the vFOV and dFOV of two lenses were measured and plotted. The first was a SIGMA DC 18–125mm, 1:3.5–5.6 D zoom lens, set to 18 mm. This lens lies in the achievable distortion range of the inversion camera model.

The second lens was a Nikkor AF Fisheye 10.5 mm, 1:2.8 G ED lens. This is a corrected fisheye lens where the image does not appear circular but fills the whole image area. This is achieved by obtaining a 180° FOV only along the diagonal and compressing the image more along the vertical direction than the horizontal. As can be seen from Fig. 7.10, a 10.5 mm lens cannot be represented correctly by the inversion camera model. This is also due to a general property of fisheye lenses discussed below. In the pose estimation experiments presented in Sect. 8.4, it turned out, however, that the inversion camera model approximated the 10.5 mm lens well enough to achieve good pose estimation results.

It is important to note that the above analysis is only an indicator of whether a lens may be representable by the inversion camera model, since the actual lens distortion will in general be a more complex function. However, it has already been shown in [71] that the division model, which is equivalent to the inversion model in the case of lens distortion, provides a sufficiently good approximation of lens distortion.

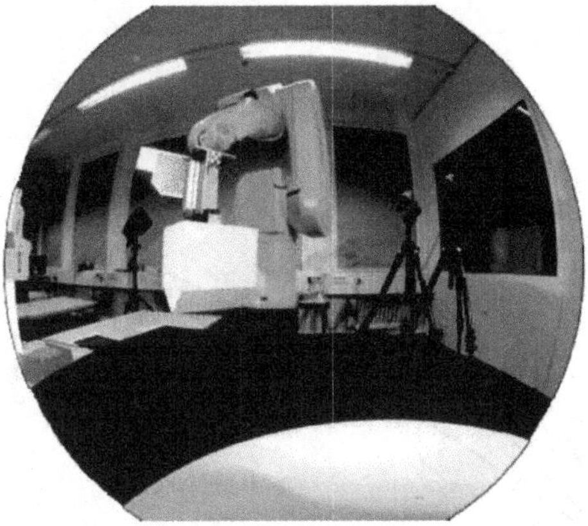

Fig. 7.12 Example of an image taken with a fisheye lens. The top and bottom are cut off, since the CCD chip used was too small for the chosen lens

7.4 Fisheye Lenses

Figure 7.12 shows an image taken with a fisheye lens. Ideal fisheye lenses implement a different type of projection from ideal thin lenses. While the latter can be modeled by central perspective projection, fisheye lenses conserve the angular separation of projection rays. That is, in the fisheye projection model the angle between a projection ray and the optical axis is proportional to the distance between the image center and the corresponding image point. This implies that incident light rays that are perpendicular to the optical axis and pass through the focal point are mapped to infinity in the central perspective projection model, whereas in the fisheye projection model they are mapped to a finite position on the image plane.

It turns out that the inversion camera model cannot represent the projection of an ideal fisheye lens. Figure 7.13 shows the relation between the image position relative to the image center and the projection ray angle for three selected configurations of the inversion camera model. The units of the image position are arbitrary, and the projection ray angle is measured in degrees relative to the optical axis. The relation between the image position and the projection angle for the ideal fisheye model is represented by the straight lines.

The configurations of the three inversion camera models shown in Fig. 7.13 differ only by the position of the inversion sphere, i.e. the value of τ_s. The radius of the inversion sphere is $r = 1$, the initial focal length is $\tau_f = 1$, and the three values of τ_s are, from left to right, 1.524, 1.399, and 1.182. The image position in Fig. 7.13 is measured in the same units as r, τ_f, and τ_s.

Out of the three configurations shown in Fig. 7.13, the middle one approximates the fisheye model best. In fact, it is a good approximation to

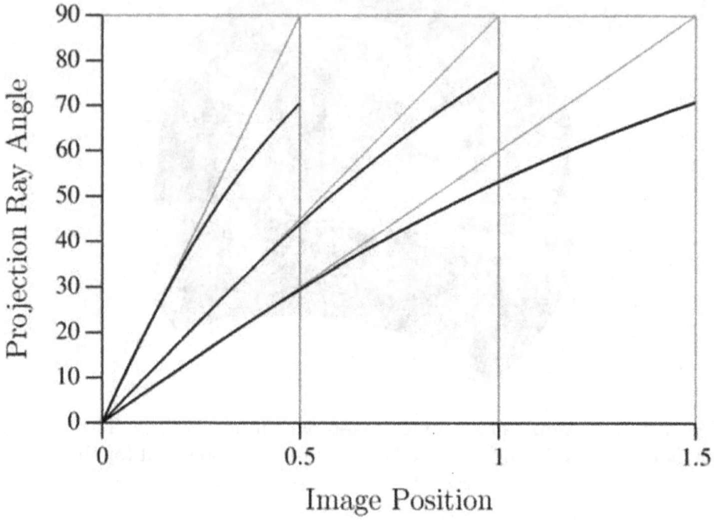

Fig. 7.13 Image position vs. projection ray angle for inversion model and ideal fisheye model

the fisheye model for about half of the total image size. This also becomes apparent in the pose estimation experiments of Sect. 8.4.

7.5 Catadioptric Camera

In this section, the application of the inversion camera model to catadioptric cameras with parabolic mirrors is discussed. Figure 7.1 shows an example of an image taken with such an imaging system. In this case, the tip of the parabola is pointing upwards and the camera looks down on it.

Figure 7.14 shows the geometric setup of this catadioptric imaging system. As was indicated in Fig. 7.3(d), this is one of the possible configurations of the inversion camera model. The generation of an image point $\boldsymbol{Y}$ from a world point $\boldsymbol{X}$ via reflection in the parabolic mirror can be represented mathematically by projecting $\boldsymbol{X}$ onto the sphere $\boldsymbol{S}_{img}$ followed by an inversion in the sphere $\boldsymbol{S}_{inv}$. In contrast to the inversion model setup of Sect. 7.3, the focal point $\boldsymbol{F}$ lies in the image plane in this case.

The relation of the physical parabolic mirror to the mathematical setup is also indicated in Fig. 7.14. In a standard setup, the sphere $\boldsymbol{S}_{img}$ has a unit radius and is centered on the focal point $\boldsymbol{F}$ of the parabolic mirror. The corresponding parabolic mirror then has to pass through the intersection points of $\boldsymbol{S}_{img}$ with the image plane $\boldsymbol{P}_{img}$. The inversion sphere $\boldsymbol{S}_{inv}$ has to

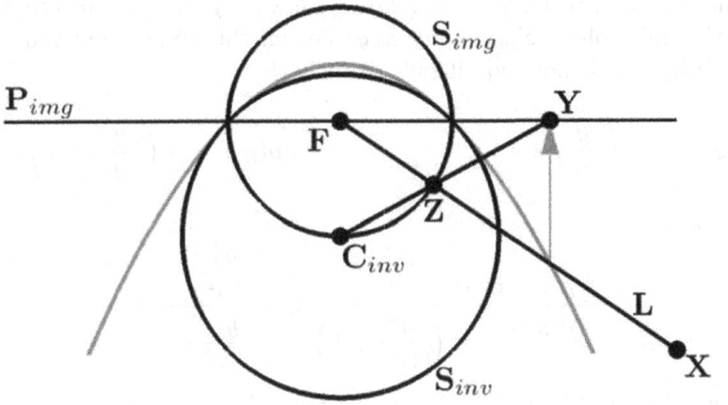

Fig. 7.14 Representation of a catadioptric camera with a parabolic mirror

be centered on $\boldsymbol{C}_{inv}$ and has to pass through the intersection points of $\boldsymbol{S}_{img}$ with $\boldsymbol{P}_{img}$. This fixes the radius of $\boldsymbol{S}_{inv}$ as $\sqrt{2}$.

If the location and radius of the inversion sphere $\boldsymbol{S}_{inv}$ are fixed, the only free parameter left in the inversion camera model represented by (7.6) is the dilation, i.e. the scaling of the image. In the following, the relation between the dilation parameter τ_d, and correspondingly $d = (1+\tau_d)/(1-\tau_d)$, and the focal length of the parabolic mirror μ is derived. The same coordinate system as before is assumed, where $\boldsymbol{F}$ lies at the origin, $\boldsymbol{e}_1$ and $\boldsymbol{e}_2$ span the image plane $\boldsymbol{P}_{img}$, and $\boldsymbol{e}_3$ points from $\boldsymbol{F}$ to $\boldsymbol{C}_{inv}$ in Fig. 7.14.

The mapping of an image point $\boldsymbol{y}$ by the inversion camera model is still given by (7.16). Substituting the values $\tau_s = 1$ and $r = \sqrt{2}$ into (7.16) results in the following expression:

$$\boldsymbol{z}(\boldsymbol{y}) = \frac{2\,d}{d^2\,\boldsymbol{y}^2 + 1}\,\boldsymbol{y} + \frac{d^2\,\boldsymbol{y}^2 - 1}{d^2\,\boldsymbol{y}^2 + 1}\,\boldsymbol{e}_3. \tag{7.31}$$

The parametric expression for a parabola with focal length μ and its focal point at the origin is given by

$$\boldsymbol{p}(\boldsymbol{y}) = \boldsymbol{y} + \left(\frac{\boldsymbol{y}^2}{4\,\mu} - \mu\right)\,\boldsymbol{e}_3. \tag{7.32}$$

Note that if d is chosen correctly, $\boldsymbol{p}(\boldsymbol{y})$ lies at the intersection point of the projection ray $\boldsymbol{L}$ and the parabola in Fig. 7.14. The goal is now to find the relation between d and μ such that $\boldsymbol{z}(\boldsymbol{y})$ and $\boldsymbol{p}(\boldsymbol{y})$ actually do lie on the same projection ray from the focal point $\boldsymbol{F}$. That such a relation must exist has already been proven by Geyer and Daniilidis [78].

One approach to finding the relation between μ and d is to project $\boldsymbol{p}(\boldsymbol{y})$ onto the unit sphere $\boldsymbol{S}_{img}$, since $\boldsymbol{z}(\boldsymbol{y})$ lies on this sphere and thus $\boldsymbol{z}(\boldsymbol{y}) = \boldsymbol{p}(\boldsymbol{y})/\|\boldsymbol{p}(\boldsymbol{y})\|$. It is not too difficult to show that

$$\boldsymbol{p}(\boldsymbol{y})^2 = \mu^2 \left(\frac{\boldsymbol{y}^2}{4\,\mu^2} + 1\right)^2 \qquad \Longleftrightarrow \qquad \|\boldsymbol{p}(\boldsymbol{y})\| = \mu \left(\frac{\boldsymbol{y}^2}{4\,\mu^2} + 1\right). \tag{7.33}$$

Therefore,

$$\frac{\boldsymbol{p}(\boldsymbol{y})}{\|\boldsymbol{p}(\boldsymbol{y})\|} = \frac{1}{\mu \left(\dfrac{\boldsymbol{y}^2}{4\,\mu^2} + 1\right)}\,\boldsymbol{y} + \frac{\dfrac{\boldsymbol{y}^2}{4\,\mu^2} - 1}{\dfrac{\boldsymbol{y}^2}{4\,\mu^2} + 1}\,\boldsymbol{e}_3. \tag{7.34}$$

If we compare (7.31) with (7.34), it can easily be seen that

$$\boldsymbol{z}(\boldsymbol{y}) = \frac{\boldsymbol{p}(\boldsymbol{y})}{\|\boldsymbol{p}(\boldsymbol{y})\|} \qquad \Longleftrightarrow \qquad d = \frac{1}{2\,\mu}. \tag{7.35}$$

To summarize, the imaging process of a parabolic catadioptric camera as shown in Fig. 7.14, whose parabolic mirror has a focal length μ, can be represented by the transformation operator $\boldsymbol{K} = \boldsymbol{S}_{inv}\,\boldsymbol{D}$ in CGA, where

$$\boldsymbol{S}_{inv} = \boldsymbol{e}_3 - \frac{1}{2}\,\boldsymbol{e}_\infty + \boldsymbol{e}_o, \qquad \boldsymbol{D} = 1 + \frac{2\,\mu - 1}{2\,\mu + 1}\,\boldsymbol{E}. \tag{7.36}$$

7.6 Extensions

The inversion camera model was presented in its most basic form in the previous sections. In this section, extensions of this basic model are presented. However, a detailed analysis of these extensions is a subject for future research.

In the previous sections it was always assumed that the optical axis passes through the image center. For real cameras, this is in general not the case. The basic inversion camera model can be extended as follows to incorporate such deviations:

$$\boldsymbol{K} = \boldsymbol{T}_S\,\boldsymbol{S}\,\boldsymbol{T}_S\,\boldsymbol{D}\,\boldsymbol{T}_C, \tag{7.37}$$

where $\boldsymbol{S}$ is a sphere of radius r centered at the origin, $\boldsymbol{T}_S$ is a translator along $\boldsymbol{e}_3$, $\boldsymbol{D}$ is a dilator, and $\boldsymbol{T}_C$ is a translator in the image plane. That is, $\boldsymbol{T}_C$ translates the image points in the image plane before they are transformed. The effect of this additional translation is shown in Fig. 7.15(b); the initial image is shown in Fig. 7.15(a).

In addition to translation of the image plane, the inversion sphere can also be translated parallel to the image plane. That is, $\boldsymbol{T}_S$ translates not only along e_3 but also along e_1 and e_2. The effect of translating the inversion sphere parallel to the image plane is shown in Fig. 7.15(c).

Combining a translation of the image plane with a translation of the inversion sphere can generate an image such as that shown in Fig. 7.15(d). The combination of these two extensions therefore seems to allow for a translation of the viewpoint.

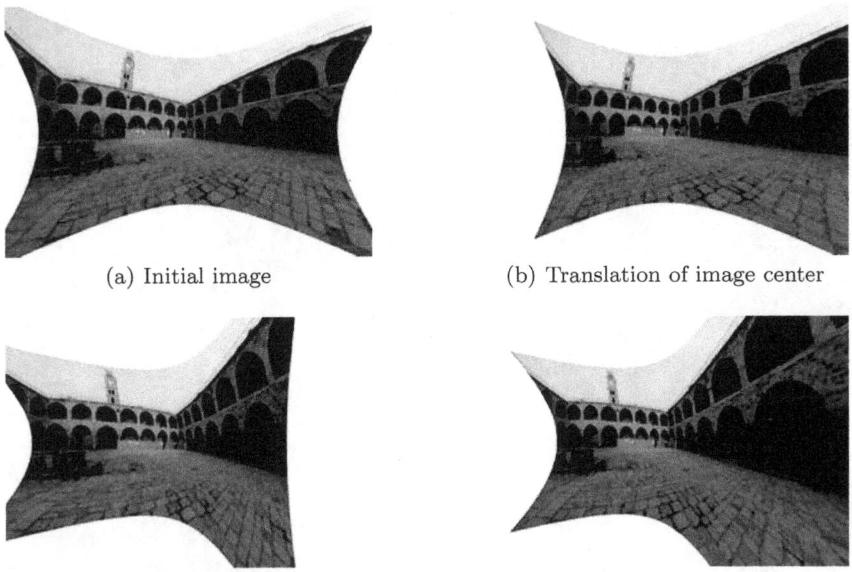

(a) Initial image

(b) Translation of image center

(c) Translation of inversion sphere parallel to image plane

(d) Translation of both image center and inversion sphere

Fig. 7.15 Effect of various extensions to the basic inversion camera model

Chapter 8
Monocular Pose Estimation

Monocular pose estimation is a more or less well-known subject area, which was first treated in the 1980s, for example by Lowe [122, 123]. Overviews of the various approaches to pose estimation can be found in [115, 150]. The various algorithms differ in the type of information that is known about the object whose pose is to be estimated and in the mathematical formalism that is used to represent the pose itself.

The data that is assumed to be given in all monocular pose estimation algorithms is some geometrical information about the object and the corresponding appearance of the object in the image taken. For example, some typical approaches are to assume knowledge of the location of points and/or lines on an object, and the location of their appearance in an image. Finding these correspondences is known as the *correspondence problem*, which is by no means trivial. It is usually only tractable if either unique markers are placed on an object or a tracking assumption is made. In the latter case it is assumed that the pose of the object is roughly known and only has to be adapted slightly. Such an assumption is particularly necessary if only a contour model of the object is given [155].

The mathematical framework used for pose estimation is typically matrix algebra. However, there have been approaches using *dual quaternions* [39, 173], which are isomorphic to motors in conformal geometric algebra. The drawback of using dual quaternions is that within this framework the only representable geometric entities are lines. In the geometric-algebra framework used in this text, any representable geometric entity can be used.

The work most closely related to this text is that by Rosenhahn et al. on pose estimation, they also employed geometric algebra. Their work treats many aspects of pose estimation, such as pose estimation with point and line correspondences [153], pose estimation with free-form contours in 2D [155, 156] and 3D [158, 159], pose estimation of kinematic chains [152, 157], and combinations thereof for estimation of human motion [151].

The aim of this chapter is to revisit the constraint equation of geometric algebra that lies at the foundation of all of the above applications, and to

C. Perwass, *Geometric Algebra with Applications in Engineering.*
Geometry and Computing.

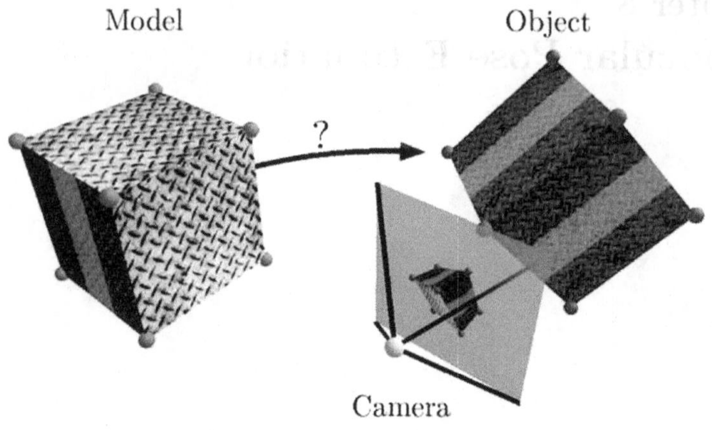

Fig. 8.1 Basic setup of monocular pose estimation problem

apply the numerical approach presented in Chap. 5 to it. This offers the following main advantages over the approach of Rosenhahn:

1. No "small-angle" approximation of the transformation operator is necessary. Instead, the linear representation of the Euclidean group in conformal geometric algebra is used directly.
2. Simultaneous estimation of pose, focal length, and lens distortion for standard lens systems and parabolic catadioptric cameras is possible.
3. Incorporation of uncertain data and estimation of the uncertainty of a pose is possible.
4. Immediate application of optimization algorithms to constraint equations through a tensor representation is possible.

In addition to the advanced treatment of the geometric-algebra constraint equation, a novel algorithm for the automatic estimation of the initial pose is presented, such that no tracking assumption has to be made if point correspondences are known.

The particular pose estimation problem that will be treated here is drawn schematically in Fig. 8.1. It is assumed that the model of the object whose pose is to be estimated is known. In this text, only object models given by a set of points and/or lines are treated.

The object is observed by a camera, and correspondences between the image points and the object model are found. Finding such correspondences is a non-trivial problem that belongs to the field of image processing. If point correspondences have to be found, the correspondence problem is currently tractable only if markers are attached to the object, or if the object per se has strong, distinguishing, pointlike features.

Given the object model, an image of the object taken with a camera with a known orientation and internal parameters, and correspondences between the image and the object model, the task of pose estimation is to estimate the transformation that the object model has to undergo to generate the same image as the object. This transformation is then called the *object pose*. Alternatively, if the location and orientation of the object are known, the camera pose can be estimated in this way.

The plan of this chapter is to first discuss a simple algorithm that gives a rough initial estimate of the object pose in Sect. 8.1. Next, Sect. 8.2 introduces the representation of the monocular pose estimation problem in geometric algebra, which is followed by a presentation of the solution method in Sect. 8.3, using the numerical approach introduced in Chap. 5. The validity and quality of the solution method are finally verified in Sect. 8.4 in an experiment with real data, generated by movements of a robotic arm. The conclusions in Sect. 8.5 summarize the main features of the approach.

8.1 Initial Pose

The goal of the algorithm presented in this section is to give a very rough initial pose estimate which removes the ambiguities in the actual pose estimation algorithm presented later. One of these ambiguities is that in the mathematical model of a pinhole camera, an image of an object can be generated either by an object in front of the camera or by an object placed upside down *behind* the camera Fig. 8.2.

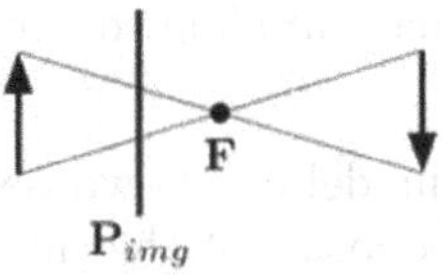

Fig. 8.2 Schematic representation of ambiguity in pose estimation

For example, assume that the observed object appears upside down in a camera image, compared with its orientation in the object model coordinate system, which is typically regarded as identical to the camera coordinate system. Depending on the pose estimation algorithm used, the "closest" numerical minimum may be achieved by placing the object model behind the camera instead of turning the object model upside down and placing it in front of the camera, which would be the correct possibility.

Another ambiguity in the pose estimation algorithm presented later on is encountered when the pose, focal length, and lens distortion are estimated simultaneously. It occurs because the inversion camera model can generate a reflection of a point in the center of the image. With respect to the above

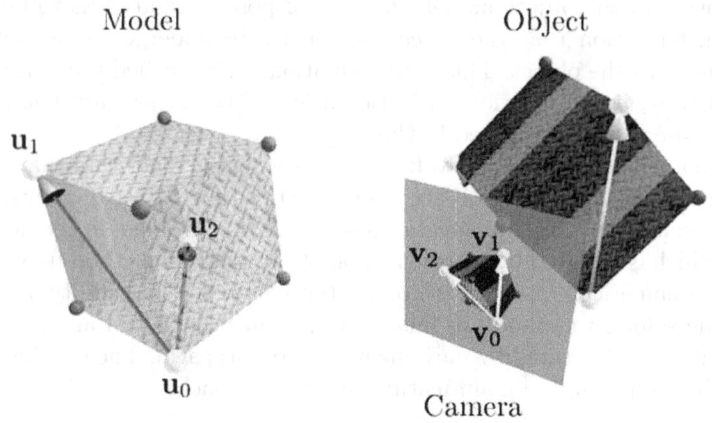

Fig. 8.3 Selection of image point triplet and corresponding model point triplet

example, this means that the algorithm may place the object model upright in front of the camera and the inversion camera model is set to generate an upside-down image.

In addition to the above-mentioned ambiguities, iterative pose estimation algorithms will also need fewer iterations when starting from a fairly sensible estimate.

The goal of the initial pose estimation algorithm is therefore to ensure that the above ambiguities are lifted. This is achieved in the following way, where it is assumed that the object model is given as a set of points and that the correspondences to the image are known. The algorithm consists of three main steps:

1. Find a set of three object model points whose corresponding image points generate a well-conditioned basis of the image plane. That is, if one of the points is chosen as the origin, the vectors to the remaining two points should be close to perpendicular and as long as possible.
2. Calculate the transformation that maps the basis given by the object model point triplet to the corresponding basis vectors of the image point triplet.
3. Calculate the translation along the projection ray of one of the image points of the point triplet so that the extent of the object model point triplet after projection is approximately that of the image point triplet.

These steps will now be discussed in some detail. In the following, the model points are denoted by $\{\boldsymbol{x}_i\}$ and the corresponding image points by $\{\boldsymbol{y}_i\}$. It is assumed that the image point vectors are given in the same coordinate system as the model points. The image center will be denoted by $\boldsymbol{c}$.

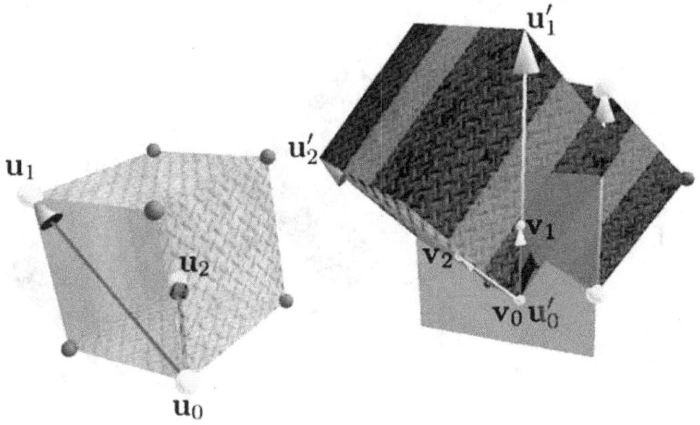

Fig. 8.4 First step of initial transformation of object model

From the set of image points three points have to be selected which generate a well-conditioned basis of the image plane. This is done in the following three steps:

1. The image point furthest away from the image center is found via

$$\boldsymbol{v}_0 := \arg \max_{\boldsymbol{x} \in \{\boldsymbol{y}_i\}} \Big(\|\boldsymbol{x} - \boldsymbol{c}\| \Big). \tag{8.1}$$

2. The image point furthest away from $\boldsymbol{v}_0$ is found, i.e.

$$\boldsymbol{v}_1 := \arg \max_{\boldsymbol{x} \in \{\boldsymbol{y}_i\}} \Big(\|\boldsymbol{x} - \boldsymbol{v}_0\| \Big). \tag{8.2}$$

3. The last image point chosen is the one whose perpendicular distance from the line through $\boldsymbol{v}_0$ and $\boldsymbol{v}_1$ is largest, i.e.

$$\boldsymbol{v}_2 := \arg \max_{\boldsymbol{x} \in \{\boldsymbol{y}_i\}} \Big(\|(\boldsymbol{v}_1 - \boldsymbol{v}_0) \wedge (\boldsymbol{x} - \boldsymbol{v}_0)\| \Big). \tag{8.3}$$

The object model points corresponding to the $\{\boldsymbol{v}_i\}$ are denoted by $\{\boldsymbol{u}_i\}$. An example of the result of this procedure is shown in Fig. 8.3. For the purpose of the next steps, the following vectors are defined for $i \in \{1, 2\}$:

$$\Delta\boldsymbol{v}_i := \frac{\boldsymbol{v}_i - \boldsymbol{v}_0}{\|\boldsymbol{v}_i - \boldsymbol{v}_0\|} \quad \text{and} \quad \Delta\boldsymbol{u}_i := \frac{\boldsymbol{u}_i - \boldsymbol{u}_0}{\|\boldsymbol{u}_i - \boldsymbol{u}_0\|}. \tag{8.4}$$

In the next step the rotor $\boldsymbol{R} \in \mathbb{G}_3^+$ is estimated, which satisfies

$$\Delta\boldsymbol{v}_i = \boldsymbol{R}\, \Delta\boldsymbol{u}_i\, \widetilde{\boldsymbol{R}}, \qquad \forall\, i \in \{1, 2\}. \tag{8.5}$$

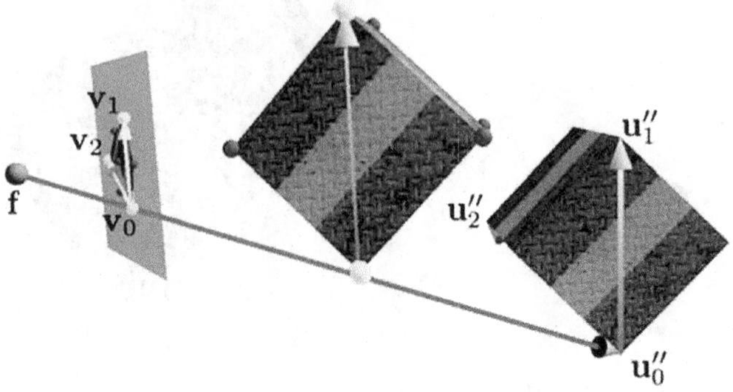

Fig. 8.5 Translation of model to front of camera

To estimate the rotor $\boldsymbol{R}$, the algorithms described in Sects. 5.2.2 and 6.2.6 can be applied. The model points $\{\boldsymbol{x}_i\}$ are then transformed as follows:

$$\boldsymbol{x}'_i := \boldsymbol{R}\,(\boldsymbol{x}_i - \boldsymbol{u}_0)\,\widetilde{\boldsymbol{R}} + \boldsymbol{v}_0, \qquad \forall\, i. \tag{8.6}$$

Figure 8.4 shows the result of this transformation. The plane in the object model defined by the $\{\boldsymbol{u}_i\}$ now lies in the image plane, and the vectors $\{\Delta\boldsymbol{u}_i\}$ are aligned with $\{\Delta\boldsymbol{v}_i\}$. This transformation ensures that the model is oriented appropriately with respect to the image.

The last step is to translate the points $\{\boldsymbol{x}'_i\}$ along the projection ray through $\boldsymbol{v}_0$, which ensures that the object comes to lie in front of the camera. The amount by which the points are translated is given by the ratio $\|\boldsymbol{u}_1 - \boldsymbol{u}_0\|/\|\boldsymbol{v}_1 - \boldsymbol{v}_0\|$. That is, the larger the distance between the model points with respect to the distance between the corresponding image points, the further the model is pushed away from the camera. The $\{\boldsymbol{x}'_i\}$ are thus transformed as follows to give the initial pose estimate:

$$\boldsymbol{x}''_i := \boldsymbol{x}'_i + \frac{\boldsymbol{v}_0 - \boldsymbol{f}}{\|\boldsymbol{v}_0 - \boldsymbol{f}\|}\,\frac{\|\boldsymbol{u}_1 - \boldsymbol{u}_0\|}{\|\boldsymbol{v}_1 - \boldsymbol{v}_0\|}, \tag{8.7}$$

where $\boldsymbol{f}$ denotes the optical center. The effect of this transformation is shown in Fig. 8.5. The complete initial transformation is thus given by

$$\boldsymbol{x}''_i := \boldsymbol{R}\,(\boldsymbol{x}_i - \boldsymbol{t}_1)\,\widetilde{\boldsymbol{R}} + \boldsymbol{t}_2, \tag{8.8}$$

where

$$\boldsymbol{t}_1 := \boldsymbol{u}_0 \qquad \text{and} \qquad \boldsymbol{t}_2 := \boldsymbol{v}_0 + \frac{\boldsymbol{v}_0 - \boldsymbol{f}}{\|\boldsymbol{v}_0 - \boldsymbol{f}\|}\,\frac{\|\boldsymbol{u}_1 - \boldsymbol{u}_0\|}{\|\boldsymbol{v}_1 - \boldsymbol{v}_0\|}\,. \tag{8.9}$$

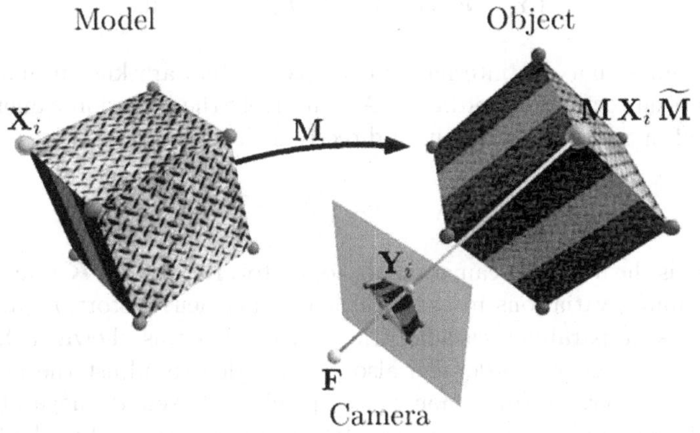

Fig. 8.6 Representation of monocular pose estimation problem in CGA

In CGA, this transformation can be represented by a single operator, which can be found as follows. Let $\boldsymbol{R}$ represent the rotor defined above, only this time embedded in $\mathbb{G}_{4,1}$. Furthermore, let $\boldsymbol{T}_1$ and $\boldsymbol{T}_2$ denote the translators that translate by $\boldsymbol{t}_1$ and $\boldsymbol{t}_2$, respectively. The initial transformation operator is then given by

$$\boldsymbol{M} = \boldsymbol{T}_2\, \boldsymbol{R}\, \widetilde{\boldsymbol{T}}_1. \tag{8.10}$$

If $\{\boldsymbol{X}_i\}$ denotes the object model points embedded in $\mathbb{G}_{4,1}$, then the initial transformation of the model points is evaluated via $\boldsymbol{M}\, \boldsymbol{X}_i\, \widetilde{\boldsymbol{M}}$.

This method of estimating an initial pose works particularly well for the example presented in Figs. 8.3, 8.4, and 8.5. In general, however, the pose estimated in this way can be considerably off. Nevertheless, the method ensures that the object lies in front of the camera and is oriented roughly correctly. In Sect. 8.4, it will be seen that this method works quite well. It even works for images taken by a catadioptric camera.

8.2 Formulation of the Problem in CGA

The problem of monocular pose estimation can be formulated in a straight forward manner in CGA. The geometric configuration is sketched in Fig. 8.6. The goal of the pose estimation algorithm is to estimate the transformation operator $\boldsymbol{M}$ such that the transformation of any model point $\boldsymbol{X}_i$, resulting in $\boldsymbol{M}\, \boldsymbol{X}_i\, \widetilde{\boldsymbol{M}}$, lies on the projection ray $\boldsymbol{L}_i$ of the corresponding image point $\boldsymbol{Y}_i$, where $\boldsymbol{L}_i := \boldsymbol{Y}_i \wedge \boldsymbol{F} \wedge \boldsymbol{e}_\infty$. The constraint equation for $\boldsymbol{M}$ is thus

$$\Big(\boldsymbol{Y}_i \wedge \boldsymbol{F} \wedge \boldsymbol{e}_\infty\Big) \wedge \Big(\boldsymbol{M}\,\boldsymbol{X}_i\,\widetilde{\boldsymbol{M}}\Big) = 0. \tag{8.11}$$

This assumes that the internal camera parameters are known and the image has been rectified accordingly. Alternatively the inversion camera model described in Sect. 7.2 can be used directly in (8.11). That is,

$$\Big((\boldsymbol{K}\,\boldsymbol{Y}_i\,\widetilde{\boldsymbol{K}}) \wedge \boldsymbol{F} \wedge \boldsymbol{e}_\infty\Big) \wedge \Big(\boldsymbol{M}\,\boldsymbol{X}_i\,\widetilde{\boldsymbol{M}}\Big) = 0, \tag{8.12}$$

where $\boldsymbol{K}$ is the inversion camera model operator. Recall that $\boldsymbol{K}$ is an operator that combines variations in the focal length and lens distortion, and it can also represent parabolic catadioptric cameras. As was shown in Sect. 7.6, the inversion camera model can also be extended to adjust the position of the image center. However, non-square pixels and skewed image plane axes cannot be represented in this model. It is therefore assumed in the following that these types of distortion have already been rectified before the pose estimation is applied.

Estimating the pose from correspondences between model points and image points is not the only possibility. Correspondences between model lines and image points can also be used. In this case, typically a number of image points are selected that correspond to the model line. Let $\boldsymbol{U}_i$ denote a model line and $\boldsymbol{Y}_i$ a corresponding image point; then the constraint equation for $\boldsymbol{M}$ becomes

$$\Big((\boldsymbol{K}\,\boldsymbol{Y}_i\,\widetilde{\boldsymbol{K}}) \wedge \boldsymbol{F}\Big) \wedge \Big(\boldsymbol{M}\,\boldsymbol{U}_i\,\widetilde{\boldsymbol{M}}\Big) = 0. \tag{8.13}$$

Note that the left bracket does not contain the $\wedge \boldsymbol{e}_\infty$ anymore since $\boldsymbol{U}_i$ is already a line (cf. section 4.3.5). An example of the application of this type of constraint to robot navigation can be found in [76].

If the inversion camera model is not used, there are even more constraints that can be used in addition. The two most interesting additional constraints are correspondences between model points and image lines, and between model and image lines. The inversion camera model cannot be applied in these cases, since the operator $\boldsymbol{K}$ maps a line into a circle. It would then be necessary to represent an ellipsoidal cone that passes through the focal point and this circle, which is not possible in CGA.

These two additional constraints have the following mathematical formulation. Let $\boldsymbol{V}_i$ denote an image line and $\boldsymbol{X}_i$ a model point that lies on the corresponding model line. The constraint equation is then given by

$$\Big(\boldsymbol{V}_i \wedge \boldsymbol{F}\Big) \wedge \Big(\boldsymbol{M}\,\boldsymbol{X}_i\,\widetilde{\boldsymbol{M}}\Big) = 0\,. \tag{8.14}$$

Note that the expression $\boldsymbol{V}_i \wedge \boldsymbol{F}$ represents a projection plane. If $\boldsymbol{U}_i$ denotes a model line that corresponds to the image line $\boldsymbol{V}_i$, then the constraint becomes

$$\Big(\boldsymbol{V}_i \wedge \boldsymbol{F}\Big) \veebar \Big(\boldsymbol{M}\,\boldsymbol{U}_i\,\widetilde{\boldsymbol{M}}\Big) = 0\,. \tag{8.15}$$

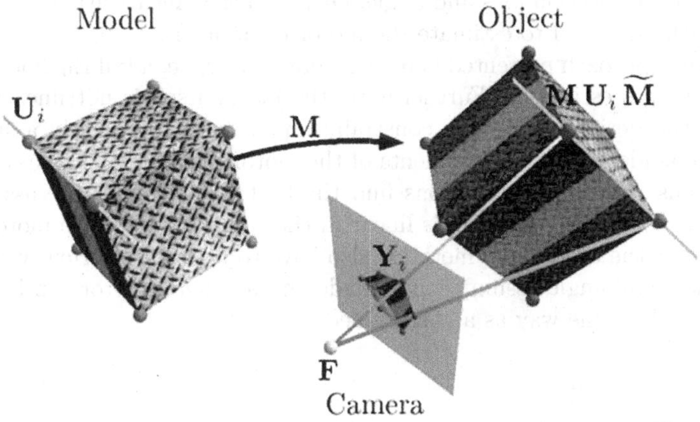

Fig. 8.7 Representation of monocular pose estimation problem using correspondences between model lines and image points

Note that this is one of the containment relations presented in Table 4.6.

8.3 Solution Method

In the following a numerical solution method for monocular pose estimation using the point–line constraint expressed by (8.12) is discussed. The goal is to estimate not only the pose but also the camera parameters contained in the inversion camera model operator $\boldsymbol{K}$. The most straightforward way to estimate the motor and the camera model operator is probably to differentiate (8.12) and to solve for the components of $\boldsymbol{M}$ and $\boldsymbol{K}$ with a standard non-linear minimization approach. However, if the uncertainty of the measured image points is to be taken into account and the covariance of the resultant motor is to be estimated, a combination of Gauss–Markov and Gauss–Helmert estimations is a good choice.

A solution method for the constraint equation (8.11), where a full internal camera calibration is assumed, has been presented by Rosenhahn in [150]. In [150], a motor of the form $\boldsymbol{M} := \boldsymbol{T}\boldsymbol{R}\widetilde{\boldsymbol{T}}$, i.e. a general rotation, is assumed. This implies that the resultant equation systems will have a singularity if the relation between the object model and the observed object is a pure translation. Although this is almost never the case in real scenarios, the equations may become numerically sensitive if a pose is nearly a pure translation. To generate the equation system, the motor $\boldsymbol{M}$ is linearized under the assumption that the rotation angle is small, and the constraint equation is made linear in the rotation axis and angle. Owing to this linearization, an iterative approach is needed to estimate the motor components.

In the approach presented here, a general motor, i.e. a full Euclidean transformation, is estimated. Furthermore, the motor itself is not linearized. Instead, the derivatives of the constraint equation are implicitly assumed to be independent of the components of the motor. That is, the Gauss–Markov and Gauss–Helmert estimations find the best motor under the assumption that the constraint equation is linear in the components of the motor. Since this is not the case, these methods also have to be iterated. However, in this way, no small-angle assumption is made and a general motor can be treated in much the same way as a constrained motor.

8.3.1 Tensor Form

We recall the constraint equation (8.12),

$$\left((\boldsymbol{K}\, \boldsymbol{Y}_n\, \widetilde{\boldsymbol{K}}) \wedge \boldsymbol{F} \wedge \boldsymbol{e}_\infty \right) \wedge \left(\boldsymbol{M}\, \boldsymbol{X}_n\, \widetilde{\boldsymbol{M}} \right) = 0\,.$$

To apply Gauss–Markov and Gauss–Helmert estimation, this constraint equation has to be expressed in tensor form, which is done in a number of steps.

We define the necessary component vectors as $\mathsf{k} := \mathcal{K}(\boldsymbol{K}) \in \mathbb{R}^4$, $\mathsf{x}_n := \mathcal{K}(\boldsymbol{X}_n) \in \mathbb{R}^5$, $\mathsf{y}_n := \mathcal{K}(\boldsymbol{Y}_n) \in \mathbb{R}^5$, $\mathsf{z}_n := \mathcal{K}(\boldsymbol{Z}_n) \in \mathbb{R}^5$, $\mathsf{f} := \mathcal{K}(\boldsymbol{F}) \in \mathbb{R}^5$, $\mathsf{m} := \mathcal{K}(\boldsymbol{M}) \in \mathbb{R}^8$, and $\mathsf{e}_\infty := \mathcal{K}(\boldsymbol{e}_\infty) \in \mathbb{R}^5$. The camera model component vector k may be regarded as a component vector. However, it is advantageous to use the multivectors that generate $\boldsymbol{K}$ as parameter vectors. From (7.10), it follows that $\boldsymbol{K}$ can be written as $\boldsymbol{K} = \boldsymbol{S}_{inv}\, \boldsymbol{D}$, where

$$\boldsymbol{S}_{inv} = \tau_s\, \boldsymbol{e}_3 + \frac{1}{2}\,(\tau_s^2 - r^2)\, \boldsymbol{e}_\infty + \boldsymbol{e}_o \quad \text{and} \quad \boldsymbol{D} = 1 - \tau_d\, \boldsymbol{E}\,,$$

with $\boldsymbol{E} := \boldsymbol{e}_\infty \wedge \boldsymbol{e}_o$. If we defining the component vectors $\mathsf{s} := \mathcal{K}(\boldsymbol{S}_{inv}) \in \mathbb{R}^3$ and $\mathsf{d} := \mathcal{K}(\boldsymbol{D}) \in \mathbb{R}^2$, the component vector k is given by

$$k^r = s^i\, d^j\, \Gamma^r{}_{ij}\,. \tag{8.16}$$

The transformation of the image point $\boldsymbol{Y}_n$ by the camera model operator $\boldsymbol{K}$ becomes

$$\boldsymbol{Z}_n := \boldsymbol{K}\, \boldsymbol{Y}_n\, \widetilde{\boldsymbol{K}} \quad \Longleftrightarrow \quad \mathsf{z}_n^r = y_n^{i_1}\, k^{i_2}\, k^{i_3}\, K^r{}_{i_1\, i_2\, i_3}\,, \tag{8.17}$$

where $K^r{}_{i_1\, i_2\, i_3} \in \mathbb{R}^{5\times 5\times 4\times 4}$ is defined as

$$K^r{}_{i_1\, i_2\, i_3} := \Gamma^{j_1}{}_{i_2 i_1}\, \Gamma^r{}_{j_1 j_2}\, R^{j_2}{}_{i_3}\,.$$

Note that the algebraic product and operator tensors are assumed to be reduced to the minimal dimensions necessary. When implementing the tensor form of (8.17), it is advantageous *not* to calculate the tensor ${K^r}_{i_1\,i_2\,i_3}$ in advance, but to contract the constituent product tensors as soon as possible with the data vectors. In this way, the computational load is considerably reduced. However, the analytic presentation of the formulas is much clearer if the product tensors are combined.

The expression for the transformed model point $\boldsymbol{X}_n$ is very similar:

$$\boldsymbol{W}_n := \boldsymbol{M}\,\boldsymbol{X}_m\,\widetilde{\boldsymbol{M}} \quad\Longleftrightarrow\quad \mathsf{w}_n^r = x_n^{i_1}\; m^{i_2}\; m^{i_3}\;\; {M^r}_{i_1\,i_2\,i_3}\,, \tag{8.18}$$

where $\mathsf{w}_n := \mathcal{K}(\boldsymbol{W}_n)$, and ${M^r}_{i_1\,i_2\,i_3} \in \mathbb{R}^{5\times5\times8\times8}$ is defined as

$${M^r}_{i_1\,i_2\,i_3} := {\Gamma^{j_1}}_{i_2 i_1}\; {\Gamma^r}_{j_1 j_2}\; {R^{j_2}}_{i_3}\,.$$

The projection ray $\boldsymbol{L}_n := \boldsymbol{Z}_n \wedge \boldsymbol{F} \wedge \boldsymbol{e}_\infty$ is expressed in tensor form as

$$\boldsymbol{L}_n := \boldsymbol{Z}_n \wedge \boldsymbol{F} \wedge \boldsymbol{e}_\infty \quad\Longleftrightarrow\quad \ell_n^r = z_n^{i_1}\; f^{i_2}\;\; {L^r}_{i_1\,i_2}\,, \tag{8.19}$$

where ${L^r}_{i_1\,i_2} \in \mathbb{R}^{6\times5\times5}$ is defined as

$${L^r}_{i_1\,i_2} = e_\infty^{j_2}\;\; {\Lambda^{j_1}}_{i_1 i_2}\; {\Lambda^r}_{j_1 j_2}\,.$$

The complete constraint is therefore given by

$$\begin{aligned} \boldsymbol{Q} &:= \Big(\big(\boldsymbol{K}\,\boldsymbol{Y}_n\,\widetilde{\boldsymbol{K}}\big) \wedge \boldsymbol{F} \wedge \boldsymbol{e}_\infty\Big) \wedge \Big(\boldsymbol{M}\,\boldsymbol{X}_n\,\widetilde{\boldsymbol{M}}\Big) \\ \Longleftrightarrow\quad q_n^r &= \ell_n^{i_1}\; w_n^{i_2}\;\; {\Lambda^r}_{i_1 i_2}\,, \end{aligned} \tag{8.20}$$

where $\mathsf{q} \in \mathbb{R}^4$, because $\boldsymbol{Q}$ represents a plane (see Table 4.3). Substituting the expressions for $\ell_n^{i_1}$ and $w_n^{i_2}$ into (8.20) gives

$$\begin{aligned} q_n^r = {}& y_n^{i_1}\; k^{i_2}\; k^{i_3}\;\; x_n^{p_1}\; m^{p_2}\; m^{p_3}\;\; f^{j_2} \\ & \times {K^{j_1}}_{i_1\,i_2\,i_3}\; {L^{q_1}}_{j_1\,j_2}\; {M^{q_2}}_{p_1\,p_2\,p_3}\; {\Lambda^r}_{q_1 q_2}\,. \end{aligned} \tag{8.21}$$

This may also be written as

$$q_n^r = y_n^{i_1}\;\; k^{i_2}\; k^{i_3}\;\; m^{p_2}\; m^{p_3}\;\; {Q^r}_{n\,i_1\,i_2\,i_3\,p_2\,p_3}\,, \tag{8.22}$$

where ${Q^r}_{n\,i_1\,i_2\,i_3\,p_2\,p_3}$ is defined as

$${Q^r}_{n\,i_1\,i_2\,i_3\,p_2\,p_3} := x_n^{p_1}\; f^{j_2}\;\; {K^{j_1}}_{i_1\,i_2\,i_3}\; {L^{q_1}}_{j_1\,j_2}\; {M^{q_2}}_{p_1\,p_2\,p_3}\; {\Lambda^r}_{q_1 q_2}\,.$$

Here x_n and f are assumed to be known exactly and are thus multiplied into the tensor ${Q^r}_{n\,i_1\,i_2\,i_3\,p_2\,p_3}$. If they also have associated covariance matrices, they have to be regarded as data vectors and as part of the constraint function.

Note again that it is computationally much less expensive to calculate q_n^r step by step and not to precalculate the tensor $Q^r{}_{n\,i_1\,i_2\,i_3\,p_2\,p_3}$. The form of (8.22) is merely helpful for seeing that the constraint function that is to be satisfied is quadratic in the parameter vectors k and m. That is, the pose estimation problem can be expressed as a set of coupled quadratic equations.

8.3.2 Jacobi Matrices

To apply the Gauss–Markov and Gauss–Helmert estimation methods, the Jacobi matrices of (8.22) have to be evaluated. In the estimation procedure, the vectors s, d and m are regarded as the parameter vectors and the $\{\,\mathsf{y}_n\,\}$ as the data vectors. To apply the formulas for the construction of the normal equations of the Gauss–Markov and Gauss–Helmert estimation methods, the three parameter vectors have to be combined into a single vector $\mathsf{p} := \left[\,\mathsf{m}^\mathsf{T}\,,\ \mathsf{s}^\mathsf{T}\,,\ \mathsf{d}^\mathsf{T}\,\right]^\mathsf{T}$. The Jacobi matrix $\partial_\mathsf{p}\,\mathsf{q}_n$ is then given by

$$\mathsf{U}_n := \partial_\mathsf{p}\,\mathsf{q}_n = \left[\,\partial_\mathsf{m}\,\mathsf{q}_n\,,\ \partial_\mathsf{s}\,\mathsf{q}_n\,,\ \partial_\mathsf{d}\,\mathsf{q}_n\,\right]\,. \tag{8.23}$$

The evaluation of the separate Jacobi matrices in this expression is most easily done step by step. First of all, from (8.16) it follows that

$$\frac{\partial\, k^r}{\partial\, s^p} = d^j\, \Gamma^r{}_{pj} \qquad \text{and} \qquad \frac{\partial\, k^r}{\partial\, d^p} = s^i\, \Gamma^r{}_{ip}\,, \tag{8.24}$$

where r indicates the row and p the column of the Jacobi matrices. The Jacobi matrices of z_n can be derived from (8.17) as follows:

$$\begin{aligned}
\frac{\partial\, z_n^r}{\partial\, y_n^p} &= k^{i_2}\, k^{i_3}\, K^r{}_{p\,i_2\,i_3}\,,\\
\frac{\partial\, z_n^r}{\partial\, s^p} &= y_n^{i_1}\, \frac{\partial\, k^{i_2}}{\partial\, s^p}\, k^{i_3}\, \left(K^r{}_{i_1\,i_2\,i_3} + K^r{}_{i_1\,i_3\,i_2}\right)\,,\\
\frac{\partial\, z_n^r}{\partial\, d^p} &= y_n^{i_1}\, \frac{\partial\, k^{i_2}}{\partial\, d^p}\, k^{i_3}\, \left(K^r{}_{i_1\,i_2\,i_3} + K^r{}_{i_1\,i_3\,i_2}\right)\,.
\end{aligned} \tag{8.25}$$

The Jacobi matrices of $\boldsymbol{\ell}_n$ then follow directly from (8.19):

$$\begin{aligned}
\frac{\partial\, \ell_n^r}{\partial\, y_n^p} &= \frac{\partial\, z_n^{i_1}}{\partial\, y_n^p}\, f^{i_2}\, L^r{}_{i_1\,i_2}\,,\\
\frac{\partial\, \ell_n^r}{\partial\, s^p} &= \frac{\partial\, z_n^{i_1}}{\partial\, s^p}\, f^{i_2}\, L^r{}_{i_1\,i_2}\,,\\
\frac{\partial\, \ell_n^r}{\partial\, d^p} &= \frac{\partial\, z_n^{i_1}}{\partial\, d^p}\, f^{i_2}\, L^r{}_{i_1\,i_2}\,.
\end{aligned} \tag{8.26}$$

The Jacobi matrices of w_n are similar to those of z_n and are given by

$$\frac{\partial\, w_n^r}{\partial\, m^p} = x_n^{i_1}\, m^{i_2}\, \left(M^r{}_{i_1\, i_2\, p} + M^r{}_{i_1\, p\, i_2}\right)\,. \tag{8.27}$$

Finally, the Jacobi matrices of q_n^r with respect to the parameter vectors are given by

$$\begin{aligned}
\frac{\partial\, q_n^r}{\partial\, m^p} &= \ell_n^{i_1}\, \frac{\partial\, w_n^{i_2}}{\partial\, m^p}\, \Lambda^r{}_{i_1 i_2}\,, \\
\frac{\partial\, q_n^r}{\partial\, s^p} &= \frac{\partial\, \ell_n^{i_1}}{\partial\, s^p}\, w_n^{i_2}\, \Lambda^r{}_{i_1 i_2}\,, \\
\frac{\partial\, q_n^r}{\partial\, d^p} &= \frac{\partial\, \ell_n^{i_1}}{\partial\, d^p}\, w_n^{i_2}\, \Lambda^r{}_{i_1 i_2}\,.
\end{aligned} \tag{8.28}$$

For the Gauss–Helmert estimation method, the Jacobi matrix of q_n with respect to the data vector y_n also has to be known. This is given by

$$\mathsf{V}_n := \partial_{\mathsf{y}_n}\, \mathsf{q}_n = \frac{\partial\, q_n^r}{\partial\, y_n^p} = \frac{\partial\, \ell_n^{i_1}}{\partial\, y_n^p}\, w_n^{i_2}\, \Lambda^r{}_{i_1 i_2}\,. \tag{8.29}$$

8.3.3 Constraints on Parameters

Two different types of constraints on the parameter vectors are employed here. There are two constraints of the first type:

1. A constraint function $\boldsymbol{H}(\boldsymbol{M})$, which ensures that $\boldsymbol{M}$ is a motor.
2. A constraint function $\boldsymbol{G}(\boldsymbol{S}_{inv})$, which fixes the radius of $\boldsymbol{S}_{inv}$ at unity.

The second type of constraint is a regularization term that fixes part of the parameter vectors $\boldsymbol{S}_{inv}$ and $\boldsymbol{D}$. This regularization term is expressed as a covariance matrix for the parameters as discussed in Sects. 5.8.5 and 5.9.5.

The constraint on $\boldsymbol{M}$ is given by

$$\boldsymbol{H}(\boldsymbol{M}) = \boldsymbol{M}\, \widetilde{\boldsymbol{M}} - 1\,, \tag{8.30}$$

which ensures that $\boldsymbol{M}$ is a motor and removes the degree of freedom associated with the arbitrary scale of $\boldsymbol{M}$. In tensor form, this constraint reads

$$h^r = m^{i_1}\, m^{i_2}\, \Gamma^r{}_{i_1 j}\, R^j{}_{i_2} - \delta^r{}_1 \tag{8.31}$$

where $\delta^r{}_1$ is the Kronecker delta, and it is assumed that the first parameter refers to the scalar component. The corresponding Jacobi matrix is

$$\mathsf{H}^r{}_p := \frac{\partial\, h^r}{\partial\, m^p} = m^{i_1}\, \left(R^j{}_{i_1}\, \Gamma^r{}_{pj} + R^j{}_p\, \Gamma^r{}_{i_1 j}\right)\,. \tag{8.32}$$

The first constraint on $\boldsymbol{S}_{inv}$, given by

$$\boldsymbol{G}(\boldsymbol{S}) = \boldsymbol{S}_{inv}\, \boldsymbol{S}_{inv} - 1\,, \tag{8.33}$$

ensures, together with the second constraint on $\boldsymbol{S}_{inv}$ discussed below, that the radius of $\boldsymbol{S}_{inv}$ is unity. This follows directly from (4.44), p. 153. In tensor form, this constraint reads

$$g^r = s^{i_1}\, s^{i_2}\, \Gamma^r{}_{i_1 i_2} - \delta^r{}_1\,, \tag{8.34}$$

where $\delta^r{}_1$ is defined as before. The corresponding Jacobi matrix is

$$\mathsf{G}^r{}_p := \frac{\partial\, g^r}{\partial\, s^p} = s^i\,(\Gamma^r{}_{pi} + \Gamma^r{}_{ip})\,. \tag{8.35}$$

The second type of constraint on $\boldsymbol{S}_{inv}$ and $\boldsymbol{D}$ ensures that their constant components stay constant. That is, in

$$\boldsymbol{S}_{inv} = \tau_s\, \boldsymbol{e}_3 + \frac{1}{2}\,(\tau_s^2 - r^2)\, \boldsymbol{e}_\infty + \boldsymbol{e}_o \qquad \text{and} \qquad \boldsymbol{D} = 1 - \tau_d\, \boldsymbol{E}\,,$$

the $\boldsymbol{e}_o$ component of $\boldsymbol{S}_{inv}$ and the scalar component of $\boldsymbol{D}$ are constant. This is enforced by introducing a diagonal covariance matrix for the parameter vector p, whose inverse has zeros on the diagonal at those places that relate to free parameters. Diagonal entries that relate to fixed parameters are set to high values, for example 1×10^8. In this way, it is also possible to fix $\boldsymbol{S}_{inv}$ completely and vary only $\boldsymbol{D}$, which is equivalent to estimating only the focal length. Similarly, $\boldsymbol{M}$ can be constrained further if it is known, for example, that the Euclidean transformations have to lie in one of the planes of the coordinate axes.

Let ν^i_{m}, ν^i_{s}, and ν^i_{d} denote the inverse variances (regularization values) of m^i, s^i, and d^i, respectively. Since the full parameter vector p is given by $\mathsf{p} = \left[\mathsf{m}^\mathsf{T}, \mathsf{s}^\mathsf{T}, \mathsf{d}^\mathsf{T}\right]^\mathsf{T}$, the corresponding inverse covariance (regularization) matrix $\Sigma^{-1}_{\mathsf{p},\mathsf{p}}$ is given by

$$\Sigma^{-1}_{\mathsf{p},\mathsf{p}} = \begin{bmatrix} \nu^1_{\mathsf{m}} &&&&&&& \\ & \ddots &&&&&& \\ && \nu^8_{\mathsf{m}} &&&&& \\ &&& \nu^1_{\mathsf{s}} &&&& \\ &&&& \nu^2_{\mathsf{s}} &&& \\ &&&&& \nu^3_{\mathsf{s}} && \\ &&&&&& \nu^1_{\mathsf{d}} & \\ &&&&&&& \nu^2_{\mathsf{d}} \end{bmatrix}. \tag{8.36}$$

Free parameters have a corresponding inverse variance that is zero, i.e. the variance is infinite, and fixed parameters have a large corresponding inverse variance, i.e. a small variance.

8.3.4 Iterative Estimation

The estimation of the parameter vector p is done in two main steps. In the first step, the Gauss-Markov estimation is iterated until it converges. In the second step, the Gauss-Helmert estimation, which takes account of the covariances of the data vectors is applied once. The idea here is that the Gauss-Markov estimation finds the minimum under the assumption that the data vectors are certain. If the data covariances are not too large, this minimum should be close to the minimum when the data covariances are taken into account. One step with the Gauss–Helmert method should then suffice to estimate the appropriate minimum.

8.3.4.1 Setup

The image is assumed to lie in the $\boldsymbol{e}_1$–$\boldsymbol{e}_2$ plane, where the image center lies at the origin. The horizontal image axis pointing to the right from the image center is $\boldsymbol{e}_1$, and the vertical image axis pointing *upwards* from the image center is $\boldsymbol{e}_2$. Therefore, the optical axis pointing away from the camera has a direction $-\boldsymbol{e}_3$.

It is assumed that the image resolution in pixels and the size of the CCD chip in millimeters are known. The image points $\boldsymbol{Y}_n$ that correspond to the model points $\boldsymbol{X}_n$ are then expressed in terms of millimeter positions relative to the image center. Furthermore, the image point coordinates are scaled isotropically, such that the extent of the CCD chip away from the image center is of order 1. The model points are typically not scaled accordingly, since this would result in very large differences in magnitude between the image and model points, which would lead to numerical problems in the estimation procedure. Instead, the model points are scaled isotropically such that the extent of the model is not more than one order of magnitude larger than the chip size.

The model itself is placed at the origin, where one model point is chosen as the model's origin. Alternatively, the model's center may be placed at the origin.

8.3.4.2 Initial Values

The initial estimate of the parameter vector m of the motor $\boldsymbol{M}$ is obtained from the initial pose estimation described in Sect. 8.1. The initial parameter vectors s and d of the inversion sphere $\boldsymbol{S}_{inv}$ and the dilator $\boldsymbol{D}$ are set either to the values of the pinhole setup for images with lens distortion (cf. Sect. 7.3) or to the values for the catadioptric setup for a parabolic mirror with unit focal length (cf. Sect. 7.5).

- *Lens distortion.* $\boldsymbol{S}_{inv}$ represents a sphere of radius 1 centered on the optical center and $\boldsymbol{D}$ is a unit dilator. The optical center or focal point $\boldsymbol{F}$ is placed at the point $\boldsymbol{e}_3$. This implies an initial focal length of 1, which relates to an initial angular field of view of approximately 45 degrees, because of the scaling of the CCD chip. Hence,

$$\boldsymbol{F} = \boldsymbol{e}_3 + \frac{1}{2}\boldsymbol{e}_\infty + \boldsymbol{e}_o \iff \mathsf{f} = \left[0\,,\,0\,,\,1\,,\,\tfrac{1}{2}\,,\,1\right]^\mathsf{T}\,,$$
$$\boldsymbol{S}_{inv} = \boldsymbol{e}_3 + \boldsymbol{e}_0 \iff \mathsf{s} = [1\,,\,0\,,\,1]^\mathsf{T}\,,$$
$$\boldsymbol{D} = 1 \iff \mathsf{d} = [1\,,\,0]^\mathsf{T}\,.$$

- *Parabolic mirror.* The focal point $\boldsymbol{F}$ is now placed at the origin, i.e. the same point as the image center. $\boldsymbol{S}_{inv}$ represents a sphere of radius $\sqrt{2}$ centered on $\boldsymbol{e}_3$, and $\boldsymbol{D}$ is a unit dilator. This represents the setup shown in Fig. 7.14. Note that only $\boldsymbol{S}_{inv}$ is kept fixed in the minimization procedure. Hence,

$$\boldsymbol{F} = \boldsymbol{e}_o \iff \mathsf{f} = [0\,,\,0\,,\,0\,,\,0\,,\,1]^\mathsf{T}\,,$$
$$\boldsymbol{S}_{inv} = \boldsymbol{e}_3 - \frac{1}{2}\boldsymbol{e}_\infty + \boldsymbol{e}_0 \iff \mathsf{s} = \left[1\,,\,-\frac{1}{2}\,,\,1\right]^\mathsf{T}\,,$$
$$\boldsymbol{D} = 1 \iff \mathsf{d} = [1\,,\,0]^\mathsf{T}\,.$$

If the camera model parameters are known, for example if rectified images are used, then these can be set, and the corresponding parameter vectors are fixed by setting the corresponding entries on the diagonal of $\Sigma^{-1}_{\mathsf{p},\mathsf{p}}$ to large values.

In the following, λ_I and λ_M denote the image and model point scales, respectively. If the focal length τ_f of a standard camera is known and the image is rectified, then the focal-point parameter vector f is set to

$$\mathsf{f} = \left[0\,,\,0\,,\,\lambda_I\,\tau_f\,,\,\frac{1}{2}\lambda_I^2\,\tau_f^2\,,\,1\right]\,,$$

and s and d are regarded as fixed parameters.

If the focal length τ_f of a parabolic mirror is known, then the dilator parameter vector d is set to

$$\mathsf{d} = \left[1\,,\,\frac{2\,\lambda_M\,\tau_f - 1}{2\,\lambda_M\,\tau_f + 1}\right]^\mathsf{T}\,.$$

The parameter vectors s and d are then both kept fixed in the minimization.

8.3.4.3 Iteration

Given the Jacobi matrices and parameter covariance $\Sigma_{\mathsf{p},\mathsf{p}}$ as derived in Sects. 8.3.2 and 8.3.3, the normal equations of the generalized Gauss–Markov method (cf. Sect. 5.8.5) can be generated and solved for the adjustment vector $\Delta\mathsf{p}$ of the parameter vector p. If $\mathsf{p}^{(i)}$ and $\Delta\mathsf{p}^{(i)}$ denote the parameter and adjustment vectors of the ith iteration, then

$$\mathsf{p}^{(i+1)} = \mathsf{p}^{(i)} + \Delta\mathsf{p}^{(i)} .$$

The Jacobi matrices are then reevaluated with the updated parameter vector $\mathsf{p}^{(i+1)}$, and a new adjustment vector $\Delta\mathsf{p}^{(i+1)}$ is estimated. This iterative procedure is stopped when the magnitude of the adjustment vector $\|\Delta\mathsf{p}^{(i)}\|$ is sufficiently small.

Note that this procedure is very similar to the Levenberg–Marquardt method, with the main difference that the adjustment vector is not scaled before adding it to the current estimate. The update direction is, however, evaluated in exactly the same way.

A problem in minimizing the given constraint function for the motor and camera model parameters is that the parameters are of different scales. That is, an adjustment that is small in terms of the motor parameters can have a very large effect when applied to the camera model parameters. This has the effect that the adjustment of the camera model parameters is often overestimated, which can lead to an oscillation or a convergence to a local minimum in the minimization procedure. The situation can be remedied by setting the components in $\Sigma^{-1}_{\mathsf{p},\mathsf{p}}$ that correspond to the camera model parameters not to zero but to small values. The exact magnitude that these components should have is difficult to find, however. Instead, the following approach has been implemented, which leads to good results.

With respect to (8.36), the regularization values for m and $\{\,\nu^i_{\mathsf{m}}\,\}$, are set to zero. The motor parameters are therefore treated as free parameters. The regularization values $\{\,\nu^i_{\mathsf{s}}\,\}$ and $\{\,\nu^i_{\mathsf{d}}\,\}$ are initially set to a large value such as 10^7. In the lens distortion setup, the only camera model parameters that have to be adjusted are s^1, s^2, and d^2. Keeping the remaining camera model parameters constant also fixes the scale of s and d.

Owing to this initialization, the camera model parameters are not adjusted in the first iteration step, which helps to guide the motor update in the right "direction". From now on, the regularization terms are set depending on the corresponding parameter adjustment of the previous step as follows:

$$\nu^1_{\mathsf{s}} = 2\,\|s^1\|\,, \quad \nu^2_{\mathsf{s}} = 2\,\|s^2\|\,, \quad \nu^2_{\mathsf{d}} = 2\,\|d^2\|\,. \tag{8.37}$$

The idea is that if a large adjustment step is made in one iteration, then in the next iteration only a small step is made. For example, suppose that in one step a parameter is adjusted by too much, such that the derivative at the adjusted position is even larger in magnitude in the opposite direction.

(a) Setup of a standard camera above a catadioptric camera with a parabolic mirror

(b) Setup of cameras relative to robot arm

Fig. 8.8 Experimental setup

In this case the minimization will oscillate and not reach a minimum. The variation of the regularization terms has a damping effect, such that the step in the opposite direction is kept small. If the minimization routine reaches an area close to the minimum in this way, only small steps are necessary and damping is removed.

In this way, good and stable results were obtained in experiments. In the parabolic-mirror setup, the $\{\nu_s^i\}$ are, of course, left at their large initial values, because only d^2 has to be adjusted.

8.4 Experiments

To test the pose estimation method presented in the previous section, experiments with a number of different imaging systems were performed. The basic idea was to have a robot arm move an object by well-defined steps, to generate exact and repeatable movements. These were observed by standard projective cameras with different lenses and with a parabolic mirror. Because the robot movements were known exactly, they could be used as ground truth for the pose estimation from the images.

In the remainder of this section, first the experimental setup and then the results are discussed.

8.4.1 Setup

Figure 8.8 shows the experimental setup. A model of a house was attached to the end effector of a robot arm and moved in front of a stationary camera. Since the robot movements had a positioning uncertainty of below 1 *mm*, the positions could be used as ground truth. The model was not rotated, since an exact calibration of the rotation center with respect to the model was not available. The model was translated within an area of approximately 50 *cm* parallel to the image plane and 35 *cm* perpendicular to the image plane. The exact positions in the robot coordinate system are given in Table 8.1 in millimeters and are visualized in Fig. 8.9. With respect to the image in Fig. 8.8(b), $\boldsymbol{e}_1$ points towards the desk in the foreground, $\boldsymbol{e}_2$ to the right, and $\boldsymbol{e}_3$ upwards. The closest approach of the object to the camera was approximately 10 *cm* for the catadioptric and fisheye camera setups.

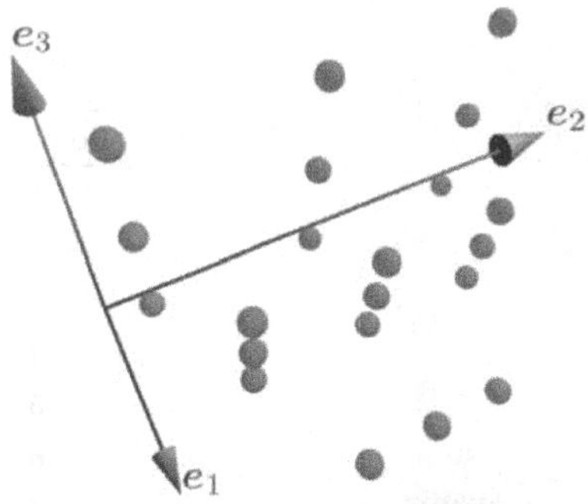

Fig. 8.9 Visualization of the 21 robot positions given in Table 8.1

Table 8.1 Robot positions in millimeters in robot coordinate system

	1	*2*	*3*	*4*	*5*	*6*	*7*	*8*	*9*	*10*	
$\boldsymbol{e}_1$	427	593	758	427	593	758	427	593	758	427	
$\boldsymbol{e}_2$	−141	−61.5	18	116	119	123	372	300	227	−144	.
$\boldsymbol{e}_3$	43	43	43	43	43	43	43	43	43	198	
	11	*12*	*13*	*14*	*15*	*16*	*17*	*18*	*19*	*20*	*21*
$\boldsymbol{e}_1$	593	427	593	427	593	427	593	427	593	427	593
$\boldsymbol{e}_2$	−63	114	118	372	300	−147	−64.5	113	118	372	300
$\boldsymbol{e}_3$	120	198	120	198	120	352	198	352	198	352	198

Three different cameras were used. A Nikon D70 digital SLR camera with a CCD chip size of 23.7 × 15.6 mm and a resolution of 3008 × 2000 pixels was used to take pictures with three different lenses: the SIGMA DC 18–125 mm, 1:3.5–5.6 D zoom lens, set to 18 mm and 50 mm, a Nikkor AF Fisheye 10.5 mm, 1:2.8 G ED lens, and a Sigma 8 mm 1:4.0 EX DG Circular-Fisheye lens.

The two remaining cameras were used with a parabolic-mirror catadioptric imaging system. One camera was a Sony analog camera with a resolution of 768 × 576 pixels and a CCD chip size of 8.8 × 6.6 mm. The other camera was a LogLux camera-link camera with a resolution of 1280 × 1024 pixels and a CCD chip size of 8.576 × 6.861 mm.

Neither an internal nor an external calibration of the cameras was carried out before the pose estimation experiments. However, the size of the CCD chip in millimeters and its resolution in pixels were known and it was assumed that the optical axis passed at a right angle through the center of the CCD chip.

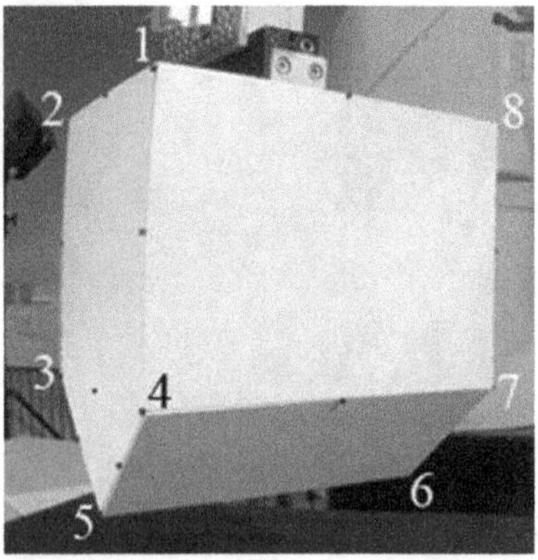

(a) House model, where the markers used are indicated by numbers

Point No.	e_1	e_2	e_3
1	0	0	0
2	0	0	148
3	0	145	148
4	0	145	0
5	0	207	74
6	210	207	74
7	210	145	0
8	210	0	0

(b) Coordinates of model points of house model shown in (a) in millimeters

Fig. 8.10 House model used in the experiments

Eight markers at the visible corners of the house model shown in Fig. 8.10(a) were used as model points. The corresponding model coordinates in millimeters are given in Fig. 8.10(b). Figure 8.11 shows how the house model appeared in the different imaging systems for robot position 3. Note that for the 18 mm and the 50 mm lens, the D70 SLR camera had to be placed further away from the robot arm than shown in Fig. 8.8 so that the house model could be seen for all robot positions. The correspondences between these model points and their apparent positions in the images were found manually.

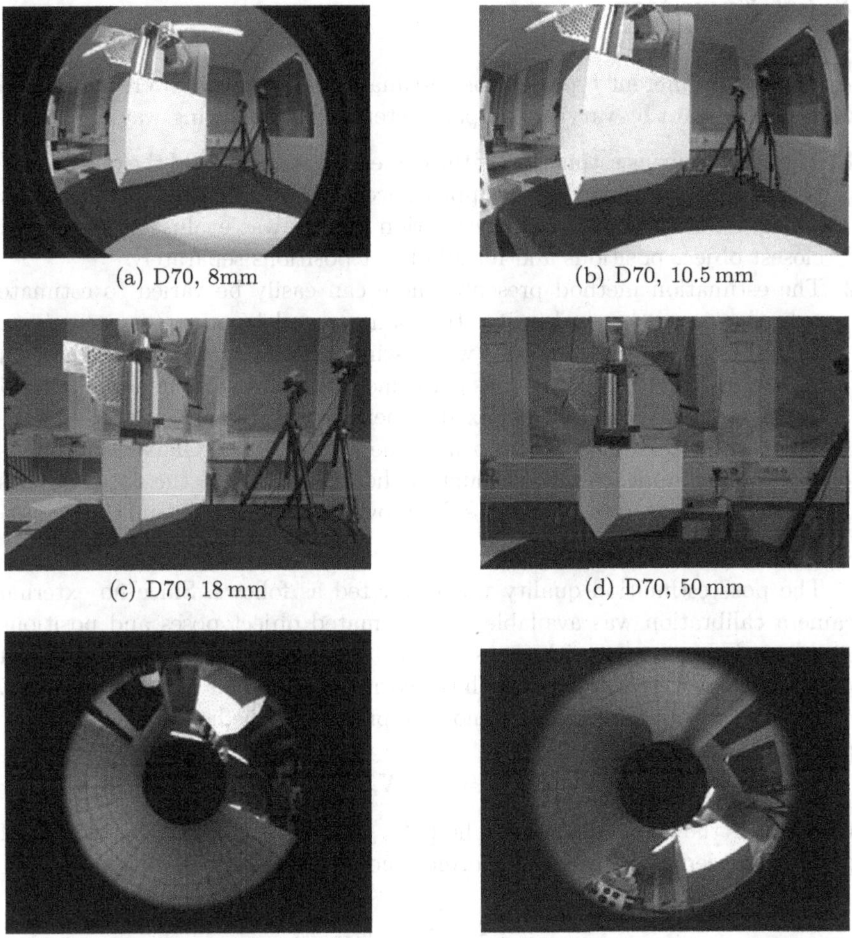

(a) D70, 8mm (b) D70, 10.5 mm
(c) D70, 18 mm (d) D70, 50 mm
(e) LogLux, parabolic mirror (f) Sony, parabolic mirror

Fig. 8.11 Images of house model for robot position 3, for each imaging system

The algorithm was implemented in CLUScript, an interpreted programming language, and was executed with CLUCalc (see Chap. 2). The software ran on a 1.6 GHz Pentium M processor. An optimized implementation in C++ would be expected to increase the execution speed by at least a factor of 10.

8.4.2 Execution

A number of different types of pose estimation experiments were performed on the same data by varying the parameters in the following ways:

1. The simultaneous estimation of the camera parameters and the object pose is only well conditioned if the appearance of the object in an image is large enough. Therefore, the pose estimation quality was evaluated for the six closest object positions and for all object positions separately.
2. The estimation method presented here can easily be varied to estimate only the pose when values for the camera model parameters are given. The camera model parameters were estimated with a separate calibration object, and pose estimation was performed with a camera model calibrated in this way and also without fixed camera model parameters.
3. The Gauss–Helmert estimation uses the results of the Gauss–Markov estimation as input, to take account of the uncertainty in the data. Results are therefore given for the Gauss–Markov estimation alone and the Gauss–Helmert estimation.

The pose estimation quality was evaluated as follows. Since no external camera calibration was available, the estimated object poses and positions could not be compared with the robot positions directly. Instead, all pairs of pose estimates were compared with the corresponding pair of robot positions.

Let $\boldsymbol{M}_n$ denote the estimated motor representing the house pose for robot position n, and define

$$\boldsymbol{W}_{n\,i} := \boldsymbol{M}_n\, \boldsymbol{X}_i\, \widetilde{\boldsymbol{M}}_n \;,$$

where $\boldsymbol{X}_i$ is the ith model point. The $\{\, \boldsymbol{W}_{n\,i} \,\}$ are thus the transformed model points. The center of these transformed model points was evaluated as

$$\boldsymbol{w}_n := \frac{1}{M} \sum_{i=1}^{M} \mathcal{C}^{-1}(\boldsymbol{W}_{n\,i}) \;,$$

where $\mathcal{C}^{-1}(\boldsymbol{W}_{n\,i})$ maps $\boldsymbol{W}_{n\,i}$ to Euclidean space and M is the number of model points. The difference between the estimated transformed model centers for robot positions i and j, namely $\boldsymbol{w}_i$ and $\boldsymbol{w}_j$, is denoted by

$$\Delta\boldsymbol{w}_{ij} := \boldsymbol{w}_i - \boldsymbol{w}_j \;.$$

Similarly, the difference vector between the true robot positions i and j, namely $\boldsymbol{r}_i$ and $\boldsymbol{r}_j$, is given by

$$\Delta\boldsymbol{r}_{ij} := \boldsymbol{r}_i - \boldsymbol{r}_j \;.$$

These difference vectors were calculated for all pairs of pose estimates. From the pairs $\Delta\boldsymbol{w}_{ij}$ and $\Delta\boldsymbol{r}_{ij}$, the best rotor (rotation operator) $\boldsymbol{R}$ can be estimated that rotates the $\{\, \Delta\boldsymbol{w}_{ij} \,\}$ into the corresponding $\{\, \Delta\boldsymbol{r}_{ij} \,\}$, which is part of the external calibration. The RMS error of the pose estimate is then given by

$$\Delta := \sqrt{\frac{1}{N} \sum_{(i,j)} \left\| \Delta r_{ij} - \boldsymbol{R}\, \Delta \boldsymbol{w}_{ij}\, \widetilde{\boldsymbol{R}} \right\|^2} , \tag{8.38}$$

where the sum goes over all pairs (i, j) and N is the number of these pairs. For example, for the six closest robot positions, $N = 15$, and if all 21 robot positions are considered, then $N = 210$.

Additionally, a relative RMS error was calculated as the ratio of the RMS error given by (8.38) and the mean distance of the estimated model positions relative to the camera origin. That is,

$$\Delta_{Rel} := \frac{\Delta}{\frac{1}{N} \sum_{n=1}^{N} \|\boldsymbol{w}_n\|} . \tag{8.39}$$

8.4.3 Results

The results of the experiments when the six closest object positions were considered are shown in Table 8.2, and the results when all 21 object positions were considered are shown in Table 8.3. In these tables, Δ is given in units of millimeters, Δ_{Rel} in percent, the number of iterations (Iter.) without units, and the total execution time (Time) in seconds. Note again that the uncalibrated results were obtained with only a knowledge of the size of the CCD chip and its resolution. No additional calibration information was given. A number of conclusions can be drawn from this data.

In Table 8.2, it may be surprising that the pose estimation for the 8 mm and 10.5 mm lenses is more accurate in the uncalibrated than in the calibrated case. The reason is that the inversion camera model cannot represent the distortion of these fisheye lenses correctly. Therefore, in the calibrated case the best mean calibration was used for all images. In the uncalibrated, case the best calibration for the current position of the house object in the image was used, which was much better. However, it is also apparent from Table 8.3 that the simultaneous calibration and pose estimation fails or at least contains a large error if the object is too far away from the camera. Here the pose estimated using a calibrated camera model is better.

It is also interesting to see in Table 8.2 that, apparently, the pose estimates in the uncalibrated case are better for the 8 mm and 10.5 mm lenses than for the 18 mm lens. The reason seems to be that the images taken with the 18 mm lens do not constrain the camera model as well as those taken with the 8 mm and 10.5 mm lenses. Also, the camera was placed further away from the robot arm, such that the relative RMS error Δ_{Rel} stayed approximately the same.

For the 50 mm lens, a simultaneous estimation of the pose and camera model parameters was too badly conditioned to give stable results, which is why no results are listed. The main reason is that in this case variations in focal length and lens distortion hardly change the appearance of the object in the image and can also be partly recovered by a different pose. In particular, as the projection of the object becomes more orthographic, variations in the focal length and in the depth position result in virtually the same image. In other words, with increasing focal length, the object dimensions also have to increase to allow a stable estimation of the camera parameters.

In the comparison between the two catadioptric imaging systems, the LogLux camera gave much better results than did the Sony camera; this was due to the higher resolution of the former. For the six closest object positions, the LogLux camera gave equally good results in the calibrated and the uncalibrated case, even though more iterations were necessary in the latter case. This shows that the camera model parameters were estimated stably and a good calibration could be achieved for the close object positions. When all object positions are considered, though, the calibrated pose estimation is much better. Again, if the object is too far away, its extension in space does not suffice to generate well-conditioned constraint equations.

The additional use of the Gauss–Helmert method after the initial Gauss–Markov estimation leads partly to slightly better and partly to slightly worse results. In all cases a standard deviation of two pixels was assumed for the image points. It is not necessarily to be expected that the Gauss–Helmert estimation will give better results in the Euclidean sense, since it regards the data covariance matrices as local metrics instead of using the standard Euclidean metric. The result of the Gauss–Helmert method is thus optimal with respect to these metrics. This should be of particular importance if different image points can be associated with different covariance matrices, reflecting the uncertainty of the correspondence. For example, if an image-processing algorithm that detects markers on the object also returns a covariance matrix reflecting the spatial precision of the detection, this could be used advantageously in the Gauss–Helmert estimation.

8.5 Conclusions

In this chapter, a monocular pose estimation algorithm was presented that incorporates the inversion camera model. In this way, the same constraint equation can be used for a pinhole camera, a camera with lens distortion, and a catadioptric camera with a parabolic mirror. It was shown that when the extension of the object whose pose is to be estimated is large enough in the image, camera calibration is possible and leads to good results. In particular, it is easy to change the estimation method from a calibration method to a pure pose estimation method by varying the covariance matrix

Table 8.2 Experimental results for pose estimation using the six closest positions of the house model to the camera. Δ is given in millimeters, Δ_{Rel} in percent, and the execution time in seconds. GM, Gauss–Markov method; GH, Gauss–Helmert method; Iter., number of interations

	Calibrated				Uncalibrated			
	GM & GH		GM		GM & GH		GM	
Camera	Δ	Iter.	Δ	Iter.	Δ	Iter.	Δ	Iter.
Lens	Δ_{Rel}	Time	Δ_{Rel}	Time	Δ_{Rel}	Time	Δ_{Rel}	Time
D70	**7.19**	*4.67*	**7.48**	*3.67*	**2.64**	*7.83*	**2.65**	*6.83*
8 mm	2.61	0.41	2.72	0.30	0.96	0.63	0.96	0.53
D70	**7.74**	*4.00*	**8.19**	*3.00*	**2.36**	*6.50*	**2.37**	*5.50*
10.5 mm	2.80	0.34	2.96	0.24	0.85	0.53	0.85	0.43
D70	**5.56**	*5.50*	**5.30**	*3.50*	**6.28**	*6.50*	**6.28**	*5.50*
18 mm	0.87	0.47	0.83	0.26	0.87	0.53	0.87	0.43
D70	**12.88**	*5.00*	**12.86**	*4.00*				
50 mm	0.74	0.41	0.74	0.30				
LogLux	**8.06**	*6.00*	**7.98**	*5.00*	**8.00**	*10.83*	**7.89**	*9.83*
Parabolic	1.81	0.48	1.79	0.37	1.82	0.83	1.80	0.72
Sony	**13.63**	*6.83*	**14.26**	*5.17*	**31.12**	*10.67*	**31.49**	*9.00*
Parabolic	3.73	0.58	3.92	0.41	8.48	0.86	8.81	0.69

of the parameters, which acts as a regularization matrix. Furthermore, the pose estimation algorithm has the following advantageous aspects:

1. No initial pose estimate is needed.
2. No approximations of the Euclidean transformation operator, the motor, are made, such as a small-angle approximation or an assumption that the transformation is a general rotation, which excludes pure translations.
3. The constraint equation is quadratic in the components of the motor and the camera model that are to be estimated.
4. Covariance matrices can be obtained for the motor and the camera model operator.

For pose estimation alone, the approach presented is certainly very useful; it is also useful because it can be extended to estimating the pose from line matches as well. When the camera model is included, it is well suited for catadioptric cameras with a parabolic mirror, lens systems with a slight distortion, and pinhole cameras, i.e. rectified images. For fisheye lenses, the pose estimation quality is limited, because fisheye lenses cannot be modeled to a high accuracy.

Table 8.3 Experimental results for pose estimation using all 21 positions of the house model. Δ is given in millimeters, Δ_{Rel} in percent, and the execution time in seconds. GM, Gauss–Markov method; GH, Gauss–Helmert method; Iter., number of interations

	Calibrated				Uncalibrated			
	GM & GH		GM		GM & GH		GM	
Camera / Lens	Δ / Δ_{Rel}	Iter. / Time	Δ / Δ_{Rel}	Iter. / Time	Δ / Δ_{Rel}	Iter. / Time	Δ / Δ_{Rel}	Iter. / Time
D70	**18.87**	*4.90*	**17.83**	*3.76*	**40.97**	*12.57*	**40.50**	*11.43*
8 mm	4.29	0.42	4.06	0.30	9.36	1.01	9.23	0.89
D70	**20.12**	*4.33*	**19.06**	*3.33*	**36.62**	*11.91*	**31.88**	*10.86*
10.5 mm	4.55	0.36	4.32	0.26	7.30	0.95	7.13	0.85
D70	**10.23**	*5.62*	**10.10**	*3.67*	**27.67**	*7.00*	**27.59**	*6.00*
18 mm	1.35	0.50	1.33	0.30	3.19	0.57	3.19	0.47
D70	**23.32**	*5.00*	**23.18**	*4.00*				
50 mm	1.26	0.42	1.25	0.32				
LogLux	**13.75**	*6.05*	**13.64**	*4.95*	**39.42**	*12.48*	**39.44**	*11.43*
Parabolic	2.22	0.49	2.20	0.38	6.57	0.96	6.57	0.86
Sony	**18.15**	*6.52*	**18.10**	*5.00*	**85.42**	*12.76*	**85.35**	*11.24*
Parabolic	3.36	0.55	3.35	0.40	16.35	1.03	16.38	0.88

Chapter 9
Versor Functions

In this chapter, several different types of versor functions are discussed, which demonstrate interesting relationships between Fourier series of complex-valued functions, coupled twists, space curves in n dimensions, and a special type of polynomial curve, *Pythagorean-hodograph* (PH) curves. All of these stem from a fundamental versor function $\boldsymbol{F} : \mathbb{R} \to \mathbb{G}_{p,q}$ defined as

$$\boldsymbol{F} : t \mapsto \boldsymbol{A}(t)\, \boldsymbol{N}\, \widetilde{\boldsymbol{A}}(t)\,, \qquad \boldsymbol{N} \in \mathbb{G}_{p,q}\,, \quad \boldsymbol{A} : \mathbb{R} \to \mathfrak{G}_{p,q}\,, \tag{9.1}$$

where $\mathfrak{G}_{p,q}$ denotes the Clifford group of $\mathbb{G}_{p,q}$ (cf. Sect. 3.3).

The chapter consists of two main parts: a discussion of coupled motors and a discussion of PH curves. Coupled motors occur quite naturally in the treatment of robotic arms and kinematic chains in general [150, 161]. It is shown how this is related to cycloidal curves and Fourier descriptors of planar curves. The discussion of coupled motors is concluded with a brief discussion of space curves in higher dimensions generated through coupled motors.

The second main part of this chapter is a discussion of PH curves. PH curves were introduced by Farouki and Sakkalis for $\mathbb{R}^2$ [64] and $\mathbb{R}^3$ [65], and are now well-known polynomial functions. Nevertheless, this section introduces some new aspects of spatial PH curves, which were developed by the author in collaboration with Farouki [135]. The main result is that cubic and quintic PH curves, which are typically represented in terms of quaternions, can be represented equivalently in a vector form in geometric algebra. This form has the advantages that it is dimension-independent, it gives the free parameters a geometric meaning, and it lends itself well to further analysis. Another interesting new result is the analysis of PH curves of constant length in the setting of Hermite interpolation. This could be interesting for applications, since the PH curves vary in shape while keeping their length constant. The section ends with a brief discussion of PH curves in $\mathbb{R}^n$ and a summary of the results.

C. Perwass, *Geometric Algebra with Applications in Engineering.* Geometry and Computing.

9.1 Coupled Motors

A coupled motor system describes the curve that is generated by coupling together a number of general Euclidean transformations, which are called motors or screws (cf. Sect. 4.3.11). A motor describes a rotation about an arbitrary axis with a simultaneous translation along that axis. In conformal space $\mathbb{G}_{4,1}$, such an operator can be represented as follows. Let $\boldsymbol{R}, \boldsymbol{T}_R, \boldsymbol{T}_A \in \mathfrak{S}_{4,1}$ denote a rotor about an axis through the origin and two translators, respectively, where $\mathfrak{S}_{4,1}$ denotes the spin group of $\mathbb{G}_{4,1}$. A motor $\boldsymbol{M}$ can then be written as

$$\boldsymbol{M} := \boldsymbol{T}_A\, \boldsymbol{T}_R\, \boldsymbol{R}\, \widetilde{\boldsymbol{T}}_R\,, \tag{9.2}$$

where $\boldsymbol{T}_R$ translates the rotation axis of $\boldsymbol{R}$ and $\boldsymbol{T}_A$ is a translation in the direction of the rotation axis. If we define a motor function $\boldsymbol{M} : \mathbb{R} \to \mathfrak{S}_{4,1}$, then a curve can be generated by the function $\boldsymbol{F} : \mathbb{R} \to \mathbb{G}^1_{p,q}$, where

$$\boldsymbol{F} : t \mapsto \boldsymbol{M}(t)\, \boldsymbol{N}\, \widetilde{\boldsymbol{M}}(t)\,, \tag{9.3}$$

and where $\boldsymbol{N} := \mathcal{C}(\boldsymbol{n}) \in \mathbb{G}^1_{4,1}$ represents a point in conformal space and $\boldsymbol{n} \in \mathbb{G}^1_3$ (see (4.36), p. 150). A coupled motor curve is generated by the combination of a number of motor functions $\boldsymbol{M}_i : \mathbb{R} \to \mathfrak{S}_{4,1}$, as

$$\boldsymbol{F} : t \mapsto \boldsymbol{M}_k(t) \cdots \boldsymbol{M}_1(t)\, \boldsymbol{N}\, \widetilde{\boldsymbol{M}}_1(t) \cdots \widetilde{\boldsymbol{M}}_k(t)\,. \tag{9.4}$$

9.1.1 Cycloidal Curves

A particular subset of coupled motor functions in $\mathbb{R}^2$ generates the well-known cycloidal curves. An overview of these curves can be found in [113, 128, 150, 166]. They are of particular interest here because they have a direct relation to Fourier series of complex-valued functions, which was first shown in [154, 155]. Before the connection to the Fourier series is shown, the representation of cycloidal curves in the conformal space $\mathbb{G}_{3,1}$ of $\mathbb{R}^2$ will be discussed.

A rotor function $\boldsymbol{R} : \mathbb{R} \to \mathfrak{S}_{3,1}$ representing a rotation about the origin can be defined as

$$\boldsymbol{R} : t \mapsto \exp(-\pi\, \mu\, t\, \boldsymbol{U})\,, \tag{9.5}$$

where $\mu \in \mathbb{R}$ is the rotation frequency and $\boldsymbol{U} := \boldsymbol{e}_1\, \boldsymbol{e}_2$ is the unit bivector representing the rotation plane. The $\{\boldsymbol{e}_i\} := \overline{\mathbb{R}}^{3,1}$ denote the canonical basis of $\mathbb{R}^{3,1}$.

A rotation about an arbitrary point in $\mathbb{R}^2$ can be represented by a motor function $\boldsymbol{M} : \mathbb{R} \to \mathfrak{S}_{3,1}$, defined in terms of the above rotor function $\boldsymbol{R}(t)$ as

$$\boldsymbol{M} : t \mapsto \boldsymbol{T}\, \boldsymbol{R}(t)\, \widetilde{\boldsymbol{T}}\,, \tag{9.6}$$

where $\boldsymbol{T} := 1 - \frac{1}{2}\,\boldsymbol{d}\,\boldsymbol{e}_\infty \in \mathfrak{S}_{3,1}$ is a constant translator.

If $\boldsymbol{N} := \mathcal{C}(\boldsymbol{n}) \in \mathbb{G}^1_{3,1}$ represents the point $\boldsymbol{n} \in \mathbb{R}^2$ in conformal space, then a function $\boldsymbol{F} : \mathbb{R} \to \mathbb{G}^1_{3,1}$, defined as

$$\boldsymbol{F} : t \mapsto \boldsymbol{M}(t)\,\boldsymbol{N}\,\widetilde{\boldsymbol{M}}(t)\,, \tag{9.7}$$

represents a curve in $\mathbb{R}^2$. In this case, this curve will simply be a circle, because the point $\boldsymbol{N}$ is rotated about the point $\boldsymbol{T}\,\boldsymbol{e}_o\,\widetilde{\boldsymbol{T}}$, i.e. the origin translated by $\boldsymbol{T}$.

Suppose a set of motor functions $\boldsymbol{M}_i : \mathbb{R} \to \mathfrak{S}_{3,1}$ are defined as

$$\boldsymbol{M}_i : t \mapsto \boldsymbol{T}_i\,\boldsymbol{R}_i(t)\,\widetilde{\boldsymbol{T}}_i\,, \tag{9.8}$$

with

$$\boldsymbol{T}_i := 1 - \frac{1}{2}\,\boldsymbol{d}_i\,\boldsymbol{e}_\infty\,, \quad \boldsymbol{R}_i : t \mapsto \exp(-\pi\,\mu_i\,t\,\boldsymbol{U})\,. \tag{9.9}$$

The $\{\,\boldsymbol{d}_i\,\} \subset \mathbb{R}^2 \subset \mathbb{G}^1_{3,1}$ define a set of 2D translation vectors, and the $\{\,\mu_i\,\} \subset \mathbb{R}$ a set of rotation frequencies. The curve generated by the coupling of these motor functions is represented by a function $\boldsymbol{F} : \mathbb{R} \to \mathbb{G}^1_{3,1}$, defined as

$$\boldsymbol{F} : t \mapsto \boldsymbol{M}_k(t) \cdots \boldsymbol{M}_1(t)\,\boldsymbol{N}\,\widetilde{\boldsymbol{M}}_1(t) \cdots \widetilde{\boldsymbol{M}}_k(t)\,. \tag{9.10}$$

Various sets of rotation axis positions $\{\,\boldsymbol{d}_i\,\}$ and rotation frequencies $\{\,\mu_i\,\}$ generate the well-known *cycloidal curves.*

9.1.2 Fourier Series

The connection of (9.10) to a Fourier series of a complex-valued function can be most easily shown by expressing (9.10) in $\mathbb{G}_2$, the geometric algebra of $\mathbb{R}^2$.

The rotor $\boldsymbol{R}(t) \in \mathfrak{S}_2$ has the form

$$\boldsymbol{R} : t \mapsto \exp(-\pi\,\mu\,t\,\boldsymbol{U})\,, \tag{9.11}$$

where $\boldsymbol{U} := \boldsymbol{e}_1\,\boldsymbol{e}_2$ is in this case the pseudoscalar of $\mathbb{G}_2$. However, there exists no translation operator. Instead, the translations are generated by addition. The function $\boldsymbol{F}(t)$ in (9.7) can therefore be expressed as a function $\boldsymbol{f} : \mathbb{R} \to \mathbb{G}^1_2$ as follows:

$$\boldsymbol{f} : t \mapsto \boldsymbol{R}\,(\boldsymbol{n} - \boldsymbol{d})\,\widetilde{\boldsymbol{R}} + \boldsymbol{d}\,. \tag{9.12}$$

Since $\boldsymbol{n} - \boldsymbol{d} \in \mathbb{G}^1_2$ is a grade 1 vector and the rotation plane $\boldsymbol{U} \in \mathbb{G}^2_2$ is the pseudoscalar, it follows from the rules of the inner product that

$$\boldsymbol{U}\,(\boldsymbol{n}-\boldsymbol{d}) = \boldsymbol{U}\cdot(\boldsymbol{n}-\boldsymbol{d}) = -(\boldsymbol{n}-\boldsymbol{d})\cdot\boldsymbol{U} = -(\boldsymbol{n}-\boldsymbol{d})\,\boldsymbol{U}\ .$$

Since

$$\boldsymbol{R} = \exp(-\pi\,\mu\,t\,\boldsymbol{U}) = \cos(\pi\mu\,t) - \sin(\pi\mu\,t)\,\boldsymbol{U}\ ,$$

it follows that $\boldsymbol{R}\,(\boldsymbol{n}-\boldsymbol{d}) = (\boldsymbol{n}-\boldsymbol{d})\,\widetilde{\boldsymbol{R}}$. Hence, $\boldsymbol{f}(t)$ can be written as

$$\boldsymbol{f}(t) = (\boldsymbol{n}-\boldsymbol{d})\,\widetilde{\boldsymbol{R}}^2(t) + \boldsymbol{d}\ ,$$

where $\widetilde{\boldsymbol{R}}^2(t) = \exp(2\pi\,\mu\,t\,\boldsymbol{U})$.

The function $\boldsymbol{F}(t)$ in (9.10) may thus be written in $\mathbb{G}_2$ as a function $\boldsymbol{f}\,:\,\mathbb{R}\to\mathbb{G}_2^1$ defined as

$$\boldsymbol{f}\,:\,t\mapsto\cdots\,\boldsymbol{R}_2(t)\left(\boldsymbol{R}_1(t)\,(\boldsymbol{d}_0-\boldsymbol{d}_1)\,\widetilde{\boldsymbol{R}}_1(t)+\boldsymbol{d}_1-\boldsymbol{d}_2\right)\widetilde{\boldsymbol{R}}_2(t)+\boldsymbol{d}_2-\cdots\ , \tag{9.13}$$

where $\boldsymbol{d}_0 := \boldsymbol{n}$ is the initial vector mentioned earlier. This can also be written as

$$\begin{aligned}\boldsymbol{f}(t) = &\prod_{i=1}^{k}\left((\boldsymbol{d}_0-\boldsymbol{d}_1)\,\widetilde{\boldsymbol{R}}_i^2(t)\right) + \prod_{i=2}^{k}\left((\boldsymbol{d}_1-\boldsymbol{d}_2)\,\widetilde{\boldsymbol{R}}_i^2(t)\right) + \dots\\ &+(\boldsymbol{d}_{k-1}-\boldsymbol{d}_k)\,\widetilde{\boldsymbol{R}}_k^2(t) + \boldsymbol{d}_k\ .\end{aligned} \tag{9.14}$$

If we define $\boldsymbol{p}_i := \boldsymbol{d}_{k-i}-\boldsymbol{d}_{k-i+1}$ for $i\in\{1,\dots,k\}$, $\boldsymbol{p}_0 := \boldsymbol{d}_k$, and

$$\widetilde{\boldsymbol{V}}_j(t) := \prod_{i=k-j+1}^{k}\widetilde{\boldsymbol{R}}_i^2(t)\ ,$$

(9.13) can be written as

$$\boldsymbol{f}(t) = \boldsymbol{p}_0 + \sum_{i=1}^{k}\boldsymbol{p}_i\,\widetilde{\boldsymbol{V}}_i(t) = \boldsymbol{p}_0 + \sum_{i=1}^{k}\boldsymbol{p}_i\,\exp(2\,\pi\,\nu_i\,t\,\boldsymbol{U})\ . \tag{9.15}$$

The definition of the $\{\,\widetilde{\boldsymbol{V}}_j\,\}$ implies that the frequencies $\{\nu_i\}$ are given in terms of the $\{\mu_i\}$ as $\nu_j := \sum_{i=k-j+1}^{k}\mu_i$. If $\mu_i = 1$ for all $i\in\{\,1,\dots,k\,\}$, then $\nu_i = i$, and (9.15) represents a finite Fourier series

$$\boldsymbol{f}(t) = \sum_{i=0}^{k}\boldsymbol{p}_i\,\exp(2\,\pi\,i\,t\,\boldsymbol{U})\ . \tag{9.16}$$

The relation to a complex-valued Fourier series becomes clear when we note that the even subalgebra $\mathbb{G}_2^+\subset\mathbb{G}_2$ is isomorphic to the complex numbers, as shown in Sect. 3.8.2. Consider a vector $\boldsymbol{x}\in\mathbb{G}_2^1$, where $\boldsymbol{x} := x^1\,\boldsymbol{e}_1 + x^2\,\boldsymbol{e}_2$; then

$$Z := e_1 x = x^1 + x^2 e_1 e_2 = x^1 + x^2 U .$$

Since $U^2 = -1$, Z is isomorphic to a complex number, where x^1 is the real part and x^2 the imaginary part. Therefore, $e_1 f(t)$ is isomorphic to a complex-valued function $f : \mathbb{R} \to \mathbb{C}$ defined as

$$f : t \mapsto \sum_{i=0}^{k} p_i \exp(2\pi i t \mathrm{i}) , \tag{9.17}$$

where $\mathrm{i} := \sqrt{-1}$ is the imaginary unit and the $\{ p_i \} \in \mathbb{C}$ are complex numbers isomorphic to $\{ e_1 p_i \} \in \mathbb{G}_2^+$. Equation (9.17) is the standard definition of a finite Fourier series. The discrete Fourier transform of $f(t)$ is defined as

$$F_j := \int_0^1 f(t) \exp(-2\pi j t \mathrm{i}) \, \mathrm{d}t . \tag{9.18}$$

When applied to (9.17), this results in

$$F_j = \sum_{i=0}^{k} p_i \int_0^1 \exp(2\pi (i-j) t \mathrm{i}) \, \mathrm{d}t = p_j . \tag{9.19}$$

In geometric-algebra terms, the discrete Fourier transform of a 1-vector-valued function $f : \mathbb{R} \to \mathbb{G}_2^1$ is defined as

$$F_j := \int_0^1 f(t) \exp(-2\pi j t U) \, \mathrm{d}t . \tag{9.20}$$

Applied to (9.16), this results in

$$F_j = \sum_{i=0}^{k} p_i \int_0^1 \exp(2\pi (i-j) t U) \, \mathrm{d}t = p_j . \tag{9.21}$$

The amplitude spectrum of $f(t)$ is therefore given by the magnitudes $\{ \|p_i\| \}$ of the phase vectors $\{ p_i \}$, and the phase angles are given by $\{ \arctan(p_i \cdot e_2 / p_i \cdot e_1) \}$. The phase vectors are called *Fourier descriptors*. It is also possible to construct *affine-invariant* Fourier descriptors [10, 80] from the phase vectors. These are elements that stay invariant under affine transformations of the closed, planar curve that they describe; such descriptord have been used extensively for object recognition (see e.g. [11, 70, 169]).

9.1.3 Space Curves

The representation of a curve generated by coupled motors as a Fourier series was only possible because all rotors $\{\, \boldsymbol{R}_i(t)\, \}$ represented rotations in the same plane $\boldsymbol{U}$. The $\{\, \boldsymbol{R}_i(t)\, \}$ therefore commute, i.e. $\boldsymbol{R}_i(t)\, \boldsymbol{R}_j(t) = \boldsymbol{R}_j(t)\, \boldsymbol{R}_i(t)$. This means that any cycloidal curve that lies in some plane in a higher-dimensional embedding space can be represented as a Fourier series. The only element that changes is the bivector $\boldsymbol{U}$ representing the plane.

Another interesting type of cycloidal curve is those that do not lie in a plane. In this case the rotors $\{\, \boldsymbol{R}_i(t)\, \}$ in the coupled-motor representation of (9.13) do not all share the same rotation plane. One effect of this is that the rotors do not commute, in general, and thus the equation cannot be reduced to the form of (9.15). An interesting subject for future research would be to find a transformation that extracts the rotation frequencies and axes from a given space curve function in this case.

9.2 Pythagorean-Hodograph Curves

PH curves are a particularly interesting type of polynomial curve. Their name describes their distinguishing property, namely that their hodograph satisfies the Pythagorean condition. The *hodograph* of a parametric curve $\boldsymbol{r} : \mathbb{R} \to \mathbb{R}^n$ is the locus defined by its derivative $\boldsymbol{r}'(t) := (\partial_t\, \boldsymbol{r})(t)$, considered as a parametric curve in its own right. The Pythagorean condition refers to the fundamental equation $c^2 = a^2 + b^2$. In terms of the polynomial curve $\boldsymbol{r}(t)$, its hodograph is Pythagorean if there exists a polynomial $\sigma : \mathbb{R} \mapsto \mathbb{R}$ such that

$$\big(\sigma(t)\big)^2 = \big(\boldsymbol{r}'(t)\big)^2 \quad \iff \quad \sigma(t) = \|\boldsymbol{r}'(t)\| \,.$$

The main effect of this constraint is that the cumulative arc length $s(t)$ of the curve $\boldsymbol{r}(t)$ is given by a polynomial, since

$$s(t) = \int_0^t \|\boldsymbol{r}'(u)\|\, \mathrm{d}u = \int_0^t \sigma(u)\, \mathrm{d}u \,.$$

Owing to this particular structure, PH curves offer significant computational advantages over "ordinary" polynomial parametric curves. They find application in computer graphics, computer-aided design, motion control, robotics, and related fields [55].

PH curves in the Euclidean spaces $\mathbb{R}^2$ and $\mathbb{R}^3$ have been thoroughly investigated. Planar PH curves [64] can be conveniently expressed in terms of a complex-variable formulation [53], which facilitates key construction and analysis algorithms [5, 60, 63, 66, 101, 125]. Spatial PH curves were first studied in [65], although the characterization used there constitutes only a *sufficient* condition for a polynomial hodograph $\boldsymbol{r}'(t) \in \mathbb{R}^3$ to be Pythagorean. A sufficient-and-necessary condition for the sum of squares of three polynomials to yield the perfect square of a polynomial was presented in a different context in [43], and this was subsequently interpreted in terms of a quaternion formulation [32, 59] for spatial PH curves. PH curves have also been constructed under the Minkowski metric, facilitating exact boundary recovery of a planar region from its medial axis transform [124]. Choi et al. [32] have thoroughly investigated, from a Clifford algebra perspective, the diverse algebraic structures that are evident among Pythagorean hodographs residing in Euclidean and Minkowski spaces. For a thorough discussion of the various aspects of PH curves, see [56].

Among the useful features that distinguish a (Euclidean) PH curve $\boldsymbol{r}(t)$ from an "ordinary" polynomial curve are the following:

- The cumulative arc-length function $s(t)$ is a *polynomial* in the parameter t, and the total arc length S can be computed *exactly* – i.e. without numerical quadrature – by rational arithmetic on the coefficients of the curve [52].
- Shape measures such as the *elastic energy* (the integral of the squared curvature with respect to arc length) are amenable to exact closed-form evaluation [66].
- PH curves admit *real-time CNC interpolator algorithms* that allow computer-numerical-control machines to accurately traverse curved paths with speeds dependent on time, arc length, or curvature [61, 66, 171].
- The *offsets* (or *parallels*) of planar PH curves admit rational parameterizations – and likewise for the "tubular" *canal surfaces* that have a given spatial PH curve as their "spine" curve [52, 64, 65].
- Spatial PH curves allow exact derivation of *rotation-minimizing frames*, which avoid the "unnecessary" rotation of the Frenet frame in the curve normal plane [54]. These frames incur logarithmic terms; efficient rational approximations are available [57, 102] as alternatives.
- PH curves typically yield "fair" interpolants of discrete data, exhibiting more even curvature distributions, compared with "ordinary" polynomial splines or Hermite interpolants [5, 63, 65, 66, 125].

The focus in this section is on elucidating further the intrinsic structure of *spatial* PH curves, and their construction by interpolation of discrete data. Although the quaternion representation is rotation-invariant [59] and proves useful in a variety of construction and analysis algorithms [31, 54, 58, 57, 62], it nevertheless entails certain open problems. The quaternion form amounts to generating a hodograph $\boldsymbol{r}'(t)$ by a continuous sequence of scalings/rotations of a fixed unit "reference" vector $\hat{\boldsymbol{n}}$, and the significance of the choice of $\hat{\boldsymbol{n}}$ needs clarification. Also, for any prescribed $\hat{\boldsymbol{n}}$, two free parameters arise in constructing spatial PH quintic interpolants to first-order Hermite data [58], whose meaning and proper choice remain to be understood.

The plan of this section is as follows. First, the relation to the general versor equation is discussed in Sect. 9.2.1. Then, the vector form of spatial Pythagorean hodographs is discussed in Sect. 9.2.2. A thorough analysis of the relation between the vector and the quaternion form of spatial PH curves is presented in Sect. 9.2.3. The first-order Hermite interpolation problem for spatial PH quintics is then addressed in Sect. 9.2.4, in terms of both the vector and the quaternion formulation, and some novel aspects of the interpolants are identified in Sects. 9.2.5 and 9.2.6. The generalization to Pythagorean hodographs in $\mathbb{R}^n$ is briefly discussed in Sect. 9.2.7.

9.2.1 Relation to Versor Equation

To recognize the relation of PH curves to the general versor equation (9.1), it is helpful to consider the Pythagorean condition from a more general point of view. Consider first of all a general, not necessarily polynomial, function $\boldsymbol{F} : \mathbb{R} \to \mathbb{R}^3$ with a hodograph $\boldsymbol{f} : \mathbb{R} \to \mathbb{R}^3$. Then there always exists a function $\phi : \mathbb{R} \to \mathbb{R}$ such that

$$\boldsymbol{f}^2(t) = \phi^2(t) \qquad \Longleftrightarrow \qquad \|\boldsymbol{f}(t)\| = \phi(t)\,. \tag{9.22}$$

If the goal is, given a function $\phi(t)$, to find the set of functions $\boldsymbol{f}(t)$ that satisfy this equation (9.22), the trivial solution is clearly

$$\boldsymbol{f} : t \mapsto \phi(t)\,\hat{\boldsymbol{n}}\,, \qquad \hat{\boldsymbol{n}} \in \mathbb{S}^2 \subset \mathbb{R}^3\,,$$

where $\mathbb{S}^2$ denotes the unit sphere. However, this solution is not unique. In fact, at a particular time t_0, the sphere of radius $\phi(t_0)$ is the solution set for $\boldsymbol{f}(t_0)$ that satisfies the equation $\boldsymbol{f}^2(t_0) = \phi^2(t_0)$.

Hence, the set of functions $\boldsymbol{f}(t)$ that satisfy (9.22) is generated by

$$\boldsymbol{f} : t \mapsto \boldsymbol{R}(t)\,(\phi(t)\,\hat{\boldsymbol{n}})\,\widetilde{\boldsymbol{R}}(t)\,, \tag{9.23}$$

where $\boldsymbol{R} : \mathbb{R} \to \mathfrak{G}_3$ is an arbitrary rotor function that satisfies $\boldsymbol{R}(t)\,\widetilde{\boldsymbol{R}}(t) = 1$, i.e. it represents a rotation. Rotating $\hat{\boldsymbol{n}}$ does not change its magnitude. Therefore, independent of the function $\boldsymbol{R}(t)$, it follows that

$$\boldsymbol{f}^2(t) = \phi^2(t)\,\boldsymbol{R}(t)\,\hat{\boldsymbol{n}}\,\widetilde{\boldsymbol{R}}(t)\,\boldsymbol{R}(t)\,\hat{\boldsymbol{n}}\,\widetilde{\boldsymbol{R}}(t) = \phi^2(t)\,.$$

PH curves are just those polynomial curves whose hodograph satisfies (9.22), where $\phi(t)$ is also a polynomial. If the general functions $\boldsymbol{f}(t)$ and $\phi(t)$ are constrained to be polynomial functions $\boldsymbol{r}(t)$ and $\sigma(t)$, then the $\boldsymbol{r}(t)$ and $\sigma(t)$ that satisfy (9.22) have to form a subset of the set of functions generated by (9.23). Hence, the necessary and sufficient condition for $\boldsymbol{r}(t)$ to be the hodograph of a PH curve is that

$$\boldsymbol{r} \,:\, t \mapsto \boldsymbol{A}(t)\, (\nu(t)\, \hat{\boldsymbol{n}})\, \widetilde{\boldsymbol{A}}(t) \,, \tag{9.24}$$

where $\boldsymbol{A} \,:\, \mathbb{R} \to \mathbb{G}_3^+ \cong \mathbb{H}$ and $\nu \,:\, \mathbb{R} \to \mathbb{R}$ are polynomials. $\boldsymbol{A}(t)$ represents a scaling–rotation polynomial function, which can be expressed in terms of a general rotor function $\boldsymbol{R} \,:\, \mathbb{R} \to \mathfrak{S}_3$ and a general scaling function $\alpha \,:\, \mathbb{R} \to \mathbb{R}$ as

$$\boldsymbol{A}(t) = \alpha(t)\, \boldsymbol{R}(t) \,, \qquad \boldsymbol{R} \,:\, t \mapsto \frac{\boldsymbol{A}(t)}{\sqrt{\boldsymbol{A}(t)\, \widetilde{\boldsymbol{A}}(t)}} \,, \qquad \alpha \,:\, t \mapsto \sqrt{\boldsymbol{A}(t)\, \widetilde{\boldsymbol{A}}(t)} \,.$$

Therefore,

$$\boldsymbol{r}(t) = \alpha^2(t)\, \nu(t)\; \boldsymbol{R}(t)\, \hat{\boldsymbol{n}}\, \widetilde{\boldsymbol{R}}(t) \,,$$

such that

$$\boldsymbol{r}^2(t) = \alpha^4(t)\, \nu^2(t) \,.$$

The condition expressed in (9.22) is therefore satisfied if

$$\alpha^2(t)\, \nu(t) = \sigma(t) \,.$$

Since $\alpha^2(t)$ and $\nu(t)$ are polynomial, $\sigma(t)$ is polynomial as desired.

Note that for PH curves in $\mathbb{R}^2$, the rotors in (9.23) are isomorphic to complex numbers, and in $\mathbb{R}^3$ they are isomorphic to quaternions. In higher dimensions, the rotors represent the appropriate rotation operators to rotate $\hat{\boldsymbol{n}} \in \mathbb{R}^n$ to any position on $\mathbb{S}^{n-1}$. In geometric algebra, these rotors can still be generated by the geometric product of two vectors. However, there is an added complication in dimensions higher than three. Whereas the sum of two quaternions still represents a scaling–rotation, this is not necessarily the case for scaling–rotation operators in higher dimensions. In other words, the Clifford group $\mathfrak{G}_n$ is only a subalgebra of $\mathbb{G}_n$ for $1 \leq n \leq 3$. A possible method to generate polynomials of the type $\boldsymbol{A} \,:\, \mathbb{R} \to \mathfrak{G}_n$ for $n > 3$ is through the geometric product of a number of polynomials of the type $\boldsymbol{a}_i \,:\, \mathbb{R} \to \mathbb{G}_n^1$. For example,

$$\boldsymbol{A} \,:\, t \mapsto \boldsymbol{a}_1(t) \cdots \boldsymbol{a}_n(t)$$

is a Clifford-group-valued polynomial in any dimension n.

9.2.2 Pythagorean-Hodograph Curves

The distinguishing property of a PH curve $\boldsymbol{r}(t)$ is that $\|\boldsymbol{r}'(t)\| = \sigma(t)$, where $\sigma(t)$ is a *polynomial* in t. For brevity, only *regular* PH curves, which satisfy $\boldsymbol{r}'(t) \neq \boldsymbol{0}$ for all t, are considered here, and the phrase "PH curve" is used to mean a regular PH curve.

The standard way of expressing spatial PH curves is by writing them in terms of quaternion-valued polynomials, as follows. Let $\boldsymbol{A} \,:\, \mathbb{R} \to \mathbb{H}$ be a quaternion-valued polynomial, and let $\hat{\boldsymbol{n}} \in \mathbb{H}$ be a unit pure quaternion (i.e. $\hat{\boldsymbol{n}}$ has no scalar part). The hodograph $\boldsymbol{r}' : \mathbb{R} \to \mathbb{H}$ of a PH curve $\boldsymbol{r} : \mathbb{R} \to \mathbb{H}$ is then given by

$$\boldsymbol{r}'(t) = \boldsymbol{A}(t)\, \hat{\boldsymbol{n}}\, \boldsymbol{A}^*(t)\,, \tag{9.25}$$

whence $\boldsymbol{r}'(t)$ is always a pure quaternion. Identifying pure quaternions with vectors in $\mathbb{R}^3$ then gives a spatial Pythagorean hodograph. It can be shown [32, 43, 59] that the form (9.25) is necessary and sufficient for $\boldsymbol{r}(t) = \int \boldsymbol{r}'(t)\, dt$ to be a spatial PH curve.

Equation (9.25) can be expressed equivalently in geometric algebra by invoking the isomorphism between $\mathbb{H}$ and $\mathbb{G}_3^+$. Let $\boldsymbol{A} : \mathbb{R} \to \mathbb{G}_3^+$ again denote a polynomial, and let $\hat{\boldsymbol{n}} \in \mathbb{R}^3$ be a unit vector. The hodograph $\boldsymbol{r}' : \mathbb{R} \to \mathbb{R}^3$ of a PH curve $\boldsymbol{r} : \mathbb{R} \to \mathbb{R}^3$ is then given by

$$\boldsymbol{r}'(t) = \boldsymbol{A}(t)\, \hat{\boldsymbol{n}}\, \widetilde{\boldsymbol{A}}(t)\,. \tag{9.26}$$

The difference from (9.25) is that here the isomorphic embedding of quaternions in geometric algebra is used and, instead of using pure quaternions to represent points in $\mathbb{R}^3$, vectors of $\mathbb{R}^3$ are used. The product between the entities is, of course, the *geometric product.* It may be shown [32] that (9.26) is also a necessary and sufficient condition for generating a PH curve.

The magnitude $\|\boldsymbol{r}'(t)\|$ can be evaluated from (9.26) as follows. First, note that $\boldsymbol{A}(t)$ defines a scaling rotation, which means that $\boldsymbol{A}(t)\, \widetilde{\boldsymbol{A}}(t) = \widetilde{\boldsymbol{A}}(t)\, \boldsymbol{A}(t) = \|\boldsymbol{A}(t)\|^2$. Therefore,

$$\left[\boldsymbol{r}'(t)\right]^2 = \boldsymbol{A}(t)\, \hat{\boldsymbol{n}}\, \widetilde{\boldsymbol{A}}(t)\, \boldsymbol{A}(t)\, \hat{\boldsymbol{n}}\, \widetilde{\boldsymbol{A}}(t) = \left[\boldsymbol{A}(t)\, \widetilde{\boldsymbol{A}}(t)\right]^2 = \|\boldsymbol{A}(t)\|^4\,,$$

and hence $\|\boldsymbol{r}'(t)\| = \boldsymbol{A}(t)\, \widetilde{\boldsymbol{A}}(t)$, which is polynomial if $\boldsymbol{A}(t)$ is polynomial.

Spatial Pythagorean hodographs can also be generated using scaling reflections, rather than scaling rotations. Let $\boldsymbol{a} : \mathbb{R} \to \mathbb{R}^3$ be a vector-valued polynomial, let $\hat{\boldsymbol{n}} \in \mathbb{R}^3$ be a unit vector, and let a hodograph $\boldsymbol{r}' : \mathbb{R} \to \mathbb{R}^3$ be given by

$$\boldsymbol{r}'(t) = \boldsymbol{a}(t)\, \hat{\boldsymbol{n}}\, \boldsymbol{a}(t)\,. \tag{9.27}$$

It then follows immediately that $[\boldsymbol{r}'(t)]^2 = [\boldsymbol{a}(t)]^4$, and thus $\|\boldsymbol{r}'(t)\| = [\boldsymbol{a}(t)]^2$, which is a polynomial. If $\boldsymbol{A}(t) := \boldsymbol{a}(t)\, \hat{\boldsymbol{n}}\, \mathsf{rot}(\phi, \hat{\boldsymbol{n}})$, then $\boldsymbol{A}(t)\, \hat{\boldsymbol{n}}\, \widetilde{\boldsymbol{A}}(t) \equiv \boldsymbol{a}(t)\, \hat{\boldsymbol{n}}\, \boldsymbol{a}(t)$ for all $\phi \in [0, 2\pi)$, since $\mathsf{rot}(\phi, \hat{\boldsymbol{n}})\, \hat{\boldsymbol{n}} = \hat{\boldsymbol{n}}\, \mathsf{rot}(\phi, \hat{\boldsymbol{n}})$ and $\hat{\boldsymbol{n}}^2 = 1$.

In the rotation representation (9.26), the choice of $\hat{\boldsymbol{n}}$ is immaterial, as can be seen by choosing $\hat{\boldsymbol{n}} = \boldsymbol{S}\, \hat{\boldsymbol{m}}\, \widetilde{\boldsymbol{S}}$, where $\boldsymbol{S}$ is a rotor, and observing that

$$\boldsymbol{r}'(t) = \boldsymbol{A}(t)\, \hat{\boldsymbol{n}}\, \widetilde{\boldsymbol{A}}(t) = \boldsymbol{A}(t)\, \boldsymbol{S}\, \hat{\boldsymbol{m}}\, \widetilde{\boldsymbol{S}}\, \widetilde{\boldsymbol{A}}(t) = \boldsymbol{B}(t)\, \hat{\boldsymbol{m}}\, \widetilde{\boldsymbol{B}}(t)\,,$$

where $\boldsymbol{B}(t) := \boldsymbol{A}(t)\, \boldsymbol{S}$ also defines a scaling rotation. Hence, for any unit vector $\hat{\boldsymbol{m}}$, it is always possible to find a corresponding $\boldsymbol{B}(t)$ such that $\boldsymbol{B}(t)\, \hat{\boldsymbol{m}}\, \widetilde{\boldsymbol{B}}(t)$ defines exactly the same Pythagorean hodograph as $\boldsymbol{A}(t)\, \hat{\boldsymbol{n}}\, \widetilde{\boldsymbol{A}}(t)$. However, for the reflection representation (9.27) this is not, in general, true.

Instead,

$$\boldsymbol{r}'(t) = \boldsymbol{a}(t)\,\hat{\boldsymbol{n}}\,\boldsymbol{a}(t) = \boldsymbol{a}(t)\,\boldsymbol{S}\,\hat{\boldsymbol{m}}\,\widetilde{\boldsymbol{S}}\,\boldsymbol{a}(t)\,,$$

and the quantity $\boldsymbol{a}(t)\,\boldsymbol{S}$ is vector-valued only if $\boldsymbol{a}(t)$ lies in the rotation plane of $\boldsymbol{S}$. Hence, it is *not* always possible to find a vector-valued polynomial $\boldsymbol{b}(t)$ to replace $\boldsymbol{a}(t)$ so that $\boldsymbol{b}(t)\,\hat{\boldsymbol{m}}\,\boldsymbol{b}(t)$ defines exactly the same hodograph as $\boldsymbol{a}(t)\,\hat{\boldsymbol{n}}\,\boldsymbol{a}(t)$.

Nevertheless, both of these representations of a hodograph are rotation-invariant. If we note that $\boldsymbol{S}\,\widetilde{\boldsymbol{S}} = \widetilde{\boldsymbol{S}}\,\boldsymbol{S} = 1$ for any rotor $\boldsymbol{S}$, upon rotating the hodograph (9.26), we obtain

$$\boldsymbol{S}\,\boldsymbol{A}(t)\,\hat{\boldsymbol{n}}\,\widetilde{\boldsymbol{A}}(t)\,\widetilde{\boldsymbol{S}} = (\boldsymbol{S}\,\boldsymbol{A}(t)\,\widetilde{\boldsymbol{S}})\,(\boldsymbol{S}\,\hat{\boldsymbol{n}}\,\widetilde{\boldsymbol{S}})\,(\boldsymbol{S}\,\widetilde{\boldsymbol{A}}(t)\,\widetilde{\boldsymbol{S}})\,,$$

and similarly, for (9.27),

$$\boldsymbol{S}\,\boldsymbol{a}(t)\,\hat{\boldsymbol{n}}\,\boldsymbol{a}(t)\,\widetilde{\boldsymbol{S}} = (\boldsymbol{S}\,\boldsymbol{a}(t)\,\widetilde{\boldsymbol{S}})\,(\boldsymbol{S}\,\hat{\boldsymbol{n}}\,\widetilde{\boldsymbol{S}})\,(\boldsymbol{S}\,\boldsymbol{a}(t)\,\widetilde{\boldsymbol{S}})\,.$$

Hence, the rotated hodograph is obtained by applying the same rotor $\boldsymbol{S}$ to both $\hat{\boldsymbol{n}}$ and either $\boldsymbol{A}(t)$ or $\boldsymbol{a}(t)$, as appropriate.

In terms of component functions, the relationship between the representations (9.26) and (9.27) is as follows. Since $\hat{\boldsymbol{n}}$ can be chosen freely in the rotation form (9.26), let us choose $\hat{\boldsymbol{n}} = \boldsymbol{e}_1$. Furthermore, let $\boldsymbol{A}(t) := A_0(t) + A_1(t)\,\boldsymbol{e}_{23} + A_2(t)\,\boldsymbol{e}_{31} + A_3(t)\,\boldsymbol{e}_{12}$, where $\boldsymbol{e}_{ij} \equiv \boldsymbol{e}_i\,\boldsymbol{e}_j$. It can then be shown that

$$\begin{aligned}\boldsymbol{r}'(t) = \boldsymbol{A}(t)\,\boldsymbol{e}_1\,\widetilde{\boldsymbol{A}}(t) = & \big[A_0^2(t) + A_1^2(t) - A_2^2(t) - A_3^2(t)\big]\,\boldsymbol{e}_1 \\ & + 2\,\big[A_1(t)\,A_2(t) - A_0(t)\,A_3(t)\big]\,\boldsymbol{e}_2 \\ & + 2\,\big[A_0(t)\,A_2(t) + A_1(t)\,A_3(t)\big]\,\boldsymbol{e}_3\,. \end{aligned} \tag{9.28}$$

Using the same $\hat{\boldsymbol{n}}$ for the reflection form (9.27) and writing $\boldsymbol{a}(t) := a_1(t)\,\boldsymbol{e}_1 + a_2(t)\,\boldsymbol{e}_2 + a_3(t)\,\boldsymbol{e}_3$, it follows that

$$\begin{aligned}\boldsymbol{r}'(t) = \boldsymbol{a}(t)\,\boldsymbol{e}_1\,\boldsymbol{a}(t) = & \big[a_1^2(t) - a_2^2(t) - a_3^2(t)\big]\,\boldsymbol{e}_1 \\ & + 2\,a_1(t)\,a_2(t)\,\boldsymbol{e}_2 \\ & + 2\,a_1(t)\,a_3(t)\,\boldsymbol{e}_3\,. \end{aligned}$$

The reflection representation is thus equivalent to the initial formulation of spatial PH curves given in [65], which is sufficient but not necessary for a hodograph to be Pythagorean. The rotor representation introduced in [32, 59], which is equivalent to (9.28), on the other hand, is a sufficient-and-necessary form.

The relation between the representations can be elucidated further by noting that

$$\boldsymbol{A}(t)\,\boldsymbol{I} = A_0(t)\,\boldsymbol{I} + A_1(t)\,\boldsymbol{e}_1 + A_2(t)\,\boldsymbol{e}_2 + A_3(t)\,\boldsymbol{e}_3\,.$$

Therefore, if $a_i(t) = A_i(t)$ for $i \in \{1, 2, 3\}$, then

$$\boldsymbol{A}(t)\,\boldsymbol{I} \;=\; A_0(t)\,\boldsymbol{I} + \boldsymbol{a}(t) \quad\iff\quad \boldsymbol{A}(t) = A_0(t) + \boldsymbol{a}^*(t)\,,$$

where $\boldsymbol{a}^*(t)$ denotes the dual of $\boldsymbol{a}(t)$, namely $\boldsymbol{a}^*(t) = \boldsymbol{a}(t)\,\boldsymbol{I}^{-1}$. Note that the dual of a vector in $\mathbb{R}^3$ is isomorphic to a pure quaternion.

9.2.3 Relation Between the Rotation and Reflection Forms

In this subsection the relation between the rotation and the reflection form of a PH curve is investigated in more detail. The main result is that cubic and quintic PH curves can be represented equivalently in rotation and reflection form.

In general, the preimage of a Pythagorean hodograph in $\mathbb{R}^3$ can be written as

$$\boldsymbol{A}(t) \;=\; \sum_{i=0}^{N} f_i(t)\,\boldsymbol{A}_i\,, \tag{9.29}$$

where $\{\boldsymbol{A}_i\} \subset \mathbb{G}_3^+$, and $\{f_i((t)\}$ is a polynomial basis. The hodographs of all PH curves of *odd* degree M in $\mathbb{R}^3$ can be expressed [32, 43, 59] in the form $\boldsymbol{r}'(t) = \boldsymbol{A}(t)\,\hat{\boldsymbol{n}}\,\widetilde{\boldsymbol{A}}(t)$, where $\hat{\boldsymbol{n}} \in \mathbb{S}^2$ and the preimage $\boldsymbol{A}(t) \in \mathbb{G}_3^+$ is a polynomial of degree $N = \dfrac{1}{2}(M-1)$. The corresponding hodograph $\boldsymbol{r}'(t)$ is then given by

$$\boldsymbol{r}'(t) \;=\; \boldsymbol{A}(t)\,\hat{\boldsymbol{n}}\,\widetilde{\boldsymbol{A}}(t) \;=\; \sum_{i=0}^{N}\sum_{j=0}^{N} f_{ij}(t)\,\boldsymbol{r}_{ij}\,,$$

where $f_{ij}(t) := f_i(t)\,f_j(t)$ and

$$\boldsymbol{r}_{ij} \;:=\; \boldsymbol{A}_i\,\hat{\boldsymbol{n}}\,\widetilde{\boldsymbol{A}}_j\,. \tag{9.30}$$

For any given basis $\{f_i(t)\}$, the hodograph $\boldsymbol{r}'(t)$ depends only on the coefficients $\{\boldsymbol{r}_{ij}\}$. The rotor representation can be transformed into the reflection representation if there exists a rotor $\boldsymbol{S} \in \mathbb{G}_3^+$ such that

$$\boldsymbol{A}(t) = \boldsymbol{a}(t)\,\boldsymbol{I}^{-1}\,\boldsymbol{S}\,, \qquad \boldsymbol{a} : \mathbb{R} \to \mathbb{R}^3\,. \tag{9.31}$$

Recall that $\boldsymbol{a}^*(t) = \boldsymbol{a}(t)\,\boldsymbol{I}^{-1}$ is the dual of $\boldsymbol{a}(t)$, which is isomorphic to a pure quaternion. The hodograph then becomes

$$\boldsymbol{r}'(t) \;=\; \boldsymbol{A}(t)\,\hat{\boldsymbol{n}}\,\widetilde{\boldsymbol{A}}(t) \;=\; \boldsymbol{a}(t)\,\boldsymbol{I}^{-1}\,\boldsymbol{S}\,\hat{\boldsymbol{n}}\,\widetilde{\boldsymbol{S}}\,\widetilde{\boldsymbol{I}}^{-1}\,\boldsymbol{a}(t) \;=\; \boldsymbol{a}(t)\,\hat{\boldsymbol{m}}\,\boldsymbol{a}(t)\,, \tag{9.32}$$

where $\hat{\boldsymbol{m}} := \boldsymbol{S}\,\hat{\boldsymbol{n}}\,\widetilde{\boldsymbol{S}}$. Furthermore, since $\widetilde{\boldsymbol{I}}^{-1} = \boldsymbol{I}$ and $\boldsymbol{I}$ commutes with all elements of $\mathbb{G}_3$,

$$\boldsymbol{I}^{-1}\,\hat{\boldsymbol{m}}\,\widetilde{\boldsymbol{I}}^{-1} = \hat{\boldsymbol{m}}\,\boldsymbol{I}^{-1}\,\boldsymbol{I} = \hat{\boldsymbol{m}}\,.$$

Writing $\boldsymbol{a}(t)$ in terms of the $\{f_i(t)\}$ as

$$\boldsymbol{a}(t) = \sum_{i=0}^{N} f_i(t)\,\boldsymbol{a}_i\,, \qquad \{\boldsymbol{a}_i\} \subset \mathbb{R}^3\,, \tag{9.33}$$

implies that the components $\{\boldsymbol{r}_{ij}\}$ of the hodograph $\boldsymbol{r}'(t)$ are given by

$$\boldsymbol{r}_{ij} = \boldsymbol{A}_i\,\hat{\boldsymbol{n}}\,\boldsymbol{A}_j = \boldsymbol{a}_i\,\hat{\boldsymbol{m}}\,\boldsymbol{a}_j\,, \tag{9.34}$$

if $\boldsymbol{A}(t) = \boldsymbol{a}(t)\,\boldsymbol{I}^{-1}\,\boldsymbol{S}$. The rotor representation can therefore be replaced by the reflection representation if there exists a rotor $\boldsymbol{S} \in \mathbb{G}_3^+$ together with a set $\{\boldsymbol{a}_i\} \subset \mathbb{R}^3$ such that

$$\boldsymbol{A}_i = \boldsymbol{a}_i\,\boldsymbol{I}^{-1}\,\boldsymbol{S}\,. \tag{9.35}$$

In the following, the conditions when such a relation can be found are derived.

Lemma 9.1. *Given two elements $\boldsymbol{A}_i$, $\boldsymbol{A}_j \in \mathbb{G}_3^+$, there then exist three vectors $\boldsymbol{a}_i, \boldsymbol{a}_j, \boldsymbol{b} \in \mathbb{R}^3$ such that*

$$\boldsymbol{A}_i = \boldsymbol{a}_i\,\boldsymbol{b} \qquad \text{and} \qquad \boldsymbol{A}_j = \boldsymbol{a}_j\,\boldsymbol{b}\,.$$

Proof. Since $\boldsymbol{A}_i$ and $\boldsymbol{A}_j$ represent scaling rotations about the origin in $\mathbb{R}^3$, their rotation planes have to intersect in at least a line. We choose $\boldsymbol{b} \in \mathbb{R}^3$ to lie in the intersection of the rotation planes. Since all scaling rotations can be represented by two consecutive scaling reflections, there have to exist vectors $\boldsymbol{a}_i, \boldsymbol{a}_j \in \mathbb{R}^3$ such that the proposition is satisfied. □

Lemma 9.2. *Given three elements $\boldsymbol{A}_i$, $\boldsymbol{A}_j$, $\boldsymbol{A}_k \in \mathbb{G}_3^+$, then in general there* do not *exist four vectors $\boldsymbol{a}_i, \boldsymbol{a}_j, \boldsymbol{a}_k, \boldsymbol{b} \in \mathbb{R}^3$ such that*

$$\boldsymbol{A}_i = \boldsymbol{a}_i\,\boldsymbol{b}\,, \qquad \text{and} \qquad \boldsymbol{A}_j = \boldsymbol{a}_j\,\boldsymbol{b}\,, \qquad \text{and} \qquad \boldsymbol{A}_k = \boldsymbol{a}_k\,\boldsymbol{b}\,.$$

Proof. Since $\boldsymbol{A}_i$, $\boldsymbol{A}_j$, and $\boldsymbol{A}_k$ represent scaling rotations about the origin in $\mathbb{R}^3$, their rotation planes need not intersect in a single line, and thus they need not share a reflection about some vector $\boldsymbol{b} \in \mathbb{R}^3$. □

Lemma 9.3. *Let $\boldsymbol{A}_0\,\boldsymbol{A}_1 \in \mathbb{G}_3^+$; there then exist a rotor $\boldsymbol{S} \in \mathbb{G}_3^+$ and $\boldsymbol{a}_0, \boldsymbol{a}_1 \in \mathbb{R}^3$ such that*

$$\boldsymbol{A}_i = \boldsymbol{a}_i\,\boldsymbol{I}^{-1}\,\boldsymbol{S}\,, \quad \forall\, i \in \{0, 1\}\,.$$

Proof. From Lemma 9.1, it follows that there exist vectors $\boldsymbol{a}_0, \boldsymbol{a}_1, \boldsymbol{b} \in \mathbb{R}^3$ such that $\boldsymbol{A}_i = \boldsymbol{a}_i\,\boldsymbol{b}$, for $i \in \{0, 1\}$. Hence, we choose $\boldsymbol{S} = \boldsymbol{I}\,\boldsymbol{b}$, whence

$$\boldsymbol{A}_i = \boldsymbol{a}_i\,\boldsymbol{I}^{-1}\,\boldsymbol{I}\,\boldsymbol{b} = \boldsymbol{a}_i\,\boldsymbol{b}\,.$$

□

Lemma 9.4. *Given* $\boldsymbol{A}_0, \boldsymbol{A}_1, \boldsymbol{A}_2 \in \mathbb{G}_3^+$*, there then exist a rotor* $\boldsymbol{S} \in \mathbb{G}_3^+$ *and vectors* $\boldsymbol{a}_0, \boldsymbol{a}_1, \boldsymbol{a}_2 \in \mathbb{R}^3$ *such that*

$$\boldsymbol{A}_i = \boldsymbol{a}_i\, \boldsymbol{I}^{-1}\, \boldsymbol{S}\,, \quad \forall\, i \in \{0,1,2\}\,.$$

Proof. Without loss of generality, we choose $\boldsymbol{S} = \boldsymbol{I}\,\boldsymbol{a}_0^{-1}\,\boldsymbol{A}_0$. The proposition is satisfied if there exist vectors $\boldsymbol{a}_0, \boldsymbol{a}_1, \boldsymbol{a}_2 \in \mathbb{R}^3$ such that

$$\boldsymbol{A}_i \;=\; \boldsymbol{a}_i\, \boldsymbol{I}^{-1}\, \boldsymbol{S} \;=\; \boldsymbol{a}_i\, \boldsymbol{I}^{-1}\, \boldsymbol{I}\, \boldsymbol{a}_0^{-1}\, \boldsymbol{A}_0 \;=\; \boldsymbol{a}_i\, \boldsymbol{a}_0^{-1}\, \boldsymbol{A}_0\,, \quad i \in \{1,2\}\,.$$

For $i \in \{1,2\}$,

$$\boldsymbol{A}_i = \boldsymbol{a}_i\, \boldsymbol{a}_0^{-1}\, \boldsymbol{A}_0 \quad \iff \quad \boldsymbol{A}_i\, \boldsymbol{A}_0^{-1} = \boldsymbol{a}_i\, \boldsymbol{a}_0^{-1}\,.$$

The elements $\boldsymbol{A}_{i0} := \boldsymbol{A}_i\, \boldsymbol{A}_0^{-1}$ again represent scaling rotations. Therefore, it follows from Lemma 9.1 that vectors $\boldsymbol{a}_0$, $\boldsymbol{a}_1$ and $\boldsymbol{a}_2$ have to exist for arbitrary elements $\boldsymbol{A}_{10}$ and $\boldsymbol{A}_{20}$ and thus for arbitrary $\boldsymbol{A}_0$, $\boldsymbol{A}_1$, and $\boldsymbol{A}_2$. The vector $\boldsymbol{a}_0^{-1}$ can be found from $\boldsymbol{A}_{10}$ and $\boldsymbol{A}_{20}$ by intersecting their rotation planes. Once $\boldsymbol{a}_0^{-1}$ is known, the vectors $\boldsymbol{a}_1$ and $\boldsymbol{a}_2$ can be found. □

The proof of Lemma 9.4 implies that given $\boldsymbol{A}_0, \boldsymbol{A}_1, \ldots, \boldsymbol{A}_N \in \mathbb{G}_3^+$, with $N > 2$, there exist a rotor $\boldsymbol{S} \in \mathbb{G}_3^+$ and vectors $\boldsymbol{a}_0, \boldsymbol{a}_1, \ldots, \boldsymbol{a}_N \in \mathbb{R}^3$, such that $\boldsymbol{A}_i = \boldsymbol{a}_i\, \boldsymbol{I}^{-1}\, \boldsymbol{S}$ for $i \in \{0, \ldots, N\}$, only if the rotation planes of the $\{\boldsymbol{A}_1, \ldots, \boldsymbol{A}_N\}$ intersect in a single line, i.e. $\boldsymbol{A}_i = \boldsymbol{a}_i\, \boldsymbol{a}_0^{-1}\, \boldsymbol{A}_0$ for $i \in \{1, \ldots, k\}$.

Theorem 9.1. *Given the preimage* $\boldsymbol{A} : \mathbb{R} \to \mathbb{G}_3^+$ *of a hodograph in the form*

$$\boldsymbol{A}(t) = \sum_{i=0}^{N} f_i(t)\, \boldsymbol{A}_i\,,$$

with $0 \leq N \leq 2$ *and* $\{\boldsymbol{A}_i\} \subset \mathbb{G}_3^+$*, and where* $\{f_i(t)\}$ *is a polynomial basis, then there exist a function* $\boldsymbol{a} : \mathbb{R} \to \mathbb{R}^3$ *defined by*

$$\boldsymbol{a}(t) = \sum_{i=0}^{N} f_i(t)\, \boldsymbol{a}_i\,,$$

with $\{\boldsymbol{a}_i\} \subset \mathbb{R}^3$*, and vectors* $\hat{\boldsymbol{n}}, \hat{\boldsymbol{m}} \in \mathbb{S}^2$ *such that*

$$\boldsymbol{A}(t)\, \hat{\boldsymbol{n}}\, \widetilde{\boldsymbol{A}}(t) = \boldsymbol{a}(t)\, \hat{\boldsymbol{m}}\, \boldsymbol{a}(t)\,.$$

Proof. Since the polynomial bases of the functions $\boldsymbol{A}(t)$ and $\boldsymbol{a}(t)$ are identical, the proposition is proven if it can be shown that

$$\boldsymbol{A}_i\, \hat{\boldsymbol{n}}\, \boldsymbol{A}_j \;=\; \boldsymbol{a}_i\, \hat{\boldsymbol{m}}\, \boldsymbol{a}_j\,, \qquad \forall\, i,j \in \{0, \ldots, N\}\,.$$

This condition is satisfied if there exist a rotor $\boldsymbol{S} \in \mathbb{G}_3^+$ and a set $\{\boldsymbol{a}_i\} \subset \mathbb{R}^3$ such that $\boldsymbol{A}_i = \boldsymbol{a}_i \, \boldsymbol{I}^{-1} \, \boldsymbol{S}$. Furthermore, $\hat{\boldsymbol{m}}$ has to be defined in terms of $\hat{\boldsymbol{n}}$ as $\hat{\boldsymbol{m}} := \boldsymbol{S} \, \hat{\boldsymbol{n}} \, \widetilde{\boldsymbol{S}}$. From Lemmas 9.3 and 9.4, it follows that such a rotor $\boldsymbol{S}$ and set $\{\boldsymbol{a}_i\}$ exist for $0 \leq N \leq 2$. □

Theorem 9.1 implies that cubic and quintic PH curves can be represented equivalently in rotation and reflection form.

9.2.4 Pythagorean-Hodograph Quintic Hermite Interpolation

We desire a spatial PH quintic $\boldsymbol{r}(t)$, $t \in [0,1]$, with prescribed points $\boldsymbol{r}(0) = \boldsymbol{p}_0$ and $\boldsymbol{r}(1) = \boldsymbol{p}_2$ and end derivatives $\boldsymbol{r}'(0) = \boldsymbol{d}_0$ and $\boldsymbol{r}'(1) = \boldsymbol{d}_2$. Beginning with the rotation representation, the hodograph is written as

$$\boldsymbol{r}'(t) = \boldsymbol{A}(t) \, \hat{\boldsymbol{n}} \, \widetilde{\boldsymbol{A}}(t) ,$$

where $\boldsymbol{A} : \mathbb{R} \to \mathbb{G}_3^+$ is a quadratic polynomial, and $\hat{\boldsymbol{n}} \in \mathbb{R}^3$ is a unit vector. We express $\boldsymbol{A}(t)$ in Bernstein–Bézier form as

$$\boldsymbol{A}(t) = \sum_{i=0}^{2} f_i(t) \, \boldsymbol{A}_i , \quad f_0(t) := (1-t)^2, \; f_1(t) := 2\,(1-t)\,t, \; f_2(t) := t^2 , \tag{9.36}$$

with $\boldsymbol{A}_0, \boldsymbol{A}_1, \boldsymbol{A}_2 \in \mathbb{G}_3^+$. The hodograph is thus given by

$$\boldsymbol{r}'(t) = \sum_{i,j} f_{ij}(t) \, \boldsymbol{r}_{ij} , \quad f_{ij}(t) := f_i(t) \, f_j(t), \quad \boldsymbol{r}_{ij} := \boldsymbol{A}_i \, \hat{\boldsymbol{n}} \, \widetilde{\boldsymbol{A}}_j .$$

Now, $\boldsymbol{r}_{ij}$ is not necessarily a vector for $i \neq j$, but one can verify that $(\boldsymbol{r}_{ij} + \boldsymbol{r}_{ji}) \in \mathbb{R}^3$. $\boldsymbol{A}_0$ and $\boldsymbol{A}_2$ can be evaluated immediately from the end derivatives $\boldsymbol{d}_0$ and $\boldsymbol{d}_2$ as

$$\boldsymbol{A}_0 = \mathsf{ref}(\hat{\boldsymbol{n}}, \boldsymbol{d}_0) \, \hat{\boldsymbol{n}} \, \mathsf{rot}(\phi_0, \hat{\boldsymbol{n}}) , \quad \boldsymbol{A}_2 = \mathsf{ref}(\hat{\boldsymbol{n}}, \boldsymbol{d}_2) \, \hat{\boldsymbol{n}} \, \mathsf{rot}(\phi_2, \hat{\boldsymbol{n}}) ,$$

where ϕ_0 and ϕ_2 are free parameters. To evaluate $\boldsymbol{A}_1$, a constraint involving the end points is invoked:

$$\Delta \boldsymbol{p} = \int_0^1 \boldsymbol{r}'(t) \, \mathrm{d}t = \sum_{i,j} F_{ij} \, \boldsymbol{r}_{ij} , \quad F_{ij} := \int_0^1 f_{ij}(t) \, dt ,$$

where $\Delta \boldsymbol{p} := \boldsymbol{p}_2 - \boldsymbol{p}_0$. To determine $\boldsymbol{A}_1$ from this constraint, the following approach is used. Let $\boldsymbol{V} := \sum_i v_i \, \boldsymbol{A}_i$, where the $\{v_i\}$ are scalars. Then

$$\boldsymbol{V}\,\hat{\boldsymbol{n}}\,\widetilde{\boldsymbol{V}} = \sum_{i,j} V_{ij}\,\boldsymbol{r}_{ij} = \sum_{i,j} \left(V_{ij} - F_{ij}\right)\boldsymbol{r}_{ij} + \sum_{i,j} F_{ij}\,\boldsymbol{r}_{ij}\,,$$

where $V_{ij} := v_i\,v_j$. If the $\{v_i\}$ can be chosen so that the sum $\sum_{i,j}(V_{ij}-F_{ij})\,\boldsymbol{r}_{ij}$ is independent of $\boldsymbol{A}_1$, the right-hand side

$$\boldsymbol{u} := \sum_{i,j} \left(V_{ij} - F_{ij}\right)\boldsymbol{r}_{ij} + \sum_{i,j} F_{ij}\,\boldsymbol{r}_{ij}$$

of the above equation can be evaluated, since $\sum_{i,j} F_{ij}\;\boldsymbol{r}_{ij} = \Delta\boldsymbol{p}$ and $\boldsymbol{A}_0$ and $\boldsymbol{A}_2$ are known. Furthermore, if $\boldsymbol{u}$ can be evaluated, then $\boldsymbol{V} = \mathsf{ref}(\hat{\boldsymbol{n}},\,\boldsymbol{u})\,\hat{\boldsymbol{n}}\,\mathsf{rot}(\phi_1,\,\hat{\boldsymbol{n}})$, where ϕ_1 is a free parameter, and $\boldsymbol{A}_1 = (\boldsymbol{V} - v_0\,\boldsymbol{A}_0 - v_2\,\boldsymbol{A}_2)/v_1$. A real solution for the $\{v_i\}$ can be found if $F_{11} > 0$, and is given by

$$v_1 = \sqrt{F_{11}}\,,\quad v_0 = F_{01}/v_1\,,\quad v_2 = F_{02}/v_1\,,$$

since the F_{ij} define a symmetric matrix. Using the specific form (9.36) of the basis functions $\{f_i(t)\}$, it is found, in accordance with [63], that

$$\boldsymbol{A}_1 = \frac{1}{4}\left(\boldsymbol{V} - 3\,\boldsymbol{A}_0 - 3\,\boldsymbol{A}_2\right),\qquad \boldsymbol{V} = \mathsf{ref}(\hat{\boldsymbol{n}},\,\boldsymbol{u})\,\hat{\boldsymbol{n}}\,\mathsf{rot}(\phi_1,\,\hat{\boldsymbol{n}})\,,$$

where

$$\boldsymbol{u} = 120\,\Delta\boldsymbol{p} - 15\,(\boldsymbol{d}_0 + \boldsymbol{d}_2) + 5\,(\boldsymbol{r}_{02} + \boldsymbol{r}_{20})\,. \tag{9.37}$$

The same derivation can also be done for the reflection representation of the hodograph. In this case

$$\boldsymbol{r}'(t) = \boldsymbol{a}(t)\,\hat{\boldsymbol{n}}\,\boldsymbol{a}(t)\,,$$

where $\boldsymbol{a} : \mathbb{R} \to \mathbb{R}^3$ is given by $\boldsymbol{a}(t) := \sum_{i=1}^{3} f_i(t)\,\boldsymbol{a}_i$ with $\boldsymbol{a}_1, \boldsymbol{a}_2, \boldsymbol{a}_3 \in \mathbb{R}^3$. Given the same data as before and setting $\boldsymbol{v} = \mathsf{ref}(\hat{\boldsymbol{n}},\,\boldsymbol{u})$, we obtain the result

$$\boldsymbol{a}_0 = \mathsf{ref}(\hat{\boldsymbol{n}},\,\boldsymbol{d}_0)\,,\quad \boldsymbol{a}_2 = \mathsf{ref}(\hat{\boldsymbol{n}},\,\boldsymbol{d}_2)\,,\quad \boldsymbol{a}_1 = \frac{1}{4}\left(\boldsymbol{v} - 3\,\boldsymbol{a}_0 - 3\,\boldsymbol{a}_2\right).$$

9.2.5 Degrees of Freedom

While both the rotation and the reflection representation give solutions to the Hermite interpolation problem, the degrees of freedom are encoded differently. In the case of the rotation form, the set of PH curves is generated by varying the parameters ϕ_0, ϕ_1, and ϕ_2, whereas in the case of the reflection form, varying $\hat{\boldsymbol{n}}$ generates most of the possible PH curves. Why not all PH curves can be generated by varying $\hat{\boldsymbol{n}}$ is discussed later on. First of all, the rotation form will be investigated in some more detail.

To simplify the formulas somewhat, the following definitions are made. The reflectors and rotors are written as functions of $\hat{\boldsymbol{n}}$, i.e.

$$\boldsymbol{b}_i(\hat{\boldsymbol{n}}) := \mathsf{ref}(\hat{\boldsymbol{n}}, \boldsymbol{d}_i) \qquad \text{and} \qquad \boldsymbol{R}_i(\hat{\boldsymbol{n}}) := \mathsf{rot}(\phi_i, \hat{\boldsymbol{n}}) \,.$$

The preimage components can be brought into a consistent form by defining $\boldsymbol{B}_0 := \boldsymbol{A}_0$, $\boldsymbol{B}_1 := \boldsymbol{V}$, $\boldsymbol{B}_2 := \boldsymbol{A}_2$, $\boldsymbol{d}_1 := \boldsymbol{u}$ and $g_0(t) := f_0(t) - \frac{3}{4} f_1(t)$, $g_1(t) := f_1(t)$, $g_2(t) := f_2(t) - \frac{3}{4} f_1(t)$, whence

$$\boldsymbol{A}(t) \equiv \boldsymbol{B}(t) := \sum_{i=0}^{2} g_i(t)\, \boldsymbol{B}_i \,, \qquad \boldsymbol{B}_i = \boldsymbol{b}_i(\hat{\boldsymbol{n}})\, \hat{\boldsymbol{n}}\, \boldsymbol{R}_i(\hat{\boldsymbol{n}}) \,.$$

Hence, $\boldsymbol{B}_0\, \hat{\boldsymbol{n}}\, \boldsymbol{B}_0 = \boldsymbol{d}_0$, $\boldsymbol{B}_2\, \hat{\boldsymbol{n}}\, \boldsymbol{B}_2 = \boldsymbol{d}_2$ and $\boldsymbol{B}_1\, \hat{\boldsymbol{n}}\, \boldsymbol{B}_1 = \boldsymbol{d}_1$. In this parameterization, the hodograph $\boldsymbol{r}'(t)$ takes the form

$$\boldsymbol{r}'(t) = \sum_{i=0}^{2} \sum_{j=0}^{2} g_{ij}(t)\, \boldsymbol{r}_{ij} \,,$$

where $g_{ij}(t) := g_i(t)\, g_j(t)$ and $\boldsymbol{r}_{ij} = \boldsymbol{B}_i\, \hat{\boldsymbol{n}}\, \widetilde{\boldsymbol{B}}_j$. Expanding the $\boldsymbol{r}_{ij}$ gives

$$\boldsymbol{r}_{ij} = \boldsymbol{b}_i(\hat{\boldsymbol{n}})\, \hat{\boldsymbol{n}}\, \boldsymbol{R}_i(\hat{\boldsymbol{n}})\, \hat{\boldsymbol{n}}\, \widetilde{\boldsymbol{R}}_j(\hat{\boldsymbol{n}})\, \hat{\boldsymbol{n}}\, \boldsymbol{b}_j(\hat{\boldsymbol{n}}) = \boldsymbol{b}_i(\hat{\boldsymbol{n}})\, \hat{\boldsymbol{n}}\, \boldsymbol{R}_{ij}(\hat{\boldsymbol{n}})\, \boldsymbol{b}_j(\hat{\boldsymbol{n}}) \,,$$

where $\boldsymbol{R}_{ij}(\hat{\boldsymbol{n}}) := \mathsf{rot}(\phi_i - \phi_j, \hat{\boldsymbol{n}})$. This implies that

$$\boldsymbol{r}_{ii} = \boldsymbol{b}_i(\hat{\boldsymbol{n}})\, \hat{\boldsymbol{n}}\, \boldsymbol{b}_i(\hat{\boldsymbol{n}}) = \boldsymbol{d}_i \,.$$

Whereas $\boldsymbol{d}_0$ and $\boldsymbol{d}_2$ are constants, $\boldsymbol{d}_1 = \boldsymbol{u}$ depends on $\boldsymbol{r}_{02} + \boldsymbol{r}_{20}$ (see (9.37)) and thus on $\phi_2 - \phi_0$. Hence, the $\{\boldsymbol{r}_{ij}\}$ depend only on the differences between the $\{\phi_i\}$, which means that the hodograph $\boldsymbol{r}'(t)$ has only two degrees of freedom. Different choices of $\hat{\boldsymbol{n}}$ do not generate new sets of hodographs in this case.

Theorem 9.1 guarantees that every hodograph in the rotation form can also be expressed in the reflection form. However, some care has to be taken when using the reflection form directly. The two degrees of freedom of the hodograph in the reflection form are contained in the choice of $\hat{\boldsymbol{n}} \in \mathbb{S}^2$. The hodograph is given in this case by

$$\boldsymbol{r}'(t) = \boldsymbol{b}(t)\, \hat{\boldsymbol{n}}\, \boldsymbol{b}(t) \,, \quad \boldsymbol{b}(t) := \sum_{i=0}^{2} g_i(t)\, \boldsymbol{b}_i(\hat{\boldsymbol{n}}) \,, \quad \boldsymbol{b}_i(\hat{\boldsymbol{n}}) := \mathsf{ref}(\hat{\boldsymbol{n}}, \boldsymbol{d}_i) \,. \tag{9.38}$$

A restriction of the parameterization in terms of $\hat{\boldsymbol{n}}$ is that the $\{\boldsymbol{b}_i(\hat{\boldsymbol{n}})\}$ are not well defined if $\hat{\boldsymbol{n}} = -\hat{\boldsymbol{d}}_i$ for $i \in \{0, 1, 2\}$. In this case, there does not exist a unique reflector. Instead, all unit vectors in the plane perpendicular to $\boldsymbol{d}_i$ are valid reflectors. Hence, for each $\hat{\boldsymbol{n}} = -\hat{\boldsymbol{d}}_i$ there is a one-parameter family of reflectors that generate different hodographs.

9.2.6 Curves of Constant Length

In the previous subsection, the degrees of freedom of PH curves that satisfy a given Hermite interpolation were discussed. In this subsection, it is shown that there exist subsets of PH curves that are of equal length but different shape.

Using the definitions of the previous subsection, the arc length L of the PH curve

$$\boldsymbol{r}(t) = \boldsymbol{p}_0 + \int_0^t \sum_{i=0}^{2} \sum_{j=0}^{2} g_{ij}(s)\, \boldsymbol{r}_{ij}\, \mathrm{d}s$$

is given by

$$L = \int_0^1 \|\boldsymbol{r}'(t)\|\, \mathrm{d}t = \int_0^1 \boldsymbol{b}^2(t)\, \mathrm{d}t = \sum_{i=0}^{2} \sum_{j=0}^{2} G_{ij}\, \boldsymbol{b}_i(\hat{\boldsymbol{n}})\, \boldsymbol{b}_j(\hat{\boldsymbol{n}}) \,,$$

where $G_{ij} := \int_0^1 g_{ij}(t)\, dt$. For the particular functions $g_i(t)$ used above, it may be shown that

$$G_{ij} = \frac{1}{120} \begin{bmatrix} 15 & 0 & -5 \\ 0 & 1 & 0 \\ -5 & 0 & 15 \end{bmatrix} .$$

From the definitions of the $\boldsymbol{b}_i(\hat{\boldsymbol{n}})$, it is clear that $\boldsymbol{b}_i^2(\hat{\boldsymbol{n}}) = \|\boldsymbol{d}_i\|$, where $\boldsymbol{d}_1 = \boldsymbol{u}$ as defined in (9.37). Hence,

$$120\, L = 15\, (\|\boldsymbol{d}_0\| + \|\boldsymbol{d}_2\|) + \|\boldsymbol{u}\| - 10\, \boldsymbol{b}_0(\hat{\boldsymbol{n}}) \cdot \boldsymbol{b}_2(\hat{\boldsymbol{n}}), \qquad (9.39)$$

where the identity

$$\boldsymbol{b}_i(\hat{\boldsymbol{n}})\, \boldsymbol{b}_j(\hat{\boldsymbol{n}}) + \boldsymbol{b}_j(\hat{\boldsymbol{n}})\, \boldsymbol{b}_i(\hat{\boldsymbol{n}}) = 2\, \boldsymbol{b}_i(\hat{\boldsymbol{n}}) \cdot \boldsymbol{b}_j(\hat{\boldsymbol{n}}) \,,$$

which follows from Lemma 3.11 has been used. For given $\boldsymbol{p}_0$, $\boldsymbol{p}_2$, $\boldsymbol{d}_0$, $\boldsymbol{d}_2$, it follows that L is a function of $\hat{\boldsymbol{n}}$. Note that $\boldsymbol{u}$ is also a function of $\hat{\boldsymbol{n}}$, since it depends on $\boldsymbol{b}_0(\hat{\boldsymbol{n}})$ and $\boldsymbol{b}_2(\hat{\boldsymbol{n}})$. In the following it is shown that two different families of PH curves of constant length can be generated by rotating $\hat{\boldsymbol{n}}$ in the plane spanned by $\boldsymbol{b}_0(\hat{\boldsymbol{n}})$ and $\boldsymbol{b}_2(\hat{\boldsymbol{n}})$. The first step is the following lemma.

Lemma 9.5. *Let $\hat{\boldsymbol{b}}_i(\hat{\boldsymbol{n}}) := \mathsf{ref}(\hat{\boldsymbol{n}}, \hat{\boldsymbol{d}}_i)$ for some $\hat{\boldsymbol{n}} \neq -\hat{\boldsymbol{d}}_i$. Also, let $\boldsymbol{Q} = \mathsf{rot}(\alpha, \hat{\boldsymbol{q}})$, $\alpha \in [0, \pi)$, be a rotor, such that $\hat{\boldsymbol{b}}_i(\hat{\boldsymbol{n}})$ lies in its rotation plane. There then exists an angle $\beta \in (0, 2\pi)$ such that $\mathsf{rot}(\beta, \hat{\boldsymbol{q}})$ rotates $\hat{\boldsymbol{n}}$ into $-\hat{\boldsymbol{d}}_i$, and*

(a) if $2\alpha < \beta$, then $\hat{\boldsymbol{b}}_i(\boldsymbol{Q}^2\, \hat{\boldsymbol{n}}\, \widetilde{\boldsymbol{Q}}^2) = \boldsymbol{Q}\, \hat{\boldsymbol{b}}_i(\hat{\boldsymbol{n}})\, \widetilde{\boldsymbol{Q}}$;

(b) if $2\alpha > \beta$, then $\hat{\boldsymbol{b}}_i(\boldsymbol{Q}^2\, \hat{\boldsymbol{n}}\, \widetilde{\boldsymbol{Q}}^2) = -\boldsymbol{Q}\, \hat{\boldsymbol{b}}_i(\hat{\boldsymbol{n}})\, \widetilde{\boldsymbol{Q}}$.

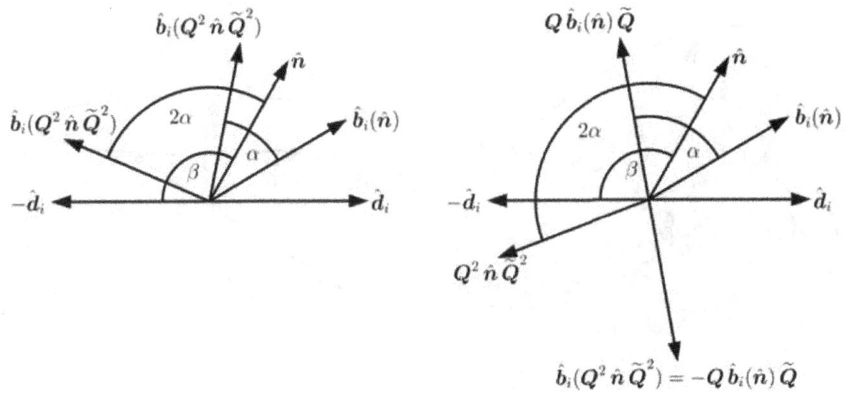

(a) For $2\alpha < \beta$ (the angle between $\hat{\boldsymbol{n}}$ and $-\hat{\boldsymbol{d}}_i$), the reflector is simply rotated by an angle α

(b) For $2\alpha > \beta$, the reflector is rotated by α and then *reversed*

Fig. 9.1 Behavior of the reflector $\hat{\boldsymbol{b}}_i(\hat{\boldsymbol{n}})$ of $\hat{\boldsymbol{n}}$ into $\hat{\boldsymbol{d}}_i$, as $\hat{\boldsymbol{n}}$ is rotated by an angle 2α into $\boldsymbol{Q}^2\hat{\boldsymbol{n}}\widetilde{\boldsymbol{Q}}^2$ by a rotor $\boldsymbol{Q} = \mathsf{rot}(\alpha,\, \hat{\boldsymbol{q}})$, where $\hat{\boldsymbol{q}}$ is orthogonal to $\hat{\boldsymbol{n}}$ and $\hat{\boldsymbol{d}}_i$

Proof. We defer the formal proof, which is rather technical, to Sect. 9.2.9. However, the geometrical significance of this lemma is easily grasped. Let β be the acute angle between $\hat{\boldsymbol{n}}$ and $-\hat{\boldsymbol{d}}_i$. When $\hat{\boldsymbol{n}}$ is rotated by an angle $2\alpha < \beta$ about $\hat{\boldsymbol{q}}$, the reflector $\mathsf{ref}(\hat{\boldsymbol{n}},\, \hat{\boldsymbol{d}}_i)$ is rotated by an angle α in the same sense, as shown in Fig. 9.1(a). However, when $\hat{\boldsymbol{n}}$ is rotated by an angle $2\alpha > \beta$ about $\hat{\boldsymbol{q}}$, the reflector $\mathsf{ref}(\hat{\boldsymbol{n}},\, \hat{\boldsymbol{d}}_i)$ is rotated by an angle α and then *reversed*, as shown in Fig. 9.1(b). □

Lemma 9.6. *Let* $\boldsymbol{W}(\hat{\boldsymbol{n}}) := \mathbf{b}_0(\hat{\boldsymbol{n}}) \wedge \mathbf{b}_2(\hat{\boldsymbol{n}})$*, which represents the plane spanned by* $\mathbf{b}_0(\hat{\boldsymbol{n}})$ *and* $\mathbf{b}_2(\hat{\boldsymbol{n}})$*. Rotating* $\hat{\boldsymbol{n}}$ *in the plane* $\boldsymbol{W}(\hat{\boldsymbol{n}})$ *leaves* $\boldsymbol{W}(\hat{\boldsymbol{n}})$ *unchanged, independent of whether* $\hat{\boldsymbol{n}}$ *lies in the plane* $\boldsymbol{W}(\hat{\boldsymbol{n}})$ *or not.*

Proof. Let $\boldsymbol{Q}$ denote a rotor that rotates in the plane $\boldsymbol{W}(\hat{\boldsymbol{n}})$; then it follows from Lemma 9.5 that

$$\boldsymbol{b}_i(\boldsymbol{Q}^2\,\hat{\boldsymbol{n}}\,\widetilde{\boldsymbol{Q}}^2) = \pm\,\boldsymbol{Q}\,\boldsymbol{b}_i(\hat{\boldsymbol{n}})\,\widetilde{\boldsymbol{Q}}\,, \qquad i \in \{0, 2\}\,.$$

It may be shown with the help of Lemma 3.11 that

$$\begin{aligned} \boldsymbol{W}(\boldsymbol{Q}^2\,\hat{\boldsymbol{n}}\,\widetilde{\boldsymbol{Q}}^2) &= \mathbf{b}_0(\boldsymbol{Q}^2\,\hat{\boldsymbol{n}}\,\widetilde{\boldsymbol{Q}}^2) \,\wedge\, \mathbf{b}_0(\boldsymbol{Q}^2\,\hat{\boldsymbol{n}}\,\widetilde{\boldsymbol{Q}}^2) \\ &= \pm\,(\boldsymbol{Q}\,\mathbf{b}_0(\hat{\boldsymbol{n}})\,\widetilde{\boldsymbol{Q}}) \,\wedge\, (\boldsymbol{Q}\,\mathbf{b}_2(\hat{\boldsymbol{n}})\,\widetilde{\boldsymbol{Q}}) \\ &= \pm\,\boldsymbol{Q}\,(\mathbf{b}_0(\hat{\boldsymbol{n}}) \wedge \mathbf{b}_2(\hat{\boldsymbol{n}}))\,\widetilde{\boldsymbol{Q}} \\ &= \pm\,\boldsymbol{Q}\,\boldsymbol{W}(\hat{\boldsymbol{n}})\,\widetilde{\boldsymbol{Q}}\,. \end{aligned} \tag{9.40}$$

Since $\boldsymbol{W}(\hat{\boldsymbol{n}})$ is the rotation plane of $\boldsymbol{Q}$, $\boldsymbol{Q}\,\boldsymbol{W}(\hat{\boldsymbol{n}}) = \boldsymbol{W}(\hat{\boldsymbol{n}})\,\boldsymbol{Q}$ and thus

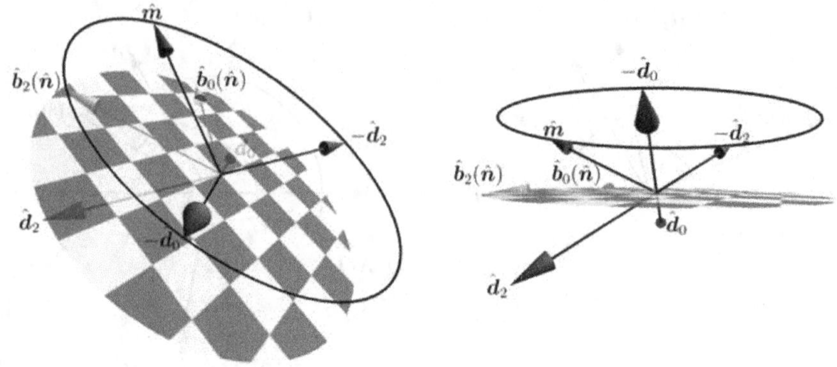

Fig. 9.2 Two views of the rotation of a vector $\hat{\boldsymbol{m}}$ in the plane spanned by $\hat{\boldsymbol{b}}_0(\hat{\boldsymbol{n}})$ and $\hat{\boldsymbol{b}}_2(\hat{\boldsymbol{n}})$

$$\boldsymbol{Q}\,\boldsymbol{W}(\hat{n})\,\widetilde{\boldsymbol{Q}} = \boldsymbol{W}(\hat{n})\,\boldsymbol{Q}\,\widetilde{\boldsymbol{Q}} = \boldsymbol{W}(\hat{n})\,,$$

which proves the proposition. Note that changing the sign of $\boldsymbol{W}(\hat{\boldsymbol{n}})$ has no effect on the plane it represents. □

By Lemma 9.5, it follows that there exist rotations of $\hat{\boldsymbol{n}}$ in the plane $\boldsymbol{U} := \boldsymbol{b}_0(\hat{\boldsymbol{n}}) \wedge \boldsymbol{b}_2(\hat{\boldsymbol{n}})$ that map it into $-\boldsymbol{d}_0$ and $-\boldsymbol{d}_2$. This construction is shown in Fig. 9.2. Let $\hat{\boldsymbol{m}}(\alpha)$, $\alpha \in [0, 2\pi)$, denote the rotation of $\hat{\boldsymbol{n}}$ by an angle α in the plane $\boldsymbol{W}(\hat{\boldsymbol{n}})$, and let $\beta_0, \beta_2 \in [0, 2\pi)$ denote the angles such that $\hat{\boldsymbol{m}}(\beta_0) = -\boldsymbol{d}_0$ and $\hat{\boldsymbol{m}}(\beta_2) = -\boldsymbol{d}_2$. Without loss of generality, it may be assumed that $\beta_0 < \beta_2$ (when $\beta_2 < \beta_0$, everything that follows remains valid with interchanged indices for β_0, β_2).

The dependence of $\boldsymbol{u}$ on $\hat{\boldsymbol{m}}(\alpha)$ is given by the expression

$$\boldsymbol{b}_0(\hat{\boldsymbol{m}})\,\hat{\boldsymbol{m}}\,\boldsymbol{b}_2(\hat{\boldsymbol{m}}) + \boldsymbol{b}_2(\hat{\boldsymbol{m}})\,\hat{\boldsymbol{m}}\,\boldsymbol{b}_0(\hat{\boldsymbol{m}})\,. \tag{9.41}$$

The length $L(\alpha)$ therefore depends on this expression and on

$$2\,\boldsymbol{b}_0(\hat{\boldsymbol{m}}) \cdot \boldsymbol{b}_2(\hat{\boldsymbol{m}}) \;=\; \boldsymbol{b}_0(\hat{\boldsymbol{m}})\,\boldsymbol{b}_2(\hat{\boldsymbol{m}}) + \boldsymbol{b}_2(\hat{\boldsymbol{m}})\,\boldsymbol{b}_0(\hat{\boldsymbol{m}})\,. \tag{9.42}$$

In much the same way as in the case of (9.40), it may be shown with the help of Lemma 9.5 that the above two expressions change at most in sign for different values of $\alpha \in [0, 2\pi)\backslash\{\beta_0, \beta_2\}$. In the case $\alpha = 0$, the expressions (9.41) and (9.42) change their sign when $\beta_0 < \alpha < \beta_2$. There are therefore two different lengths $L(\alpha) = L_1$ for $\alpha < \beta_0$ and $\alpha > \beta_2$, and $L(\alpha) = L_2$ for $\beta_0 < \alpha < \beta_2$.

If $\hat{\boldsymbol{m}}(\alpha)$ does not describe a rotation in the plane $\boldsymbol{W}(\hat{\boldsymbol{n}})$ but one in the plane bisecting $\boldsymbol{d}_0$ and $\boldsymbol{d}_2$, then PH curves of all possible lengths are

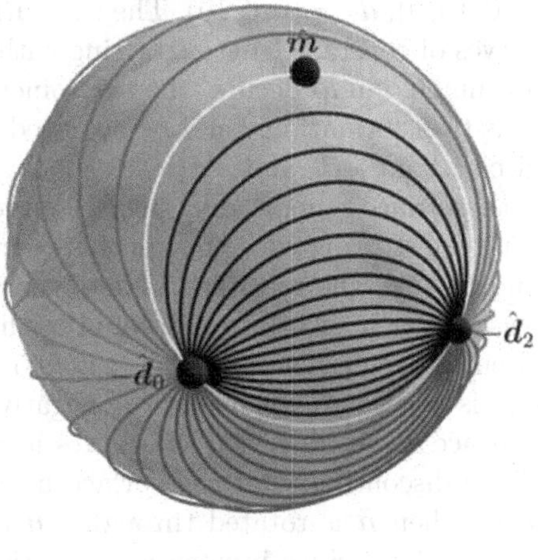

Fig. 9.3 Moving $\hat{\boldsymbol{m}}$ on a circle generates PH curves of constant length but different shape, while different circles generate PH curves of different lengths

generated. The positions of $\hat{\boldsymbol{m}}(\alpha)$ that generate PH curves of constant length for all lengths therefore constitute a "common points" system of coaxial circles on the unit sphere, as illustrated in Fig. 9.3.

Fig. 9.4 Two views of a pair of surfaces, where each of the surfaces is generated by a one-parameter family of spatial PH quintic Hermite interpolants of constant arc length

Although the arc length remains constant within each range of α, the PH curves generated by different values of α within these ranges differ in shape.

Figure 9.4 shows an example of this for the Hermite data $\boldsymbol{p}_0 = (0,0,0)$, $\boldsymbol{p}_2 = (1,0,0)$, $\boldsymbol{d}_0 = (0,10,0)$, $\boldsymbol{d}_2 = (0,0,10)$. The two surfaces shown are each generated by PH curves of constant length, starting with $\hat{\boldsymbol{n}}$ as the bisector of $\boldsymbol{d}_0$ and $\boldsymbol{d}_2$. A discontinuous jump occurs when $\hat{\boldsymbol{n}}$ coincides with either $-\boldsymbol{d}_0$ or $-\boldsymbol{d}_2$. A continuous transformation could be achieved by rotating $\hat{\boldsymbol{n}}$ on an infinitesimal half-circle about $-\boldsymbol{d}_0$ and $-\boldsymbol{d}_2$.

Figure 9.5 shows the surface generated by the PH curves as $\hat{\boldsymbol{n}}$ rotates in the plane bisecting $\boldsymbol{d}_0$ and $\boldsymbol{d}_2$, starting with an $\hat{\boldsymbol{n}}$ in this plane. This rotation of $\hat{\boldsymbol{n}}$ is in a plane perpendicular to the rotation plane of constant arc length, and PH curves of all possible arc lengths are generated in this way. The shortest and the longest PH curve are drawn in black in Fig. 9.5. In fact, in this case, the longest PH curve is generated when $\hat{\boldsymbol{n}}$ is the negative bisector of $\boldsymbol{d}_0$ and $\boldsymbol{d}_2$. Note that the surface spanned by the PH curves in Fig. 9.4 contains the longest PH curve. The discontinuity of the surface in Fig. 9.5, close to the shortest curve, occurs when $\hat{\boldsymbol{n}}$ is rotated through $-\boldsymbol{u}$ (this discontinuity is particular to the chosen data, since in general a rotation of $\hat{\boldsymbol{n}}$ in the plane bisecting $\boldsymbol{d}_0$ and $\boldsymbol{d}_2$ does not pass through $-\boldsymbol{u}$).

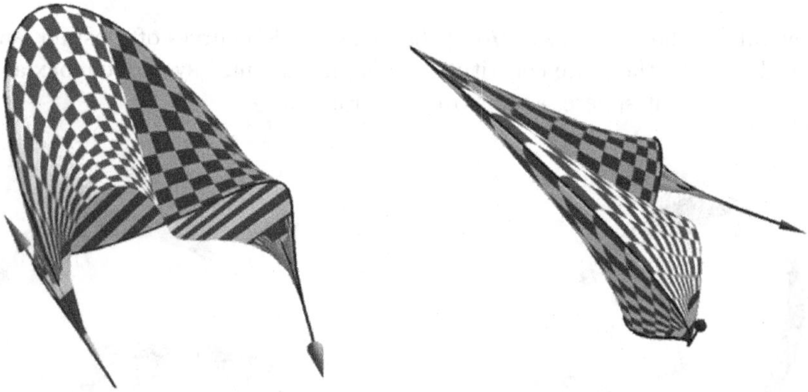

Fig. 9.5 Two views of the surface generated by rotating $\hat{\boldsymbol{n}}$ in the plane bisecting $\boldsymbol{d}_0$ and $\boldsymbol{d}_2$, using the same Hermite data as in Fig. 9.4. This generates PH curves of all possible arc lengths. The shortest and longest PH curves are shown in black

Note that, in the special case $\boldsymbol{d}_0 = \boldsymbol{d}_2$, a rotation of $\hat{\boldsymbol{m}}$ in any plane passing through $\mathsf{ref}(\hat{\boldsymbol{m}}, \hat{\boldsymbol{d}}_0)$ will leave the arc length of the PH curve unchanged. Since this set of rotation planes allows the rotation of $\hat{\boldsymbol{m}}$ into any point on the unit sphere, all PH curve interpolants have the same length in this case.

9.2.7 Pythagorean-Hodograph Curves in $\mathbb{R}^n$

The reflection representation of a hodograph can easily be extended to dimensions higher than three, in which case it always provides a sufficient condition for a hodograph to be Pythagorean – i.e., for a polynomial $\boldsymbol{a} : \mathbb{R} \to \mathbb{R}^n$ and a unit vector $\hat{\boldsymbol{n}} \in \mathbb{R}^n$, integration of the hodograph $\boldsymbol{r}' : \mathbb{R} \to \mathbb{R}^n$ given by $\boldsymbol{r}'(t) = \boldsymbol{a}(t)\, \hat{\boldsymbol{n}}\, \boldsymbol{a}(t)$ yields a PH curve in $\mathbb{R}^n$. With the rotation representation, on the other hand, elements of $\mathbb{G}_n^+$ for $n > 3$ are not always grade-preserving, which means that if $\boldsymbol{A} \in \mathbb{G}_n^+$ and $\hat{\boldsymbol{n}} \in \mathbb{R}^n$, $\boldsymbol{A}\, \hat{\boldsymbol{n}}\, \widetilde{\boldsymbol{A}}$ is not necessarily a vector. Equivalently, one can say that $\boldsymbol{A}\, \widetilde{\boldsymbol{A}}$ is not always a scalar.

In fact, the set of algebraic entities that have to be considered are not the elements of the even subalgebra $\mathbb{G}_n^+$ of $\mathbb{G}_n$, but the elements of the *Clifford group*. The Clifford group is defined as the set of those multivectors that can be generated by the geometric product of a number of grade 1 vectors [148], i.e. by the concatenation of reflections. The elements of the Clifford group are also called *versors*. For a set of vector-valued polynomials $\{\boldsymbol{a}_i\} : \mathbb{R} \to \mathbb{R}^n$, it therefore follows that $\boldsymbol{A}(t) := \boldsymbol{a}_1(t)\, \boldsymbol{a}_2(t) \ldots \boldsymbol{a}_k(t)$ is a polynomial in the Clifford group. However, whether the complete set of versor-valued polynomials can be generated in this way remains to be seen.

9.2.8 Summary

The main result of this section is the proposal of a novel formulation for spatial PH curves, based on the geometric product of vectors in geometric algebra. This formulation encompasses the initial (sufficient) characterization [65] of spatial Pythagorean hodographs as a special case, and is equivalent to the sufficient-and-necessary quaternion form [32, 59] for cubic and quintic PH curves. Whereas the quaternion form corresponds to generating a hodograph by a continuous sequence of scalings/rotations of a fixed unit vector $\hat{\boldsymbol{n}}$, the vector form amounts to a sequence of scalings/reflections of $\hat{\boldsymbol{n}}$. The quaternion form entails "hidden" angular variables in the coefficients of the preimage curve $\boldsymbol{A}(t)$, but the preimage curve $\boldsymbol{a}(t)$ of the vector form contains no free variables in addition to $\hat{\boldsymbol{n}}$ – instead, maintaining flexibility in the choice of $\hat{\boldsymbol{n}}$ ensures coverage of the complete space of spatial PH curves, modulo exceptional cases.

Variation of the unit vector $\hat{\boldsymbol{n}}$ offers a geometrically more intuitive access to the intrinsic freedoms of the shape of spatial PH curves than manipulating the free angular variables of the quaternion coefficients. As an illustration of this, the vector form reveals a decomposition of the two-dimensional space of spatial PH quintics that interpolate given first-order Hermite data into a product of two one-parameter spaces: one of the parameters controls the total arc length of the interpolants, while the other alters the shape of the interpolants of given total arc length.

The simpler and more intuitive nature of the vector formulation suggests that it could be fruitfully employed in other PH curve construction and analysis problems. One interesting subject for future research is to use the vector representation for defining PH curves in $\mathbb{R}^N$, and for extending existing PH spline algorithms [5, 60] from planar to spatial PH curves.

9.2.9 Proof of Lemma 9.5

Since $\hat{\boldsymbol{b}}_i(\hat{\boldsymbol{n}})$ lies in the rotation plane of $\boldsymbol{Q}$, it follows that $\boldsymbol{Q}\,\hat{\boldsymbol{b}}_i(\hat{\boldsymbol{n}}) = \hat{\boldsymbol{b}}_i(\hat{\boldsymbol{n}})\,\widetilde{\boldsymbol{Q}}$ (see Lemma 4.2, p. 131). Hence,

$$\begin{aligned}\hat{\boldsymbol{d}}_i &= \hat{\boldsymbol{b}}_i(\hat{\boldsymbol{n}})\,\hat{\boldsymbol{n}}\,\hat{\boldsymbol{b}}_i(\hat{\boldsymbol{n}})\\ &= \hat{\boldsymbol{b}}_i(\hat{\boldsymbol{n}})\,\widetilde{\boldsymbol{Q}}^2\;\;\boldsymbol{Q}^2\,\hat{\boldsymbol{n}}\,\widetilde{\boldsymbol{Q}}^2\;\;\boldsymbol{Q}^2\,\hat{\boldsymbol{b}}_i(\hat{\boldsymbol{n}})\\ &= \boldsymbol{Q}\,\hat{\boldsymbol{b}}_i(\hat{\boldsymbol{n}})\,\widetilde{\boldsymbol{Q}}\;\;\boldsymbol{Q}^2\,\hat{\boldsymbol{n}}\,\widetilde{\boldsymbol{Q}}^2\;\;\boldsymbol{Q}\,\hat{\boldsymbol{b}}_i(\hat{\boldsymbol{n}})\,\widetilde{\boldsymbol{Q}}\,.\end{aligned}$$

Therefore, $\boldsymbol{Q}\,\hat{\boldsymbol{b}}_i(\hat{\boldsymbol{n}})\,\widetilde{\boldsymbol{Q}}$ must be either the positive or the negative reflector of $\hat{\boldsymbol{d}}_i$, and $\hat{\boldsymbol{m}} := \boldsymbol{Q}^2\,\hat{\boldsymbol{n}}\,\widetilde{\boldsymbol{Q}}^2$. Furthermore, the rotation plane of $\boldsymbol{Q}$ has to intersect the plane perpendicular to $\hat{\boldsymbol{d}}_i$, which implies that there exists an angle $\alpha_0 \in [0,\pi)$ such that $\boldsymbol{Q}_0 := \mathsf{rot}(\alpha_0,\,\hat{\boldsymbol{q}})$ rotates $\hat{\boldsymbol{b}}_i(\hat{\boldsymbol{n}})$ into this plane. Since $\boldsymbol{Q}_0\,\hat{\boldsymbol{b}}_i(\hat{\boldsymbol{n}})\,\widetilde{\boldsymbol{Q}}_0$ is perpendicular to $\boldsymbol{d}_i$, it is a bisector of $\boldsymbol{d}_i$ and $-\boldsymbol{d}_i$. Thus, $\boldsymbol{Q}_0^2\,\hat{\boldsymbol{n}}\,\widetilde{\boldsymbol{Q}}_0^2 = -\hat{\boldsymbol{d}}_i$ and $\beta = 2\,\alpha_0$.

In order to investigate the other parts of the lemma, we need to express reflectors in a particular form. Let $\hat{\boldsymbol{x}}$ and $\hat{\boldsymbol{y}}$ be two perpendicular unit vectors in the rotation plane of some rotor $\boldsymbol{R}(\alpha) = \mathsf{rot}(\alpha,\,(\hat{\boldsymbol{x}}\wedge\hat{\boldsymbol{y}})^*)$, where α is the rotation angle. Since

$$\boldsymbol{R}(\alpha)\,\hat{\boldsymbol{x}}\,\widetilde{\boldsymbol{R}}(\alpha) = \cos\alpha\,\hat{\boldsymbol{x}} + \sin\alpha\,\hat{\boldsymbol{y}}\,,\qquad \widetilde{\boldsymbol{R}}(\alpha)\,\hat{\boldsymbol{x}}\,\boldsymbol{R}(\alpha) = \cos\alpha\,\hat{\boldsymbol{x}} - \sin\alpha\,\hat{\boldsymbol{y}}\,,$$

it follows that

$$\boldsymbol{R}(\alpha)\,\hat{\boldsymbol{x}}\,\widetilde{\boldsymbol{R}}(\alpha) - \widetilde{\boldsymbol{R}}(\alpha)\,\hat{\boldsymbol{x}}\,\boldsymbol{R}(\alpha) = 2\sin\alpha\,\hat{\boldsymbol{y}}\,.$$

Hence,

$$\mathsf{ref}\big(\boldsymbol{R}(\alpha)\,\hat{\boldsymbol{x}}\,\widetilde{\boldsymbol{R}}(\alpha),\,-\widetilde{\boldsymbol{R}}(\alpha)\,\hat{\boldsymbol{x}}\,\boldsymbol{R}(\alpha)\big) = \begin{cases}\hat{\boldsymbol{y}},\ \alpha\in(0,\pi)\,,\\ -\hat{\boldsymbol{y}},\ \alpha\in(-\pi,0)\,.\end{cases} \tag{9.43}$$

Conversely, this also implies that for any vector $\hat{\boldsymbol{x}}$ in the rotation plane of $\boldsymbol{R}(\alpha)$, $\mathsf{ref}\big(\boldsymbol{R}(\alpha)\,\hat{\boldsymbol{x}}\,\widetilde{\boldsymbol{R}}(\alpha),\,-\widetilde{\boldsymbol{R}}(\alpha)\,\hat{\boldsymbol{x}}\,\boldsymbol{R}(\alpha)\big)$ is perpendicular to $\hat{\boldsymbol{x}}$.

We now return to the initial problem. Let $\boldsymbol{Q}_0 := \mathsf{rot}(\frac{1}{2}\beta,\, \hat{\boldsymbol{q}})$, such that $\boldsymbol{Q}_0^2\, \hat{\boldsymbol{n}}\, \widetilde{\boldsymbol{Q}}_0^2 = -\hat{\boldsymbol{d}}_i$, and let $\boldsymbol{Q}_\Delta := \mathsf{rot}(\Delta\alpha,\, \hat{\boldsymbol{q}})$, such that $\boldsymbol{Q}^2 = \boldsymbol{Q}_0^2\, \boldsymbol{Q}_\Delta^2$, whence $\Delta\alpha = \alpha - \frac{1}{2}\beta$. Note that $\boldsymbol{Q}_0^2\, \boldsymbol{Q}_\Delta^2 = \boldsymbol{Q}_\Delta^2\, \boldsymbol{Q}_0^2$, since the two rotors have the same rotation plane. It may be shown by straightforward calculation that

$$\boldsymbol{Q}\, \hat{\boldsymbol{b}}(\hat{\boldsymbol{n}})\, \widetilde{\boldsymbol{Q}} = \boldsymbol{Q}\, \mathsf{ref}\big(\boldsymbol{Q}_0\, \hat{\boldsymbol{x}}_i\, \widetilde{\boldsymbol{Q}}_0,\; -\widetilde{\boldsymbol{Q}}_0\, \hat{\boldsymbol{x}}_i\, \boldsymbol{Q}_0\big)\, \widetilde{\boldsymbol{Q}},$$

where $\hat{\boldsymbol{x}}_i := \widetilde{\boldsymbol{Q}}_0\, \hat{\boldsymbol{d}}_i\, \boldsymbol{Q}_0$. Similarly, we find that

$$\hat{\boldsymbol{b}}_i(\boldsymbol{Q}^2\, \hat{\boldsymbol{n}}\, \widetilde{\boldsymbol{Q}}^2) = -\boldsymbol{Q}\, \mathsf{ref}\big(\boldsymbol{Q}_\Delta\, \hat{\boldsymbol{x}}_i\, \widetilde{\boldsymbol{Q}}_\Delta,\; -\widetilde{\boldsymbol{Q}}_\Delta\, \hat{\boldsymbol{x}}_i\, \boldsymbol{Q}_\Delta\big)\, \widetilde{\boldsymbol{Q}}.$$

Thus, if we define $\hat{\boldsymbol{y}} := \mathsf{ref}\big(\boldsymbol{Q}_0\, \hat{\boldsymbol{x}}_i\, \widetilde{\boldsymbol{Q}}_0,\; -\widetilde{\boldsymbol{Q}}_0\, \hat{\boldsymbol{x}}_i\, \boldsymbol{Q}_0\big)$, it follows from (9.43), that

$$-\mathsf{ref}\big(\boldsymbol{Q}_\Delta\, \hat{\boldsymbol{x}}_i\, \widetilde{\boldsymbol{Q}}_\Delta,\; -\widetilde{\boldsymbol{Q}}_\Delta\, \hat{\boldsymbol{x}}_i\, \boldsymbol{Q}_\Delta\big) = \begin{cases} \hat{\boldsymbol{y}}, \; \Delta\alpha \in [-\alpha_0,\, 0) \\ -\hat{\boldsymbol{y}}, \; \Delta\alpha \in (0,\, \pi - \alpha_0). \end{cases}$$

Since $\frac{1}{2}\beta = \alpha_0 \in [0, \pi)$, we have $\Delta\alpha \in [-\alpha_0,\, 0)$ if $\alpha \in [0,\, \alpha_0)$ and $\Delta\alpha \in (0,\, \pi - \alpha_0)$ if $\alpha \in (\alpha_0,\, \pi)$, and therefore

$$\boldsymbol{Q}\, \hat{\boldsymbol{b}}_i(\hat{\boldsymbol{n}})\widetilde{\boldsymbol{Q}} = \begin{cases} \hat{\boldsymbol{b}}_i(\boldsymbol{Q}^2\, \hat{\boldsymbol{n}}\, \boldsymbol{Q}^2),\; \alpha \in [0, \beta/2)\,, \\ -\hat{\boldsymbol{b}}_i(\boldsymbol{Q}^2\, \hat{\boldsymbol{n}}\, \boldsymbol{Q}^2),\; \alpha \in (\beta/2, \pi)\,. \end{cases} \qquad \square$$

Chapter 10
Random-Variable Space

The purpose of this chapter is to give an example of a geometric algebra over a space other than a real vector space. From Axiom 3.1 of geometric algebra, it follows that a geometric algebra can also be formed over a finite-dimensional Hilbert space. The particular example considered here is the Hilbert space of random variables. Another example could be the Hilbert space of the basis functions of a finite Fourier series. In all cases, the concepts of blades, null spaces, intersections, and combinations of subspaces are still valid in the geometric algebra over the Hilbert space, even though they may not have the same geometric meaning. However, a geometric meaning may be given to otherwise abstract operations in this way. This may help us to gain additional insights into various fields and to draw parallels between different fields.

While the Hilbert space of random variables is well known, the geometric algebra over this Hilbert space has not so far been treated. The main results of this chapter are about how the variance, the co-variance, and the the Cauchy–Schwarz inequality follow directly from operations on blades of random variables. Furthermore, an equation for the correlation coefficient between an arbitrary number of random variables is derived.

The plan of this chapter is as follows. In Sect. 10.1, some basic properties of random variables are introduced and operations between random variables and their corresponding density functions are defined. In Sect. 10.2, the Hilbert space of random variables is developed using the previously defined notation. Here, also, the Dirac delta random variable is defined, and it is shown that it can be interpreted as the "direction" of the expectation value of a random variable. In this way, a *homogeneous* random-variable Hilbert space can be developed. After this initial work, the geometric algebra over the homogeneous random-variable space is introduced in Sect. 10.3. Note that Sects. 10.1 and 10.2 give a formal treatment of the Hilbert space of random variables. Readers who are interested mainly in the treatment of the geometric algebra over the Hilbert space of random variables may go straight to Sect. 10.3, with a quick look at Sect. 10.2.3.

C. Perwass, *Geometric Algebra with Applications in Engineering.*
Geometry and Computing.

10.1 A Random-Variable Vector Space

10.1.1 Probability Space

Without rederiving the basics of statistics, some fundamental definitions are given in the following, mainly to introduce the nomenclature used and also to stress some concepts that not every reader may be familiar with. A probability space consists of three things:

1. A sample space $\mathbb{S}$.
2. A collection $\mathbb{A}$ of subsets of $\mathbb{S}$ which form a σ-algebra.
3. A probability measure $\mathcal{P}$ with $\mathcal{P} : \mathbb{A} \to [0, 1]$.

A definition of a σ-algebra, or Borel algebra, can be found, for example, in [82]. What is important to know at this point is that $\mathbb{S}$ contains all fundamental events and $\mathbb{A}$ contains all those events of interest. The probability measure $\mathcal{P}$ then gives a measure of how likely an event in $\mathbb{A}$ is. A probability space is therefore denoted by $(\mathbb{S}, \mathbb{A}, \mathcal{P})$.

10.1.2 Continuous Random Variables

It is important to understand that a random variable is a *function* that maps elements of the space $\mathbb{A}$ to values in the reals $\mathbb{R}$. That is, a random variable $\underline{X}$ is a map $\mathbb{A} \to \mathbb{R}$. A random variable is therefore a special type of the more general class of functions that map $\mathbb{S} \to \mathbb{R}$ and is thus specifically called an "$\mathbb{A}$-measurable function".

Since, for each element of $\mathbb{A}$, the measure $\mathcal{P}$ gives the likelihood that it occurs, one can also evaluate the likelihood that a random variable takes a particular value. For every random variable $\underline{X}$, one can therefore evaluate a *cumulative distribution function* (cdf) $F_{\underline{X}}$, which is a map $\mathbb{R} \to [0, 1]$ and is defined as

$$F_{\underline{X}} : x \mapsto \mathcal{P}\Big(\{a \in \mathbb{A} : \underline{X}(a) < x\} \Big). \tag{10.1}$$

Note that one property of the measure $\mathcal{P}$ on the σ-algebra is that if

$$\{A_1, A_2, \ldots, A_n\} \subseteq \mathbb{A}$$

are disjoint subsets of $\mathbb{A}$, then

$$\mathcal{P}\Big(\bigcup_{i=1}^{n} A_i\Big) = \sum_{i=1}^{n} \mathcal{P}(A_i).$$

$F_{\underline{X}}(x)$ gives the probability that $\underline{X}$ takes a value smaller than x. A cdf has to satisfy the conditions

$$\lim_{x\to-\infty} F_{\underline{X}}(x) = 0 \qquad \text{and} \qquad \lim_{x\to\infty} F_{\underline{X}}(x) = 1\,,$$

which implies that

$$\int_{-\infty}^{\infty} dF_{\underline{X}} = 1.$$

We now define $f_{\underline{X}}(x) := \dfrac{d}{dx} F_{\underline{X}}(x)$. Then the above integral can also be written as

$$\int_{-\infty}^{\infty} dF_{\underline{X}} = \int_{-\infty}^{\infty} f_{\underline{X}}(x)\, dx = 1.$$

The function $f_{\underline{X}}$ is called the *probability density function* (pdf) of $\underline{X}$. This is the function that will be used most often here to discuss properties of a random variable. Note that from the definition of $F_{\underline{X}}$, it follows that $f_{\underline{X}}$ is a map $\mathbb{R} \to [0,\infty)$.

A particularly important measure on random variables is the *expectation value*. The expectation value operator will be written as $\mathcal{E}$ and is defined as

$$\mathcal{E}(\underline{X}) := \int_{-\infty}^{\infty} \underline{X}\, dF_{\underline{X}} = \int_{-\infty}^{\infty} x\, f_{\underline{X}}(x)\, dx. \tag{10.2}$$

More generally, one can define the moment operator $\mathcal{M}$. The kth moment of a random variable $\underline{X}$ is defined as

$$\mathcal{M}_k(\underline{X}) := \mathcal{E}(\underline{X}^k) = \int_{-\infty}^{\infty} \underline{X}^k\, dF_{\underline{X}} = \int_{-\infty}^{\infty} x^k\, f_{\underline{X}}(x)\, dx, \quad k \in \mathbb{N} \geq 0.$$

Since random variables are functions, it would be interesting to define a norm on them such that they formed a Banach space. One general norm that may be useful is the following:

$$\|\underline{X}\|_k := \left(\int_{-\infty}^{\infty} |x|^k\, f_{\underline{X}}(x)\, dx \right)^{\frac{1}{k}}, \quad k \in \mathbb{N} > 0. \tag{10.3}$$

If k is even, it clearly follows that

$$\mathcal{M}_k(\underline{X}) = \Big(\|\underline{X}\|_k \Big)^k, \quad k \in \mathbb{N},\, k > 0,\, k \text{ even}, \tag{10.4}$$

and, in particular,

$$\|\underline{X}\|_2 = \sqrt{\mathcal{E}(\underline{X}^2)}.$$

Random variables which have a finite second moment are of particular interest here. The space of all random variables defined on $(\mathbb{S}, \mathbb{A}, \mathcal{P})$ with a finite second moment will be denoted by $L_2(\mathbb{S}, \mathbb{A}, \mathcal{P})$. It may be shown that

$L_2(\mathbb{S}, \mathbb{A}, \mathcal{P})$ (using the norm $\|.\|_2$) is a Banach space. In the following, L_2 will be used as a shorthand to mean $L_2(\mathbb{S}, \mathbb{A}, \mathcal{P})$.

10.1.3 Multiple Random Variables

Let $\mathbb{L}^n \subset L_2$ be a finite-dimensional set of random variables with a finite second moment. Also, let $\{\underline{X}_1, \underline{X}_2, \ldots, \underline{X}_n\} \equiv \mathbb{L}^n$ denote the separate random variables. Each random variable has associated with it a cumulative distribution function denoted by $F_{\underline{X}_i}$. Since the various random variables may have a functional dependence on each other, there will in general be a *joint* cdf for all random variables. That is,

$$F_{\underline{X}_1,\underline{X}_2,\ldots,\underline{X}_n} : (x_1, x_2, \ldots, x_n) \mapsto \mathcal{P}\Big(\{a \in \mathbb{A} : \underline{X}_1(a) < x_1\}, \ldots, \{a \in \mathbb{A} : \underline{X}_n(a) < x_n\} \Big). \tag{10.5}$$

This joint cdf is directly related to the separate cdfs via

$$F_{\underline{X}_k}(x_k) = \lim_{x_1 \to \infty} \ldots \lim_{x_{k-1} \to \infty} \lim_{x_{k+1} \to \infty} \ldots \lim_{x_n \to \infty} \Big(F_{\underline{X}_1,\underline{X}_2,\ldots,\underline{X}_n}(x_1, \ldots, x_n) \Big). \tag{10.6}$$

In terms of the joint pdf, this becomes

$$f_{\underline{X}_k}(x_k) = \int_{-\infty}^{\infty} \cdots \int_{-\infty}^{\infty} f_{\underline{X}_1,\ldots,\underline{X}_n}(x_1, \ldots, x_n)\, dx_1 \ldots dx_{k-1}\, dx_{k+1} \ldots dx_n. \tag{10.7}$$

Note that two random variables $\underline{X}_i$ and $\underline{X}_j$ are said to be *statistically independent* if $F_{\underline{X}_1,\underline{X}_2} = F_{\underline{X}_1} F_{\underline{X}_2}$.

With the above definitions, the joint cdf $F_{\underline{X}_1,\underline{X}_2,\ldots,\underline{X}_n}$ may also be written as $F_{\mathbb{L}^n}$. We denote by $\mathbb{F}^n(\mathbb{L}^n)$ the set of cdfs of $\mathbb{L}^n$. Furthermore, we denote by $\mathbb{F}_n(\mathbb{L}^n)$ the set of all cdfs of the power set of $\mathbb{L}^n$. For example, given $\mathbb{L}^3 := \{\underline{X}_1, \underline{X}_2, \underline{X}_3\}$, then

$$\mathbb{F}_3(\mathbb{L}^3) = \Big\{ F_\emptyset,\, F_{\underline{X}_1},\, F_{\underline{X}_2},\, F_{\underline{X}_3},\, F_{\underline{X}_1,\underline{X}_2},\, F_{\underline{X}_1,\underline{X}_3},\, F_{\underline{X}_2,\underline{X}_3},\, F_{\underline{X}_1,\underline{X}_2,\underline{X}_3} \Big\},$$

where $F_\emptyset = 1$. Note that a cdf is independent of the order of its indices; for example, $F_{\underline{X}_1,\underline{X}_3} = F_{\underline{X}_3,\underline{X}_1}$. In the following, $\mathbb{F}^n$ and $\mathbb{F}_n$ will be written instead of $\mathbb{F}^n(\mathbb{L}^n)$ and $\mathbb{F}_n(\mathbb{L}^n)$ if it is clear which set $\mathbb{L}^n$ is meant. Let $\otimes$ denote a product on $\mathbb{F}_n$ that maps $\mathbb{F}_n \times \mathbb{F}_n \to \mathbb{F}_n$. For two sets $\mathbb{U} \subseteq \mathbb{L}^n$ and $\mathbb{V} \subseteq \mathbb{L}^n$, this product is defined as

$$F_{\mathbb{U}} \otimes F_{\mathbb{V}} := F_{\mathbb{U} \cup \mathbb{V}}. \tag{10.8}$$

For example,

$$F_{\underline{X}_1} \otimes F_{\underline{X}_1} = F_{\underline{X}_1} \quad \text{and} \quad F_{\underline{X}_1} \otimes F_{\underline{X}_2} = F_{\underline{X}_2} \otimes F_{\underline{X}_1} = F_{\underline{X}_1,\underline{X}_2}.$$

Furthermore, for some $\mathbb{U} \subseteq \mathbb{L}^n$, let $F_{\mathbb{U}}^c$ denote the complement of $F_{\mathbb{U}}$, which is defined as

$$F_{\mathbb{U}}^c := F_{\mathbb{V}}, \quad \mathbb{V} := \mathbb{L}^n \setminus \mathbb{U}, \tag{10.9}$$

where $\mathbb{L}^n \setminus \mathbb{U}$ denotes the set $\mathbb{L}^n$ without the elements of $\mathbb{U}$. Together with the definition of the complement, $(\mathbb{F}_n, \otimes)$ forms a σ-algebra.

One may also define a product, denoted by $\otimes$, and a sum, denoted by $\oplus$, on the set $\mathbb{L}^n$ which are simply the standard product, and sum, respectively, in $\mathbb{R}$ of two random variables. That is, for $\underline{X}_i, \underline{X}_j \in \mathbb{L}^n$,

$$\left(\underline{X}_i \otimes \underline{X}_j\right)(a) := \underline{X}_i(a)\, \underline{X}_j(a), \quad \forall a \in \mathbb{A}, \tag{10.10}$$

and

$$\left(\underline{X}_i \oplus \underline{X}_j\right)(a) := \underline{X}_i(a) + \underline{X}_j(a), \quad \forall a \in \mathbb{A}. \tag{10.11}$$

We denote by $(\mathbb{L}^n, \mathbb{F}^n)$ the set of corresponding pairs of random variables and cdfs. That is,

$$(\mathbb{L}^n, \mathbb{F}^n) = \left\{ (\underline{X}_1, F_{\underline{X}_1}), (\underline{X}_2, F_{\underline{X}_2}), \ldots, (\underline{X}_n, F_{\underline{X}_n}) \right\}. \tag{10.12}$$

This set can be regarded as the basis of a vector space over the reals by first defining the $\oplus$-sum on elements of $(\mathbb{L}^n, \mathbb{F}^n)$. Let $(\underline{X}, F_{\underline{X}}), (\underline{Y}, F_{\underline{Y}}) \in (\mathbb{L}^n, \mathbb{F}^n)$; then

$$(\underline{X}, F_{\underline{X}}) \oplus (\underline{Y}, F_{\underline{Y}}) := (\underline{X} \oplus \underline{Y}, F_{\underline{X}} \otimes F_{\underline{Y}}). \tag{10.13}$$

For example, for $(\underline{X}_1, F_{\underline{X}_1}), (\underline{X}_2, F_{\underline{X}_2}) \in (\mathbb{L}^n, \mathbb{F}^n)$,

$$(\underline{X}_1, F_{\underline{X}_1}) \oplus (\underline{X}_2, F_{\underline{X}_2}) = (\underline{X}_1 \oplus \underline{X}_2, F_{\underline{X}_1,\underline{X}_2}).$$

Furthermore, a product between elements of $\mathbb{R}$ and elements of $(\mathbb{L}^n, \mathbb{F}^n)$ is defined as follows. Let $\alpha, \beta \in \mathbb{R}$ and $(\underline{X}, F_{\underline{X}}) \in (\mathbb{L}^n, \mathbb{F}^n)$; then

1. $\alpha\,(\underline{X}, F_{\underline{X}}) = (\underline{X}, F_{\underline{X}})\,\alpha = (\alpha\,\underline{X}, F_{\underline{X}})$.
2. $(\alpha + \beta)\,(\underline{X}, F_{\underline{X}}) = \alpha\,(\underline{X}, F_{\underline{X}}) \oplus \beta\,(\underline{X}, F_{\underline{X}})$.
3. $1\,(\underline{X}, F_{\underline{X}}) = (\underline{X}, F_{\underline{X}})\,1 = (\underline{X}, F_{\underline{X}})$.

Distributivity of scalars with respect to $\oplus$ follows immediately. If we combine the above definitions, it is clear that, for example,

$$(\underline{X}_1, F_{\underline{X}_1}) \oplus (\underline{X}_1, F_{\underline{X}_1}) = (\underline{X}_1 \oplus \underline{X}_1, F_{\underline{X}_1}) = 2\,(\underline{X}_1, F_{\underline{X}_1}).$$

It may thus be shown that the set $(\mathbb{L}^n, \mathbb{F}^n)$, together with the above product with elements of $\mathbb{R}$ and the $\oplus$-sum, forms a vector space over $\mathbb{R}$. This vector space will be denoted by $(\mathbb{L}^n, \mathbb{F}^n, \oplus)$.

10.2 A Hilbert Space of Random Variables

In order to define a Hilbert space on the vector space $(\mathbb{L}^n, \mathbb{F}^n, \oplus)$, the vector space has to be shown to be a Banach space and a scalar product has to be defined. The first step therefore is to define a norm on $(\mathbb{L}^n, \mathbb{F}^n, \oplus)$. This is done in the same spirit as before. Then the expectation operator is introduced, which leads to the definition of a scalar product.

10.2.1 The Norm

Let $(\underline{X},\, F_{\underline{X}}) \in (\mathbb{L}^n,\, \mathbb{F}^n)$; then

$$\|(\underline{X},\, F_{\underline{X}})\|_k := \left(\int_{-\infty}^{\infty} |x|^k\, f_{\underline{X}}(x)\, dx\right)^{\frac{1}{k}}, \quad k \in \mathbb{N},\, k > 0. \tag{10.14}$$

Since $\underline{X} \in \mathbb{L}^n \subset L_2$, this norm exists for $k = 2$ by definition. This is in fact the same definition as before, with the only difference being that instead of writing $\|\underline{X}\|_k$ and keeping in mind that there is a cdf associated with $\underline{X}$, this dependence is made algebraically explicit. Using the algebraic properties of $(\mathbb{L}^n, \mathbb{F}^n)$, it follows that for $(\underline{X}, F_{\underline{X}}), (\underline{Y}, F_{\underline{Y}}) \in (\mathbb{L}^n, \mathbb{F}^n)$,

$$\begin{aligned}\|(\underline{X}, F_{\underline{X}}) \oplus (\underline{Y}, F_{\underline{Y}})\|_k &= \|(\underline{X} \oplus \underline{Y},\, F_{\underline{X},\underline{Y}})\|_k \\ &= \left(\int_{-\infty}^{\infty}\int_{-\infty}^{\infty} |x+y|^k\, f_{\underline{X},\underline{Y}}(x,y)\, dx\, dy\right)^{\frac{1}{k}}.\end{aligned} \tag{10.15}$$

Since $(\mathbb{L}^n, \mathbb{F}^n)$ is a basis of $(\mathbb{L}^n, \mathbb{F}^n, \oplus)$, it may be shown that $(\mathbb{L}^n, \mathbb{F}^n, \oplus)$, together with the norm $\|.\|_2$, is a Banach space.

In order to introduce a scalar product, the expectation operator is used. The expectation operator is again denoted by $\mathcal{E}$ and is defined by

$$\mathcal{E}\big((\underline{X},\, F_{\underline{X}})\big) := \int_{-\infty}^{\infty} x\, f_{\underline{X}}(x)\, dx, \tag{10.16}$$

where $(\underline{X},\, F_{\underline{X}}) \in (\mathbb{L}^n, \mathbb{F}^n)$. This definition can be extended to the whole of $(\mathbb{L}^n, \mathbb{F}^n, \oplus)$. For $(\underline{X},\, F_{\underline{X}}), (\underline{Y}, F_{\underline{Y}}) \in (\mathbb{L}^n, \mathbb{F}^n)$ and $\alpha, \beta \in \mathbb{R}$,

$$
\begin{aligned}
&\mathcal{E}\big(\alpha(\underline{X},\, F_{\underline{X}}) \oplus \beta(\underline{Y}, F_{\underline{Y}})\big) \\
&\quad = \mathcal{E}\big((\alpha\underline{X} \oplus \beta\underline{Y},\, F_{\underline{X},\underline{Y}})\big) \\
&\quad := \int_{-\infty}^{\infty}\int_{-\infty}^{\infty} (\alpha x + \beta y)\, f_{\underline{X},\underline{Y}}(x, y)\, dx\, dy \\
&\quad = \int_{-\infty}^{\infty} \alpha\, x\, f_{\underline{X}}(x)\, dx + \int_{-\infty}^{\infty} \beta\, y\, f_{\underline{Y}}(y)\, dy \\
&\quad = \alpha\,\mathcal{E}\big((\underline{X}, F_{\underline{X}})\big) + \beta\,\mathcal{E}\big((\underline{Y}, F_{\underline{Y}})\big).
\end{aligned}
\tag{10.17}
$$

That is, $\mathcal{E}$ is a linear operator on $(\mathbb{L}^n, \mathbb{F}^n, \oplus)$.

10.2.2 The Scalar Product

A scalar product is now introduced in terms of the expectation operator. In order to do this, first a product operator $\otimes$ on elements of $(\mathbb{L}^n, \mathbb{F}^n)$ has to be introduced. Let $(\underline{X}, F_{\underline{X}}), (\underline{Y}, F_{\underline{Y}}) \in (\mathbb{L}^n, \mathbb{F}^n)$; then

$$
(\underline{X}, F_{\underline{X}}) \otimes (\underline{Y}, F_{\underline{Y}}) := (\underline{X} \otimes \underline{Y},\, F_{\underline{X}} \otimes F_{\underline{Y}}). \tag{10.18}
$$

This definition implies that

$$
(\underline{X}, F_{\underline{X}}) \otimes (\underline{Y}, F_{\underline{Y}}) = (\underline{Y}, F_{\underline{Y}}) \otimes (\underline{X}, F_{\underline{X}}).
$$

For example,

$$
(\underline{X}, F_{\underline{X}}) \otimes (\underline{Y}, F_{\underline{Y}}) = (\underline{X} \otimes \underline{Y},\, F_{\underline{X},\underline{Y}})
$$

and

$$
(\underline{X}, F_{\underline{X}}) \otimes (\underline{X}, F_{\underline{X}}) = (\underline{X} \otimes \underline{X},\, F_{\underline{X}}).
$$

The definition of $\otimes$ can be extended to the whole of $(\mathbb{L}^n, \mathbb{F}^n, \oplus)$ as follows. Let $\alpha \in \mathbb{R}$ and $(\underline{X}, F_{\underline{X}}), (\underline{Y}, F_{\underline{Y}}), (\underline{Z}, F_{\underline{Z}}) \in (\mathbb{L}^n, \mathbb{F}^n)$. We then make the following definitions:

1. $$
\begin{aligned}
&(\underline{X}, F_{\underline{X}}) \otimes \Big((\underline{Y}, F_{\underline{Y}}) \oplus (\underline{Z}, F_{\underline{Z}})\Big) \\
&\quad = \Big((\underline{X}, F_{\underline{X}}) \otimes (\underline{Y}, F_{\underline{Y}})\Big) \oplus \Big((\underline{X}, F_{\underline{X}}) \otimes (\underline{Z}, F_{\underline{Z}})\Big) \\
&\quad = \Big((\underline{Y}, F_{\underline{Y}}) \oplus (\underline{Z}, F_{\underline{Z}})\Big) \otimes (\underline{X}, F_{\underline{X}}).
\end{aligned}
$$

2. $$\alpha\Big((\underline{X},F_{\underline{X}})\otimes(\underline{Y},F_{\underline{Y}})\Big)$$
$$=\Big(\alpha\,(\underline{X},F_{\underline{X}})\Big)\otimes(\underline{Y},F_{\underline{Y}})$$
$$=(\underline{X},F_{\underline{X}})\otimes\Big(\alpha\,(\underline{Y},F_{\underline{Y}})\Big)$$
$$=\Big((\underline{X},F_{\underline{X}})\otimes(\underline{Y},F_{\underline{Y}})\Big)\,\alpha\,.$$

After these preliminaries, a scalar product denoted by $*$ is defined for two elements $(\underline{X},F_{\underline{X}}),(\underline{Y},F_{\underline{Y}})\in(\mathbb{L}^n,\mathbb{F}^n)$ by

$$\begin{aligned}(\underline{X},F_{\underline{X}})*(\underline{Y},F_{\underline{Y}}) &:= \mathcal{E}\big((\underline{X},F_{\underline{X}})\otimes(\underline{Y},F_{\underline{Y}})\big)\\ &= \mathcal{E}\big((\underline{X}\otimes\underline{Y},F_{\underline{X},\underline{Y}})\big)\\ &= \int_{-\infty}^{\infty}\int_{-\infty}^{\infty} x\,y\,f_{\underline{X},\underline{Y}}(x,y)\,dx\,dy.\end{aligned}\tag{10.19}$$

Using the properties of the respective operators, this definition can be extended directly to the whole of $(\mathbb{L}^n,\mathbb{F}^n,\oplus)$. This scalar product is also related to the norm, as required for a Hilbert space. That is,

$$\begin{aligned}(\underline{X},F_{\underline{X}})*(\underline{X},F_{\underline{X}}) &= \mathcal{E}\big((\underline{X},F_{\underline{X}})\otimes(\underline{X},F_{\underline{X}})\big)\\ &= \mathcal{E}\big((\underline{X}\otimes\underline{X},F_{\underline{X}})\big)\\ &= \int_{-\infty}^{\infty} x^2\,f_{\underline{X}}(x)\,dx\\ &= \|(\underline{X},F_{\underline{X}})\|_2^2.\end{aligned}\tag{10.20}$$

This may also be written as

$$\|(\underline{X},F_{\underline{X}})\|_2=\sqrt{(\underline{X},F_{\underline{X}})*(\underline{X},F_{\underline{X}})}.\tag{10.21}$$

Therefore, the vector space $(\mathbb{L}^n,\mathbb{F}^n,\oplus)$ together with the norm $\|.\|_2$ and the scalar product $*$ is a Hilbert space. This Hilbert space will be denoted by $H^n:=(\mathbb{L}^n,\mathbb{F}^n,\oplus,*)$.

10.2.3 The Dirac Delta Distribution

Since H^n is a Hilbert space, elements of H^n can be treated just like the vectors in $\mathbb{R}^n$. One particularly useful operation is the orthogonalization of a set of elements of H^n. As in $\mathbb{R}^n$, two elements of H^n are said to be orthogonal if their scalar product is zero. For the sake of brevity, elements of H^n will no longer be denoted by a pair $(\underline{X}, F_{\underline{X}})$ in this chapter but simply by the respective random variable. That is, $\underline{X}$ will be written instead of $(\underline{X}, F_{\underline{X}})$. Furthermore, the $\oplus$-sum of elements of H^n will now be denoted by the standard plus sign "+".

The (Dirac) delta distribution plays an important role in the context of the Hilbert space of random variables, since it allows the evaluation of the expectation value of a random variable via the scalar product. The delta random variable of expectation $\alpha \in \mathbb{R}$ will be denoted by $\underline{D}_\alpha$ and is defined on $(\mathbb{S}, \mathbb{A}, \mathcal{P})$ as

$$\underline{D}_\alpha : a \mapsto \alpha. \tag{10.22}$$

That is, for all events in $\mathbb{A}$, $\underline{D}_\alpha$ has the same result, α. The cdf of $\underline{D}_\alpha$ is therefore

$$F_{\underline{D}_\alpha} : x \mapsto \begin{cases} 0 : x \leq \alpha\,, \\ 1 : x > \alpha\,. \end{cases} \tag{10.23}$$

Since the pdf $f_{\underline{D}_\alpha}$ of $\underline{D}_\alpha$ is the derivative of $F_{\underline{D}_\alpha}$, $f_{\underline{D}_\alpha}$ is the Dirac delta distribution at the position α, i.e.

$$f_{\underline{D}_\alpha} : x \mapsto \delta(x - \alpha).$$

Recall that the Dirac delta distribution has the following properties.

1. $\displaystyle\int_{-\infty}^{\infty} \delta(x - \alpha)\, dx = 1\,.$

2. $\displaystyle\int_{-\infty}^{\infty} g(x)\, \delta(x - \alpha)\, dx = g(\alpha),$

where $g : \mathbb{R} \to \mathbb{R}$ is some function. The expectation value of $\underline{D}_\alpha$ is

$$\mathcal{E}(\underline{D}_\alpha) = \int_{-\infty}^{\infty} x\, \delta(x - \alpha)\, dx = \alpha,$$

as was assumed initially. In the following, the identification $\underline{D} \equiv \underline{D}_1$ will be made. Suppose now that $\underline{D}, \underline{X} \in H^n$ are statistically independent. Recall that this implies that $F_{\underline{X},\underline{D}}(x, y) = F_{\underline{X}}(x)\, F_{\underline{D}}(y)$. It therefore follows that

$$\begin{aligned} \underline{X} * \underline{D} &= \int_{-\infty}^{\infty} \int_{-\infty}^{\infty} x\, y\, f_{\underline{X}}(x)\, \delta(y - 1)\, dx\, dy \\ &= \int_{-\infty}^{\infty} x\, f_{\underline{X}}(x)\, dx \\ &= \mathcal{E}(\underline{X}). \end{aligned}$$

This can be interpreted by saying that $\underline{D}$ "points in the direction" of the expectation of a random variable. In fact, $\underline{D}$ may be called the *expectation dimension.*

The expectation value is an important feature of a random variable. By extending the random-variable space H^n by the statistically independent delta distribution random variable, expectation values, as well as (co)variances of random variables, can be evaluated with the scalar product. For normally distributed random variables, this can be regarded as representing the expectation values and variances in a single matrix. This is much like the homogeneous representation of ellipses described in Sect. 4.4.5, where the translation of the origin can be identified with the expectation value.

The process of extending H^n by the delta distribution $\underline{D}$ will be called the *homogenization* of H^n, and the corresponding homogeneous random-variable space will be denoted by H^n_h. If $\{\underline{X}_1, \ldots, \underline{X}_n\}$ is a basis of H^n, then $\{\underline{X}_1, \ldots, \underline{X}_n, \underline{D}\}$ is a basis of H^n_h.

10.3 Geometric Algebra over Random Variables

As noted in the previous section, expectation values of random variables in the homogeneous random-variable space H^n_h are represented by an "expectation dimension", spanned by the delta distribution. Therefore, the expectation value of a random variable can be evaluated by means of the scalar product with the delta distribution. Since the norm and the scalar product are the only operations that allow the evaluation of scalar features of random variables in H^n_h, the homogeneous representation of the random-variable space is essential. The geometric algebra will therefore be constructed over H^n_h.

A geometric algebra may be defined over any vector space on which a quadratic form is defined. Since the Hilbert space H^n_h already defines a scalar product, and thus a quadratic form, on the elements of the vector space, a geometric algebra can also be defined over H^n_h. Everything that has been shown for geometric algebras over real-valued vector spaces in Chap. 3 is therefore also valid for $\mathbb{G}(H^n_h)$, which justifies the simpler notation $\mathbb{G}_n$ for $\mathbb{G}(H^n_h)$.

The confinement to the scalar product and the norm when one is evaluating scalar features of random variables implies that, at most, moments of order two can be evaluated. This makes the algebra well suited for spaces of normally distributed random variables, since the first two moments completely describe their probability density functions. In addition, orthogonality (i.e. uncorrelatedness) of normally distributed random variables with zero mean in H^n_h implies statistical independence, which is not necessarily the case for other probability distributions. Any basis of normally distributed random variables can therefore be transformed into a basis of statistically independent random variables using an orthogonalization scheme. Only when this is possible can all statistical properties of and between random variables be captured by the vector components on such an orthogonal basis.

The algebra $\mathbb{G}(H_h^n)$ allows more than the evaluation of the first two moments of random variables, since subspaces of random variables and operators on random variables can also be represented. It is important to note here that this treatment of random variables is quite different from that of uncertain multivectors in Sect. 5.3. Whereas in Sect. 5.3 the normally distributed random variables are the scalar-valued components of multivectors, here the basis elements themselves are random variables.

In the following, first some properties of $\mathbb{G}(H_h^n)$ for random variables with arbitrary pdfs will be presented. Later, the algebra over normally distributed random variables will be treated in some more detail.

10.3.1 The Norm

Before we treat some general properties of $\mathbb{G}(H_h^n)$, the norm of H_h^n has to be extended to the whole algebra. This is done in just the same way as for $\mathbb{G}(\mathbb{R}^n)$. For any $\underline{A} \in \mathbb{G}_n$, the norm of $\underline{A}$ is defined as

$$\|\underline{A}\|_2 := \sqrt{\underline{A} * \widetilde{\underline{A}}}\,.$$

To simplify the notation somewhat, the symbol $\|\cdot\|$ will be used to denote $\|\cdot\|_2$ in the following.

Consider the geometric product of two random variables $\underline{A} := \underline{X}\,\underline{Y}$, with $\underline{X}, \underline{Y} \in \mathbb{G}_n^1$. The norm of $\underline{A}$ is

$$\|\underline{A}\|^2 = (\underline{X}\,\underline{Y}) * (\underline{Y}\,\underline{X}) = \langle(\underline{X}\,\underline{Y})\,(\underline{Y}\,\underline{X})\rangle_0 = \langle\underline{X}\,(\underline{Y}\,\underline{Y})\,\underline{X}\rangle_0 = (\underline{X} * \underline{X})\,(\underline{Y} * \underline{Y})\,,$$

where the last step follows from $\underline{X}\,\underline{X} = \underline{X} * \underline{X}$, which is the defining axiom of the algebra. Similarly, the norm of the blade $\underline{X} \wedge \underline{Y} \in \mathbb{G}_n^2$ is given by

$$\|\underline{X} \wedge \underline{Y}\|^2 = (\underline{X} \wedge \underline{Y}) \cdot (\underline{Y} \wedge \underline{X}) = \underline{X}^2\,\underline{Y}^2 - (\underline{X} \cdot \underline{Y})^2\,.$$

Just as was shown in Lemma 3.8 for the geometric algebra over a vector space $\mathbb{R}^{p,q}$, the blade $\underline{X} \wedge \underline{Y}$ can be replaced by $\underline{X} \wedge \underline{Y}'$, where $\underline{X}$ and $\underline{Y}'$ are orthogonal. That is,

$$\|\underline{X} \wedge \underline{Y}\|^2 = \|\underline{X} \wedge \underline{Y}'\|^2 = \|\underline{X}\,\underline{Y}'\|^2 = (\underline{X} * \underline{X})\,(\underline{Y}' * \underline{Y}')\,.$$

Since $\underline{X} * \underline{X} \geq 0$ and $\underline{Y}' * \underline{Y}' \geq 0$ by definition, it follows that the magnitude of the blade $\underline{X} \wedge \underline{Y}$ is positive semidefinite, i.e. $\|\underline{X} \wedge \underline{Y}\|^2 \geq 0$. In general, it holds that for $\underline{X}_1, \ldots, \underline{X}_k \in \mathbb{G}_n^1$, $k \leq n$, the magnitude obeys the inequality $\|\underline{X}_1 \wedge \ldots \wedge \underline{X}_k\| \geq 0$. From $\|\underline{X} \wedge \underline{Y}\| \geq 0$, it follows that

$$(\underline{X} \cdot \underline{Y})^2 \leq \underline{X}^2\,\underline{Y}^2\,.$$

In standard notation, this is usually written as

$$\mathcal{E}(\underline{X}\underline{Y})^2 \leq \mathcal{E}(\underline{Y}^2)\,\mathcal{E}(\underline{X}^2)\,, \tag{10.24}$$

and is known as the *Cauchy–Schwarz inequality* (see [82]). Just as in a standard proof, the inequality follows here from the fact that $\|\underline{X}\|^2 \geq 0$ and $\|\underline{Y}\|^2 \geq 0$. Inequalities for k random variables can be derived in exactly the same way, using $\|\underline{X}_1 \wedge \underline{X}_2 \wedge \cdots \wedge \underline{X}_k\| \geq 0$.

10.3.2 General Properties

Recall that two random variables $\underline{X}$ and $\underline{Y}$ are statistically independent if their pdfs satisfy $f_{\underline{X},\underline{Y}}(x,y) = f_{\underline{X}}(x)\, f_{\underline{Y}}(y)$. Statistical independence therefore implies that $\underline{X} * \underline{Y} = \mathcal{E}(\underline{X})\,\mathcal{E}(\underline{Y})$. The converse, however, is not true in general. If $\underline{X} * \underline{Y} = \mathcal{E}(\underline{X})\,\mathcal{E}(\underline{Y})$, the random variables $\underline{X}$ and $\underline{Y}$ are said to be *uncorrelated*, but they are not necessarily statistically independent.

The important feature of the homogeneous random-variable space H_h^n in contrast to the corresponding H^n is that it contains the delta distribution $\underline{D}$, which is assumed to be statistically independent of all elements of H^n. The expectation value of $\underline{X} \in H_h^n$ can then be evaluated as

$$\mathcal{E}(\underline{X}) = \underline{D} * \underline{X} = \underline{D} \cdot \underline{X}\,.$$

Recall that in $\mathbb{G}_n$, the scalar product of vectors (grade 1) is equivalent to their inner product. The delta distribution $\underline{D}$ may therefore be regarded as the expectation direction. To measure the *variance* of a random variable, the "linear dependence" on the expectation dimension has to be removed, i.e.

$$\check{\underline{X}} := \underline{X} - \mathcal{E}(\underline{X})\,\underline{D}\,.$$

Clearly, $\check{\underline{X}}$ is the expectation-free version of $\underline{X}$, since $\check{\underline{X}} * \underline{D} = 0$. The norm squared of $\check{\underline{X}}$ is then the variance $\mathcal{V}(\underline{X})$ of $\underline{X}$,

$$\check{\underline{X}} \cdot \check{\underline{X}} = \underline{X} \cdot \underline{X} - \mathcal{E}(\underline{X})^2 \;=: \mathcal{V}(\underline{X}).$$

Given two random variables $\underline{X}, \underline{Y} \in H_h^n$, with corresponding expectation-free variables $\check{\underline{X}}$ and $\check{\underline{Y}}$, their *covariance* $\mathcal{C}(\underline{X}, \underline{Y})$ can be evaluated as

$$\check{\underline{X}} \cdot \check{\underline{Y}} = \underline{X} \cdot \underline{Y} - \mathcal{E}(\underline{X})\,\mathcal{E}(\underline{Y}) =: \mathcal{C}(\underline{X}, \underline{Y})\,.$$

If $\underline{X}$ and $\underline{Y}$ are statistically independent (or even only uncorrelated), then $\underline{X} \cdot \underline{Y} = \mathcal{E}(\underline{X})\,\mathcal{E}(\underline{Y})$ and thus $\mathcal{C}(\underline{X}, \underline{Y}) = 0$.

The variance and covariance can also be recovered from the magnitudes of appropriate blades. Consider the blade $\underline{X} \wedge \underline{D} \in \mathbb{G}_n^2$, with a magnitude squared equal to

$$\|\underline{X} \wedge \underline{D}\|^2 = (\underline{X} \wedge \underline{D}) \cdot (\underline{D} \wedge \underline{X}) = \underline{X}^2 - \mathcal{E}(\underline{X})^2 = \mathcal{V}(\underline{X}) .$$

This is not too surprising, since $\underline{D} \wedge \underline{D} = 0$ and thus

$$\underline{X} \wedge \underline{D} = (\check{\underline{X}} + \mathcal{E}(\underline{X})\, \underline{D}) \wedge \underline{D} = \check{\underline{X}} \wedge \underline{D} = \check{\underline{X}}\, \underline{D} .$$

The covariance of $\underline{X}, \underline{Y} \in \mathbb{G}^1_n$ is given by

$$(\underline{X} \wedge \underline{D}) \cdot (\underline{D} \wedge \underline{Y}) = \underline{X} \cdot \underline{Y} - \mathcal{E}(\underline{X})\, \mathcal{E}(\underline{Y}) = \mathcal{C}(\underline{X}, \underline{Y}) .$$

10.3.3 Correlation

Consider a blade $\check{\underline{X}} \wedge \check{\underline{Y}}$ of two expectation-free random variables $\check{\underline{X}}, \check{\underline{Y}} \in \mathbb{G}^1_n$. Its magnitude squared is given by

$$(\check{\underline{X}} \wedge \check{\underline{Y}}) \cdot (\check{\underline{Y}} \wedge \check{\underline{X}}) = \check{\underline{X}}^2\, \check{\underline{Y}}^2 - (\check{\underline{X}} \cdot \check{\underline{Y}})^2 = \mathcal{V}(\check{\underline{X}})\, \mathcal{V}(\check{\underline{Y}}) - \mathcal{C}(\check{\underline{X}}, \check{\underline{Y}})^2 ,$$

which is equivalent to the determinant of the covariance matrix of $\check{\underline{X}}$ and $\check{\underline{Y}}$ if these variables are normally distributed. The lower bound on $\|\check{\underline{X}} \wedge \check{\underline{Y}}\|^2$ is zero; the upper bound is obtained if $\check{\underline{X}}$ and $\check{\underline{Y}}$ are orthogonal. In this case

$$\|\check{\underline{X}} \wedge \check{\underline{Y}}\| = \|\check{\underline{X}}\, \check{\underline{Y}}\| = \|\check{\underline{X}}\|\, \|\check{\underline{Y}}\| .$$

Thus,

$$0 \leq \|\check{\underline{X}} \wedge \check{\underline{Y}}\|^2 \leq \mathcal{V}(\check{\underline{X}})\, \mathcal{V}(\check{\underline{Y}}) .$$

If $\mathcal{V}(\check{\underline{X}}) > 0$ and $\mathcal{V}(\check{\underline{Y}}) > 0$, it follows that

$$0 \leq \frac{\mathcal{C}(\check{\underline{X}}, \check{\underline{Y}})^2}{\mathcal{V}(\check{\underline{X}})\, \mathcal{V}(\check{\underline{Y}})} \leq 1 .$$

This ratio describes the correlation between $\check{\underline{X}}$ and $\check{\underline{Y}}$: it is zero if they are uncorrelated and unity if they are completely dependent. The correlation coefficient $\rho(\underline{X}, \underline{Y})$ for random variables $\underline{X}, \underline{Y} \in \mathbb{G}^1_n$ is defined as

$$\rho(\underline{X}, \underline{Y}) := \frac{\mathcal{C}(\underline{X}, \underline{Y})}{\sqrt{\mathcal{V}(\underline{X})\, \mathcal{V}(\underline{Y})}} .$$

This concept can be extended to more than two random variables using the same approach. For example, for $\check{\underline{X}}, \check{\underline{Y}}, \check{\underline{Z}} \in \mathbb{G}^1_n$,

$$\begin{aligned} \|\check{\underline{X}} \wedge \check{\underline{Y}} \wedge \check{\underline{Z}}\|^2 &= (\check{\underline{X}} \wedge \check{\underline{Y}} \wedge \check{\underline{Z}}) \cdot (\check{\underline{Z}} \wedge \check{\underline{Y}} \wedge \check{\underline{X}}) \\ &= \check{\underline{X}}^2\, \check{\underline{Y}}^2\, \check{\underline{Z}}^2 + 2\, (\check{\underline{X}} \cdot \check{\underline{Y}})\, (\check{\underline{Y}} \cdot \check{\underline{Z}})\, (\check{\underline{Z}} \cdot \check{\underline{X}}) \\ &\quad - \check{\underline{X}}^2\, (\check{\underline{Y}} \cdot \check{\underline{Z}})^2 - \check{\underline{Y}}^2\, (\check{\underline{Z}} \cdot \check{\underline{X}})^2 - \check{\underline{Z}}^2\, (\check{\underline{X}} \cdot \check{\underline{Y}})^2. \end{aligned} \tag{10.25}$$

Just as before,

$$0 \le \|\check{\underline{X}} \wedge \check{\underline{Y}} \wedge \check{\underline{Z}}\|^2 \le \mathcal{V}(\underline{X})\, \mathcal{V}(\underline{Y})\, \mathcal{V}(\underline{Z}) \,.$$

Hence,

$$\begin{aligned} 0 \;\le\; &\mathcal{V}(\underline{X})\, \mathcal{C}(\underline{Y},\underline{Z})^2 + \mathcal{V}(\underline{Y})\, \mathcal{C}(\underline{Z},\underline{X})^2 + \mathcal{V}(\underline{Z})\, \mathcal{C}(\underline{X},\underline{Y})^2 \\ &- 2\, \mathcal{C}(\underline{X},\underline{Y})\, \mathcal{C}(\underline{Y},\underline{Z})\, \mathcal{C}(\underline{Z},\underline{X}) \;\le\; \mathcal{V}(\underline{X})\, \mathcal{V}(\underline{Y})\, \mathcal{V}(\underline{Z}) \,. \end{aligned} \tag{10.26}$$

Dividing by the expression on the right-hand side gives

$$0 \;\le\; \rho(\underline{X},\underline{Y})^2 + \rho(\underline{Y},\underline{Z})^2 + \rho(\underline{Z},\underline{X})^2 - 2\, \rho(\underline{X},\underline{Y})\, \rho(\underline{Y},\underline{Z})\, \rho(\underline{Z},\underline{X}) \;\le\; 1 \,.$$

The correlation coefficient for three random variables $\underline{X}, \underline{Y}, \underline{Z} \in \mathbb{G}^1_n$ is therefore defined as

$$\rho(\underline{X},\underline{Y},\underline{Z})^2 := \rho(\underline{X},\underline{Y})^2 + \rho(\underline{Y},\underline{Z})^2 + \rho(\underline{Z},\underline{X})^2 - 2\, \rho(\underline{X},\underline{Y})\, \rho(\underline{Y},\underline{Z})\, \rho(\underline{Z},\underline{X}) \,.$$

Recall that the magnitude of a blade is the determinant of the matrix constructed from the constituent vectors (see 3.66). Geometrically, this magnitude gives the (hyper)volume of the parallelepiped spanned by the vectors (see Fig. 1.1). The smaller the components that are mutually perpendicular, the smaller this volume is. Since correlation is directly related to the concept of orthogonality, the "volume" spanned by a set of random variables is related to the correlation coefficient.

The relation between the correlation coefficient of a number of random variables $\underline{X}_1, \ldots, \underline{X}_k \in \mathbb{G}^1_n$, $k \le n$, and the magnitude of the blade $\check{\underline{X}}_1 \wedge \cdots \wedge \check{\underline{X}}_k$ is given by

$$1 - \rho(\underline{X}_1, \ldots, \underline{X}_k)^2 := \frac{\left\|(\underline{X}_1 - \mathcal{E}(\underline{X}_1)\, \underline{D}) \wedge \cdots \wedge (\underline{X}_k - \mathcal{E}(\underline{X}_k)\, \underline{D})\right\|^2}{\mathcal{V}(\underline{X}_1) \cdots \mathcal{V}(\underline{X}_k)} \,.$$

The correlation between two random variables can also be expressed as an angle, which follows directly from the relation between vectors in Euclidean space. That is,

$$\begin{aligned} &\check{\underline{X}} \cdot \check{\underline{Y}} = \|\check{\underline{X}}\|\, \|\check{\underline{Y}}\|\, \cos(\gamma) \\ \iff &\cos(\gamma) = \frac{\mathcal{C}(\check{\underline{X}},\check{\underline{Y}})}{\sqrt{\mathcal{V}(\check{\underline{X}})\, \mathcal{V}(\check{\underline{Y}})}} = \rho(\check{\underline{X}},\check{\underline{Y}}) \,. \end{aligned} \tag{10.27}$$

The magnitude of the blade $\check{\underline{X}} \wedge \check{\underline{Y}}$ is also related to the angle γ (cf. section 4.1.1):

$$\|\check{\underline{X}} \wedge \check{\underline{Y}}\| = \|\check{\underline{X}}\|\,\|\check{\underline{Y}}\|\,\sin(\gamma)$$
$$\iff \sin(\gamma) = \frac{\|\check{\underline{X}} \wedge \check{\underline{Y}}\|}{\sqrt{\mathcal{V}(\check{\underline{X}})\,\mathcal{V}(\check{\underline{Y}})}} = \sqrt{1 - \rho(\check{\underline{X}}, \check{\underline{Y}})^2}\,. \tag{10.28}$$

The geometric product of two random variables $\check{\underline{X}}, \check{\underline{Y}} \in \mathbb{G}^1_n$ can therefore be written as

$$\check{\underline{X}}\,\check{\underline{Y}} = \sqrt{\mathcal{V}(\check{\underline{X}})\,\mathcal{V}(\check{\underline{Y}})}\,\big(\cos(\gamma) + \sin(\gamma)\,\widehat{\underline{U}}_{\langle 2\rangle}\big) = \sqrt{\mathcal{V}(\check{\underline{X}})\,\mathcal{V}(\check{\underline{Y}})}\;e^{\gamma\,\widehat{\underline{U}}_{\langle 2\rangle}}\,,$$

where $\widehat{\underline{U}}_{\langle 2\rangle} := (\check{\underline{X}} \wedge \check{\underline{Y}})/\|\check{\underline{X}} \wedge \check{\underline{Y}}\|$ is a unit bivector.

10.3.4 Normal Random Variables

As mentioned before, normally distributed random variables are particularly well suited for a representation in H^n_h. Their pdfs are completely described by the first two moments, which can be evaluated with the use of the scalar product. This also implies that if two normally distributed random variables are uncorrelated, they are also statistically independent. Orthogonalization of a basis $\{\,\underline{D}, \underline{X}_1, \dots, \underline{X}_n\,\}$ of H^n_h thus results in a basis of statistically independent distributions. Variances and correlations between random variables can then be represented by the components of linear combinations of this orthogonal basis.

10.3.4.1 Canonical Basis

The equivalence of uncorrelatedness and statistical independence can be derived directly from the normal distribution. Let $\underline{\mathsf{X}} := (\underline{X}_1, \dots, \underline{X}_n)$ denote a multivariate random variable with a multivariate normal distribution $N(\mathsf{m}, \mathsf{V})$, where m is the expectation vector and V is the covariance matrix. $N(\mathsf{m}, \mathsf{V})$ is the joint density function of the set $\{\,\underline{X}_1, \dots, \underline{X}_n\,\}$, given by

$$f(\mathsf{x}) = \frac{1}{\sqrt{(2\pi)^n\,|\mathsf{V}|}}\,\exp\left(-\frac{1}{2}\,(\mathsf{x} - \mathsf{m})^{\mathsf{T}}\,\mathsf{V}^{-1}\,(\mathsf{x} - \mathsf{m})\right).$$

The expectation dimension is already statistically independent of all other distributions in H^n_h by definition. The important orthogonalization is therefore that of the corresponding set of zero-mean random variables $\{\,\check{\underline{X}}_1, \dots, \check{\underline{X}}_n\,\}$. The joint density function of this set is $N(0, \mathsf{V})$, given by

$$f(\mathsf{x}) = \frac{1}{\sqrt{(2\pi)^n\,|\mathsf{V}|}}\,\exp\left(-\frac{1}{2}\,\mathsf{x}^{\mathsf{T}}\,\mathsf{V}^{-1}\,\mathsf{x}\right).$$

The covariance matrix V is a positive definite symmetric matrix of full rank, and may thus be orthogonalized. Let D denote the matrix that orthogonalizes V; that is, $\mathsf{D}^{\mathsf{T}}\mathsf{V}\mathsf{D} = \Lambda$, where Λ is a diagonal matrix. We define $\mathsf{y} := \mathsf{D}\,\mathsf{x}$, such that $\mathsf{x} = \mathsf{D}^{-1}\,\mathsf{y}$; it may then be shown (see [82]) that the joint density function of $\underline{\mathsf{Y}} = \mathsf{D}\,\underline{\mathsf{X}}$ is given by

$$f(\mathsf{y}) = \frac{1}{\sqrt{(2\pi)^n\,|\Lambda|}}\,\exp\left(-\frac{1}{2}\,\mathsf{y}^{\mathsf{T}}\,\Lambda^{-1}\,\mathsf{y}\right) = \prod_{i=1}^{n}\frac{1}{\sqrt{2\pi\,\sigma_i^2}}\,\exp\left(-\frac{1}{2}\left(\frac{y_i}{\sigma_i}\right)^2\right),$$

where y_i is the ith component of y and σ_i^2 is the corresponding component in the diagonal matrix Λ. This can also be written as $f(\mathsf{y}) = \prod_{i=1}^{n} f_{\underline{\mathsf{Y}}_i}(y_i)$, which is the condition for statistical independence.

From this it follows that for any H_h^n of normally distributed random variables, an orthonormal basis of n statistically independent random variables of unit variance can be found, which we call a canonical basis of H_h^n. Given an arbitrary basis of H_h^n, a canonical basis can also be constructed using the Gram–Schmidt orthogonalization.

The canonical basis of H_h^n is denoted by $\{\,\underline{e}_0, \underline{e}_1, \ldots, \underline{e}_n\,\}$, where the $\{\,\underline{e}_1, \ldots, \underline{e}_n\,\}$ represent statistically independent, zero-mean, unit-variance random variables. The basis element $\underline{e}_0$ represents the Dirac delta distribution, which is statistically independent of all other random variables. Hence, $\{\,\underline{e}_0, \underline{e}_1, \ldots, \underline{e}_n\,\}$ is an orthonormal basis of H_h^n; that is, $\underline{e}_i * \underline{e}_j = \delta_{ij}$.

10.3.4.2 Representation of Random Variables

A general random variable $\underline{X} \in H_h^n$ can be written as $\underline{X} = \sum_{i=1}^{n} x^i\,\underline{e}_i$ or simply $\underline{X} = x^i\,\underline{e}_i$, using the Einstein summation convention. This is possible because a sum of normally distributed random variables is also normally distributed. The component x^0 gives the expectation value of $\underline{X}$, while the $\{\,x^1, \ldots, x^n\,\}$ give the standard deviations along the corresponding statistical dependencies. Consider for example two random variables $\underline{\check{X}}, \underline{\check{Y}} \in H_h^2$ defined as $\underline{\check{X}} := x^1\,\underline{e}_1 + x^2\,\underline{e}_2$ and $\underline{\check{Y}} := y^1\,\underline{e}_1 + y^2\,\underline{e}_2$, where $x^1, x^2, y^1, y^2 \in \mathbb{R}$. All statistical properties of these two random variables are now captured in the components $\{\,x^i\,\}$ and $\{\,y^i\,\}$. Their variances are

$$\mathcal{V}(\underline{\check{X}}) = \underline{\check{X}}\cdot\underline{\check{X}} = (x^1)^2 + (x^2)^2\,, \qquad \mathcal{V}(\underline{\check{Y}}) = \underline{\check{Y}}\cdot\underline{\check{Y}} = (y^1)^2 + (y^2)^2\,,$$

and their covariance is

$$\mathcal{C}(\underline{\check{X}}, \underline{\check{Y}}) = \underline{\check{X}}\cdot\underline{\check{Y}} = x^1\,y^1 + x^2\,y^2\,.$$

Typically, for a set of random variables $\{\,\underline{X}_1, \ldots, \underline{X}_n\,\}$ that form a basis of H_h^n, the expectation values and the covariance matrix V are given. To perform numerical calculations in this space with the random variables $\{\,\underline{X}_i\,\}$, a representation of them in terms of the canonical basis is needed. This can be evaluated as follows. First, the expectation values give the corresponding dependencies on $\underline{e}_0$. The dependencies on the remaining dimensions can be derived from the covariance matrix V. For this purpose, the matrix D which diagonalizes the covariance matrix V, such that $\mathsf{D}^{\mathsf{T}}\,\mathsf{V}\,\mathsf{D}$ is diagonal, has to be found. Then the relation between the multivariate distribution $\underline{\check{\mathsf{X}}} := (\underline{\check{X}}_1, \ldots, \underline{\check{X}}_n)$ and the canonical multivariate distribution $\underline{\check{\mathsf{E}}} := (\underline{e}_1, \ldots, \underline{e}_n)$ is $\underline{\check{\mathsf{X}}} = \mathsf{D}^{-1}\,\underline{\check{\mathsf{E}}}$. That is, the rows of D^{-1} give the components of the $\{\underline{\check{X}}_i\}$ in terms of the canonical basis elements.

10.3.4.3 Further Properties

In general, the sum $\underline{\check{X}} + \underline{\check{Y}}$ and the difference $\underline{\check{X}} - \underline{\check{Y}}$ have variances

$$
\begin{aligned}
\mathcal{V}(\underline{\check{X}} + \underline{\check{Y}}) &= (\underline{\check{X}} + \underline{\check{Y}})^2 \\
&= \underline{\check{X}}^2 + \underline{\check{Y}}^2 + 2\,\underline{\check{X}} \cdot \underline{\check{Y}} \\
&= \mathcal{V}(\underline{\check{X}}) + \mathcal{V}(\underline{\check{Y}}) + 2\,\mathcal{C}(\underline{\check{X}}, \underline{\check{Y}})\,,
\end{aligned}
$$

and, similarly,

$$
\mathcal{V}(\underline{\check{X}} - \underline{\check{Y}}) = \mathcal{V}(\underline{\check{X}}) + \mathcal{V}(\underline{\check{Y}}) - 2\,\mathcal{C}(\underline{\check{X}}, \underline{\check{Y}})\,.
$$

The sum and difference are statistically independent if

$$
(\underline{\check{X}} + \underline{\check{Y}}) \cdot (\underline{\check{X}} - \underline{\check{Y}}) = 0 \quad \Longleftrightarrow \quad \underline{\check{X}}^2 = \underline{\check{Y}}^2 \quad \Longleftrightarrow \quad \mathcal{V}(\underline{\check{X}}) = \mathcal{V}(\underline{\check{Y}})\,.
$$

For random variables of H_h^n that are not expectation-free, the split between the expectation dimension and the "variance dimensions" can be made explicit through the projection on and rejection from (cf. section 3.2.10) the expectation dimension $\underline{e}_0$. That is, for $\underline{X} \in H_h^n$ with $\underline{X} = x^i\,\underline{e}_i$,

$$
\underline{X} \;=\; \mathcal{P}_{\underline{e}_0}(\underline{X}) \;+\; \mathcal{P}^{\perp}_{\underline{e}_0}(\underline{X}) \;=\; \mathcal{E}(\underline{X})\,\underline{e}_0 \;+\; \underline{\check{X}}\,,
$$

where $\underline{\check{X}} = \mathcal{P}^{\perp}_{\underline{e}_0}(\underline{X})$ is the zero-mean version of $\underline{X}$. The second moment of $\underline{X}$ is thus

$$
\underline{X} * \underline{X} = \mathcal{E}(\underline{X})^2 + \underline{\check{X}} * \underline{\check{X}} = (x^0)^2 + \sum_{i=1}^{n} (x^i)^2 \;=\; \mathcal{E}(\underline{X})^2 + \mathcal{V}(\underline{X})\,,
$$

where $\mathcal{E}(\underline{X}) = x^0$ and $\mathcal{V}(\underline{X}) = \underline{\check{X}} * \underline{\check{X}} = \sum_{i=1}^{n} (x^i)^2$.

10.3.4.4 Geometric Algebra

The algebraic basis of $\mathbb{G}(H_h^n)$ has dimension $2^{(n+1)}$ and can be constructed by the appropriate algebraic products of the basis elements $\{\,\underline{e}_i\,\}$ (cf. section 3.1). For example, given H^{2+1} with a basis $\{\,\underline{e}_0, \underline{e}_1, \underline{e}_2\,\}$, the canonical algebraic basis of $\mathbb{G}(H^{2+1})$ is given by

$$\{\,1,\, \underline{e}_0,\, \underline{e}_1,\, \underline{e}_2,\, \underline{e}_0\,\underline{e}_1,\, \underline{e}_0\,\underline{e}_2,\, \underline{e}_1\,\underline{e}_2,\, \underline{e}_0\,\underline{e}_1\,\underline{e}_2\,\}\,.$$

While $\underline{e}_1$ and $\underline{e}_2$ represent statistically independent random variables, the basis element $\underline{e}_{12} \equiv \underline{e}_1\,\underline{e}_2$ can be regarded as representing the random-variable subspace spanned by $\underline{e}_1$ and $\underline{e}_2$. All operations that are possible in the geometric algebra of Euclidean space, such as projection, reflection, and rotation, can now also be applied to normally distributed random variables.

Another example is the treatment of random-variable subspaces in much the same way as random variables. That is, if $\underline{\breve{A}}, \underline{\breve{B}}, \underline{\breve{C}}, \underline{\breve{D}} \in \mathbb{G}_n^1$, then $\underline{\breve{A}} \wedge \underline{\breve{B}}$ and $\underline{\breve{C}} \wedge \underline{\breve{D}}$ are elements of the vector space $\mathbb{G}_n^2$. The variance of the subspace $\underline{\breve{A}} \wedge \underline{\breve{B}}$ may now be defined as

$$\|\underline{\breve{A}} \wedge \underline{\breve{B}}\|^2 = \underline{\breve{A}}^2\, \underline{\breve{B}}^2 - (\underline{\breve{A}} \cdot \underline{\breve{B}})^2\,.$$

The covariance between the subspaces $\underline{\breve{A}} \wedge \underline{\breve{B}}$ and $\underline{\breve{C}} \wedge \underline{\breve{D}}$ may be defined as

$$(\underline{\breve{A}} \wedge \underline{\breve{B}}) \cdot (\underline{\breve{D}} \wedge \underline{\breve{C}}) = (\underline{\breve{A}} \cdot \underline{\breve{C}})\,(\underline{\breve{B}} \cdot \underline{\breve{D}}) - (\underline{\breve{A}} \cdot \underline{\breve{D}})\,(\underline{\breve{B}} \cdot \underline{\breve{C}})\,.$$

The meet and join operators can also be used to intersect and combine random-variable subspaces.

Although we can easily apply all operations defined in geometric algebra, such as reflection and rotation, to elements of random-variable spaces, the significance of these operations, in the context of random variables, is currently not clear. However, as demonstrated in the derivation of the Cauchy–Schwartz inequality, and the derivation of the generalized correlation coefficient, the geometric concepts of geometric algebra can lead to new insights. Future research in this field, may therefore be quite fruitful.

Notation

The aim of the notation chosen was to allow a distinction between different types of elements through their fonts. Here is a list of the most important elements of the notation:

$\mathbb{A}$	A general set of arbitrary entities
$\mathbb{R}$	The real numbers
$\mathbb{S}^n$	The unit sphere in $\mathbb{R}^{n+1}$
$\mathbb{C}$	The complex numbers
$\mathbb{H}$	The quaternions
$\mathbb{R}^n$	A vector space of dimension n over the field $\mathbb{R}$ with a Euclidean signature
$\mathbb{R}^{p,q}$	A vector space of dimension $n = p+q$ over the field $\mathbb{R}$ with signature (p,q)
$\mathbb{R}^{m \times n}$	The direct product $\mathbb{R}^m \otimes \mathbb{R}^n$
$\mathbb{G}_n$	The geometric algebra over $\mathbb{R}^n$
$\mathbb{G}_{p,q}$	The geometric algebra over $\mathbb{R}^{p,q}$
$\mathbb{G}^k_{p,q}$	The k-vector space of $\mathbb{G}_{p,q}$
$\mathbb{G}^{\circ k}_{p,q}$	The set of grade-k null-blades of $\mathbb{G}_{p,q}$
$\mathbb{G}^{\varnothing k}_{p,q}$	The set of grade-k non-null-blades of $\mathbb{G}_{p,q}$
$\mathfrak{G}_{p,q}$	The Clifford group of $\mathbb{G}_{p,q}$

a, A	Scalar elements
$\boldsymbol{a}, \boldsymbol{A}$	Multivectors of geometric algebra
a, A	A column vector and a matrix in matrix algebra
$A^k{}_{ij}$	A 3-valence tensor in $\mathbb{R}^{p \times q \times r}$
$\mathcal{F}$	A general function
$\boldsymbol{A}_{\langle k \rangle}$	A blade of grade k
$\underline{\boldsymbol{a}}, \underline{\boldsymbol{A}}$	Random multivector variables
$\underline{\mathsf{a}}, \underline{\mathsf{A}}$	Random matrix variables
$\bar{\boldsymbol{a}}, \bar{\boldsymbol{A}}$	The expectation value of a random multivector variable
$\bar{\mathsf{a}}, \bar{\mathsf{A}}$	The expectation value of a random matrix variable

The following is a list of the various operator symbols used in this text:

$\boldsymbol{A}\,\boldsymbol{B}$	Geometric product of $\boldsymbol{A}$ and $\boldsymbol{B}$
$\boldsymbol{A} * \boldsymbol{B}$	Scalar product of $\boldsymbol{A}$ and $\boldsymbol{B}$
$\boldsymbol{A} \star \boldsymbol{B}$	Euclidean scalar product of $\boldsymbol{A}$ and $\boldsymbol{B}$
$\boldsymbol{A} \cdot \boldsymbol{B}$	Inner product of $\boldsymbol{A}$ and $\boldsymbol{B}$
$\boldsymbol{A} \wedge \boldsymbol{B}$	Outer product of $\boldsymbol{A}$ and $\boldsymbol{B}$
$\boldsymbol{A} \vee \boldsymbol{B}$	Meet of $\boldsymbol{A}$ and $\boldsymbol{B}$
$\boldsymbol{A} \dot{\wedge} \boldsymbol{B}$	Join of $\boldsymbol{A}$ and $\boldsymbol{B}$
$\boldsymbol{A} \triangledown \boldsymbol{B}$	Regressive product of $\boldsymbol{A}$ and $\boldsymbol{B}$
$\boldsymbol{A}^{-1}$	Inverse of $\boldsymbol{A}$
$\langle \boldsymbol{A} \rangle_k$	Projection of $\boldsymbol{A}$ onto grade k
$\boldsymbol{A}^*$	Dual of $\boldsymbol{A}$
$\widetilde{\boldsymbol{A}}$	Reverse of $\boldsymbol{A}$
$\boldsymbol{A}^\dagger$	Conjugate of $\boldsymbol{A}$
$\lVert \boldsymbol{A} \rVert$	Norm of $\boldsymbol{A}$
$\mathcal{P}_{\boldsymbol{B}_{\langle l \rangle}}(\boldsymbol{A}_{\langle k \rangle})$	Projection of $\boldsymbol{A}_{\langle k \rangle}$ onto $\boldsymbol{B}_{\langle l \rangle}$
$\mathcal{P}^{\perp}_{\boldsymbol{B}_{\langle l \rangle}}(\boldsymbol{A}_{\langle k \rangle})$	Rejection of $\boldsymbol{A}_{\langle k \rangle}$ from $\boldsymbol{B}_{\langle l \rangle}$
$\partial_{\boldsymbol{A}}$	Multivector differentiation operator with respect to $\boldsymbol{A}$

References

1. Ablamowicz, R.: Clifford algebra computations with Maple. In: W.E. Baylis (ed.) Clifford (Geometric) Algebras with Applications in Physics, Mathematics and Engineering, pp. 463–501. Birkhäuser, Boston (1996)
2. Ablamowicz, R., Fauser, B.: The CLIFFORD home page. http://math.tntech.edu/rafal/cliff9/ (2006)
3. Ablamowicz, R., Lounesto, P., Parra, J.M.: Clifford Algebras with Numeric and Symbolic Computations. Birkhäuser, Boston, MA (1996)
4. Ablamowicz, R., Sobczyk, G. (eds.): Lectures on Clifford (Geometric) Algebras and Applications. Birkhäuser (2004)
5. Albrecht, G., Farouki, R.T.: Construction of C^2 Pythagorean-hodograph interpolating splines by the homotopy method. Adv. Comp. Math. **5**, 417–442 (1996)
6. Angles, P.: Construction de revêtements du groupe conforme d'un espace vectoriel muni d'une métrique de type (p, q). Ann. l'I.H.P., Section A **33**(1), 33–51 (1980)
7. Angles, P.: Géométrie spinorielle conforme orthogonale triviale et groupes de spinorialité conformes. In: Report HTKK Math., vol. A, pp. 1–36. Helsinki University of Technology (1982)
8. Angles, P.: Algèbres de Clifford c+(r, s) des espaces quadratiques pseudo-Euclidiens standards er, s et structures correspondantes sur les espaces de spineurs associés; plongements naturels des quadriques projectives associées. In: J. Chisholm, A. Common (eds.) Clifford algebras and their applications in mathematical physics, *C*, vol. 123, pp. 79–91. Reidel, Dordrecht (1986)
9. Angles, P.: Conformal Groups in Geometry and Spin Structures. Progress in Mathematical Physics. Birkhäuser (2006)
10. Arbter, K.: Affine-invariant Fourier descriptors. In: J.C. Simon (ed.) From Pixels to Features, pp. 153–164. Elsevier Science (1989)
11. Arbter, K., Snyder, W.E., Burkhardt, H., Hirzinger, G.: Application of affine-invariant Fourier descriptors to recognition of 3-D objects. IEEE Trans. Pattern Anal. Mach. Intell. **12**(7), 640–647 (1990)
12. Arun, K.S., Huang, T.S., Blostein, S.D.: Least-squares fitting of two 3-d point sets. IEEE Trans. Pattern Anal. Mach. Intell. **9**(5), 698–700 (1987)
13. Baeza, R.: Quadratic Forms over Semilocal Rings, *Lecture Notes in Mathematics*, vol. 655. Springer, Berlin, Heidelberg (1978)
14. Banarer, V., Perwass, C., Sommer, G.: Design of a multilayered feed-forward neural network using hypersphere neurons. In: N. Petkov, M.A. Westenberg (eds.) Proceedings of the 10th International Conference on Computer Analysis of Images and Patterns, CAIP 2003, Groningen, The Netherlands, August 2003,

Lecture Notes in Computer Science, vol. 2756, pp. 571–578. Springer, Berlin, Heidelberg (2003)
15. Banarer, V., Perwass, C., Sommer, G.: The hypersphere neuron. In: Proceedings of the 11th European Symposium on Artificial Neural Networks, ESANN 2003, Bruges, pp. 469–474. d-side Publications, Evere, Belgium (2003)
16. Baylis, W.E. (ed.): Clifford (Geometric) Algebras with Applications to Physics, Mathematics and Engineering. Birkhäuser, Boston (1996)
17. Bayro-Corrochano, E., Buchholz, S.: Geometric neural networks. In: G. Sommer, J.J. Koenderink (eds.) Algebraic Frames for the Perception–Action Cycle, *Lecture Notes in Computer Science*, vol. 1315, pp. 379–394. Springer, Heidelberg (1997)
18. Bayro-Corrochano, E., Lasenby, J., Sommer, G.: Geometric algebra: A framework for computing point and line correspondences and projective structure using n uncalibrated cameras. In: International Conference on Pattern Recognition (ICPR '96) (1996)
19. Bayro-Corrochano, E., Sommer, G.: Object modelling and collision avoidance using Clifford algebra. In: V. Hlavac, R. Sara (eds.) Computer Analysis of Images and Patterns, Proceedings of CAIP'95, Prague, *Lecture Notes in Computer Science*, vol. 970, pp. 699–704. Springer (1995)
20. Berberich, E., Eigenwillig, A., Hemmer, M., Hert, S., Mehlhorn, K., Schömer, E.: A computational basis for conic arcs and boolean operations on conic polygons. In: 10th European Symposium on Algorithms, *Lecture Notes in Computer Science*, vol. 2461, pp. 174–186. Springer, Heidelberg (2002)
21. Bie, H.D., Sommen, F.: Correct rules for Clifford calculus on superspace. Advances in Applied Clifford Algebras **17**(3) (2007)
22. Board, O.A.R.: OpenGL Programming Guide, 2nd edn. Addison-Wesley Developers Press (1997)
23. Board, O.A.R.: OpenGL Reference Manual, 2nd edn. Addison-Wesley Developers Press (1997)
24. Brackx, F., Delanghe, R., Sommen, F.: Clifford Analysis, *Research Notes in Mathematics*, vol. 76. Pitman, London (1982)
25. Brannan, D.A., Esplen, M.F., Gray, J.J.: Geometry. Cambridge University Press (1999)
26. Bronstein, I.N., Semendjajew, K.A., Musiol, G., Mühlig, H.: Taschenbuch der Mathematik, 4th edn. Verlag Harri Deutsch (1999)
27. Buchholz, S.: A theory of neural computation with Clifford algebra. Ph.D. thesis, Christian-Albrechts-Universität zu Kiel (2005)
28. Buchholz, S., Bihan, N.L.: Optimal separation of polarized signals by quaternionic neural networks. In: 14th European Signal Processing Conference, EUSIPCO 2006, September 4–8, Florence, Italy (2006)
29. Buchholz, S., Sommer, G.: Introduction to Neural Computation in Clifford Algebra, pp. 291–314. Springer, Heidelberg (2001)
30. Buchholz, S., Sommer, G.: On averaging in Clifford groups. In: H. Li, P.J. Olver, G. Sommer (eds.) Computer Algebra and Geometric Algebra with Applications, *Lecture Notes in Computer Science*, vol. 3519, pp. 229–238. Springer, Berlin, Heidelberg (2005)
31. Choi, H.I., Han, C.Y.: Euler–Rodrigues frames on spatial Pythagorean-hodograph curves. Comput. Aided Geom. Design **19**, 603–620 (2002)
32. Choi, H.I., Lee, D.S., Moon, H.P.: Clifford algebra, spin representation, and rational parameterization of curves and surfaces. Adv. Comp. Math. **17**, 5–48 (2002)
33. Claus, D., Fitzgibbon, A.W.: A rational function lens distortion model for general cameras. In: IEEE Computer Society Conference on Computer Vision and Pattern Recognition, vol. 1, pp. 213–219 (2005)

34. Clifford, W.K.: Preliminary sketch of bi-quaternions. In: Proc. Lond. Math. Soc., vol. 4, pp. 381–395 (1873)
35. Clifford, W.K.: Applications of Grassmann's Extensive Algebra, pp. 266–276. Macmillan, London (1882)
36. Clifford, W.K.: Mathematical Papers. Macmillan, London (1882)
37. Clifford, W.K.: On the Classification of Geometric Algebras, pp. 397–401. Macmillan, London (1882)
38. Cox, D., Little, J., O'Shea, D.: Ideals, Varieties and Algorithms. Springer, New York (1998)
39. Daniilidis, K.: Hand–eye calibration using dual quaternions. Int. J. Robot. Res. **18**, 286–298 (1999)
40. Daniilidis, K.: Using the algebra of dual quaternions for motion alignment. In: G. Sommer (ed.) Geometric Computing with Clifford Algebras, pp. 489–500. Springer, Berlin, Heidelberg (2001)
41. Delanghe, R.: Clifford analysis: History and perspective. In: Computational Methods and Function Theory, vol. 1, pp. 107–153. Heldermann (2001)
42. Delanghe, R., Sommen, F., Soucek, V.: Clifford Algebra and Spinor-Valued Functions. Kluwer, Dordrecht (1992)
43. Dietz, R., Hoschek, J., Jüttler, B.: An algebraic approach to curves and surfaces on the sphere and on other quadrics. Comput. Aided Geom. Design **10**, 211–229 (1993)
44. Differ, A.: The Clados home page. http://sourceforge.net/projects/clados/ (2005)
45. Doran, C., Lasenby, A.: Geometric Algebra for Physicists. Cambridge University Press (2003)
46. Doran, C.J.L., Lasenby, A.N., Gull, S.F.: Gravity as a gauge theory in the spacetime algebra. In: F. Brackx, R. Delanghe (eds.) Clifford Algebras and Their Applications in Mathematical Physics, pp. 375–385. Kluwer Academic, Dordrecht (1993)
47. Dorst, L.: Honing Geometric Algebra for Its Use in the Computer Sciences, pp. 127–152. Springer (2000)
48. Dorst, L., Doran, C., Lasenby, J. (eds.): Applications of Geometric Algebra in Computer Science and Engineering. Birkhäuser, Boston, Basel, Berlin (2002)
49. Dorst, L., Fontijne, D.: An algebraic foundation for object-oriented euclidean geometry. In: E. Hitzer (ed.) Proceedings of RIMS Symposium–Innovative Teaching in Mathematics with Geometric Algebra, Nov. 20–23, 2003, Kyoto, pp. 138–153. Research Institute for Mathematical Sciences, Kyoto, Japan (2004)
50. Dorst, L., Fontijne, D., Mann, S.: Geometric Algebra for Computer Science. Morgan Kaufmann, San Francisco (2007)
51. Dorst, L., Mann, S., Bouma, T.: GABLE: A MatLab tutorial for Geometric Algebra. http://staff.science.uva.nl/~leo/GABLE/ (2002)
52. Farouki, R.T.: Pythagorean-hodograph curves in practical use. In: R.E. Barnhill (ed.) Geometry Processing for Design and Manufacturing, pp. 3–33. SIAM (1992)
53. Farouki, R.T.: The conformal map $z \rightarrow z^2$ of the hodograph plane. Comput. Aided Geom. Design **11**, 363–390 (1994)
54. Farouki, R.T.: Exact rotation-minimizing frames for spatial Pythagorean-hodograph curves. Graph. Models **64**, 382–395 (2002)
55. Farouki, R.T.: Pythagorean-Hodograph Curves, pp. 405–427. North-Holland, Amsterdam (2002)
56. Farouki, R.T.: Pythagorean-Hodograph Curves, *Geometry and Computing*, vol. 1. Springer (2008)

57. Farouki, R.T., Han, C.Y.: Rational approximation schemes for rotation-minimizing frames on Pythagorean-hodograph curves. Comput. Aided Geom. Design **20**, 435–454 (2003)
58. Farouki, R.T., al Kandari, M., Sakkalis, T.: Hermite interpolation by rotation-invariant spatial Pythagorean-hodograph curves. Adv. Comp. Math. **17**, 369–383 (2002)
59. Farouki, R.T., al Kandari, M., Sakkalis, T.: Structural invariance of spatial Pythagorean hodographs. Comput. Aided Geom. Design **19**, 395–407 (2002)
60. Farouki, R.T., Kuspa, B.K., Manni, C., Sestini, A.: Efficient solution of the complex quadratic tridiagonal system for C^2 PH quintic splines. Numer. Algorithms **27**, 35–60 (2001)
61. Farouki, R.T., Manjunathaiah, J., Nicholas, D., Yuan, G.F., Jee, S.: Variable feedrate CNC interpolators for constant material removal rates along Pythagorean-hodograph curves. Comput. Aided Design **30**, 631–640 (1998)
62. Farouki, R.T., Manni, C., Sestini, A.: Spatial C^2 PH quintic splines. In: T. Lyche, M.L. Mazure, L.L. Schumaker (eds.) Curve and Surface Design: Saint Malo 2002, pp. 147–156. Nashboro Press, Brentwood (2003)
63. Farouki, R.T., Neff, C.A.: Hermite interpolation by Pythagorean-hodograph quintics. Math. Comp. **64**, 1589–1609 (1995)
64. Farouki, R.T., Sakkalis, T.: Pythagorean hodographs. IBM J. Res. Develop. **34**, 736–752 (1990)
65. Farouki, R.T., Sakkalis, T.: Pythagorean-hodograph space curves. Adv. Comp. Math. **2**, 41–66 (1994)
66. Farouki, R.T., Shah, S.: Real-time CNC interpolators for Pythagorean-hodograph curves. Comput. Aided Geom. Design **13**, 583–600 (1996)
67. Faugeras, O.: Stratification of three dimensional vision: Projective, affine and metric representations. J. Opt. Soc. Am. A **12**(3), 465–484 (1995)
68. Faugeras, O., Mourrain, B.: On the geometry and algebra of the point and line correspondences between n images. In: Proceedings of ICCV'95, pp. 951–956 (1995)
69. Faugeras, O., Papadopoulo, T.: Grassmann–Cayley algebra for modelling systems of cameras and the algebraic equations of the manifold of trifocal tensors. Phil. Trans. R. Soc. Lond. A **356**(1740), 1123–1152 (1998)
70. Fenske, A.: Affin-invariante erkennung von grauwertmustern mit Fourierdeskriptoren. In: Mustererkennung 1993, pp. 75–83. Springer (1993)
71. Fitzgibbon, A.W.: Simultaneous linear estimation of multiple view geometry and lens distortion. In: IEEE Computer Society Conference on Computer Vision and Pattern Recognition, vol. 1, pp. 125–132 (2001)
72. Fontijne, D.: The Gaigen home page. http://www.science.uva.nl/ga/gaigen/ (2005)
73. Fontijne, D., Dorst, L.: The GAViewer home page. http://www.science.uva.nl/ga/viewer/content_viewer.html (2006)
74. Förstner, W., Brunn, A., Heuel, S.: Statistically testing uncertain geometric relations. In: G. Sommer, N. Krüger, C. Perwass (eds.) Mustererkennung 2000, Informatik Aktuell, pp. 17–26. Springer, Berlin (2000)
75. Gallier, J.: Geometric Methods and Applications for Computer Science and Engineering, *Texts in Applied Mathematics*, vol. 38. Springer (2001)
76. Gebken, C., Tolvanen, A., Perwass, C., Sommer, G.: Perspective pose estimation from uncertain omnidirectional image data. In: International Conference on Pattern Recognition (ICPR), vol. I, pp. 793–796 (2006)
77. Geyer, C., Daniilidis, K.: Catadioptric camera calibration. In: 7th International Conference on Computer Vision, vol. 1, pp. 398–404 (1999)
78. Geyer, C., Daniilidis, K.: Catadioptric projective geometry. Int. J. Comput. Vis. **45**, 223–243 (2001)

79. Gilbert, J.E., Murray, M.A.M.: Clifford Algebras and Dirac Operators in Harmonic Analysis. Cambridge University Press (1991)
80. Granlund, G.: Fourier preprocessing for hand print character recognition. IEEE Trans. Comput. **21**, 195–201 (1972)
81. Grassmann, H.: Die lineale Ausdehnungslehre, ein neuer Zweig der Mathematik, dargestellt und durch Anwendungen auf die übrigen Zweige der Mathematik, wie auch auf die Statik, Mechanik, die Lehre vom Magnetismus und die Krystallonomie erläutert von Hermann Grassmann. O. Wigand, Leipzig (1844)
82. Grimmet, G., Stirzaker, D.: Probability and Random Processes, 3rd edn. Oxford University Press, Oxford (2001)
83. Gull, S.F., Lasenby, A.N., Doran, C.J.L.: Imaginary numbers are not real – the geometric algebra of space time. Found. Phys. **23**(9), 1175 (1993)
84. Haddon, J.A., Forsyth, D.A.: Noise in bilinear problems. In: International Conference on Computer Vision, vol. 2, pp. 622–627. IEEE Computer Society, Vancouver (2001)
85. Hartley, R.I., Zissermann, A.: Multiple View Geometry in Computer Vision, 2nd edn. Cambridge University Press, Cambridge (2003)
86. Helmert, F.R.: Die Ausgleichsrechnung nach der Methode der kleinsten Quadrate. Teubner, Leipzig (1872)
87. Hestenes, D.: Space-Time Algebra. Gordon & Breach (1966)
88. Hestenes, D.: New Foundations for Classical Mechanics. Kluwer (1986)
89. Hestenes, D.: Point groups and space groups in geometric algebra. In: L. Doerst, C. Doran, J. Lasenby (eds.) Applications of Geometric Algebra with Applications in Computer Science and Engineering, pp. 3–34. Birkhäuser (2002)
90. Hestenes, D.: Geometric calculus. http://modelingnts.la.asu.edu/ (2006)
91. Hestenes, D., Sobczyk, G.: Clifford Algebra to Geometric Calculus: A Unified Language for Mathematics and Physics. Kluwer (1984)
92. Hestenes, D., Ziegler, R.: Projective geometry with Clifford algebra. Acta Applicandae Mathematicae **23**, 25–63 (1991)
93. Heuel, S.: Uncertain Projective Geometry, *LNCS*, vol. 3008. Springer (2004)
94. Hildenbrand, D.: Homepage. http://www.gris.informatik.tu-darmstadt.de/~dhilden/ (2006)
95. Hildenbrand, D., Bayro-Corrochano, E., Zamora, J.: Advanced geometric approach for graphics and visual guided robot object manipulation. In: International Conference on Robotics and Automation. Barcelona, Spain (2005)
96. Hildenbrand, D., Zamora, J., Bayro-Corrochano, E.: Inverse kinematics computation in computer graphics and robotics using conformal Geometric Algebra. In: 7th International Conference on Clifford Algebras and Their Applications. Toulouse, France (2005)
97. Hitzer, E., Perwass, C.: Crystal cells in Geometric Algebra. In: Proceedings of the International Symposium on Advanced Mechanical Engineering, November 2004, Fukui, Japan, pp. 290–295. Department of Mechanical Engineering, University of Fukui, Japan (2004)
98. Hitzer, E.M.S., Perwass, C.: Crystal cell and space lattice symmetries in Clifford geometric algebra. In: T.E. Simos, G. Sihoyios, C. Tsitouras (eds.) International Conference on Numerical Analysis and Applied Mathematics, ICNAAM 2005, pp. 937–941. Wiley-VCH, Weinheim (2005)
99. Hitzer, E.M.S., Perwass, C.: Full geometric description of all symmetry elements of crystal space groups by the suitable choice of only three vectors for each Bravais cell or crystal family. In: Proceedings of the International Symposium on Advanced Mechanical Engineering, between the University of Fukui (Japan), Pukyong National University (Korea) and University of Shanghai for Science and Technology (China), 23–26 Nov. 2005, pp. 19–25. Pukyong National University (2005)

100. Hitzer, E.M.S., Perwass, C.: Three vector generation of crystal space groups in Geometric Algebra. Bull. Soc. Sci. Form **1**(21), 55–56 (2006)
101. Jüttler, B.: Hermite interpolation by Pythagorean hodograph curves of degree seven. Math. Comp. **70**, 1089–1111 (2001)
102. Jüttler, B., Mäurer, C.: Rational approximation of rotation minimizing frames using Pythagorean-hodograph cubics. J. Geom. Graph. **3**, 141–159 (1999)
103. Kilpelä, E.: Compensation of systematic errors of image and model coordinates. Int. Arch. Photogramm. **XXIII**(B9), 407–427 (1980)
104. Koch, K.R.: Parameter Estimation and Hypothesis Testing in Linear Models, 3rd edn. Springer (1999)
105. Lasenby, A.: Recent applications of conformal geometric algebra. In: H. Li, P.J. Olver, G. Sommer (eds.) Computer Algebra and Geometric Algebra with Applications, *Lecture Notes in Computer Science*, vol. 3519, pp. 298–328 (2004)
106. Lasenby, A.N., Doran, C.J.L., Gull, S.F.: Cosmological consequences of a flat-space theory of gravity. In: F. Brackx, R. Delanghe (eds.) Clifford Algebras and Their Applications in Mathematical Physics, pp. 387–396. Kluwer Academic, Dordrecht (1993)
107. Lasenby, J., Bayro-Corrochano, E.: Computing 3D projective invariants from points and lines. In: G. Sommer, K. Daniilidis, J. Pauli (eds.) Computer Analysis of Images and Patterns, CAIP'97, Kiel, *Lecture Notes in Computer Science*, vol. 1296, pp. 82–89. Springer (1997)
108. Lasenby, J., Bayro-Corrochano, E., Sommer, G.: A new framework to the formation of invariants and multiple-view. In: Proceedings of ICIP'96, Lausanne, Switzerland (1996)
109. Lasenby, J., Lasenby, A.N.: Estimating tensors for matching over multiple views. Phil. Trans. R. Soc. Lond. A **356**(1740), 1267–1282 (1998)
110. Lasenby, J., Lasenby, A.N., Doran, C.J.L.: A unified mathematical language for physics and engineering in the 21st century. Phil. Trans. R. Soc. Lond. A: Special Millennium Issue **Issue II Mathematics, Physics and Engineering** (2000)
111. Lasenby, J., Lasenby, A.N., Doran, C.J.L., Fitzgerald, W.J.: New geometric methods for computer vision – an application to structure and motion estimation. Int. J. Comput. Vis. **26**(3), 191–213 (1998)
112. Lasenby, J., Stevenson, A.: Using geometric algebra in optical motion capture. In: E. Bayro, G. Sobczyk (eds.) Geometric Algebra: A Geometric Approach to Computer Vision, Neural and Quantum Computing, Robotics and Engineering. Birkhäuser (2000)
113. Lee, X.: A visual dictionary of special plane curves. http://xahlee.org/SpecialPlaneCurves_dir/specialPlaneCurves.html (2005)
114. Leopardi, P.: The GluCat home page. http://glucat.sourceforge.net/ (2006)
115. Lepetit, V., Fua, P.: Monocular model-based 3D tracking of rigid objects: A survey. Found. Trends Comput. Graph. Vis. **1**(1), 1–89 (2005)
116. Li, H., Hestenes, D., Rockwood, A.: Generalized homogeneous coordinates for computational geometry. In: G. Sommer (ed.) Geometric Computing with Clifford Algebras, pp. 27–59. Springer (2001)
117. Li, H., Hestenes, D., Rockwood, A.: Spherical conformal geometry with geometric algebra. In: G. Sommer (ed.) Geometric Computing with Clifford Algebras, pp. 61–75. Springer (2001)
118. Li, H., Hestenes, D., Rockwood, A.: A universal model for conformal geometries of euclidean, spherical and double-hyperbolic spaces. In: G. Sommer (ed.) Geometric Computing with Clifford Algebras, pp. 77–104. Springer (2001)
119. Lounesto, P.: Clifford Algebras and Spinors. Cambridge University Press (1997)
120. Lounesto, P.: Counterexamples to theorems published and proved in recent literature on Clifford algebras, spinors, spin groups, and the exterior algebra. http://users.tkk.fi/~ppuska/mirror/Lounesto/counterexamples.htm (1997)

121. Lounesto, P.: CLICAL. http://users.tkk.fi/~ppuska/mirror/Lounesto/CLICAL.htm (2002)
122. Lowe, D.G.: Solving for the parameters of object models from image descriptions. In: ARPA Image Understanding Workshop, pp. 121–127 (1980)
123. Lowe, D.G.: Three-dimensional object recognition from single two-dimensional images. Artif. Intell. **31**(3), 355–395 (1987)
124. Moon, H.P.: Minkowski Pythagorean hodographs. Comput. Aided Geom. Design **16**, 739–753 (1999)
125. Moon, H.P., Farouki, R.T., Choi, H.I.: Construction and shape analysis of PH quintic Hermite interpolants. Comput. Aided Geom. Design **18**, 93–115 (2001)
126. Nayfeh, M.H., Brussel, M.K.: Electricity and Magnetism. Wiley, New York (1985)
127. Needham, T.: Visual Complex Analysis. Oxford University Press, Oxford (1997)
128. O'Connor, J.J., Robertson, E.F.: Famous curves index. http://www-history.mcs.st-andrews.ac.uk/history/Curves/Curves.html (2006)
129. O'Meara, O.T.: Introduction to Quadratic Forms, 2nd edn. Springer, Berlin, Heidelberg (1973)
130. Perwass, C.: Applications of geometric algebra in computer vision. Ph.D. thesis, Cambridge University (2000)
131. Perwass, C.: Analysis of local image structure using intersections of conics. Tech. Rep. 0403, Christian-Albrechts-Universität zu Kiel (2004)
132. Perwass, C.: Junction and corner detection through the extraction and analysis of line segments. In: Combinatorial Image Analysis, *Lecture Notes in Computer Science*, vol. 3322, pp. 568–582. Springer, Berlin, Heidelberg (2004)
133. Perwass, C.: CLUCalc. http://www.clucalc.info/ (2006)
134. Perwass, C., Banarer, V., Sommer, G.: Spherical decision surfaces using conformal modelling. In: B. Michaelis, G. Krell (eds.) DAGM 2003, Magdeburg, *Lecture Notes in Computer Science*, vol. 2781, pp. 9–16. Springer, Berlin, Heidelberg (2003)
135. Perwass, C., Farouki, R.T., Noakes, L.: A geometric product formulation for spatial pythagorean hodograph curves with applications to hermite interpolation. Comput. Aided Geom. Des. **24**(4), 220–237 (2007). DOI http://dx.doi.org/10.1016/j.cagd.2007.01.002
136. Perwass, C., Förstner, W.: Uncertain geometry with circles, spheres and conics. In: R. Klette, R. Kozera, L. Noakes, J. Weickert (eds.) Geometric Properties from Incomplete Data, *Computational Imaging and Vision*, vol. 31, pp. 23–41. Springer (2006)
137. Perwass, C., Gebken, C., Sommer, G.: Implementation of a Clifford algebra co-processor design on a field programmable gate array. In: R. Ablamowicz (ed.) Clifford Algebras: Application to Mathematics, Physics, and Engineering, Progress in Mathematical Physics, pp. 561–575. Birkhäuser, Boston (2003)
138. Perwass, C., Gebken, C., Sommer, G.: Estimation of geometric entities and operators from uncertain data. In: 27. Symposium für Mustererkennung, DAGM 2005, Wien, 29.8.-2.9.005, *Lecture Notes in Computer Science*, vol. 3663. Springer, Berlin, Heidelberg (2005)
139. Perwass, C., Gebken, C., Sommer, G.: Geometry and kinematics with uncertain data. In: A. Leonardis, H. Bischof, A. Pinz (eds.) Computer Vision – ECCV 2006, *Lecture Notes in Computer Science*, vol. 3951, pp. 225–237. Springer (2006)
140. Perwass, C., Hildenbrand, D.: Aspects of geometric algebra in euclidean, projective and conformal space. Tech. Rep. 0310, Christian-Albrechts-Universität zu Kiel (2003)
141. Perwass, C., Hitzer, E.: spacegroup.info. http://www.spacegroup.info/ (2006)

142. Perwass, C., Hitzer, E.M.S.: Interactive visualization of full geometric description of crystal space groups. In: Proceedings of the International Symposium on Advanced Mechanical Engineering, between the University of Fukui (Japan), Pukyong National University (Korea) and University of Shanghai for Science and Technology (China), 23–26 Nov. 2005, pp. 276–282. Pukyong National University (2005)
143. Perwass, C., Hitzer, E.M.S.: Space group visualizer for monoclinic space groups. Bull. Soc. Sci. Form **1**(21), 38–39 (2006)
144. Perwass, C., Lasenby, J.: A geometric analysis of the trifocal tensor. In: R. Klette, G. Gimel'farb, R. Kakarala (eds.) Image and Vision Computing New Zealand, IVCNZ'98, Proceedings, pp. 157–162. University of Auckland (1998)
145. Perwass, C., Lasenby, J.: A Unified Description of Multiple View Geometry, pp. 337–369. Springer, Heidelberg (2001)
146. Perwass, C., Sommer, G.: Numerical evaluation of versors with Clifford algebra. In: L. Dorst, C. Doran, J. Lasenby (eds.) Applications of Geometric Algebra in Computer Science and Engineering, pp. 341–349. Birkhäuser (2002)
147. Perwass, C., Sommer, G.: The inversion camera model. In: Pattern Recognition, *Lecture Notes in Computer Science*, vol. 4174. Springer, Berlin, Heidelberg (2006)
148. Porteous, I.R.: Clifford Algebras and the Classical Groups. Cambridge University Press, Cambridge, UK (1995)
149. Riesz, M.: Clifford Numbers and Spinors. Kluwer Academic, Dordrecht (1993)
150. Rosenhahn, B.: Pose estimation revisited. Ph.D. thesis, Institute für Informatik und Praktische Mathematik, Christian-Albrechts-Universität zu Kiel (2003)
151. Rosenhahn, B., Brox, T., Kersting, U., Smith, D., Gurney, J., Klette, R.: A system for marker-less human motion estimation. Kuenstliche Intell. **1**, 45–51 (2006)
152. Rosenhahn, B., Granert, O., Sommer, G.: Monocular pose estimation of kinematic chains. In: L. Dorst, C. Doran, J. Lasenby (eds.) Applications of Geometric Algebra in Computer Science and Engineering, pp. 373–375. Birkhäuser, Boston (2002)
153. Rosenhahn, B., Krüger, N., Rabsch, T., Sommer, G.: Tracking with a novel pose estimation algorithm. In: R. Klette, S. Peleg, G. Sommer (eds.) International Workshop RobVis 2001 Auckland, New Zealand, *Lecture Notes in Computer Science*, vol. 1998, pp. 9–18. Springer, Berlin, Heidelberg (2001)
154. Rosenhahn, B., Perwass, C., Sommer, G.: Pose estimation of 3d free-form contours. Tech. Rep. 0207, Christian-Albrechts-Universität zu Kiel (2002)
155. Rosenhahn, B., Perwass, C., Sommer, G.: Pose estimation of 3d free-form contours in conformal geometry. In: D. Kenwright (ed.) Proceedings of Image and Vision Computing, IVCNZ, Auckland, New Zealand, pp. 29–34 (2002)
156. Rosenhahn, B., Perwass, C., Sommer, G.: Free-form pose estimation by using twist representations. Algorithmica **38**, 91–113 (2003)
157. Rosenhahn, B., Perwass, C., Sommer, G.: Modeling adaptive deformations during free-form pose estimation. In: N. Petkov, M.A. Westenberg (eds.) Proceedings of the 10th International Conference on Computer Analysis of Images and Patterns 2003, Groningen, The Netherlands, August 2003, *Lecture Notes in Computer Science*, vol. 2756, pp. 664–672. Springer, Berlin, Heidelberg (2003)
158. Rosenhahn, B., Perwass, C., Sommer, G.: Pose estimation of free-form surface models. In: B. Michaelis, G. Krell (eds.) 25. Symposium für Mustererkennung, DAGM 2003, Magdeburg, *LNCS*, vol. 2781, pp. 574–581. Springer, Berlin (2003)
159. Rosenhahn, B., Perwass, C., Sommer, G.: Pose estimation of 3d free-form contours. Int. J. Comput. Vis. **62**(3), 267–289 (2005)
160. Rosenhahn, B., Sommer, G.: Pose estimation in conformal geometric algebra, part I: The stratification of mathematical spaces. J. Math. Imaging Vis. **22**, 27–48 (2005)

161. Selig, J.M.: Geometrical Methods in Robotics. Monographs in Computer Science. Springer, New York (1996)
162. Sommen, F.: An algebra of abstract vector variables. Portugaliae Math. **54**, 287–310 (1997)
163. Sommen, F.: The problem of defining abstract bivectors. Result. Math. **31**, 148–160 (1997)
164. Sommer, G. (ed.): Geometric Computing with Clifford Algebras. Springer, Berlin, Heidelberg (2001)
165. Sommer, G., Koenderink, J.J. (eds.): Algebraic Frames for the Perception–Action Cycle, *Lecture Notes in Computer Science*, vol. 1315. Springer, Heidelberg (1997)
166. Sommer, G., Rosenhahn, B., Perwass, C.: The twist representation of shape. Technical Report 0407, Christian-Albrechts-Universität zu Kiel, Institut für Informatik und Praktische Mathematik (2004)
167. Strang, G.: Introduction to Linear Algebra. Wellesley-Cambridge Press (1998)
168. Strang, G.: Lineare Algebra. Springer, Berlin, Heidelberg (2003)
169. Tello, R.: Fourier descriptors for computer graphics. IEEE Trans. Syst. Man. Cybern. **25**(5), 861–865 (1995)
170. Tolvanen, A., Perwass, C., Sommer, G.: Projective model for central catadioptric cameras using Clifford algebra. In: 27. Symposium für Mustererkennung, DAGM 2005, Wien, 29.8.-2.9.005, *LNCS*, vol. 3663, pp. 192–199. Springer, Berlin, Heidelberg (2005)
171. Tsai, Y.F., Farouki, R.T., Feldman, B.: Performance analysis of CNC interpolators for time-dependent feedrates along PH curves. Comput. Aided Geom. Design **18**, 245–265 (2001)
172. Vince, J.A.: Geometric Algebra for Computer Graphics. Springer (2008)
173. Walker, M.W., Shao, L.: Estimating 3-d location parameters using dual number quaternions. Comput. Vis. Graph. Image Process. **54**(3), 358–367 (1991)
174. White, N.L.: Grassmann–Cayley algebra and robotics. J. Intell. Robot. Syst. **11**, 91–107 (1994)
175. White, N.L.: A Tutorial on Grassmann–Cayley Algebra, pp. 93–106. Kluwer, Dordrecht (1995)
176. Ying, X., Hu, Z.: Catadioptric camera calibration using geometric invariants. IEEE Trans. Pattern Anal. Mach. Intell. **26**(10), 1260–1271 (2004)

Index

affine transformation, 136
algebraic basis, 58
algorithm
  basis orthogonalization, 106
  blade factorization into 1-vectors, 107
  join of blades, 108
  versor factorization, 108
anticommutator product, 55
  of vectors, 56
  tensor representation, 264

basis
  orthogonalization of, 106
  reciprocal, 98
basis blade, 56
bias
  error propagation bilinear function, 219
bilinear function, error propagation bias, 219
bivector, 65
  oriented plane, 125
blade, 64
  factorization into 1-vectors, 107
  projection of, 82
  pseudoinverse, 81
  rejection from, 83
  relation to spinor, 91
  relation to versor, 91

canonical algebraic basis, 58
canonical vector basis, 52
Cauchy–Schwarz inequality, 362
Clifford group, 90
CLUCalc, 25
CLUScript, 25
commutator product, 55
  of vectors, 56
  tensor representation, 264
component vectors, 199
conformal conic space, 193
  GOPNS, 194
conformal space
  analysis of circle representation, 178
  analysis of line representation, 177
  analysis of plane representation, 176
  analysis of point pair representation, 177
  analysis of sphere representation, 176
  dilation operator, 172
  embedding of Euclidean vector, 149
  general rotation operator, 170
  geometric algebra of, 150
  GIPNS, 152
  GOPNS, 158
  inner product of spheres, 174
  inner-product circles, 156
  inner-product imaginary sphere, 154
  inner-product lines, 157
  inner-product planes, 155
  inner-product point pairs, 157
  inner-product points, 152
  inner-product sphere, 153
  inversion operator, 166
  motor operator, 171
  outer-product circles, 160
  outer-product homogeneous points, 159
  outer-product lines, 159
  outer-product planes, 160
  outer-product point pairs, 158
  outer-product points, 158
  outer-product spheres, 160
  polynomial representation, 181

pose estimation, 305
reflection operator, 163
rotation operator, 170
stratification of spaces, 161
translation operator, 168
uncertain circle, 256
uncertain line, 256
uncertain reflection, 259
uncertain rotor, 261
conformal transformation, 145
conic space, 179
analysis of conic representation, 186
degenerate conic, 190
GIPNS, 184
GOPNS, 183
intersection of conics, 190
intersection of lines and conics, 189
representation of conic, 183
rotation operator, 184
types of conics, 188
uncertain conic, 258
conjugate, 59
of blades, 69
containment relations, 173
correlation coefficient, 363
covariance matrix, 209
inverse conformal mapping, 229
of bilinear function, 213
of conformal embedding, 228
of conic space embedding, 234
of function of random variable, 212
of inverse conic space mapping, 234
of projective embedding, 222
relation to blades, 230
relation to versors, 232
curl of vector-valued function, 100
cycloidal curve, 327
relation to Fourier series, 328

De Morgan's laws, 85
degenerate conic, 190
determinant
of linear function, 94
product of, 96
differentiation operator
multivector, 101
vector, 100
Dirac delta distribution, 359
direct difference, 77
direct sum, 67
divergence of vector-valued function, 100
dual
geometric meaning, 124
duality, 60

Einstein summation convention, 58
equivalence class, 134
error propagation, 210
conditioning in projective space, 224
construction of circle, 256
construction of conic, 258
construction of line, 256
construction of reflection, 259
construction of rotor, 261
non-Gaussivity, 215
summary of equations, 215
estimation
catadioptric camera, 314
Gauss–Helmert model, 239
Gauss–Markov model, 234
initial monocular pose, 302
lens distortion, 314
monocular pose, 299
of circle, 272
of general rotor, 274
of line, 263
of versor equation, 265
Plücker condition, 263
pose, 306
projective versor equation, 267
Euclidean scalar product, 70
expectation dimension, 359
expectation value, 209
inverse conformal mapping, 229
inverse projective mapping, 222
of bilinear function, 213
of conformal embedding, 228
of conic space embedding, 233
of function of random variable, 209
of inverse conic space mapping, 234
of projective embedding, 221

factorization
of blades, 107
of versors, 108
Fourier series, 328
relation to cycloidal curve, 328
fundamental matrix, 280
future research
conformal conic space, 194
Fourier series of space curves, 330
geometric algebra of random variables, 368

Gauss–Helmert model, 239
covariance matrix of estimation, 246
generalized normal equations, 248
iterative application, 249
normal equations, 242, 244

Gauss–Markov model, 234
generalized normal equations, 239
normal equations, 237
numerical estimation, 238
general multivector derivative, 102
general multivector integration, 104
geometric algebra
axiom, 52
basis of $\mathbb{G}(\mathbb{R}^3)$, 58
defining equation, 54
isomorphism to $\mathbb{C}$, 110
isomorphism to quaternions, 113
norm of random variables, 361
of conformal conic space, 194
of conformal space, 150
of conic space, 183
of Euclidean space, 121
of projective space, 135
of symmetric matrices, 183
relation to complex numbers, 110
relation to Gibbs's vector algebra, 109
relation to Grassmann algebra, 114
relation to Grassmann–Cayley algebra, 115
relation to quaternions, 111
geometric IPNS, *see* GIPNS
geometric OPNS, *see* GOPNS
geometric orthogonality, 71
geometric product
axioms, 54
Jacobi matrix, 199
matrix representation, 199
of 1-vectors, 65
of bivectors, 65
of random variables, 365
relation to inner product, 72
relation to outer product, 66
relation to scalar product, 54
tensor representation, 198, 199
GIPNS, 120
conformal circles, 156
conformal imaginary sphere, 154
conformal lines, 157
conformal planes, 155
conformal point, 152
conformal point pairs, 157
conformal sphere, 153
Euclidean line, 125
Euclidean plane, 124
Euclidean point, 127
in conformal space, 151
in conic space, 184
of polynomial, 180
projective line, 141
projective plane, 140
projective point, 141
GOPNS, 120
conformal circles, 160
conformal homogeneous points, 159
conformal lines, 159
conformal planes, 160
conformal point pairs, 158
conformal points, 158
conformal spheres, 160
Euclidean line, 122
Euclidean plane, 122
in conformal space, 151
in conic space, 183
projective line, 138
projective plane, 139
grade, 57
grade projection bracket, 61
grade-preserving function, 93
grade-preserving operator, 92
gradient
of multivector-valued function, 101
of vector-valued function, 100
Grassmann algebra, relation to geometric algebra, 114
Grassmann–Cayley algebra, relation to geometric algebra, 115

Hesse tensor, 103
Hilbert space of random variables, 356
homogeneous space, 134
homogeneous vector, 135
random variable of, 223
hypothesis testing, 275

idempotent
multivector example, 204
inner product, 62
of 1-vector and blade, 73
of blades, 72
of matrices, 181
of multivectors, 63
of spheres, 174
relation to geometric product, 72
relation to shuffle product, 117
tensor representation, 199
inner-product null space, *see* IPNS
integration
multivector, 104
inverse
linear function, 96
multivector, 204
of blade, 81
versor, 90

inversion camera model, 281
  lens distortion, 287
  mathematical formulation of catadioptric camera, 295
  mathematical representation, 284
  relation to fisheye lens, 294
inversion in conformal space, 166
IPNS, 73
  geometric, 120
  intersection of, 126
  relation to OPNS, 79

Jacobi matrix, 100, 103
  of product tensor, 199
  pose estimation, 310
join, 84
  of blades algorithm, 108

k-vector space, 58

linear function, 93
  grade preservation, 93
  inverse, 96

magnitude
  of multivectors, 71
matrix inner product, 181
meet, 84
metric
  point–circle, 271
  point–line, 269
  versor equation, 272
monocular pose estimation, 300
  representation in conformal space, 305
  tensor representation, 309
multivector, 58
  component vector representation, 199
  inverse, 204
  random variable, 208
  tensor representation, 198

norm
  of multivectors, 71
null blade, 73
  properties of, 78
null versor, 91

operator
  dilator in conformal space, 172
  general rotation in conformal space, 170
  grade preservation, 92
  inversion in conformal space, 166
  motor in conformal space, 171
  reflection in conformal space, 163
  reflection in Euclidean space, 128
  reflection in projective space, 143
  rotor in conformal space, 170
  rotor in conic space, 184
  rotor in Euclidean space, 131
  rotor in projective space, 144
  screw in conformal space, 171
  translator, 168
OPNS, 67
  geometric, 120
  relation to IPNS, 79
ordered power set, 57
orthogonality
  geometric, 71
  self-, 71
outer product, 62
  of blades, 66
  of multivectors, 63
  relation to geometric product, 66
  relation to vector cross product, 80
  tensor representation, 199
outer-product null space, *see* OPNS
outermorphism
  linear function, 94
  of projection, 82
  of reflection, 128
  of rotor, 132
  of versor, 92

PDF, 208
pin group, 90
pinhole camera model, 279
  camera matrix, 280
  fundamental matrix, 280
  trifocal tensor, 280
Plücker condition in geometric algebra, 263
plane
  orientation, 125
polynomial space, 180
probability distribution function, *see* PDF
probability space, 352
product
  anticommutator, 55
  commutator, 55
  Euclidean scalar, 70
  geometric, 54
  inner, 62
  join, 84
  meet, 84
  outer, 62
  regressive, 89

scalar, 52
shuffle, 115
tensor representation, 199
triple scalar product, 109
triple vector cross product, 110
product tensor, 199
change of basis, 203
example in $\mathbb{G}_2$, 200
subspace projection, 201
versor equation, 205
projection
of blade, 82
of general blades, 83
outermorphism of, 82
projective space, 134
projective transformation, 136
pseudoinverse
of blade, 81
pseudoscalar, 59
Pythagorean-hodograph
Hermite interpolation, 340
Pythagorean-hodograph curve, 330
in any dimension, 333, 347
necessary and sufficient condition, 332
of constant length, 342
of maximally different length, 346
preimage, 336
reflection representation, 334
relation between reflection and rotation form, 335

quadratic space, 52
quaternions, isomorphism to geometric algebra, 113

random multivector variable, 208
random variables, 352
canonical basis, 366
correlation coefficient, 363
correlation coefficient of n, 364
correlation coefficient of three, 364
correlation of, 363
Dirac delta distribution, 359
expectation dimension, 359
expectation operator, 356
Hilbert space, 356
norm, 356
scalar product, 358
statistical independence of, 362
reciprocal basis, 98
of null vector basis, 98
reflection
in Euclidean space, 128
in projective space, 143
outermorphism, 128
versor operation, 91
regressive product, 89
relation to shuffle product, 116
rejection
from blade, 83
reverse, 59
of blades, 69
rotor, 131
exponential form, 132
in conformal space, 170
in conic space, 184
in projective space, 144
mean, 133
outermorphism, 132

scalar product, 52
of blades, 68
relation to geometric product, 54
self-orthogonality, 71
shuffle product, 115
relation to inner product, 117
relation to regressive product, 116
spin group, 91
spinor, 91
stereographic projection, 146
inverse of, 147
stratification of spaces in conformal space, 161
symmetric-matrix vector space, 181

transformation
affine, 136
projective, 136
translator, 168
trifocal tensor, 280

unitary versor, 90

vector basis, 52
vector cross product
relation to outer product, 80
triple, 110
vector space
axioms, 53
versor, 90
coupled motors, 326
evaluation from data, 206
factorization, 108
function, 325
grade preservation, 92
inverse, 90
null, 91
outermophism of, 92
reflection, 91
relation to blade, 91
relation to spinor, 91
unitary, 90